# Techniques for the Analysis of Membrane Proteins

# Techniques for the Analysis of Membrane Proteins

*Edited by*

C. IAN RAGAN and RICHARD J. CHERRY

*London New York*
CHAPMAN and HALL

First published in 1986 by
Chapman and Hall Ltd,
11 New Fetter Lane, London EC4P 4EE
Published in the USA by
Chapman and Hall
29 West 35th Street, New York NY 10001



Printed in Great Britain at the University Press,
Cambridge

ISBN 0 412 24970 7



British Library Cataloguing in Publication Data

Techniques for the analysis of membrane proteins.
1. Membrane proteins—Analysis
2. Chemistry, Analytic
I. Ragan, C. I. II. Cherry, R. J.
574.87′5 QP552.M44

ISBN 0-412-24970-7

Library of Congress Cataloging in Publication Data

Techniques for the analysis of membrane proteins.
Includes bibliographies and index.
1. Membrane proteins—Analysis. 2. Membrane proteins—Purification. I. Ragan, C. I.
II. Cherry, R. J. [Richard J.] [DNLM:
1. Membrane
Protein—analysis. QU 55 T2547]
QP552.M44T43 1986 574.87′5 86-4183

ISBN 0-412-24970-7

# Contents

# Contributors

**Roberto Bisson**
CNR Unit for the Study of Physiology of Mitochondria and Laboratory of Molecular Biology and Pathology, Institute of General Pathology, University of Padova, Italy

**Richard J. Cherry**
Department of Chemistry, University of Essex, Colchester, UK

**Ian M. Fearnley**
Medical Research Council, Laboratory of Molecular Biology, Cambridge, UK

**Robert B. Freedman**
Biological Laboratory, University of Kent, Canterbury, UK

**Paul A. Hargrave**
Department of Ophthalmology, University of Florida College of Medicine, J. Hillis Miller Health Center, Gainesville, Florida, USA

**Hajime Hirata**
Department of Biochemistry, Jichi Medical School, Minamikawachi-machi, Tochigi, Japan

**Sven Hovmöller**
Department of Structural Chemistry, Arrhenius Laboratory, Stockholm, Sweden

**Kenneth A. Jacobson**
Department of Anatomy, Laboratories for Cell Biology, University of North Carolina at Chapel Hill, Chapel Hill, North Carolina, USA

**Yasuo Kagawa**
Department of Biochemistry, Jichi Medical School, Minamikawachi-machi, Tochigi, Japan

**Hans-Geurg Kapitza**
Department of Anatomy, Laboratories for Cell Biology, University of North Carolina at Chapel Hill, Chapel Hill, North Carolina, USA

**Jena K. Khodadad**
Departments of Pathology and Anatomy, Rush-Presbyterian-St Luke's Medical Center, Chicago, Illinois, USA

**Jerome M. Loew**
Department of Pathology, Rush-Presbyterian-St Luke's Medical Center, Chicago, Illinois, USA

**E. N. Clare Mills**
Biological Laboratory, University of Kent, Canterbury, UK

**Mauricio Montal**
Department of Neurosciences, Roche Institute of Molecular Biology, Nutley, New Jersey, USA

**Cesare Montecucco**
CNR Unit for the Study of Physiology of Mitochondria and Laboratory of Molecular Biology and Pathology, Institute of General Pathology, University of Padova, Italy

**C. Ian Ragan**
Department of Biochemistry, University of Southampton, UK

**Milton R. J. Salton**
Department of Microbiology, New York University School of Medicine, New York, USA

**David D. Thomas**
Department of Biochemistry, University of Minnesota Medical School, Minneapolis, Minnesota, USA

**John E. Walker**
Medical Research Council, Laboratory of Molecular Biology, Cambridge, UK

**Ronald S. Weinstein**
Department of Pathology, Rush-Presbyterian-St Luke's Medical Center, Chicago, Illinois, USA

**Yutaka Yanagita**
Department of Pathological Biochemistry, Atomic Disease Institute, Nagasaki University School of Medicine, Nagasaki, Japan

# Preface

A preface should justify the existence of the book it precedes and this is invariably done in scientific texts by reference to the explosive growth of the field since the last such volume appeared. In molecular biology, most fields can be justifiably described as growing explosively, as should be the case for a young and vigorous science, but the study of membrane proteins stands out as one which has taken giant strides in the last few years. Ignorance of the structure and function of membrane proteins at the molecular level was certainly not due to lack of interest but rather was a result of lack of appropriate techniques. It has above all been the development of new experimental methods which has wrenched membrane biochemistry out of what Anthony Martonosi fetchingly called its 'romantic phase' (i.e. lots of ideas and few facts), into an era when the determination of membrane protein structure and mechanism is a reasonable goal.

Membrane proteins are generally classified as peripheral or integral. Peripheral proteins are relatively easily dissociated from membranes by mild treatments whence their study is essentially no different to that of soluble proteins. This book therefore concentrates on integral proteins which are strongly bound to the membrane by hydrophobic interactions with lipids. A crucial step in their study is of necessity the development of methods of solubilization and purification under non-denaturing conditions. A peculiar difficulty with membrane proteins, in particular those which transport substances across membranes, is that their function may not be detectable in solution. An essential part of the investigation of such proteins is therefore the reconstitution of their function by incorporation into artificial lipid membranes. Successful reconstitution into both planar and vesicular membranes has now been reported for a wide range of membrane proteins.

For many years the powerful techniques of peptide sequencing and X ray crystallography, which have given us a highly detailed picture of the structure and molecular mechanism of many soluble proteins, made little contribution to the study of membrane proteins. Although it is eleven years since the single membrane-spanning sequence of glycophorin was first reported, it was only in 1979 and after much difficulty that the sequence of a larger membrane protein, bacteriorhodopsin, was determined. The hydrophobic nature of membrane proteins still makes direct sequencing a difficult task, but the advent of recombinant DNA technology and hence sequencing of the corresponding gene is producing dramatic advances in this area.

The method of structure determination by electron microscopy of two-dimensional crystals of membrane proteins was first applied to bacteriorhodopsin in 1975. Although a number of other membrane proteins have been subsequently examined by this method and much insight has been gained as to their molecular architecture, the structures obtained have invariably been of low resolution. Atomic resolution requires three-dimensional crystals suitable for X-ray diffraction studies and the recent success with the bacterial reaction centre is a noteworthy breakthrough. Nevertheless, general application of X-ray crystallography to membrane proteins remains a distant vista and even when achieved will not answer many important questions relating to the organization and interactions occurring within the native biological membrane. Techniques such as freeze–fracture electron microscopy and a variety of methods for studying membrane topology will continue to have an important role, while the immunological approach is making an increasing contribution to the field.

A feature of current investigations of biological macromolecules is an increasing interest in dynamic properties. This is particularly evident in the case of membrane proteins, where both rotational and lateral mobility are a consequence of the fluid nature of the lipid bilayer. The investigation of these motions has required the development of novel techniques and most of the methods currently in use have been developed during the last decade. The study of membrane protein dynamics is important for its functional implications in, for example, electron transport, receptor-mediated responses and various aspects of immunology. In addition, investigation of membrane protein dynamics is contributing significantly to our understanding of interactions between membrane components and between the membrane and cytoskeletal structures.

This book is intended as a guide to the current armoury of techniques for the purification and characterization of membrane proteins and for studying their structure and dynamics. An obvious difficulty for such a volume is that the methods range from those which can be practised in any reasonably well-equipped biochemical laboratory to those involving specialized technical expertise and sophisticated and expensive instrumentation. It has therefore not been practical to try to impose uniformity of treatment but rather we have encouraged contributors to provide a level of experimental detail appropriate to each individual subject. In particular, the authors have tried to convey through their own experience the scope and limitations of their techniques, the kinds of information which can be obtained and, overall, the way in which this information can be combined with that from other sources to penetrate the complexity of membrane protein structure and function. We hope that new researchers will find the book of help in assessing current methodology whilst more experienced practitioners may benefit by extending their knowledge of areas which lie outside their particular expertise. We would also hope that the book will convey to all its readers something of the fascination which membrane proteins exert over us.

# 1

# Analysis of membrane protein composition by gel electrophoresis

C. IAN RAGAN



## 1.1 Introduction

Electrophoretic separation of proteins in gels is a universally popular technique whose application is not, of course, restricted to membrane proteins. Nevertheless, the need in membrane protein research for techniques for the separation and analysis of complex mixtures of hydrophobic proteins has perhaps provided a major impetus for the development of improved electrophoretic methods over the last decade, particularly the separation of sodium dodecyl sulphate (SDS)–protein complexes on polyacrylamide gels. The latter technique, and the complementary use of isoelectric focusing methods (Chapter 2), are the basis of many aspects of membrane protein research. In view of this importance, I make no apology that this chapter describes electrophoretic methods which can equally be applied to soluble proteins. However, I have concentrated on those methods which have proven to be most useful for membrane proteins and highlighted particular problems which may be encountered in membrane research.

In principle, electrophoretic separation of proteins can be carried out under conditions either where the native conformation of the protein is preserved or where the protein is denatured. For membrane proteins, solubilization with preservation of structure is generally achieved by the use of mild detergents. The protein is then in a detergent or detergent/lipid micelle and can be separated from other such proteins by electrophoresis. As with many other

Techniques for the Analysis of Membrane Proteins. Edited by C. I. Ragan & R. J. Cherry.
Published in 1986 by Chapman and Hall, 11 New Fetter Lane, London EC4P 4EE

protein separation techniques, the resolution that can be achieved is less than that found with water-soluble proteins, and the method is of limited utility. Superior resolution can be achieved by electrophoresis under denaturing conditions, of which the most common denaturant is SDS. The latter is therefore the most widely used method for analysis of membrane proteins and, as a reflection of this, the bulk of this chapter is concerned with electrophoresis of SDS–protein complexes in polyacrylamide gels. Other less commonly used denaturants and electrophoresis in non-denaturing gels are given briefer coverage at the end of the chapter.

## 1.2 Gel electrophoresis in SDS

The interaction of a large number of soluble proteins with SDS has been extensively studied by Tanford, Reynolds and co-workers (Reynolds and Tanford, 1970; Fish *et al.*, 1970; Makino *et al.*, 1974). Most of these proteins were found to bind close to 1.4 g of SDS per g of protein, or one molecule of SDS per two amino acid residues (Nielson and Reynolds, 1978) and adopted a unique conformation in which the Stokes radius was a function of the molecular weight only. This property has been exploited to provide a simple and powerful method for the separation of polypeptides and measurement of their molecular weights by electrophoresis of SDS–protein complexes through polyacrylamide gels of various porosities. The fact that some proteins do not bind the usual amount of SDS or do not migrate in electrophoresis strictly according to their molecular weights does not, of course, prevent them from being separated from other proteins by this method. Therefore, I have considered separation methods and molecular weight determination as separate issues.

### 1.2.1 SOLUBILIZATION OF MEMBRANE PROTEINS BY SDS

The ability of SDS to solubilize otherwise recalcitrant membrane proteins accounts for the popularity of SDS-gel electrophoresis. The original conditions of Weber and Osborn (1969) are still widely used (viz. 1% SDS, 1% 2-mercaptoethanol, up to 1 mg of protein per ml in 0.01 M sodium phosphate, pH 7.0, 100° C for 2 min) and appear to be suitable for the majority of pròteins. The presence of 2-mercaptoethanol to reduce disulphide bonds is necessary for molecular weight measurements since it enables polypeptide chains to unfold completely. However, it is not obligatory and treatment with, say, *N*-ethylmaleimide and SDS may be desirable where interchain disulphide bonds need to be preserved (for example after cross-linking with a disulphide-containing bifunctional reagent; Wang and Richards, 1974). Weber *et al.* (1972) discussed various problems that may arise with the standard solubilization procedure. Firstly, heating may give rise to limited hydrolysis of the polypeptide. Omitting the 100° C step may solve this but may then render the sample prone to proteolytic digestion from endogenous proteases, particularly in crude preparations. Addition of protease inhibitors is then advisable (e.g. phenylmethanesulphonyl fluoride for serine proteases, *o*-phenanthroline for divalent cation-dependent proteases and *N*-ethylmaleimide for cysteine proteases). Alternative solubilization procedures such as 7–8 M guanidine hydrochloride and 2-mercaptoethanol followed by alkylation of sulphydryl groups with iodoacetamide may be used (Weber *et al.*, 1972) but these conditions are unlikely to solubilize all membrane proteins. An additional problem with membrane proteins is that of aggregation. Proteins containing a high proportion of hydrophobic amino acids may aggregate with themselves or other such proteins, even in the presence of SDS, and fail to enter the gel. Diverse conditions may promote this: high temperature, organic solvent extraction, trichloroacetic acid precipitation, and freeze–thawing (Briggs and Capaldi, 1977; Merle and Kadenbach, 1980). Aggregates are usually visible as a band on top of the gel or even in the stacking gel (when used). To obviate this problem, solubilization at ambient temperature (perhaps aided by sonication) and the presence of additional denaturants such as urea (Ludwig *et al.*, 1979) should be tried. There are no rules, but it is advisable to try several solubilization strategies with a new membrane system.

### 1.2.2 COMPOSITION OF POLYACRYLAMIDE GELS AND BUFFERS

An enormous variety of gel configurations and gel or buffer compositions is available for separation of SDS–protein complexes. A good starting point is the original procedure of Weber and Osborn (1969) which employed gels containing a fixed single concentration of acrylamide (e.g. 10% w/v) and bisacrylamide (1:37 by weight) as the cross-linker. A continuous buffer system of sodium phosphate containing SDS was used and electrophoresis was carried out on gels cast into cylindrical tubing. In the search for improved resolution and convenience, there have been developments in three areas: the shape of the gel, the buffer systems used and the composition of the gel.

#### (a) *Slab gels and cylindrical gels*

Cylindrical gels have now been largely superseded by slab gels. Under otherwise identical conditions, resolution on typical thin slab gels (e.g. 1.5 mm thick) is better than in typical cylindrical gels (e.g. 5 mm thick). An important factor is more even temperature control, particularly in those types of apparatus in which the slab gel is totally immersed in circulating buffer. A common problem with air-cooled gels is 'smiling' of the protein bands caused by overheating of the middle of the gel. This, of course, is worse the thicker the gel layer. The thinness of slab gels conveys other advantages. They can be more easily dried down for radioautography or fluorography (Section 1.2.3.c); it is easier to infiltrate other reagents for protein staining (Sections 1.2.3.a and 1.2.3.b) or immunological identification (Section 1.2.3.d) and protein bands can be transferred on to nitrocellulose paper or other transfer media for further analysis (Section 1.2.5). Molecular weight determination (Section 1.2.4) is facilitated by running several samples in parallel tracks on the same slab gel. Mobility between tracks is more reproducible than between apparently identical cylindrical gels. Finally, complex gel compositions such as gradients of acrylamide can be conveniently cast in slab form but not in cylindrical form.

The only disadvantage of thin slab gels is their lower sample capacity. For example, for analysis of radioactive proteins of low specific radioactivity, and when resolution is not paramount, cylindrical gels may be more convenient than thick slab gels and are more easily sliced and counted (Sections 1.2.3.c).

Several commercial designs for slab gel electrophoresis are available, but many workers prefer to build their own. A convenient design for simultaneous casting of several slab gels is described below (Section 1.2.2.c).

(b) *Continuous and discontinuous buffers*

Despite their added complexity, discontinuous buffer systems are now largely used in preference to continuous systems. The gel consists of two parts: a short (1–2 cm) stacking gel of large porosity and a separating gel. Different buffer compositions are used for the two gels and in the electrode compartments. The theory of such 'disc' systems has been described by Ornstein (1964) and Davis (1964) and they have been widely adopted for separation of SDS–protein complexes by the simple addition of SDS to some or all of the buffers. The samples are applied to the stacking gel in which they are concentrated to a very small volume ('stacking'). The change in buffer composition when the separating gel is reached causes 'unstacking' and the SDS–protein complexes separate as in continuous systems. The initial concentration to a small volume results in improved resolution and enables larger sample volumes to be handled with no loss of resolution. Several buffer systems have been described of which the most popular is that of Laemmli (1970). Others have been devised by Jovin *et al.* (1971) and the use of two of these is fully discussed by Neville and Glossman (1974). From my own experience they all produce very similar results. In Table 1.1, the compositions and properties of these systems are given. The drawback of disc systems is that they only work properly with scrupulous attention to the buffer composition and pH. In particular, high-ionic-strength samples do not stack properly and dilute protein samples may have to be dialysed, for example, to get them into an appropriate buffer. In gels of constant composition containing relatively low acrylamide (e.g. 10–12%), low-molecular-weight polypeptides may not unstack from the dye front. A brief inspection of published work reveals how common this is, but it is rarely referred to. The problem again lies in the buffers; particularly if these are not

**Table 1.1** Composition of some discontinuous buffer systems for SDS-gels
The systems J4179 and J3561 of Jovin *et al.* (1971) are described by Neville and Glossmann (1974). These have different buffers for the separating gel, the running gel and the two electrode compartments. The Laemmli (1970) system employs the same buffer for both electrode compartments.

| *Buffer* | *J4179* | *J3561* | *Laemmli* |
|---|---|---|---|
| Separating gel | 0.427 M Tris–HCl, pH 9.18 | 0.0716 M imidazole–HCl, pH 4.67 | 0.375 M Tris–HCl, 0.1% SDS, pH 8.8 |
| Stacking gel | 0.0541 M Tris–$H_2SO_4$, pH 6.14 | 0.0694 M imidazole–$H_2SO_4$, pH 4.72 | 0.125 M Tris–HCl 0.1% SDS, pH 6.8 |
| Cathode | 0.0406 M Tris–boric acid, 0.1% SDS, pH 8.64 | 0.0362 M Tris–imidazole, 0.1% SDS, pH 8.0 | 0.25 M Tris, 0.192 M glycine, 0.1% SDS (cathode and anode) |
| Anode | 0.427 M Tris–HCl, pH 9.5 | 0.733 M imidazole, 0.136 M HCl | |

freshly prepared. The easiest way out is to decrease the porosity of the gel so that small proteins do not migrate close to the dye front.

(c) *Composition of gels*

Polyacrylamide is a linear polymer and must be cross-linked to provide pores for separation of SDS–protein complexes. Bisacrylamide is certainly the most common cross-linker and is generally used at a concentration of one part to approximately 30 parts of acrylamide (Weber and Osborn, 1969). Other cross-linkers are described in Section 1.2.3c. Variation of the pore size of gels can be achieved by varying the amount of the acrylamide–bisacrylamide mixture (expressed usually just as the percentage by weight of acrylamide), the acrylamide to bisacrylamide ratio, or by addition of other reagents to the mixture such as urea or glycerol.

Gels which cover the widest range of polypeptide sizes are those which have a gradient of acrylamide increasing from the origin to the end of the gel. Linear or exponential gradients can be used and the concentrations and shape of the gradient are entirely a matter of choice. The effect on protein bands of moving always into a region of smaller porosity is to prevent them from spreading by diffusion and greatly increases the resolution, particularly of smaller proteins. With gradients covering a wide variation in polyacrylamide concentration, the proteins may eventually become almost stationary, but this is not necessary for a considerable enhancement in the resolution. Used with discontinuous buffer systems, gradient gels have the additional advantage that small proteins no longer fail occasionally to unstack since they inevitably encounter an acrylamide concentration where they leave the ion front. These gels are therefore more tolerant of ageing or imprecisely made up buffers.

Single-gradient slab gels can be made as described by Douglas *et al.* (1979) using a gradient mixer to deliver the monomer solutions. Gels containing 10% to 15% acrylamide separate proteins in the range 150 to 10 kDa. The acrylamide concentrations can be varied at will, but some adjustment to polymerizing reagent concentrations may also be necessary. It is advisable to ensure that the gel polymerizes from the top downwards (lower to higher concentration) to prevent convection currents disturbing the gradient. For routine use of gradient gels it is an advantage to be able to cast several at one time. This can be done using the apparatus shown in Fig. 1.1. Cassettes of the desired dimensions are constructed from glass plates, Perspex spacers and gel sealing tape of the type supplied by several manufacturers of electrophoretic equipment. Baking the assembled cassettes at 100° C overnight prevents the tape from lifting off during storage of the gels in buffer. The cassettes are stacked in the apparatus which is first filled with 5% (v/v) aqueous ethanol to act as an overlay. The monomer is then admitted from underneath via a gradient mixer and finally sucrose solution is admitted to push the acrylamide up to within 1–2 cm of the top of the cassettes. After setting, the solid block of

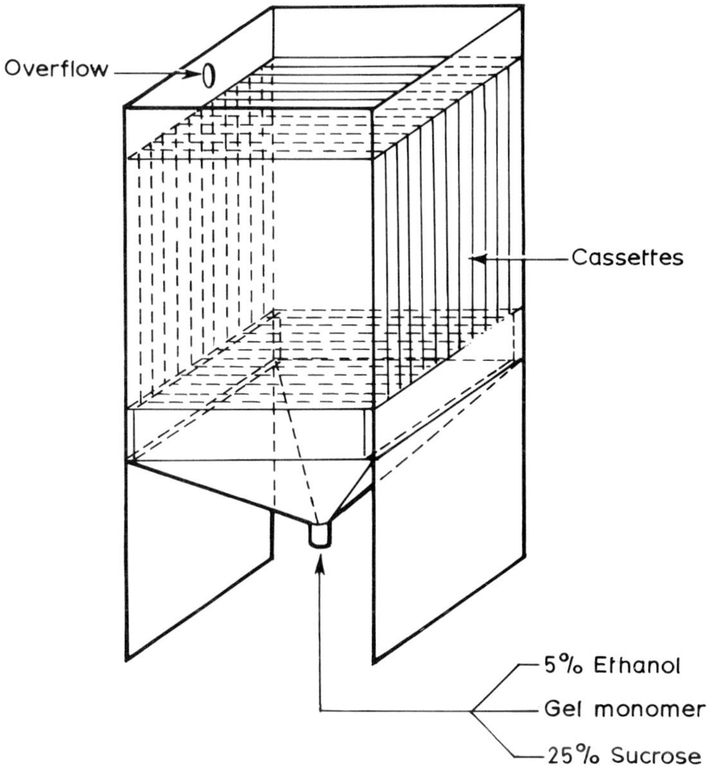


**Fig. 1.1** Apparatus for simultaneous casting of several slab gels. The apparatus is constructed out of Perspex. After filling with assembled cassettes, 5% (v/v) ethanol is admitted through the bottom. This is displaced through the overflow by acrylamide monomer from a gradient mixer followed by sucrose underlay until the acrylamide is approximately 1 cm from the top of the cassette. This space is used for a stacking gel and sample comb. Apparatus of this type is available from Universal Scientific Ltd., London, UK.

cassettes is extruded by water pressure, excess acrylamide is scraped off and the gels are stored in running gel buffer in the cold. The method of casting removes the need to use grease or clamps to seal the cassettes and spacers together. Immediately before use, a stacking gel is cast on the top of the running gel with slots for samples. These gels can be kept for weeks without any alteration in running characteristics.

While gradient gels may provide adequate resolution of small proteins (2–10 kDa), there are other methods. The first of these was that of Swank and Munkres (1971) who introduced a still popular modification of the original Weber and Osborn (1969) procedure which was to include 7–8 M urea in the sample and gel and increase the acrylamide concentration and bisacrylamide to acrylamide ratio. The urea reduces the pore size giving better separation of small proteins. Several modifications of this procedure have been published in recent years. Cabral and Schatz (1979) used a high concentration (20%) acrylamide gel containing 10% (w/v) glycerol. Merle and Kadenbach (1980) used 16% gels containing urea or urea and glycerol and a discontinuous buffer system, while Hashimoto *et al.* (1983) described the use of discontinuous gradient gels containing 7 M urea and an increased bisacrylamide to acrylamide ratio. Table 1.2 summarizes these systems.

Separation of very large proteins has not received the same amount of attention as has separation of small proteins. Large proteins, though, can be difficult to transfer efficiently on to nitrocellular or other transfer matrices (Section 1.2.5), and Elkon *et al.* (1984) report that mixed agarose–polyacrylamide gels (Warr, 1982) are helpful in this respect. Resolution was improved by inclusion of 6 M urea.

As an illustration of the improvement in resolution which can be achieved using more complex electrophoretic systems, Fig. 1.2 shows separations of the proteins of mitochondrial NADH dehydrogenase on 12.5% (w/v) acrylamide cylindrical gels using either the Weber and Osborn (1969) buffer system or the Laemmli (1970) discontinuous buffer system and on 12.5–16% (w/v) gradient slab gels using the Laemmli buffer system.

### 1.2.3 VISUALIZATION OF PROTEIN BANDS ON SDS-GELS

(a) *Coomassie Blue*

Coomassie Blue (or Brilliant Blue R250) is greatly preferred to other simple chemical stains on the grounds of sensitivity. As little as 0.1 μg of a protein can readily be detected on gels with good resolution. The early staining and destaining procedure of Weber and Osborn (1969) is still widely used. This involves staining with a solution of Coomassie Blue in aqueous acetic acid and methanol. Other procedures employ a prior stage of 'fixing' with 10% trichloroacetic acid or, for small basic proteins, formaldehyde (Steck *et al.*, 1980). Destaining can be carried out with 25% methanol and 10% acetic acid, preferably at 50° C to speed up the process.

The stain intensity is proportional to the amount of protein present (Fazekas de St. Groth *et al.*, 1963) over the range 1–15 μg, but the proportionality constant may vary considerably between different proteins. The use of relative stain intensity combined with molecular weights to estimate polypeptide stoichiometry is therefore rather risky, but nevertheless widely used, particularly when it gives the expected answer. Inevitably, membrane proteins can show an abnormal staining response to Coomassie Blue and the rate at which they take up stain can vary widely. With too short a staining period, some protein bands might be missed altogether and it is advisable to extend the staining period up to several hours, even with thin slab gels (e.g. Kadenbach

**Table 1.2** Systems for separation of low-molecular-weight proteins

| *System* | *Acrylamide concn. (%)* | *Bisacrylamide to acrylamide ratio* | *Other constituents* | *Buffers* |
|---|---|---|---|---|
| Swank & Munkres (1971) | 12.5 | 1:10 | 8 M Urea | Continuous, Tris–$P_i$* |
| Cabral & Schatz (1979) | 10/20 (step)† | 1:32 | 10% Glycerol | Discontinuous. Laemmli‡ |
| Merle & Kadenbach (1980) | 16 | 1:32 | 13% Glycerol + 3.6 M Urea | Discontinuous, Laemmli |
| Hashimoto *et al*. (1983) | 10–18 (gradient) | 1:20 | 7 M Urea | Discontinuous, Laemmli |

*0.1 M $H_3PO_4$ adjusted to pH 6.8 with Tris, 0.1% SDS.
†A 10% gel was layered on to a 20% gel to separate both large and small proteins.
‡Laemmli (1970).

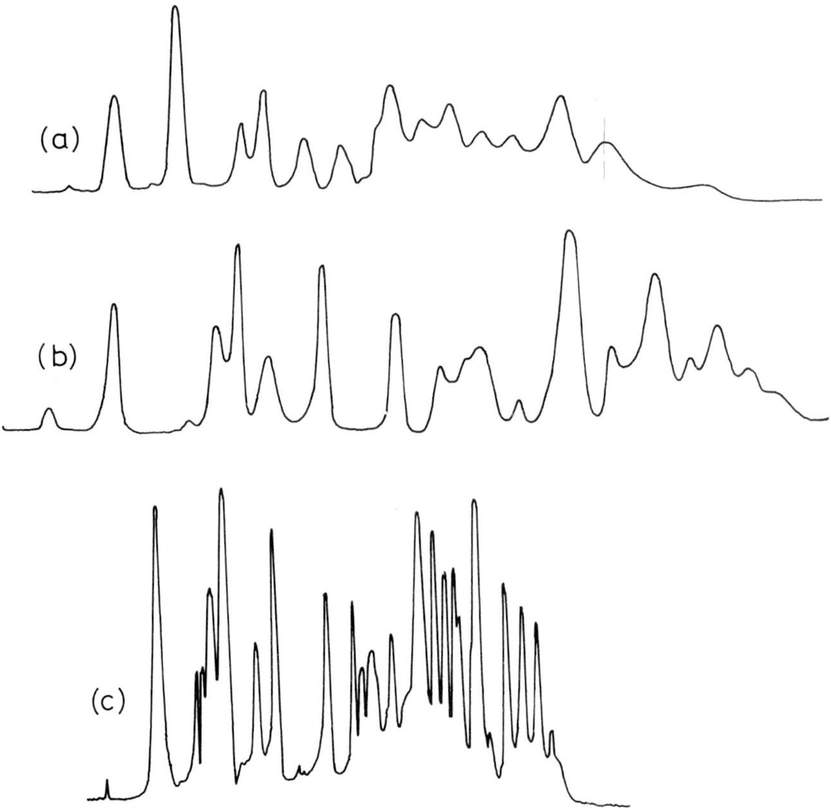


**Fig. 1.2** Separation of proteins by SDS-gel electrophoresis. Bovine heart mitochondrial NADH–ubiquinone reductase was analysed by SDS-gel electrophoresis in three different systems. In (a) electrophoresis was carried out in the continuous system of Weber and Osborn (1969) on 12.5% (w/v) acrylamide gels cast in tubes 12 cm long by 5 mm internal diameter. For (b), cylindrical gels as in (a) were used but with the discontinuous buffer system of Laemmli (1970). For (c), samples were analysed on slab gels, 1.5 mm thick by 10 cm long, containing a 12% (w/v) to 16% (w/v) concave gradient of acrylamide and using the Laemmli (1970) discontinuous buffer system. Gels were stained with Brilliant Blue and scanned in a densitometer. Origin of electrophoresis is on the left for each trace.

and Merle, 1981). In addition, glycoproteins frequently stain rather poorly with Coomassie Blue (Glossmann and Neville, 1971).

(b) *Silver stain*

Greater sensitivity can be achieved through the use of silver staining methods. Since its original introduction (Merrill *et al.*, 1979; Switzer *et al.*, 1979) the process has gone through a remarkable evolution and is now simple and cheap enough for routine use. Many modifications have arisen from the procedure of Oakley *et al.* (1980) which was considerably cheaper than earlier methods, but suffered, at least according to other workers, from problems with reproducibility. The basic procedure involves fixing of the proteins in the gel, treatment with silver nitrate solution (usually ammoniacal silver nitrate) followed by selective reduction of the silver ions to metallic silver in the presence of the proteins. The increase in sensitivity over Coomassie Blue is generally anywhere between 10- and 100-fold, depending on the exact procedure and the nature of the protein.

Fixing of the proteins is critical to the process. Glutaraldehyde, formaldehyde or methanol and acetic acid are all used, sometimes in combination. The latter reagent, as used in the procedures of Merril *et al.* (1981a,b), may fail to fix some proteins, particularly small basic ones (Steck *et al.*, 1980). Prior fixation with glutaraldehyde (Oakley *et al.*, 1980) or formaldehyde (Eschenbruck and Bürk, 1982) or methanol and acetic acid followed by glutaraldehyde (Morrissey, 1981; Porro *et al.*, 1982; Schleicher and Watterson, 1983) is therefore recommended. Irie *et al.* (1982) find more reproducible results if the gel is prestained with Coomassie Blue in the presence of methanol/acetic acid or, preferably, formaldehyde, and attribute this at least partly to enhanced retention of stained proteins in the gel. Prior staining with Coomassie Blue has the added advantage that major components and minor components can be studied independently.

Some proteins fail to stain well with certain silver procedures, and it is advisable to try more than one protocol with a new membrane system. Oakley *et al.* (1980) found that their modification visualized calmodulin which was poorly stained by earlier procedures. Schleicher and Watterson (1983) traced this to the use of glutaraldehyde for prefixing the proteins. While leaching of the protein from the gel, even after glutaraldehyde fixation, can account for some apparent loss of sensitivity to stain, this is by no means the whole story and the role of glutaraldehyde is more complex than this. Morrissey (1981), for example, found that the Oakley procedure failed to stain some proteins (despite the glutaraldehyde step) and that a soaking of the gel in dithiothreitol cured this problem as well as removing the need for illumination in the photochemical silver stain of Merril *et al.* (1981b). A mechanism for glutaraldehyde enhancement has been proposed by Dion and Pomenti (1983). A very hydrophobic subunit of mammalian cytochrome oxidase (subunit III) has been found to respond poorly to the Oakley silver stain procedure (Kadenbach *et al.*, 1983). Whether Morrissey's dithiothreitol would help is not known. This suggests that silver staining of membrane proteins may be highly variable, perhaps more so than Coomassie Blue staining.

Dark backgrounds can be a problem with silver staining, since reduction to silver can occur all over the gel, albeit at a slower rate than reduction at the protein bands. Proper washing between steps and using the correct $NH_3$ concentration can control this to a great extent (Eschenbruck and Bürk, 1982). Some procedures, however, deliberately overstain and then destain. A recent improvement (Marshall, 1984) to an earlier silver stain protocol (Marshall and Lattner, 1981; Marshall, 1983) describes a more controllable destaining procedure. Alternatively, washing between steps is largely avoided in the procedure of Porro *et al.* (1982) by using glutaraldehyde both as fixer and reducing agent.

Tasheva and Desser (1983) reported that mercaptoethanol or dithiothreitol in the sample buffer can give rise to two major artefactual bands on gels, stainable with silver (and to a lesser extent, Coomassie Blue) of apparent molecular weights 68000 and 54000. These appear to arise from skin proteins (Ochs, 1983) or from bacterial contamination of buffers and solutions (Marshall and Williams, 1984). The sensitivity of silver stain methods clearly demands very high standards of laboratory cleanliness.

(c) *Radiochemical detection*

The obvious way to assay radioactivity in protein bands is to slice the gel, extract the radioactivity and count by scintillation. For cylindrical gels, this is usually the only practical way to proceed and, for quantitative measurement, it is still the best for any kind of gel. Commercial gel slicers can provide uniform slices of less than 1 mm thick providing quite good resolution. To bring the protein into contact with the scintillant, the gel is either chemically degraded or the protein extracted from the slices. For conventional polyacrylamide/bisacrylamide gels, hydrogen peroxide can be used to dissolve the gel or, alternatively, specifically cleavable cross-linking agents can be used instead of bisacrylamide. *N,N'*-Diallyltartardiamide (Anker, 1970; Baumann and Chrambach, 1976) and *N,N'*-dihydroxyethylenebisacrylamide (O'Connell and Brady, 1977) are cross-linkers specifically cleavable with periodic acid while gels made with *N,N'*-bisacrylylcystamine (Hansen, 1977) can be dissolved with 2-mercaptoethanol. Despite the availability of these alternative cross-linkers, they have not achieved very wide popularity simply because it is not difficult to quantitatively release the radioactivity from conventional gel slices by use of hydrogen peroxide (as mentioned above) or tissue solubilizers such as NCS solubilizer from Amersham (Basch, 1968) or Protosol from New England Nuclear (Hanstein, 1979).

For many applications, quantitative estimation of radioactivity is not essential and slicing and counting may produce unacceptable loss of resolution. Under these conditions, radioautography or fluorography of slab gels is the method of choice. Recently, methods have been devised to enable radioautographic or fluorographic discrimination between different isotopes, thus bringing the power of dual-label experiments to the technique.

The commonly encountered isotopes are $^{125}I$, $^{32}P$, $^{35}S$, $^{14}C$ and $^{3}H$. The gamma and high-energy $\beta$ emissions of the first two are suitable for radioautography in which the dried gel is simply placed in contact with an appropriate X-ray film (e.g. Kodak RP X-Omat). Intensifying screens behind the film cause considerable image enhancement. Pre-exposure of the films to low light intensity and low temperature (e.g. $-70°$ C) also contribute considerably to the intensity of the image as well as giving a more linear response between image density and the number of radioactive disintegrations (Bonner and Laskey, 1974; Laskey and Mills, 1975). For generally speeding up the process, these extras are well worth the trouble.

For the low-energy $\beta$ emitters, direct radioautography is inefficient, especially for $^{3}$H. Fluorography is therefore used in which the gel is impregnated with a fluorophore (scintillant). Again combined with low temperature and preflashed film, the method gives sensitive and linear response to these isotopes and is indispensable for $^{3}$H detection (Bonner and Laskey, 1974; Laskey and Mills, 1975). Impregnation with a variety of scintillants is possible. The original procedure, still widely popular, used 2,5-diphenyloxazole (PPO) in dimethyl sulphoxide. This method smears Coomassie-stained bands somewhat and working with dimethyl sulphoxide is unpleasant. En$^{3}$Hance from New England Nuclear and Amplify from Amersham are considerably less pungent reagents which smear stained bands rather less and give very good results with $^{3}$H. Salicylic acid can also be used as a fluorophore (Chamberlain, 1979).

Double-label analysis for mixtures of $^{14}$C and $^{3}$H have been described by McConkey (1979) and Choo *et al.* (1980). The former method relies on the inability of Kodak No-screen film to detect $^{3}$H in gels, thus enabling detection of either $^{3}$H plus $^{14}$C or $^{14}$C alone. Choo *et al.* (1980) used fluorography for $^{3}$H plus $^{14}$C and radioautography through carbon paper for $^{14}$C alone. Similar procedures would no doubt allow discrimination of $^{3}$H and $^{35}$S or other isotopes with higher-energy $\beta$ or $\gamma$ emission. Cooper and Burgess (1982) describe a system for simultaneous detection of $^{35}$S and $^{32}$P. These double-label procedures are particularly useful for two-dimensional gels (Chapter 2) where conventional methods are inappropriate.

(d) *Immunological detection*

While SDS–protein complexes may be functionally inactive, they may well still be immunologically active and the identification of antigenic polypeptides in SDS-gels can be a useful experimental tool. Methods have been described for doing this involving overlaying the polyacrylamide slab gel with an antibody-containing agarose gel and allowing diffusion from the former gel to deposit precipitin lines in the agarose (e.g. Cabral *et al.*, 1978) or crossed-immunoelectrophoresis with SDS-gel electrophoresis as the first dimension (e.g. Converse and Papermaster, 1975).

The former method has been totally superseded by detection of antigen–antibody complexes in electrotransfers (Section 1.2.5) while crossed-immunoelectrophoresis is the subject of a later chapter in this volume.

(e) *Staining for glycoproteins*

Glycoproteins are common constituents of membranes, particularly plasma membranes. The classical procedure for their identification in gels is the periodic acid–Schiff stain. Gels are fixed in methanol/acetic acid or trichloroacetic acid, the carbohydrate moieties oxidized with periodic acid and stained with fuchsin–sulphite solution (McGuckin and McKenzie, 1958). Procedures are detailed by Dewald *et al.* (1974) and Neville and Glossmann (1974). This procedure is insensitive and does not provide information on the composition of the various oligosaccharide moieties present in mixtures of glycoproteins (such as those found in membranes). These problems are solved by infiltrating SDS-gels fixed in glutaraldehyde or methanol/acetic acid with various $^{125}$I-labelled lectins and determining the distribution of label by radioautography after washing (Robinson *et al.*, 1975; Tanner and Anstee, 1976; Burridge, 1976). The main drawback to the method is the time taken to infiltrate the lectin and to wash out the background (e.g. up to 6 days in all). As with immunological identification of proteins (Section 1.2.3.e), electrotransfer of the gel (Section 1.2.5) reduces the time required for processing enormously (Glass *et al.*, 1981).

### 1.2.4 DETERMINATION OF MOLECULAR WEIGHTS FROM SDS-GELS

The observation that SDS–protein complexes separate according to the protein molecular weight provides the easiest and most common technique for determination of the latter. Standard proteins, whose molecular weights are known accurately from their sequence, are used to determine the relationship between mobility and molecular size and mixtures, cold or radioactively labelled and covering various ranges, are available commercially. Accurate determination relies on the protein (standard or unknown) binding the same amount of SDS per gram of protein and adopting the same conformation as any other. This has been verified for a large number of soluble proteins whose Stokes radius in SDS is proportional to the 0.73 power of the molecular weight. However, this has been done in depth only for SDS–protein complexes in sodium phosphate buffer at pH 7.2. As Nielson and Reynolds (1978) point out, the SDS-binding characteristics and conformation of the protein may vary with ionic strength, ionic composition, pH and the presence of reagents such as urea and detergents other than SDS. Thus, the determination of molecular weight in gels run in discontinuous buffer systems or in those containing urea may not necessarily be accurate. Variations in apparent molecular weight in different gel systems are widely exploited to achieve better resolution of similar sized proteins and examples of proteins which exchange positions in different gels are known. For analyses of complex mixtures, the use of differing electrophoresis conditions can lead to glorious confusion in comparing results from different laboratories. The problem of cytochrome oxidase nomenclature is very illustrative (Kadenbach and Merle, 1981).

Membrane proteins are frequent offenders in this respect and may behave atypically under all conditions of electrophoresis. Kadenbach *et al.* (1983), for example, find that the order of separation of cytochrome oxidase subunits on SDS-gels is not strictly in accordance with the sequence-determined molecular weight and, even when it is, the apparent molecular weights may differ considerably from the real. Since, for many membrane proteins, the only measure of molecular weight is by gel electrophoresis, the possibility of appreciable error is a serious problem. Similar remarks apply to glycoproteins. Neville and Glossmann (1974) provided calibration curves relating the mobility in discontinuous gels to molecular weight for both soluble proteins and soluble glycoproteins. Deviation from theoretical behaviour is considerably more marked with the glycoproteins and they suggest a 30% margin of error is possible. However, Lambin (1978), using gradient gels, found a mean deviation from true molecular weight of only 5.5% for a wide variety of glycoproteins. It appears that the use of gradient gels may provide greater accuracy for glycoprotein molecular weights (Leach *et al.*, 1980; Poduslo, 1981). Membranous glycoproteins may therefore migrate anomalously both as a result of their hydrophobic nature and the carbohydrate moiety and may prove to be the most deviant of all.

While the degree of error cannot be estimated, except by independent measurements, it is sometimes possible to check whether a protein or glycoprotein is misbehaving by its adherence to the Ferguson equation (Ferguson, 1964) which was derived theoretically by Neville (1971), namely:

$$\log M = \log M_0 - K_R T$$

where $M$ is the electrophoretic mobility in a gel of acrylamide concentration $T$, $M_0$ is the free electrophoretic mobility and $K_R$ is the retardation coefficient. Well-behaved SDS–protein complexes have the same or very similar values for $M_0$, while $K_R$ is a uniform function of molecular weight, thus explaining the dependence of $M$ purely on molecular size. Banker and Cotman (1972) investigated this expression for a number of proteins and showed a large deviation in $M_0$ for one glycoprotein. In the absence of the true molecular weight, only variations in $M_0$ can be tested. However, it is likely that anomalous dependence of $K_R$ on molecular size can also cause spurious molecular weight determinations. The Ferguson plot would not reveal this.

Lastly, it should be pointed out that some commercial SDS preparations are seriously contaminated with tetradecyl and hexadecyl sulphates whose different binding properties can cause anomalous migration of proteins (e.g. Best *et al.*, 1981). A recent paper described the migration of a single membrane protein as two clearly separated bands when impure SDS was used (Margulies and Tiffany, 1984).

### 1.2.5 ELECTROTRANSFER FROM SDS-GELS

The electrophoretic transfer of protein bands from SDS-gels on to an adsorbent matrix (electroblotting, Western blotting) is an extremely powerful technique which renders the protein much more accessible to external reagents and therefore speeds up or increases the sensitivity of the reaction with these reagents. This can more than offset the additional time spent in making the transfer as well as offering certain unique experimental advantages. The principle of the method involves subjecting a sandwich of gel and transfer matrix to a transverse electrical field, causing the SDS–protein complexes to migrate from the gel on to the surface of the transfer matrix. Apparatus can be constructed along the lines described by Towbin *et al.* (1979) or Bittner *et al.* (1980), for example, or can be obtained commercially. The method has been reviewed by Gershoni and Palade (1983) and only a brief account of its applications is presented here.

(a) *Choice of transfer matrix*

Nitrocellulose membranes are the most widely used for protein transfers (Towbin *et al.*, 1979; Bittner *et al.*, 1980). At the pH at which electrotransfer is usually performed, both the protein and the membrane are negatively charged. Hydrophobic effects are likely to be important for binding therefore. Some proteins interact poorly with nitrocellulose, and its capacity, in any case, is not large. Nylon-based membranes may be advantageous (Gene Screen from New England Nuclear and Zetabind from AMF Cuno Division, Meridien, Connecticut). The latter is positively charged and therefore allows electrostatic interaction between protein and membrane resulting in increased capacity (Gershoni and Palade, 1982). Covalent attachment of the protein is also used to promote efficient transfer. Diazo paper (diazobenzyloxymethyl cellulose or diazophenylthioether cellulose) is the usual way of achieving this (Renart *et al.*, 1979; Reiser and Wardale, 1981). Such transfers are virtually indestructible and are extensively used for immunological identification of proteins since they can be 'erased' (i.e. the antibody removed) and used to detect another antigen.

(b) *Visualization of proteins on transfers*

Nitrocellulose membranes can be stained with the usual protein reagents, e.g. Coomassie Blue or Amido Black, of which the latter gives the greater sensitivity (Towbin *et al.*, 1979). India ink (Hancock and Tsang, 1983) is greatly superior and can give staining not far short of the intensity obtained with silver staining of gels (Moeremans *et al.*, 1985). A modified silver stain for nitrocellulose transfers has been described (Yuen *et al.*, 1982) but it results in a negative image. The most sensitive and perhaps the most convenient procedure may prove to be the use of gold or silver sols (Moeremans *et al.*, 1985). The former gives results of equal sensitivity to conventional silver staining of proteins in polyacrylamide gels and it may be easier to control and reproduce, for example, overstaining of the background does not occur. A very general procedure for detection of total protein on transfers was described by Wojtkowiak *et al.* (1983). This involves the dinitrophenylation of the transferred protein, reaction with antiserum to dinitrophenol and identification of bands containing the bound immunoglobulin.

Staining with Coomassie Blue, Amido Black and India ink is unsuitable for use with diazo papers or nylon transfer membranes because of nonspecific binding. This may also be true for the other stains as well, and, in any case, the colour of diazo paper transfers would certainly interfere with silver or gold staining. In this respect, the immunological method of Wojtkowiak *et al.* could be useful as immune complexes can be visualized by binding of $^{125}$I-labelled protein A and radioautography.

Frequently, visualization of total protein on a transfer is not needed, and, in any case, a parallel acrylamide gel can be run and stained if so desired. However, the transfer of protein to membrane or paper is not quantitative and varies with the size of the protein in particular. Methods of mitigating this problem are described by Gershoni and Palade (1983), but, in general, it is advisable to regard the transfer as suitable for qualitative studies only.

Radioactive material on transfers can be detected by radioautography or fluorography. Symington *et al.* (1981) found that the efficiency of radioautographic detection of $^{14}$C or $^{35}$S is greater on transfers than in dried acrylamide gels. Fluorographic detection is straightforward since impregnation with scintillant is very simple for transfers. However, there seems to be no advantage in detection efficiency (Erickson *et al.*, 1982).

The main attraction of transfers is their use for detection of specific ligand interactions far more rapidly and conveniently than is possible with the original gel. Generally, the transfer is first 'quenched' to prevent non-specific binding of the ligand, most commonly by using protein (e.g. bovine serum albumin, gelatin, ovalbumin, skimmed milk powder, or haemoglobin). After washing, the transfer is reacted with the ligand which is visualized, after further washing, by its radioactivity or some other specific property. Non-ionic detergents (e.g. Tween-20) are frequently used in quenching and washing buffers to reduce background. The application of the method is obvious where the determinant for the reaction is part of the amino acid sequence (e.g. antigen–antibody reaction) or some other non-denaturable group (e.g. lectin–glycoprotein interaction). However, the ability of proteins to spontaneously renature from SDS solution allows detection of other ligand-binding reactions which should depend on the secondary and tertiary structure of the protein. If the affinity of the ligand for the native protein is very high, then even a small fraction of renatured molecules will allow specific identification of the reaction. Thus, the method is suitable for identification of hormone or other receptors in membranes, nucleic acid-binding proteins, calmodulin-binding proteins, and proteins binding to whole cells. A summary of these is given in Table 1.3 and no doubt new applications will be forthcoming.

A polyclonal antiserum to a protein will generally consist of antibodies to many different determinants on the protein. Thus, not only the native enzyme, but proteolytically generated fragments can be identified in gels (Mendel-Hartvig and Nelson, 1983) or on transfers. The use of proteolytic enzymes to study membrane topology (Chapter 6) can therefore be extended to membranes containing many different proteins.

**Table 1.3** Applications of transfers

References to these applications are given in the text or in the review of Gershoni and Palade (1983)

*Types of ligand–protein interactions*

(a) DNA or RNA–nucleic acid-binding proteins
(b) Subunit–subunit interactions in multisubunit complexes
(c) Hormone–receptor interactions
(d) Antibody–antigen complex formation
(e) Glycoprotein characterization by lectin binding
(f) Identification of proteins involved in cell adhesion by binding of whole cells
(g) Calmodulin–protein interactions

*Other uses*

(h) Affinity purification of monospecific antibodies
(i) Immobilization of proteins for chemical modification
(j) Reconstitution of SDS-denatured proteins
(k) Assembly of multisubunit complexes
(l) More sensitive radioautography
(m) Erasable transfers for detection of different ligand–protein interactions

### 1.2.6 ELUTION OF PROTEINS FROM SDS-GELS

For many membrane proteins, separation on SDS-gels may be the only method for their complete purification. Proteins purified in this way, although denatured, can be used for amino acid sequence determination or for production of antisera.

More common even than papers on new silver staining procedures are papers on new methods for the elution of proteins from SDS-gels. Electrophoretic elution can give good recovery and can be applied to material which has previously been stained with, for example, Coomassie Blue. Methods fall broadly into two categories: firstly, those involving continuous elution of the protein bands from the gel and, secondly, those that involve elution from gel slices. The former methods require specialized apparatus and have not gained wide popularity. Bodhe *et al.* (1982) describe an ingenious apparatus in which the proteins are eluted transversely from the slab gel into polyurethane foam. Most methods, however, involve processing of gel slices containing the proteins of interest. The advantage is that slices from several gels, of any geometry, can be combined for isolation of milligram quantities of the protein. The slices of gel, which may have been previously stained to permit their identification, are cast into a further polyacrylamide or agarose gel and then electrophoretically eluted. The use of discontinuous buffer systems allows concentration of the protein band which can be eluted into a glycerol layer (Mendel-Hartvig, 1982) or into the interface between solutions of differing conductivities (Strälfors and Belfrage, 1983). Alternatively, the protein can be electrophoretically eluted into a dialysis bag (e.g. Walker *et al.*, 1982). The technique for visualizing proteins on gels using 4 M sodium acetate (Higgins and Dahmus, 1979) can be helpful in the slicing up of SDS-gels prior to elution.

## 1.3 Gel electrophoresis in other denaturing media

While SDS is the most commonly used denaturing detergent, there are others which may have advantages for certain aspects of membrane research.

### 1.3.1 LITHIUM DODECYL SULPHATE (LDS)

Lithium dodecyl sulphate is more soluble than SDS at lower temperatures and permits gel electrophoresis at 2° C. Under this condition, proteins may not be completely denatured and may retain non-covalently bound prosthetic groups and some degree of native conformation (Delepelaire and Chua, 1979). Guikema and Sherman (1981) showed that proteins of photosynthetic bacterial membranes could be separated on LDS-gels and retained chlorophyll and haem, thus allowing identification of chlorophyll binding proteins and cytochromes. Heating of samples in LDS led to loss of these prosthetic groups as with SDS. Similarly, Gershoni *et al.* (1982) found that identification of the toxin-binding subunit of the acetylcholine receptor was possible after separation of membrane proteins in LDS at low temperature and Guikema and Sherman (1982) have shown that iron–sulphur proteins can also be separated on LDS-gels without loss of the chromophore. LDS-gels are otherwise very similar to SDS-gels and all the previously described methods in Section 1.2 can be applied. However, because proteins are not fully dissociated, molecular weight determinations in LDS may be less reliable. Potentially, therefore, this is an extremely promising technique for membrane research and will undoubtedly attract wider attention in the future.

### 1.3.2 CHLORAL HYDRATE

A denaturing agent of quite different properties is chloral hydrate which, at high concentrations, very effectively solubilizes membranes. Several procedures have been described (Ballou *et al.*, 1974; Ballou and Smithies, 1977; Booth, 1977) of which the most recent is that of Griffin and Landon (1981). The separation on polyacrylamide gels depends not only on protein size but, to a lesser extent, on charge, and separation can therefore be altered by changes in pH (Griffin and Landon, 1981). The latter authors investigated cytochrome oxidase from bovine heart and identified fifteen components by one- and two-dimensional electrophoresis in chloral hydrate. The three components not previously identified by SDS-gel electrophoresis of this enzyme (Kadenbach and Merle, 1981) were shown to be present on these gels as highly diffuse areas. Chloral hydrate may therefore provide a useful alternative to SDS and has the added advantage that it can be easily removed from the protein by dialysis.

### 1.3.3 ACID–UREA–TRITON

Acetic acid, urea and Triton have long been used as solubilizing agents for the electrophoretic separation of histones (e.g. Franklin and Zweidler, 1977). The method permits resolution of proteins differing by only a single neutral amino acid substituent and therefore separates proteins not only according to size and charge but also to hydrophobicity. The method has been applied successfully to membrane proteins by Fernandes *et al.* (1978) and provides a degree of resolution similar to that of SDS-gel electrophoresis. The usefulness of the method lies in the fact that it may detect the products of point mutations which could be missed by isoelectric focusing and particularly by SDS-gel electrophoresis. After equilibration with SDS, acid–urea gels can be transferred on to nitrocellulose (Johnson *et al.*, 1983). Mold *et al.* (1983) found that histones separated on these gels do not silver stain very well. They attributed this either to these particular proteins lacking nucleation sites for silver deposition or to Triton masking the nucleation sites. The latter process could be important for membrane proteins in general, and Mold *et al.* report that prior staining with Amido black or 2,7-naphthalene disulphonic acid cures the problem.

### 1.3.4 CATIONIC DETERGENTS

Nozaki *et al.* (1974) found that the cationic detergent, tetradecyltrimethylammonium chloride (TMAC) interacted with proteins in a manner very similar to SDS. The TMAC–protein complexes are somewhat smaller than SDS–protein complexes but the Stokes radius is again proportional to the 0.73 power of the molecular weight. Separation of TMAC–protein complexes by electrophoresis in polyacrylamide gels should therefore be according to molecular weight, but since it is harder to saturate the sites on the protein with TMAC than with SDS, the reliability of molecular weight determinations may not be so high. Nevertheless, the method has been used with two cationic detergents, cetylpyridinium chloride (Schick, 1975) and cetyltrimethylammonium bromide (Panyim *et al.*, 1977; Mócz and Bálint, 1984). Results show a linear dependence of log (molecular weight) on mobility, and Panyim *et al.* (1977) claim that the method offers advantages for basic proteins (which may be sparingly soluble in SDS) and where nucleic acids may interfere (since these are not soluble in cetyltrimethylammonium bromide). Nevertheless, the method has not achieved wide use.

## 1.4 Gel electrophoresis under non-denaturing conditions

Electrophoresis separation of native proteins can be a useful technique since certain groups of enzymes can be identified by activity staining of the gel (e.g. Dewald *et al.*, 1974). For multisubunit enzymes the technique offers a way of assessing the purity of such proteins (a particular problem with membrane enzymes) and can, in principle, provide criteria for whether a particular polypeptide is a true subunit or an impurity. With membrane proteins, detergents are, of course, essential for solubilization. Since the detergent–protein micelle is only an approximation to the natural environment of the protein, conditions which lead to the solubilization of one enzyme without change or loss in activity may not be suitable for another, either through a failure to solubilize or through denaturation. In addition, the detergent treatment need not necessarily lead to complete dissociation of one enzyme from another while on the other hand the detergent may promote dissociation of *bona fide* subunits from a multisubunit enzyme. Non-denaturing gel electrophoresis in detergents can have its problems therefore.

Triton X-100 is the most commonly used detergent and, since it is itself neutral, separations may be expected to depend on the charge as well as on the size of the detergent–protein complexes. These complexes are much larger than the SDS–protein complexes described earlier and can only enter polyacrylamide gels of low concentration (e.g. typically 5% acrylamide). Dewald *et al.* (1974) describe a discontinuous system suitable for membrane proteins dissolved in Triton X-100. The resolution obtained does not seem to be as great as has been achieved with dissociating media such as SDS. Multisubunit enzymes may partially dissociate during electrophoresis (depending on the strength of subunit–subunit interactions) causing the enzyme to spread over an appreciable length of the gel. This has been demonstrated for cytochrome oxidase (Ludwig *et al.*, 1979) and obviously complicates the interpretation of such gels. Single-subunit enzymes, on the other hand, can migrate clearly to give sharply defined bands. An example is the identification of the mitochondrial glycerol phosphate dehydrogenase on 5% acrylamide gels by Wernette *et al.* (1981).

The anionic detergent, deoxycholate, can also be used for non-denaturing gel electrophoresis and, not unexpectedly, gives separations clearly different from those obtained with Triton X-100 (Dulaney and Touster, 1970).

## 1.5 Conclusions

The technique of resolving detergent–protein complexes by electrophoresis in polyacrylamide gels has developed from a simple method for determining homogeneity or the molecular weight of a protein into a sophisticated tool for the analysis of many diverse aspects of protein structure and function. The major drawback of the method, namely the difficulty with which access could be gained to the proteins in the gel without sacrificing yield or resolution, has now been largely overcome by the introduction of protein blotting. The possibility of renaturation of proteins following transfer could have enormous implications for membrane protein research where denaturing detergents are so frequently necessary for the initial solubilization. Further exciting developments and applications of these methods will undoubtedly appear in the near future.

## 1.6 References

Anker, H. S. (1970) *FEBS Lett.*, **7**, 293.
Ballou, B. and Smithies, O. (1977) *Anal. Biochem.*, **80**, 616–23.
Ballou, B., Sundharadas, G. and Bach, M. L. (1974) *Science*, **185**, 531–3.
Banker, G. A. and Cotman, C. W. (1972) *J. Biol. Chem.*, **247**, 5856–61.
Basch, R. S. (1968) *Anal. Biochem.*, **26**, 184–8.
Baumann, G. and Chrambach, A. (1976) *Anal. Biochem.*, **70**, 32–8.
Best, D., Watt, P. J. and Gull, K. (1981) *Anal. Biochem.*, **114**, 281–4.
Bittner, M., Kupferer, P. and Morris, C. F. (1980) *Anal. Biochem.*, **102**, 459–71.
Bodhe, A. M., Deshpande, V. V., Lakshmikantham, B. C. and Vartar, H. G. (1982) *Anal. Biochem.*, **123**, 133–42.
Bonner, W. M. and Laskey, R. A. (1974) *Eur. J. Biochem.*, **46**, 83–8.
Booth, A. G. (1977) *Biochem. J.*, **163**, 165–8.
Briggs, M. M. and Capaldi, R. A. (1977) *Biochemistry*, **16**, 73–7.
Burridge, K. (1976) *Proc. Natl. Acad. Sci. USA*, **73**, 4457–61.
Cabral, F. and Schatz, G. (1979) *Methods Enzymol.*, **56**, 602–13.
Cabral, F., Solioz, M., Rudin, Y., Schatz, G., Clavilier, L. and Slonimski, P. P. (1978) *J. Biol. Chem.*, **253**, 297–304.
Chamberlain, J. P. (1979) *Anal. Biochem.*, **98**, 132–5.
Choo, K. H., Cotton, R. G. H. and Danks, D. M. (1980) *Anal. Biochem.*, **103**, 33–8.
Converse, C. A. and Papermaster, D. S. (1975) *Science*, **189**, 469–72.
Cooper, P. C. and Burgess, A. W. (1982) *Anal. Biochem.*, **126**, 301–5.

Davis, B. J. (1964) *Ann. N.Y. Acad. Sci.*, **121**, 404–27.
Delepelaire, P. and Chua, N.-H. (1979) *Proc. Natl. Acad. Sci. USA*, **76**, 111–15.
Dewald, B., Dulaney, J. T. and Touster, O. (1974) *Methods Enzymol.*, **32**, 82–91.
Dion, A. S. and Pomenti, A. A. (1983) *Anal. Biochem.*, **129**, 490–6.
Douglas, M., Finkelstein, D. and Butow, R. A. (1979) *Methods Enzymol.*, **56**, 58–66.
Dulaney, J. T. and Touster, O. (1970) *Biochim. Biophys. Acta*, **196**, 29–34.
Elkon, K. B., Jankowski, P. W. and Chu, J.-L. (1984) *Anal. Biochem.*, **140**, 208–13.
Erickson, P. F., Minier, L. N. and Lasher, R. S. (1982) *J. Immunol. Methods*, **51**, 241–9.
Eschenbruck, M. and Bürk, R. R. (1982) *Anal. Biochem.*, **125**, 96–9.
Fazekas de St. Groth, S., Webster, R. G. and Datyner, A. (1963) *Biochim. Biophys. Acta*, **71**, 377–91.
Ferguson, K. A. (1964) *Metabolism*, **13**, 985–1002.
Fernandes, P. B., Nardi, R. V. and Franklin, S. G. (1978) *Anal. Biochem.*, **91**, 101–14.
Fish, W. W., Reynolds, J. A. and Tanford, C. (1970) *J. Biol. Chem.*, **245**, 5166–8.
Franklin, S. G. and Zweidler, A. (1977) *Nature*, **266**, 273–5.
Gershoni, J. M. and Palade, G. E. (1982) *Anal. Biochem.*, **124**, 396–405.
Gershoni, J. M. and Palade, G. E. (1983) *Anal. Biochem.*, **131**, 1–15.
Gershoni, J. M., Palade, G. E., Hawrot, E., Klimowicz, D. W. and Lentz, T. L. (1982) *J. Cell. Biol.*, **95**, 422a.
Glass, W. F., Briggs, R. C. and Hnilica, L. S. (1981) *Anal. Biochem.*, **115**, 219–24.
Glossmann, H. and Neville, D. R. (1971) *J. Biol. Chem.*, **246**, 6339–46.
Griffin, D. L. and Landon, M. (1981) *Biochem. J.*, **197**, 333–44.
Guikema, J. A. and Sherman, L. A. (1981) *Biochim. Biophys. Acta*, **637**, 189–201.
Guikema, J. and Sherman, L. A. (1982) *Biochim. Biophys. Acta*, **681**, 440–50.
Hancock, K. and Tsang, V. C. W. (1983) *Anal. Biochem.*, **133**, 157–62.
Hansen, J. N. (1977) *Anal. Biochem.*, **76**, 37–44.
Hanstein, W. G. (1979) *Methods Enzymol.*, **56**, 653–83.
Hashimoto, F., Horigome, T., Kanbayashi, M., Yoshida, K. and Sugano, M. (1983) *Anal. Biochem.*, **129**, 192–9.
Higgins, R. C. and Dahmus, M. E. (1979) *Anal. Biochem.*, **93**, 257–60.
Irie, S., Sezaki, M. and Kato, Y. (1982) *Anal. Biochem.*, **126**, 350–4.
Johnson, T. K., Yuen, K. C. L., Denell, R. E. and Consigli, R. A. (1983) *Anal. Biochem.*, **133**, 126–31.
Jovin, T. K., Dante, M. L. and Chrambach, A. (1971) Multiphasic Buffer Systems Output, Federal Scientific and Technical Information, U.S. Department of Commerce, PB 196085-196091, Springfield, VA.
Kadenbach, B., Jarausch, J., Hartmann, R. and Merle, P. (1983) *Anal. Biochem.*, **129**, 517–21.
Kadenbach, B. and Merle, P. (1981) *FEBS Lett.*, **135**, 1–11.
Laemmli, U.K. (1970) *Nature (London)*, **227**, 680–5.
Lambin, P. (1978) *Anal. Biochem.*, **85**, 114–25.
Laskey, R. A. and Mills, A. D. (1975) *Eur. J. Biochem.*, **56**, 335–41.
Leach, B. S., Collawn, J. F. and Fish, W. F. (1980) *Biochemistry*, **19**, 5734–41.
Ludwig, B., Downer, N. W. and Capaldi, R. A. (1979) *Biochemistry*, **18**, 1401–7.
Makino, S., Tanford, C. and Reynolds, J. A. (1974) *J. Biol. Chem.*, **249**, 7379–82.
Margulies, M. M. and Tiffany, H. L. (1984) *Anal. Biochem.*, **136**, 309–13.
Marshall, T. (1983) *Electrophoresis*, **4**, 269–72.

Marshall, T. (1984) *Anal. Biochem.*, **136**, 340–6.
Marshall, T. and Lattner, H. L. (1981) *Electrophoresis*, **2**, 228–35.
Marshall, T. and Williams, K. M. (1984) *Anal. Biochem.*, **139**, 502–5.
McConkey, E. J. (1979) *Anal. Biochem.*, **96**, 39–44.
McGuckin, W. F. and McKenzie, B. F. (1958) *Clin. Chem.*, **4**, 476–80.
Mendel-Hartvig, I. (1982) *Anal. Biochem.*, **121**, 215–17.
Mendel-Hartvig, I. and Nelson, B. D. (1983) *J. Bioenerg. Biomemb.*, **15**, 27–36.
Merle, P. and Kadenbach, B. (1980) *Eur. J. Biochem.*, **105**, 499–507.
Merril, C. R., Dunau, M. L. and Goldman, D. (1981a) *Anal. Biochem.*, **110**, 201–7.
Merril, C. R., Goldman, D., Sedman, S. A. and Ebert, M. H. (1981b) *Science*, **21**, 1437–8.
Merril, C. R., Switzer, R. C. and VanKeuren, M. L. (1979) *Proc. Natl. Acad. Sci. USA*, **76**, 4335–9.
Mócz, G. and Bálint, M. (1984) *Anal. Biochem.*, **143**, 283–92.
Mold, D. E., Weingart, J., Assaraf, J., Lubahn, D. B., Kelner, D. N., Shaw, B. R. and McCarty, K. S. (1983) *Anal. Biochem.*, **135**, 44–7.
Moeremans, M., Daneels, G. and DeMey, J. (1985) *Anal. Biochem.*, **145**, 315–21.
Morrissey, J. H. (1981) *Anal. Biochem.*, **117**, 307–10.
Neville, D. M. and Glossmann, H. (1974) *Methods Enzymol.*, **32**, 92–102.
Neville, D. R. (1971) *J. Biol. Chem.*, **246**, 6328–34.
Nielson, T. B. and Reynolds, J. A. (1978) *Methods Enzymol.*, **48**, 3–10.
Nozaki, Y., Reynolds, J. A. and Tanford, C. (1974) *J. Biol. Chem.*, **249**, 4452–9.
Oakley, B. R., Kirsch, D. R. and Morris, N. R. (1980) *Anal. Biochem.*, **105**, 361–3.
Ochs, D. (1983) *Anal. Biochem.*, **135**, 470–4.
O'Connell, P. B. H. and Brady, C. J. (1977) *Anal. Biochem.*, **76**, 37–44.
Ornstein, L. (1964) *Ann. N.Y. Acad. Sci.*, **121**, 321–49.
Panyim, S., Thitipongpanich, R. and Supatimusro, D. (1977) *Anal. Biochem.*, **81**, 320–7.
Poduslo, J. F. (1981) *Anal. Biochem.*, **114**, 131–9.
Porro, M., Viti, S., Antoni, G. and Saletti, M. (1982) *Anal. Biochem.*, **127**, 316–21.
Reiser, J. and Wardale, J. (1981) *Eur. J. Biochem.*, **114**, 569–75.
Renart, J., Reiser, J. and Stark, G. R. (1979) *Proc. Natl. Acad. Sci. USA*, **76**, 3116–20.
Reynolds, J. A. and Tanford, C. (1970) *J. Biol. Chem.*, **245**, 5161–5.
Robinson, P. J., Bull, F. G., Anderton, B. H. and Roitt, I. M. (1975) *FEBS Lett.*, **58**, 330–3.
Schick, M. (1975) *Anal. Biochem.*, **63**, 345–9.
Schleicher, M. and Watterson, D. M. (1983) *Anal. Biochem.*, **131**, 312–17.
Steck, G., Leuthard, P. and Bürk, R. R. (1980) *Anal. Biochem.*, **107**, 21–4.
Strälfors, P. and Belfrage, P. (1983) *Anal. Biochem.*, **128**, 7–10.
Swank, R. T. and Munkres, K. D. (1971) *Anal. Biochem.*, **39**, 462–77.
Switzer, R. C., Merril, C. R. and Shifrin, S. (1979) *Anal. Biochem.*, **98**, 231–7.
Symington, J., Green, M. and Brackmann, K. (1981) *Proc. Natl. Acad. Sci. USA*, **78**, 177–81.
Tanner, M. J. A. and Anstee, D. J. (1976) *Biochem. J.*, **153**, 265–70.
Tasheva, B. and Desser, G. (1983) *Anal. Biochem.*, **129**, 98–102.
Towbin, H., Staehelin, T. and Gordon, J. (1979) *Proc. Natl. Acad. Sci. USA*, **76**, 4350–4.
Walker, J. E., Auffret, A. D., Carne, A., Gurnett, A., Hanisch, P., Hill, D. and Saraste, M. (1982) *Eur. J. Biochem.*, **123**, 253–60.

Wang, K. and Richards, F. M. (1974) *J. Biol. Chem.*, **249**, 8005–18.
Warr, G. W. (1982) *Mol. Immunol.*, **19**, 75–81.
Weber, K. and Osborn, M. (1969) *J. Biol. Chem.*, **244**, 4406–12.
Weber, K., Pringle, J. R. and Osborn, M. (1972) *Methods Enzymol.*, **26**, 3–27.
Wernette, M. E., Ochs, R. S. and Lardy, H. A. (1981) *J. Biol. Chem.*, **256**, 12767–71.
Wojtkowiak, Z., Briggs, R. C. and Hnilica, L. S. (1983) *Anal. Biochem.*, **129**, 486–9.
Yuen, K. C. L., Johnson, T. K., Denell, R. E. and Consigli, R. A. (1982) *Anal. Biochem.*, **126**, 398–402.

# 2

# Analysis of membrane protein composition by isoelectric focusing and two-dimensional electrophoresis

ROBERT B. FREEDMAN and E. N. CLARE MILLS



## 2.1 Introduction

### 2.1.1 HISTORICAL BACKGROUND

Procedures which resolve the components of complex mixtures by successive application of two distinct separative techniques in perpendicular directions (two-dimensional techniques) are widely recognized as providing very high resolution. This appreciation dates back to the beginning of the revolution of separation techniques in biochemistry in the late 1940s and early 1950s. For example, methods for resolving metabolites or small peptides by two-dimensional separations on paper played a part in establishing the Calvin cycle of photosynthesis and in identifying the molecular defect in sickle cell haemoglobin. These successes emphasized the power of two-dimensional techniques, so that the subsequent development of successful methods for analysing protein mixtures by gel electrophoresis in the 1960s naturally led to attempts to apply the two-dimensional approach here also. However, before 1975, only one rather specialized procedure was widely and successfully exploited. This was the method developed by Kaltschmidt and Wittmann (1970) for resolving ribosomal proteins from *Escherichia coli*. The method employed (i) electrophoresis at a pH where ribosomal proteins differed from one another in charge, using a non-restrictive agarose or polyacrylamide gel,

Techniques for the Analysis of Membrane Proteins. Edited by C. I. Ragan & R. J. Cherry.
Published in 1986 by Chapman and Hall, 11 New Fetter Lane, London EC4P 4EE

followed by (ii) electrophoresis at an acid pH where all ribosomal proteins are positively charged, in a more restrictive polyacrylamide gel. The method therefore separated mainly by charge in the first dimension and mainly by size in the second, and gave a good spread of more than 50 ribosomal proteins in the second-dimension gel. Its success led to the acceptance of a uniform numbering system for ribosomal proteins (Wittmann *et al.*, 1971) and to rapid progress in studies of their function and supramolecular organization (Kurland, 1977). The method had some drawbacks: first, both charge and size played some part in the resolution in both dimensions, so that they were not entirely independent, and second, the method was devised to cope with the predominantly small and basic ribosomal proteins, and could not be easily applied to other protein mixtures.

The rapid development of two-dimensional electrophoresis as an analytical tool for proteins dates from the publication by O'Farrell (1975) of a high-resolution method employing isoelectric focusing in the first dimension and SDSth polyacrylamide gel electrophoresis (SDS-PAGE) in the second. Few fields of biochemical analysis can have had such a clear and definite starting point. Before 1975, attempts at two-dimensional resolution of proteins using these techniques were few and rather unsuccessful; in retrospect they seem half-hearted. From 1975 onwards, progress has been rapid and the methodology has been modified and refined for a great variety of applications; the original paper was cited in 60 publications in 1976, but in nearly 600 publications in 1980 (Dunn and Burghes, 1983b). This rapid and diverse application partly derives from the intrinsic advantages of the two electrophoretic methods employed. Gel isoelectric focusing is a high-capacity equilibrium technique, separating proteins on the basis of their isoelectric points and capable of resolving proteins differing in isoelectric point by $<0.01$ pH units or protein derivatives differing in only a single charge. The fact that focusing can be continued to equilibrium (in theory) eliminates any effect of protein size. SDS-PAGE, as described in Chapter 1, separates proteins essentially on the basis of polypeptide size ($M_r$ or number of residues) and resolution can be optimized by the appropriate choice of acrylamide concentration, proportion of cross-linker, shape of acrylamide gradient etc. But the impact of the O'Farrell paper also derives from its development of a comparatively simple and practical experimental protocol and its thorough discussion of key features of the system (resolving power, loading capacity, reproducibility) and of experimental factors critical for its success.

### 2.1.2 SCOPE OF THIS CHAPTER

In studies of cytosolic and soluble proteins, O'Farrell's method has been applied successfully without any modification in its essential features. However it rapidly became clear that the method was not directly applicable to membrane proteins. The problem arises from the conditions necessary for the first dimension; resolution by isoelectric focusing is dependent on the proteins in the mixture retaining their intrinsic charge properties, and thus the proteins must be in solution either free or as complexes with electrically neutral solubilizing agents. They cannot be resolved as complexes with ionic detergents, which are generally the most successful agents for solubilizing integral membrane proteins. This is the central problem in isoelectric focusing and two-dimensional electrophoresis of membrane proteins and attempts to circumvent or overcome it are discussed in a later section (2.2.4). The point to be emphasized is that this problem has prevented the development of a truly *general* method for two-dimensional electrophoresis of membrane proteins which can reliably be applied to any membrane system. The success of particular methods depends on the properties of the protein mixtures to which they are applied, so that any study of a new membrane system will inevitably involve some development work.

This chapter attempts to review the use of two-dimensional electrophoresis in studies of membrane proteins and will emphasize the minor procedural modifications which can bring success in working with membrane proteins. Since most of the essential difficulties of the technique refer to the first dimension, the separation by isoelectric focusing, and since this technique is a powerful separation technique in its own right, the theory and practice of isoelectric focusing will be considered in general terms. However, it will be assumed that the major application of isoelectric focusing will be as the first dimension of a two-dimensional separation.

Inevitably, this chapter cannot be comprehensive. Further information can be obtained from the following sources. For information on commercially available apparatus, on reagents and on the range of widely used techniques, the best sources are the publications of the major commercial suppliers in this field, in particular LKB, Pharmacia and Bio-rad. These publications are invaluable sources of practical advice and give a good presentation of the range of options available. The theory of isoelectric focusing and much early work are reviewed by Gaal *et al.* (1980); despite its publication date, this book does not really cover the expansion of the field since 1975. Theory and detailed methods are combined in the chapters by An der Lan and Chrambach (1982) and Sinclair and Rickwood (1982). An excellent review of the methodology and applications of high-resolution two-dimensional gel electrophoresis has recently appeared (Dunn and Burghes, 1983a,b); this does not specifically concentrate on membrane proteins, but it gives a complete coverage of the literature including detailed discussions of all the problems of the technique. Two methodological chapters dealing specifically with two-dimensional gel electrophoresis of *membrane* proteins have appeared (Boschetti *et al.*, 1981; Rubin and Leonardi, 1983). The latter gives a good discussion of a number of the critical problems encountered, but neither chapter presents a selection of successful procedures; each recommends a single procedure which has been successful with a number of samples.

## 2.2 Isoelectric focusing

### 2.2.1 PRINCIPLES OF ISOELECTRIC FOCUSING

A system for isoelectric focusing consists of an acidic electrolyte connected to a basic electrolyte through a medium which restricts diffusion and contains a mixture of amphoteric solutes (ampholytes). When a voltage is applied across the system such that the acidic electrolyte becomes the anode and the basic the cathode, ampholytes in the connecting medium experience both an electric field and a pH gradient; the pH gradient arises both from the nature of the electrolytes and from electrochemical reactions ($H_2O \rightarrow \frac{1}{2}O_2 + 2H^+ + 2e^-$ at the anode; $2H_2O + 2e^- \rightarrow H_2 + 2OH^-$ at the cathode). Under the influence of the electric field, charged ampholytes in the connecting medium will move; if an ampholyte has an isoelectric point (pI, ≡ pH at which the substance has net zero charge) between the pH of the anolyte and that of the catholyte, it will focus within the medium. That is, molecules initially in a region of pH > pI will be negatively charged and move towards the anode (and hence into a more acidic environment) while those in a region where pH < pI will be positive and will move towards the cathode. As a result, all the molecules of this ampholyte will collect at a point in the connecting medium where pH = pI; here they will carry no net charge and will not move in the applied electrical field. An amphoteric substance will buffer the pH of the region in which it is focused. If the connecting medium contains several such ampholytes, each with a different pI, they will focus in different regions and the pH from catholyte to anolyte will decrease in a series of steps corresponding to their pI values. When the medium contains a large number of ampholytes covering a continuous range of pI values, a continuous gradient of pH will be established and stabilized in the electrical field. The use of isoelectric focusing for the resolution of proteins depends on using low-molecular-weight ampholytes (carrier ampholytes) to establish such a pH gradient rapidly, and to stabilize it sufficiently to allow focusing of proteins which are larger ampholytes and move more slowly.

Production of a stable, continuous pH gradient requires the presence of a large number of carrier ampholytes differing marginally in pI and with good buffering capacity close to their isoelectric points; such an ampholyte mixture cannot easily be prepared from naturally occurring ampholytes. Hence the practical usefulness of isoelectric focusing as a technique really began with the first production of synthetic ampholytes by Vesterberg (1969), who coupled acrylic acid with various polyethylene polyamines in aqueous solution to form homologous, isomeric, low-molecular-weight ($M_r < 1000$) polyaminopolycarboxylic acids:

$$\begin{array}{ccccc} -CH_2- & N & -[CH_2]_n- & N & -CH_2- \\ & | & & | & \\ & [CH_2]_n & & R & \\ & | & & & \\ & NR_2 & & & \end{array}$$

where R can be H or a carboxylic acid group such as $-[CH_2]_n-COOH$, or a further substituted amino group, such as $-[CH_2]_n-NR_2$.

These derivatives were introduced by LKB and marketed under the name Ampholines. Various mixtures of components are available covering wide or narrow ranges of pI. Subsequently other synthetic carrier ampholytes have been prepared and marketed under trade names, such as Pharmalyte, Servalyt and Bio-lyte. Pharmalyte, prepared by controlled co-polymerization of selected amines with epichlorhydrin, contains –OH groups attached to the backbone, but the same ionizable amino and carboxyl groups as Ampholine; Servalyt contains additionally guanidino, sulphonate and phosphonate ionizable groups. Commercial ampholytes covering a narrow pH range ($\sim 2$ pH units) may contain more than 1000 distinct components. The various commercially available synthetic ampholytes are generally regarded as being equivalent, although there are suggestions that the different preparations differ in their success in different pH regions (Dunn and Burghes, 1983a). Ampholine is still the most widely used.

Isoelectric focusing, like other electrophoretic techniques, must be carried out in a medium which minimizes the effects of diffusion and convection, which could otherwise overwhelm the focusing process. Initially, isoelectric focusing was viewed as a preparative technique and the earliest applications involved focusing in a vertical column containing a sucrose density gradient. In addition, electrofocusing can be carried out on flat beds of a granulated gel medium such as Sephadex, or in agarose or polyacrylamide gels. Table 2.1 indicates the range of potential applications of isoelectric focusing and recommended practical techniques. Gel isoelectric focusing is preferable for high-resolution analysis and all the widely used variants of the O'Farrell procedure for two-dimensional gel electrophoresis employ isoelectric focusing in polyacrylamide gels for the first dimension.

It should be noted that Ampholines were developed before polyacrylamide gels came into common usage and were specifically designed for use in sucrose density gradients. Many problems in isoelectric focusing, particularly cathodic drift (see Section 2.2.3.b), arise partly from the use of diffusible carrier ampholytes to establish the pH gradient. An entirely new approach to the establishment of stable pH gradients has more recently been developed (Bjellquist and Ek, 1982; Righetti and Macelloni, 1982; Ek *et al.*, 1983). In this method the gradient is not established by focusing of carrier ampholytes. Instead, the buffering groups responsible for the gradient are an integral part of

**Table 2.1** Recommended isoelectric focusing techniques. IEF, isoelectric focusing; PAGIF, polyacrylamide gel IEF. (Reproduced from *Isoelectric Focusing: Principles and Methods*, Uppsala, 1982, by kind permission of Pharmacia AB.

| *Application* | *Recommended techniques* |
|---|---|
| High-resolution, proteins with molecular weights up to ca. 300 000 | PAGIF or Ultrathin |
| Peptides | Ultrathin or PAGIF |
| High-molecular-weight proteins, up to $2 \times 10^6$ | Agarose IEF |
| IEF followed by immuno-detection | Agarose IEF |
| IEF with urea | PAGIF, Agarose IEF |
| 2-D electrophoresis | IEF in gel rods |
| IEF of enzymes | Agarose IEF, PAGIF |
| IEF of tissue | Agarose IEF |
| Recovery of proteins from immuno complexes, affinity gels | Electrophoretic desorption |
| Crossed immuno IEF | Agarose IEF |
| Cells | Density gradient columns |
| Preparative IEF | Sephadex, composite gels of Sephadex/Agarose IEF<br>Agarose IEF<br>Density gradient columns |
| Physical parameters of a protein | |
| (a) pI points | PAGIF, Agarose IEF |
| (b) Titration curves | Titration curve analyses |

the polyacrylamide gel matrix; they are present as *N*-substituents on acrylamide ($CH_2{=}CH{-}CO{-}NH{-}R$) where R contains either a carboxylic acid or a tertiary amino group. These 'immobilines' are available with p*K* of 3.6 to 9.3, and a gradient is prepared by mixing an acidic immobiline (made dense by addition of glycerol) with a basic immobiline in a standard gradient mixer while pouring the gel. It is claimed that stable pH gradients as shallow as 0.01 units $cm^{-1}$ are achievable with this technique. Although the method appears promising, it has not yet been widely applied. The remainder of this chapter will deal exclusively with methods employing polyacrylamide gels and mobile carrier ampholytes.

### 2.2.2 METHODOLOGY

(a) *Systems available*

Polyacrylamide gel isoelectric focusing can be carried out in vertical slabs or cylinders of gel (as in the methods for gel electrophoresis described in Chapter 1), or in horizontal slabs of gel. Each of these methods has assets and drawbacks; horizontal slab gels are most widely used when isoelectric focusing is employed as a single-dimension analytical technique, whereas vertical cylindrical gels are most widely used for performing the first dimension of two-dimensional analyses. The major factors distinguishing the techniques are listed below:

(1) Slab gel systems allow focusing of multiple samples in parallel, and therefore give improved reproducibility and permit easier comparison of samples.
(2) Slab gels show more uniform heating during focusing and are easier to cool, which prevents distortion and allows higher voltages to be applied and hence shorter focusing times. (However, it should be noted that a typical cylindrical gel for IEF (isoelectric focusing) has a diameter of 2 mm compared to the 5 mm typical for SDS-PAGE so that heating causes fewer problems even in IEF cylindrical gels.)
(3) These advantages are particularly marked with ultrathin ($<0.5$ mm) slabs rather than conventional (1.5 mm) slabs; ultrathin slabs are also more rapidly stained and destained. However, they are difficult to take through complex development steps unless firmly bonded to a support, so in practice they have not been much used for two-dimensional techniques.
(4) Cylindrical gels have higher loading capacities; this is important when samples are dilute or contain a large number of components, such that a large total amount of protein must be loaded in order that minor components become detectable.
(5) Cylindrical gels are more economic in reagents, especially expensive carrier ampholytes, since the whole of the gel is used for resolution, with none of it wasted as empty tracks or space between tracks, as in slabs.
(6) Cylindrical gels are slightly easier to section perpendicular to the direction of migration for monitoring the pH gradient or for detecting radioactive components by scintillation counting.
(7) Cylindrical gels are slightly more convenient to extract and prepare for a second-dimension run, since without prior staining it is difficult to cut out individual tracks from a slab gel accurately.
(8) In vertical gel systems (slab or cylinder) the sample is loaded at the top of the gel (i.e. close to one extreme of pH), whereas in horizontal slab systems the sample may be loaded at any point. The end of an IEF gel is prone to a number of instabilities including pH gradient drift (see Section 2.2.3.b) and shrinkage of the gel. These effects can lead to loss of sample proteins into the electrolyte or to aggregation of the sample and subsequent streaking.

Almost all studies using two-dimensional gel electrophoresis to date have employed vertical systems to carry out the first-dimension IEF, mainly because of their convenience and because they employ the same apparatus that is used for SDS-PAGE in cylinders or slabs. However, it is possible that in the long term, the intrinsic advantages of horizontal systems in terms of stability and resolving power may lead to their being adopted more widely.

(b) *Gel composition*

Polyacrylamide gel IEF is normally carried out in gels containing a total acrylamide monomer concentration of 3–5% (w/v); higher concentrations lead to sieving effects, and hence slow migration of larger polypeptides, while lower concentrations produce gels too weak to be handled conveniently. Very-high-molecular-weight polypeptides ($M_r > 200\,000$) may be poorly focused even in low-concentration acrylamide gels, and for such polypeptides focusing in agarose gels should be considered.

Most proteins show a minimum solubility at their isoelectric point, so that precipitation of proteins during IEF is a problem even for cytosolic and other soluble proteins. To overcome this, urea (at concentrations up to 9.5 M) and non-ionic detergents (up to 2% w/v) are routinely present in IEF gels. These conditions, however, may not be sufficient to maintain membrane proteins in solution and this problem is discussed later (Section 2.2.3.c).

The resolving power of an IEF gel is dependent on the shape and range of the pH gradient established, which is in turn dependent on the ampholyte composition of the gel. Carrier ampholytes are generally present at a total concentration of 2% (w/v); this is regarded as the minimum capable of maintaining a stable gradient. Commercial ampholytes are available both as very complex mixtures covering a wide range of pI (usually 3.5–10) or as fractions covering narrower ranges (usually 2 pH units). Satisfactory pH gradients for IEF can be obtained using wide-range ampholytes alone, but it is better to optimize the ampholyte composition to match the protein composition of the sample. If the sample contains many protein components covering a narrow range of pH, e.g. 6–7, then it is best to use a preparation of wide-range ampholytes fortified with narrow-range samples covering the appropriate pH values; in this example a suitable mix might be 0.4% pH 3.5–10 ampholytes plus 0.8% pH 5–7 ampholytes plus 0.8% pH 6–8 ampholytes. If by contrast, the sample proteins are widely diverse in pI, then it is important to ensure that a wide-range pH gradient is maintained by adding narrow-range ampholytes covering the pH extremes; a suitable mix might be 1.0% pH 3.5–10 ampholytes plus 0.5% pH 4–6 ampholytes plus 0.5% pH 8–10 ampholytes. The researcher has considerable freedom in defining the optimal ampholyte composition, but it is important that this is standardized before a series of IEF or two-dimensional PAGE experiments is carried out. It should also be stressed that one standardized pH range may not resolve perfectly all the proteins in a mixture; it may therefore be necessary to sacrifice information on some components, or to run IEF in two or more conditions in order to obtain full information.

A persistent problem in gel IEF is that the full extent of the pH gradient theoretically available from the carrier ampholytes is rarely attained. This is most commonly observed as a drift of the pH gradient towards the cathode, so that in a gel containing ampholyte up to pH 10, the pH gradient at the end of focusing may not extend above pH 7–8. This problem and methods to overcome it are discussed later (Section 2.2.3.b).

(c) *Prefocusing, loading, overlays and electrolytes*

We will not discuss the practical aspects of casting a gel for IEF since the procedure is essentially that used for other forms of gel electrophoresis. However, certain particular features should be noted. Firstly, since commercial ampholytes contain amino groups they can partly replace the TEMED (*N,N,N′,N′*-tetramethylene diamine) generally included in polymerization mixtures as a radical accelerator; TEMED can therefore be present at considerably lower concentration. Secondly, the focusing will be perturbed if the gel and sample contain significant amounts of ionic species other than the proteins to be separated and the carrier ampholytes. Ionic species in the gel can arise from impurities in the acrylamide, from cyanate formed chemically from the high concentration of urea present, and from excess persulphate added as a radical initiator of polymerization. These problems are best dealt with by pretreating the acrylamide stock solution with an ion-exchange resin, and by 'prefocusing', i.e. applying a high voltage across the gel before loading the sample in order to electrophorese cyanate and persulphate out of the gel. To minimize the effects of salts in the sample it can be prepared by dialysis against sample buffer, rather than simply by dilution into this buffer.

Dilute solutions of NaOH (at the cathode) and of $H_3PO_4$ or $H_2SO_4$ (at the anode) are the most widely used electrolytes. However, serious instabilities and problems can arise if there is a marked discrepancy in either pH or osmolarity between the electrolyte and the end of the IEF gel. To obtain the appropriate pH and good buffer capacity, histidine, ethylenediamine or other amino compounds have been used as the catholyte, while glutamate, aspartate or citrate have been used as the anolyte. Other electrolytes should be considered if the pH range to be covered is unusual, e.g. if one end of the IEF gel is to be close to neutrality, a dilute HEPES buffer can be used as the electrolyte. Degassing of catholyte to remove dissolved $CO_2$ is sometimes recommended. Further discussion of the choice of electrolytes can be found in An der Lan and Chrambach (1982).

In vertical gel systems, the sample is loaded above the upper end of the gel. The electrolyte here may precipitate or chemically modify the sample proteins and so it is customary to 'overlayer' the sample with a small volume of sample buffer containing the carrier ampholytes to prevent direct contact. The sample itself is normally considerably denser than the electrolyte because of its high concentration of urea (see below Section 2.2.3.c.) so mixing and diffusion at loading are rarely a problem. Sample proteins can be included in the gel polymerization mix, making a loading step unnecessary; a detailed procedure is described by Sinclair and Rickwood (1982). Large sample volumes can be included but this method is not often used because of the dangers of chemical modification of sample proteins during polymerization.

(d) *Monitoring of the pH gradient and detection of focused proteins*

The pH gradient established at the end of an IEF run can be monitored by a number of methods. The simplest is by direct reading with a surface pH electrode; these are available with tip surface diameters of 1–3 mm, but their performance is often criticized for slow response and poor reproducibility. By far the most widely used method is the simple slicing of a focused gel into 0.5 cm or 1 cm sections, which are then allowed to equilibrate with distilled water or dilute KCl solution before the pH of the medium is read conventionally with a pH meter. This method is time-consuming, prone to errors arising from dissolution of atmospheric $CO_2$, and inevitably it involves measurement on a parallel gel (or track) rather than on the sample gel (or track). Nevertheless it requires no special apparatus and has the psychological advantage of generating numerical data, which can be expressed graphically.

Mobilities on SDS-PAGE are generally measured relative to standard proteins resolved in a parallel track. A series of pI standard proteins can provide an ideal method for monitoring the pH gradient in IEF. Such a series can be produced either (i) by mixing a series of purified proteins of carefully spaced pI, or (ii) by using a natural protein sample in which the major components have known and well separated pI, or (iii) by generating a series of derivatives of a single protein. The first possibility, a mixture of defined proteins, can be designed and prepared in the laboratory to suit a particular application, but commercial calibration kits are available; Pharmacia markets a wide-range calibration mix containing 11 markers and narrower range kits containing at least four markers. The second alternative, a natural mix, is much less expensive if an appropriate sample is available; a simpler, but less satisfactory method is to use easily recognizable internal standards, such as serum albumin, actin and tubulin, which are found in very many preparations from animal sources. The third method, derivatives of a single protein, can give very high precision in calibrating the pH gradient. Anderson and Hickman (1979) have shown that proteins can be progressively carbamoylated by heating in urea solutions; by mixing samples carbamoylated to various extents, a calibration mix can be obtained with a large number of components each differing by a single charge. Such a 'carbamoylation train' can either be run in a parallel track to calibrate a one-dimensional IEF gel, or can be included in a sample for two-dimensional electrophoresis, provided that the marker protein has an $M_r$ distinct from that of any protein of interest in the sample. Anderson and Hickman demonstrate the method with a number of proteins including haemoglobin. A small coloured protein has two additional advantages as a pI calibration marker, namely that the focusing of the marker is directly visible without staining, and that, in a second SDS-PAGE dimension, the marker protein will run close to the front, clear of the majority of sample proteins of interest. These advantages are exploited in the commercial pI calibration kit available from BDH, which consists of a mixture of various myoglobins and trifluoroacetylated myoglobin derivatives (covering the pI range 5.9–8.4), plus azurin (pI 5.6) and cytochrome *c* (pI 10.6) to extend the range. It should be noted that the pI values quoted for the components of commercial pI standard mixes are for native proteins, not those observed in urea gels.

In two-dimensional gel electrophoresis, the proteins are not usually visualized after the first-dimension separation, but a parallel track may be stained to monitor the success of the isoelectric focusing. The presence of carrier ampholytes and the possibility of rapid diffusion through the highly porous IEF gel make it vital that the proteins are well fixed and the gel extensively washed. Staining methods involving dyes (Coomassie Blue R250 or G250) or silver can be used; the Coomassie techniques currently recommended are based on the work of Vesterberg *et al.* (1977).

### 2.2.3 PROBLEMS OF THE TECHNIQUE

(a) *Artefacts*

The introduction of isoelectric focusing produced evidence of considerable heterogeneity in what were thought to be entirely pure proteins. As a result there has been extensive investigation of possible artefactual sources of heterogeneity. In general, the technique has been vindicated; most of the heterogeneity is real. Some, for example, arises from the presence of differently disulphide-linked isomers. However, some sources of artefactual heterogeneity are known, and the possibility should not be overlooked. One source is that proteins may form complexes with carrier ampholytes, the complexes with different ampholyte species then focusing at distinct pH values (see Hare *et al.*, 1978 and references therein). This phenomenon certainly occurs for some well-defined proteins, but it is not clear whether membrane proteins are likely to form such complexes (Guengerich, 1979) or whether such interactions survive in the presence of high concentrations of urea (Dunn and Burghes, 1983a).

A further source of artefactual heterogeneity is modification of amino acid side chains during sample preparation. It was noted above, that proteins readily react with cyanate on heating in concentrated urea solution, converting positively charged lysyl groups into carbamoyl groups. This process can also occur on prolonged storage of samples in urea at lower temperatures, so generating a trail of spots representing species less basic than the genuine protein. Deamidation of asparagine or glutamine side chains can also occur artefactually, producing the same consequences. Procedures for radiolabelling of sample proteins are always designed not to affect charge properties; neither iodination of tyrosyl residues, nor methylation of lysyls should modify pI. However, the conditions used for these labelling reactions may promote the artefactual modifications mentioned above.

A final source of artefacts during IEF of complex protein mixtures, such as membrane preparations, is incomplete dissociation of the components. If two polypeptides A and B interact strongly and their complex is not fully dissociated under the conditions of isoelectric focusing, the species AB will appear on the focused gel in addition to either or both A and B. Usually such a complex would be dissociated in the SDS-PAGE dimension of a two-dimensional gel, with the result that a single polypeptide such as B may give rise to two spots on the two-dimensional gel slab. This problem is seen in more extreme form if some protein components precipitate or aggregate during IEF (see Section 2.2.3c).

(b) *Instability of the pH gradient*

In theory, IEF involves the rapid establishment of a stable pH gradient by the carrier ampholytes, followed by the slower focusing of the sample proteins. In general, a focusing time of 4000–9000 volt hours is required to focus proteins in 10 cm cylindrical gels; to avoid overheating, the voltage is usually kept lower than 400 volts, so that focusing times are of the order of 10–24 h. In practice, the pH gradients established by carrier ampholytes are not entirely stable through this period, so that the pH gradient can be considerably degraded before proteins truly focus.

Decay of the pH gradient is illustrated in Fig. 2.1. A reasonable gradient is maintained for 9 h, but by this stage ampholytes have begun to disappear from the central region of the gradient. Subsequently all ampholytes migrate towards the cathode, so that the gel progressively becomes acidified. Hence a gel containing pH 3.5–10 ampholytes may be found, after a standard focusing experiment, to have a pH gradient of only 4–7. The extent of cathodic drift appears to increase with the amount of protein loaded (Cantrell *et al.*, 1981). Drift also appears to be exacerbated by the presence of $\beta$-mercaptoethanol in the sample (Righetti *et al.*, 1982). The reasons for this 'cathodic drift', which is currently the major technical problem with IEF, are not well understood. However, its effects can be minimized by a number of simple procedures.

It should be noted that the majority of known proteins have a pI in the range 5–8, so that a modest extension of the gradient towards the alkaline is often adequate. This can be achieved by reinforcing the carrier ampholytes with additional alkaline ampholytes; not only synthetic ampholytes but cheaper naturally occurring ampholytes such as lysine or arginine can also be used (Breithaupt *et al.*, 1978). Cathodic drift can also be minimized by ensuring that

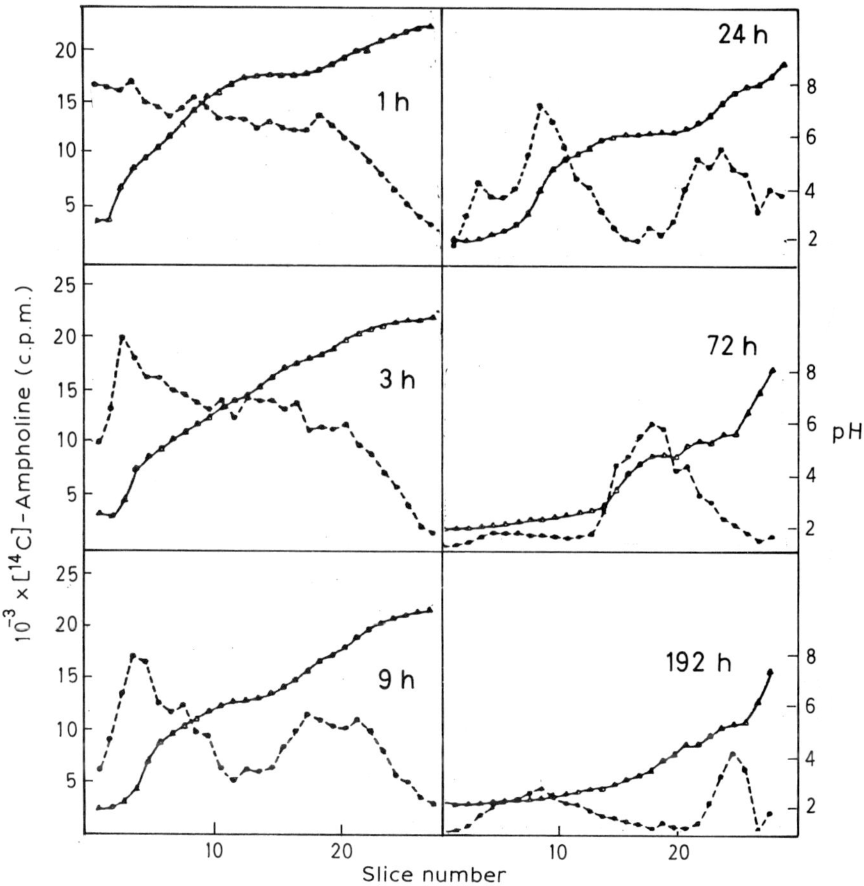


**Fig. 2.1** The formation and decay of pH gradient during electrofocusing. Gels containing [$^{14}$C]Ampholine were electrophoresed for periods of from 1 to 192 h. The pH gradients and the distribution of [$^{14}$C]Ampholine were then analysed. (From An der Lan and Chrambach, 1982).

the anolyte is not more acidic than the most acidic species in the carrier ampholytes; with carefully chosen anolyte species and concentration, gradient stability has been achieved for up to a week at 500 volts (Nguyen and Chrambach, 1977). Since cathodic drift is essentially a displacement of amphoteric species from the gel into the catholyte, an alternative preventative measure is to eliminate the catholyte and to make direct contact between the cathode and the gel (Hunter, 1978). When IEF is carried out in vertical cylindrical gels, the cathode is usually uppermost and hence the cathodic end of the gel can easily be overlayered with a solution of carrier ampholytes so that it is not in direct contact with the catholyte, which reduces cathodic drift somewhat. If for some reason the sample must be loaded from the anode, the cathode will be at the bottom and cannot readily be overlayered. In this case it is advantageous to set up the gel so that the cathode is covered by dialysis membrane containing a solution of carrier ampholytes, so that again direct contact is minimized.

The problem of cathodic drift, and the resultant difficulty of obtaining stable pH gradients covering the range pH 8–10 has made it very difficult to perform conventional IEF or two-dimensional electrophoresis of very basic proteins. Here, a more radical approach may be more successful. Non-equilibrium pH gradient electrophoresis (NEPHGE) resembles IEF in that proteins move in an electrical field and in a pH gradient set up by carrier ampholytes, but the aim of 'focusing' – achievement of equilibrium at the isoelectric point – is abandoned (O'Farrell *et al.*, 1977). Proteins are loaded at the anode and migrate towards the cathode. The voltage is applied for only 4–5 h, so that whereas acidic proteins may focus, basic proteins are still migrating when the separation is stopped. This method can resolve basic proteins such as histones or ribosomal proteins, even though most have pI greater than the maximum pH attained in the gel. Since it is not often ascertained in isoelectric focusing experiments that true focusing equilibrium has been reached, many so-called isoelectric focusing runs may actually employ the principle of NEPHGE to obtain separation.

(c) *Precipitation*

Considerable problems can be caused if proteins in the sample are not completely soluble under the conditions of isoelectric focusing. Clearly precautions can be taken to ensure that all proteins in the load sample are in solution by high speed centrifugation of the sample. But proteins may then precipitate during resolution in the IEF gel. This problem is an inevitable consequence of the poor solubility of many proteins near their isoelectric points, and is exacerbated by the use of concentrated protein loads, or by the heating which occurs during IEF. (Local hot spots can arise at points in the gel where the conductivity is low, caused by a poor choice of carrier ampholytes or by removal of ampholytes from a region of the gel through drift).

Precipitation of a protein during isoelectric focusing may produce an aggregate with non-specific affinity for other proteins, thereby limiting the resolution of many proteins in a sample. In an IEF gel this will simply be seen as a poorly resolved run; in a two dimensional run the effect may be of two kinds. Firstly, if the precipitate forms near the point of application to the IEF gel, numerous proteins will collect here, so that in the second dimension a pattern corresponding to a one-dimensional SDS-PAGE profile will be seen down one side of the two-dimensional gel. Secondly, if the protein enters the gel and then precipitates, it will usually do so over quite a wide range of pH; on resolution in a second dimension this will give rise to a horizontal 'streak' across the gel, and any proteins retained by the precipitate will also form 'streaks'. To minimize streaking in IEF gels, the general rules are to use low quantities of protein (< 50 μg for a standard cylindrical gel), to maintain the maximum urea concentration (at least 8 M) and to ensure that nucleic acids are removed from the sample (e.g. by nuclease treatment) before loading, since nucleic acids frequently precipitate in IEF gels and can entrap proteins. The requirement for a high urea concentration makes it necessary to run IEF gels of membrane proteins at room temperature, rather than in the cold room, as is sometimes recommended.

The problems of precipitation and streaking can be illustrated by the difficulties caused by the pronounced insolubility of cytochrome *P*-450 apoproteins in samples of microsomal protein (Kaderbhai and Freedman, 1980; Vlasuk and Walz, 1980). Fig. 2.2(a) shows an IEF gel of total

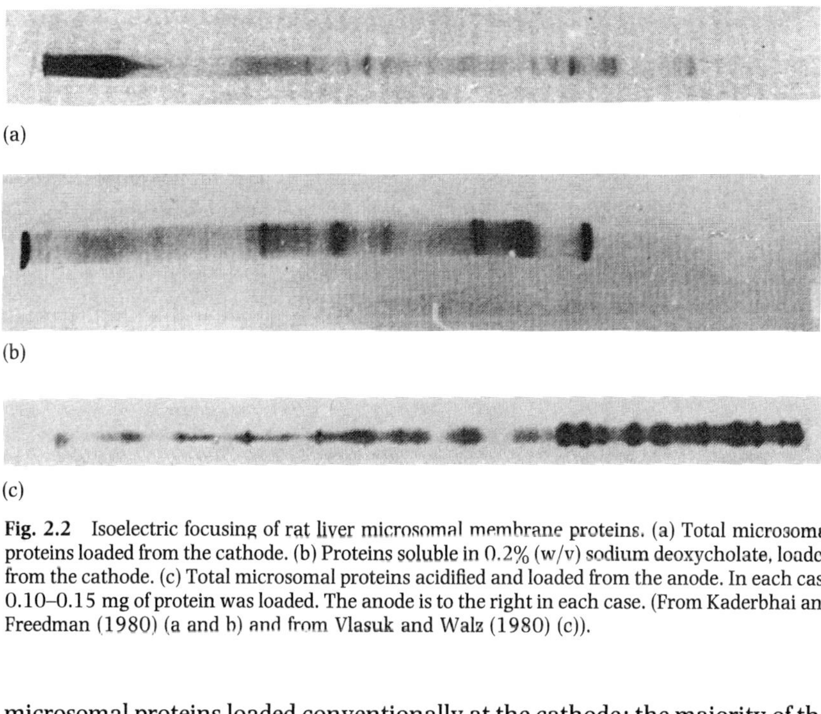


**Fig. 2.2** Isoelectric focusing of rat liver microsomal membrane proteins. (a) Total microsomal proteins loaded from the cathode. (b) Proteins soluble in 0.2% (w/v) sodium deoxycholate, loaded from the cathode. (c) Total microsomal proteins acidified and loaded from the anode. In each case 0.10–0.15 mg of protein was loaded. The anode is to the right in each case. (From Kaderbhai and Freedman (1980) (a and b) and from Vlasuk and Walz (1980) (c)).

microsomal proteins loaded conventionally at the cathode; the majority of the protein has been entrapped in a plug near the cathode, in a region of gel at pH 7–8. This problem does not arise (Fig. 2.2(b)) if the sample is a deoxycholate (0.2% w/v) extract of microsomal membranes, which does not contain the most hydrophobic peptides such as cytochrome *P*-450. Likewise, the precipitation can be prevented by acidifying the sample to the anolyte pH and applying the sample at the anode (Fig. 2.2(c)). In acidic conditions, cytochrome *P*-450s are far from their isoelectric points (pH 7–8) and migrate successfully through the gel. Because of their insolubility close to their pI, they still do not focus perfectly but by applying them at the anode, one can ensure that they do not interfere with the migration and focusing of other proteins.

### 2.2.4 SOLUBILIZATION OF MEMBRANES AND MEMBRANE PROTEINS FOR IEF

The discussion in the previous section emphasizes the importance of effective solubilization during isoelectric focusing. This presents particular problems for membrane proteins, as indicated in the Introduction. The most successful agents for solubilization of membrane proteins are ionic detergents such as SDS, which generate highly charged, water-soluble protein–detergent complexes (see Chapter 1). But such complexes mask the intrinsic charge properties of the protein and hence prevent their resolution by isoelectric focusing. Thus denaturing and solubilizing agents must be used which form complexes without conferring net charge. A thorough review of electrofocusing in detergent-containing media is given by Hjelmeland and Chrambach (1981). The protein:detergent ratio, as well as the absolute detergent concentration, are important factors. The standard approach is to use a 2% (w/v) solution of a non-ionic polyoxyethylene detergent (Nonidet P-40 or Triton X-100) in 8–10 M urea. This is perfectly adequate for many membrane proteins but fails to solubilize a significant number of major integral membrane proteins.

Three alternative approaches have overcome this problem in particular cases. The first is to use an extremely acidic or alkaline pH in addition to urea and non-ionic detergent. Several membrane preparations can be completely solubilized by the use of $K_2CO_3$ at pH 10.3 (Horst *et al.*, 1980). The use of acidic conditions to solubilize liver microsomal membranes was mentioned above; the use of acid pH has the further advantage that it permits very high concentrations of urea to be used. The second approach, and probably the most widely used, is to retain SDS as solubilizing agent but to add an excess of non-ionic detergent and to include non-ionic detergent, but not SDS, in the gel. This protocol, first suggested by O'Farrell (1975) and refined by Ames and Nikaido (1976), solubilizes the sample proteins essentially as for SDS-PAGE. Addition of excess non-ionic detergent produces complexes of protein with non-ionic detergent and mixed micelles of SDS with non-ionic detergent. When such a sample is subject to IEF, the SDS/non-ionic mixed micelles, being highly acidic, migrate rapidly to the anode, while the proteins migrate and focus as complexes with non-ionic detergent. While this approach has had considerable success, it is not infallible. Some proteins are adequately solubilized by SDS, but precipitate later in the process. This can occur either immediately after the addition of excess non-ionic detergent (in which case they can be removed by centrifugation before loading), or (more commonly) at the onset of electrofocusing, when the SDS is electrophoretically removed.

A third approach has a number of advantages in theory, but has not yet been very successful. This is to use zwitterionic detergents as solubilizing agents. Zwitterionic detergents, especially the sulphobetaines, $CH_3-[CH_2]_n-N^+[CH_3]_2-[CH_2]_3-SO_3^-$, contain both positively and negatively charged groups over the whole range of pH generally used for IEF (Hjelmeland *et al.*, 1979). Thus, although they bind to proteins and are strong denaturants, they confer no net charge on the solubilized protein. However, the sulphobetaine zwitterionic detergents, $n=14$, 16, which are effective solubilizing agents, have not yielded good results in isoelectric focusing (Booz and Travis, 1981; Burghes *et al.*, 1982; Mills and Freedman, 1983). One problem is that these detergents are not sufficiently soluble in high concentrations of urea, so that the combined effect of detergent and urea cannot be used (Burghes *et al.*, 1982; Mills and Freedman, 1983). A related zwitterionic detergent, CHAPS, in which the alkyl chain is replaced by a rigid cholyl ring system, has been described (Hjelmeland *et al.*, 1983). It is compatible with urea, and may prove useful (Perdew *et al.*, 1983).

## 2.3 Two-dimensional electrophoresis

### 2.3.1 BASIC FEATURES OF THE O'FARRELL PROCEDURE

In the method for two-dimensional electrophoresis developed by O'Farrell (1975), samples are solubilized in urea and Nonidet P-40, digested with deoxyribonuclease (DNAase), and subjected to isoelectric focusing (either immediately or after storage frozen) in vertical gel cylinders (13 cm × 2.5 mm internal diameter) containing urea and Nonidet P-40, for 5000–10 000 volt hours. The gels are extruded from the tubes and may then be equilibrated with a new buffer or applied directly to a second-dimension gel. Equilibration of the IEF gel (to remove urea and to form protein–SDS complexes for resolution in the second dimension) is required if the gel is to be frozen, but otherwise is used in some procedures but not others.

O'Farrell (1975) recommended equilibration in SDS sample buffer by shaking at room temperature for at least 30 min, and preferably 120 min. However, others, e.g. Anderson and Anderson (1978a,b), Stephenson *et al.* (1980), Rubin and Leonardi (1983), recommend equilibration for 10–30 min only, or for less than 5 min (Garrels, 1979) to minimize broadening of focused bands by diffusion and loss of proteins from the gel. Sample protein diffusion can be minimized by precipitation of the urea by cooling with dry ice (Garrels, 1979). Many workers use no equilibration step, assuming that sample proteins will be satisfactorily eluted from the IEF gel to form SDS complexes in the SDS stacking gel of the second dimension. Since many IEF gels can be run in parallel, while most laboratories are not equipped to run multiple second-dimension slab gels in parallel, freezing of gels after the first dimension is useful. Cylindrical gels can be frozen in the SDS sample buffer used for applying samples to SDS-PAGE, but this procedure is not successful for strips cut from slab IEF gels (Ames and Nikaido, 1976) possibly because of the thinness of such gels.

The IEF rod gel is then placed horizontally along the top edge of an SDS-PAGE slab gel. Any of the protocols described in Chapter 1 can be used, but discontinuous systems using Laemmli buffers and polyacrylamide gradient separating gels are by far the most common. Various methods have been developed for attaching the IEF gel cylinder to the second-dimension slab gel. In general, separating and stacking gels are poured first and the IEF gel is applied above the polymerized stacking gel. It can either be pushed down between the plates holding the slab gel, or held in the 'V' at the top if bevelled plates are used. The former method is easier if the diameter of the IEF gel is less than the thickness of the SDS-PAGE slab, but gels can be squashed. The key point is that the IEF gel must finally be positioned reproducibly, straight, horizontal and should not be stretched. The gel can be fixed in position by applying a rapid setting gel of 1% agarose in stacking gel buffer; this allows the sample gel to be well-fixed and to make good electrical contact with the slab, but avoids stretching.

The success of this basic system was established by O'Farrell (1975). Well over 1000 spots can be resolved; the average dimensions of spots were 1.5 mm in the horizontal (IEF) dimension and 0.5 mm in the vertical (SDS-PAGE) dimension so that in theory more than 10 000 spots could be resolved on a standard gel. However, this would only be possible in a mixture in which all components were present at the same concentration; in practice, a balance must be struck between loading sufficient protein to detect minor components and avoiding overloading of major components. Since resolution in the IEF dimension is greatly improved by low protein loadings (see above), the advantages of the two-dimensional technique can only really be exploited with sensitive detection methods. In a heavily loaded gel (100 $\mu$g of protein), an individual protein comprising 0.01% of the total (i.e. 0.01 $\mu$g) can just be detected by Coomassie Blue staining; however, such a loading can lead to problems of precipitation and spreading. This is easily overcome if sample proteins are radiolabelled to a reasonable specific radioactivity. A spot containing 1 c.p.m. of $^{14}C$ can be detected, so that if 100 000 c.p.m. are loaded, a protein comprising 0.001% of the total is detectable; in many cases it is possible to label proteins biosynthetically or chemically to 10 000 c.p.m./$\mu$g, so that only 10 $\mu$g of protein are required to give this level of detection.

Two-dimensional gel electrophoresis is conventionally performed with effective separating lengths of 10–15 cm in both dimensions. However, procedures have been developed for work on both larger and smaller scales. With large gels (32 cm × 40 cm), a threefold increase in resolution was observed (Voris and Young, 1980). Spot size did not increase in proportion to gel size, so that spots which overlapped in conventional gels were well resolved in large gels; furthermore, larger sample loads were possible, so that more minor components could be detected. A microscale system, permitting resolution of several hundred spots from a 0.3 $\mu$g sample, has been described by Neukirchen *et al.* (1982). Fine capillary gels are used for IEF and ultrathin gels for SDS-PAGE, so that samples 50-fold smaller than conventional samples can be resolved on gels 1–2 cm in each dimension, i.e. the size of a postage stamp.

Carrier ampholytes from the IEF gel behave like small proteins in the second-dimension SDS-PAGE gel and move close to the dye front. These species bind Coomassie Blue very strongly and are intensely stained if not previously rinsed out.

### 2.3.2 MODIFICATIONS FOR THE STUDY OF MEMBRANES AND MEMBRANE PROTEINS

Since the major difficulties associated with membrane proteins concern solubilization and maintenance in solution during isoelectric focusing, it is in these aspects that important modifications to the O'Farrell procedure have been made. The solubilization procedure of Ames and Nikaido (1976) is widely used; the sample is initially solubilized in 2% (w/v) SDS and then diluted with 2 volumes of sample buffer containing ampholytes in 9.5 M urea and 8% (w/v) Nonidet P-40. The 8-fold excess of non-ionic detergent over SDS is regarded as crucial and there must also be a 1–2-fold excess of SDS over protein.

This method was devised for investigating bacterial membranes, but close variants of it have been used to resolve a wide variety of membranes from animal (Wilson *et al.*, 1977; Rubin and Leonardi, 1983), plant (Boschetti *et al.*, 1981) and viral (Privalsky and Penhoet, 1978) sources. The different procedures differ only in details, such as whether the samples are heated in SDS for solubilization, and the temperature and time of heating if used. This variable should clearly be optimized for the given membrane sample under study. In general, samples are run from the cathode, but Privalsky and Penhoet (1978) reported that better resolution of basic viral envelope proteins was obtained by loading at the anode.

Anode loading was further developed by Vlasuk and Walz (1980), who used it to resolve liver microsomal membrane proteins. In this case, no SDS was used and the proteins were solubilized in standard O'Farrell lysis buffer containing urea and non-ionic detergent, followed by acidification and addition of further crystalline urea. An alternative approach developed by Stephenson *et al.* (1980) for mitochondrial membrane proteins involved initial solubilization in 8 M urea and SDS at an SDS: protein ratio of 2: just before loading, the sample was titrated to pH 3 with 0.1 M citric acid. In this case, non-ionic detergent was present in the IEF gel but not in the sample mix.

This study by Stephenson *et al.* is an excellent example of how standard procedures for two-dimensional electrophoresis can be modified and adapted for use with even the most challenging membrane proteins. Their purpose was to resolve the mitochondrial translation products of yeast, which consist of a small number of highly hydrophobic membrane proteins, some of which are not resolved by SDS-PAGE alone. The method finally adopted gave complete resolution of over 100 mitochondrial polypeptides, and the mitochondrial translation products could easily be recognized among the background of cytoplasmically synthesized proteins.

Some membrane proteins have, as yet, defied successful resolution on two-dimensional gels. A prominent example is spectrin from the red cell membrane. In retrospect it is unfortunate that the red cell membrane was for a long time regarded as *the* model system by membrane biochemists. Before the publication of O'Farrell's paper, Wallach and colleagues (Bhakdi *et al.*, 1974) developed two-dimensional procedures for resolution of red cell membrane proteins using IEF followed by SDS-PAGE. The results were not wholly satisfactory and did not encourage others to use the method, though the problems can now be seen to have derived from their sample material rather than their method. Rubin and Milikowski (1978) have extended the study of red cell membranes by two-dimensional electrophoresis. They found that the concentration of SDS and the SDS:protein ratio in the solubilizing buffer were crucial for successful solubilization of spectrin and Band 3 protein. Even if these proteins were successfully solubilized, they would not enter the IEF gel unless high levels of SDS were present in the sample loaded for IEF. This effect, ironically, could be used to improve resolution and detection of minor erythrocyte membrane proteins, since these were far more readily observed in two-dimensional gels from which the major proteins spectrin and Band 3 were missing. The deliberate exclusion of major hydrophobic polypeptides to improve resolution of the remainder has also been used in the case of liver microsomal membranes (Kaderbhai and Freedman, 1980), whose acidic polypeptides are best resolved by selective solubilization in 0.2% (w/v) deoxycholate before further treatment and electrophoresis by the method of Ames and Nikaido (1976).

### 2.3.3 ALTERNATIVE TWO-DIMENSIONAL PROCEDURES

Although two-dimensional electrophoresis involving IEF followed by SDS-PAGE has become the standard method for analysis of complex protein mixtures, alternative methods are useful in some circumstances. Because of the high resolving power of both IEF and SDS-PAGE, and their dependence on quite different physical properties, the most successful alternatives use the same two separation techniques in the opposite order.

Singer *et al.* (1978) introduced such a method for the analysis of genetic variants of defined proteins. Since they were interested in only one or two proteins, rather than in the whole complement of proteins in a sample, it was economic to resolve proteins in one dimension, cut out the band of interest and then resolve it alone in a second dimension. In this case, the simplicity and reproducibility of SDS-PAGE makes it more suitable for the first dimension. Samples are run on SDS-PAGE at a polyacrylamide gel concentration chosen so that the protein of interest runs close to the front, i.e. at the minimum feasible acrylamide concentration, in order to facilitate elution of the protein from the gel into the second dimension. The band of interest is cut out and placed in a sample well of a vertical IEF slab gel containing urea and Nonidet P-40. As the SDS–protein complexes enter the IEF gel from the cathode, the SDS is displaced by the excess non-ionic detergent, and the free SDS rapidly migrates to the anode. Thus corresponding proteins from many mutants can be screened for charge changes by isoelectric focusing in parallel tracks of a single IEF gel slab, rather than by more complex comparison of several two-dimensional slabs. A similar result is achieved with less convenience, in a method of Siegenthaler and Nguyen (1983) where bands from an SDS-PAGE gel are soaked to remove SDS, and then crushed to extract the proteins, which are made up in urea and Nonidet P-40 for resolution by IEF.

A genuine two-dimensional separation using SDS-PAGE preceding IEF was developed by Tuszynski *et al.* (1979). The key problem with this order, apart from the removal of SDS, is that transfer from first to second dimension involves passing the protein from a restrictive gel (at least 10% acrylamide) to a non-restrictive gel (usually 5% acrylamide). To overcome this, Tuszynski *et al.* use a careful electrophoretic elution of the sample from the SDS-PAGE gel strip into a 4.5% acrylamide gel slab in Laemmli stacking gel buffer. The proteins migrate without unstacking and this transfers the proteins into the less restrictive gel without loss of resolution (though possibly with some loss in yield). The strip containing the sample proteins is then cut from the 4.5% gel slab and applied to a *horizontal* polyacrylamide slab for isoelectric focusing in a gel containing urea and Nonidet P-40 as usual. Use of the SDS-PAGE step first, with its more comprehensive solubilizing power, has the advantage that any protein which precipitates or focuses poorly in the IEF dimension, is already resolved from most other components of the mixture and hence will not interfere with their focusing.

## 2.4 Analysis of two-dimensional gels

### 2.4.1 DETECTION AND IDENTIFICATION OF POLYPEPTIDES

Polypeptides resolved on a two-dimensional gel can be detected by any of the staining or radiochemical methods appropriate to SDS-PAGE gels. Coomassie Blue staining is widely used, but the greater sensitivity of silver staining methods, and the need to restrict the total protein loaded in order to retain good resolution, has made silver staining methods increasingly popular. Radioautography and fluorography are usually more sensitive for membranes from bacteria or from isolated cells when these can conveniently be radiolabelled biosynthetically. Methods for chemical radiolabelling of sample proteins are constantly being refined; reductive methylation of $\varepsilon$-$NH_2$ groups of lysyl residues is favoured since either $^{14}C$ (as $H^{14}CHO$) or $^{3}H$ (as $NaB^{3}H_4$) can be introduced into a very common residue with no charge modification (Finger and Choo, 1981). Two-dimensional gels can be stained for glycoproteins, etc. and can be blotted on to nitrocellulose paper for specific probing, as described in Chapter 1.

A conventional two-dimensional gel resolves 100–1000 polypeptides, so that a single gel contains a great deal of information. Much of this is meaningless unless the identity of the polypeptides on the gel is known. Even for well-studied membranes, such as the red cell membrane or the mitochondrial inner membrane, it is not yet possible to identify every polypeptide on a two-dimensional gel and for other systems the problem is enormous. Most techniques for identifying a protein on a gel require that the protein available is reasonably pure. The most valuable methods are immunological identification, peptide finger-printing and co-electrophoresis. If a membrane sample for two-dimensional electrophoresis is incubated with a specific antibody against a constituent protein, and the precipitated complex removed before the sample is applied to the IEF gel, the antigenic polypeptide will be absent from the resultant two-dimensional gel, or present in reduced quantity. This method has now been almost completely replaced by blotting methods in which the blot is developed with a specific antibody followed by a marker second antibody (see Chapter 1). A spot on a gel suspected of being identical to a known protein can be cut from the gel and subjected to protease or specific chemical cleavage in parallel with the authentic protein, and the products resolved by SDS-PAGE on a high-concentration polyacrylamide gel (Cleveland *et al.*, 1977). The only limitation to this method of identification is the sensitivity of the detection method for the final peptide fragments; if a spot has been detectable by Coomassie Blue on the original two-dimensional gel, its fragments will probably be detectable by silver staining.

The methods quoted above are essentially the same as those used for identifying a band on a one-dimensional SDS-PAGE gel. However, the greater resolution of two-dimensional electrophoresis allows a more direct co-electrophoresis method to be used with fair confidence. A membrane sample can be combined with an excess of a known membrane protein and the resultant pattern compared with that from the membrane alone; given a sufficiently pure protein and a favourable disposition of spots, the intensity of a single spot will be enhanced. (A negative result here is not necessarily conclusive since the pure protein may have undergone modification during

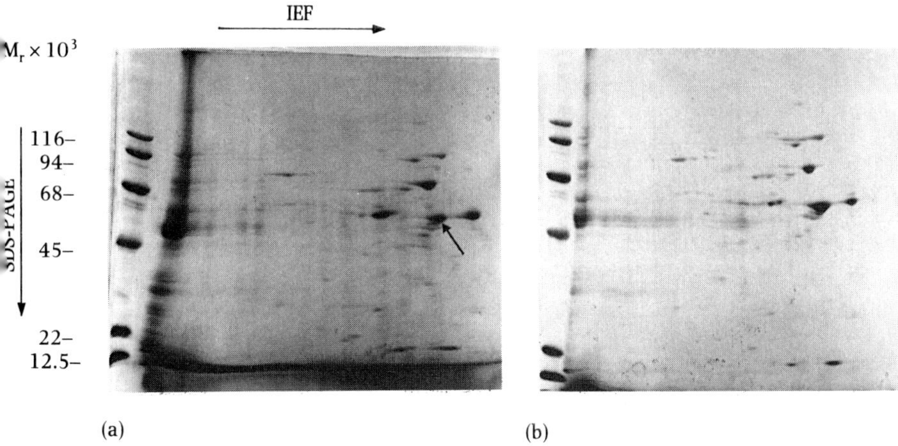


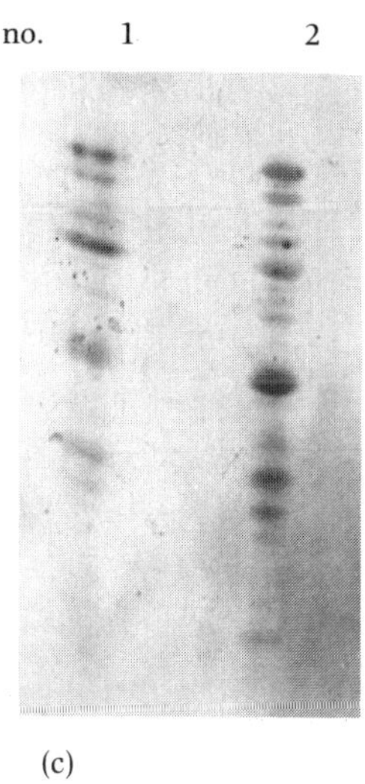


**Fig. 2.3** Identification of protein disulphide isomerase. (a) Two-dimensional electrophoresis of rat liver microsomal proteins (0.25 mg) resolved by the method of Kaderbhai and Freedman (1980). (b) As (a) but the sample contained, in addition, 0.01 mg of purified rat liver protein disulphide isomerase. (c) *Staphylococcus aureus* V8 proteinase finger-prints: track 1, authentic purified rat liver protein disulphide isomerase; track 2, gel piece containing the spot tentatively identified as protein disulphide isomerase (arrowed in a). (From Mills *et al.*, 1983.) (Reprinted by permission from Biochemical Journal, Vol. 213, pp. 245–8, © 1983, The Biochemical Society, London.)

purification). This method and the finger-printing approach are illustrated in Figs. 2.3: a,b,c.

### 2.4.2 DOUBLE-LABEL METHODS FOR COMPARING SAMPLES

In most experiments involving two-dimensional gel electrophoresis, two or more samples are compared. They may be samples from a wild-type and a mutant, from animals in different physiological states, from animals before and after a pharmacological treatment, or from samples treated in alternative ways. Comparison of two or more two-dimensional gel patterns in detail is an extremely complex task of scaling, pattern matching and quantification and is made especially complex by the difficulty of achieving reproducible running of separate gels. Reproducibility is considered in the next section (2.4.3) and developments in computerized data collection and analysis are dealt with subsequently (2.4.4). Here we consider a simpler approach to the comparison of *two* samples which circumvents the problem of reproducibility in running gels.

Differences between two samples can be highlighted if they can be mixed and run *together*, but analysed separately. This can be achieved by a number of 'double-label' methods. Essentially the samples are labelled with two distinct isotopes of different energies; the gel produced by running a mixture of the samples is then developed by methods which selectively detect one isotope, or the other, or both. In a popular procedure (McConkey, 1979), the samples are $^{3}H$- and $^{14}C$-labelled. If such a gel is developed by fluorography, both isotopes are detected; if developed by radioautography using direct contact with Kodak No-screen film, only the $^{14}C$ is detected, since this film is insensitive both to the less energetic $^{3}H$ emission and to fluorescent quanta. The radioautogram is laid over a contact negative of the fluorogram; polypeptides labelled with both isotopes appear black in the composite image, those labelled exclusively with $^{3}H$ appear white, and those labelled exclusively with $^{14}C$ appear as black spots with white halos. A variant of this procedure is to use conventional X-Omat R film for radioautography but to screen out fluorescence by placing a sheet of black carbon paper between gel and film (Choo *et al.*, 1980). These methods can be used with $^{35}S$ as the more energetic isotope in place of $^{14}C$.

A drawback of these methods is that although only one gel is run, two separate images are produced which ultimately have to be compared visually. This can be overcome by the use of colour negative film, where the different energy isotopes penetrate to different depths in the film emulsion, and hence expose the layers to different extents (Kronenberg, 1979). Using colour film suited to illumination with tungsten lights, $^{3}H$ emission does not penetrate deeper than the top (blue-sensitive) layer and hence generates a yellow image (either directly or by blue fluorescence in fluorography) while $^{14}C$ or $^{35}S$ $\beta$-particles penetrate to the green layer and produce a red image. The images can be analysed automatically by densitometer (see Section 2.4.4), but qualitative data can readily be obtained by using coloured optical filters to view the developed negative (Goldman *et al.*, 1983).

### 2.4.3 REPRODUCIBILITY AND MULTIPLE GEL SYSTEMS

The methods described above have the obvious limitation that they can only be used to compare two samples, or to compare each of a series of samples to a standard sample. For more general purposes it is inevitable that several samples run on separate gels must be compared. The exact position co-ordinates of a given polypeptide (defined in terms of distance in mm from the origin in both dimensions) will generally vary from gel to gel. Obvious reasons for this include: (i) the IEF gels may not be identical in length; (ii) the pH gradient established in IEF is most unlikely to be reproducible from one run to another (or even from one tube gel to another in a single run); (iii) the gels will suffer varying degrees of stretching and distortion during extrusion, equilibration and layering on to the second dimension; (iv) the polyacrylamide gel gradients in two different second-dimension gels are unlikely to be precisely identical; (v) the polymerization process is not fully controlled and inhomogeneities in the gel are inevitable; (vi) slab gels undergo uncontrolled distortion and shrinkage during fixing, staining and (especially) drying for radioautography.

For these reasons, workers experienced with a particular material will mentally normalize the gels they study by selecting a constellation of major spots recognizable by their relative positions and intensities, and will then identify other spots by their positions in relation to these 'internal standards'. However for processing large numbers of samples, and especially for computer analysis, more objective methods are required. To minimize the problems of normalization, it is therefore vital to maximize reproducibility in running gels. Most responses to this have involved the building of special multiple gel apparatus so that the samples to be compared can be run at the same time, on gels poured at the same time.

The first dimension presents fewer problems since it is easy to run up to 12–16 samples simultaneously on a slab gel or in a tube gel apparatus. Nevertheless, a system has been described for the simultaneous filling of up to 20 IEF tubes to precisely the same depth, their running in parallel and their rapid recovery for second-dimension electrophoresis (Anderson and Anderson, 1978a). For the second dimension, a number of multiple gel systems are possible (Sinclair and Rickwood, 1982). One possibility is to cast very wide second-dimension SDS-PAGE gels, so that several IEF gel tubes or strips can be laid on a single second-dimension gel (Garrels, 1979). Alternatively the polymerizing acrylamide solution for the second dimension can be passed through a multiple-outlet manifold so that a number of individual second-dimension gels can be poured with identical composition (Anderson and Anderson, 1978b). The ISODALT system of Anderson and Anderson (1978a,b) simultaneously casts ten polyacrylamide gradient slab gels, and runs them simultaneously in a three compartment tank. This system is commercially available.

### 2.4.4 METHODS FOR DATA COLLECTION AND COMPUTER ANALYSIS

The automated scanning of two-dimensional gels or of images produced from them, and the computer analysis of the resultant data, is a major research area which is bound to grow rapidly (Garrels, 1980). At the moment, the nature of the problem is clear, and a significant number of methods have been developed, but there is not yet sufficient experience for any one method to become standard. In any case, there will not be one ideal method, but preferred methods for different laboratories in view of the range in scale of operations (number of gels, number of spots to be analysed, expense, other possible uses for the equipment, etc.). In this section we will outline the steps involved in the automated analysis of a gel image and in the automated comparison of images. For full details, see Garrels (1979), Bossinger *et al.* (1979) or other papers cited in Dunn and Burghes (1983b).

The raw material for analysis is the distribution of stain, fluorescence or radioactivity on a gel, or of development on a film. Such data can be collected from gels or films using flat-bed *x,y* scanning densitometers which are accurate and highly resolving, but slow. More rapid data collection is possible with revolving drum densitometers, but the samples here have to be robust (i.e. radioautogram films, not gels themselves). The intensity and co-ordinate data are then filtered to remove noise and correct for background, and then some procedure is used to identify individual spots and their boundaries. A wide variety of procedures is used for this operation; the major difficulty is to distinguish minor spots close to major spots, and a refined mathematical method for obtaining such 'super-resolution' has been described (Agard *et al.*, 1981). This problem is particularly acute for membrane proteins which tend to 'streak'. After individual spots are identified and their boundaries determined, the total intensity in each peak can be integrated and then converted to d.p.m. or ng protein by calibration with standard spots. Most densitometric methods are not linear over a wide range of intensity so that it may not be possible to collect data for major and minor spots at the same time, since these may differ in intensity by at least $10^3$; it may be possible to combine data from several images of the same gel exposed for various times. At this stage in the analysis, the data are collected and expressed as a finite number of spots of defined position and total intensity (see Fig. 2.4). The next stage is to normalize the data so that different gels can be compared and corresponding spots on two gels recognized. Again a large number of procedures has been used, most of which depend on the operator identifying several spots as internal standards; the inhomogeneity of gels prevents the use of a simple scale factor in each dimension to match one gel to another. Detailed theory of gel normalization is discussed by Capel *et al.* (1979). Once normalized, the data from a large number of gels can be stored and retrieved for a variety of manipulations.

Clearly such comprehensive systems for data collection and analysis are complex and expensive, and several cheaper approaches have been developed, Mariash *et al.* (1982) describe a system whose components can be bought for less than $3500, involving a black and white video camera for data collection from fluorograms. This gives poorer spatial resolution than scanning

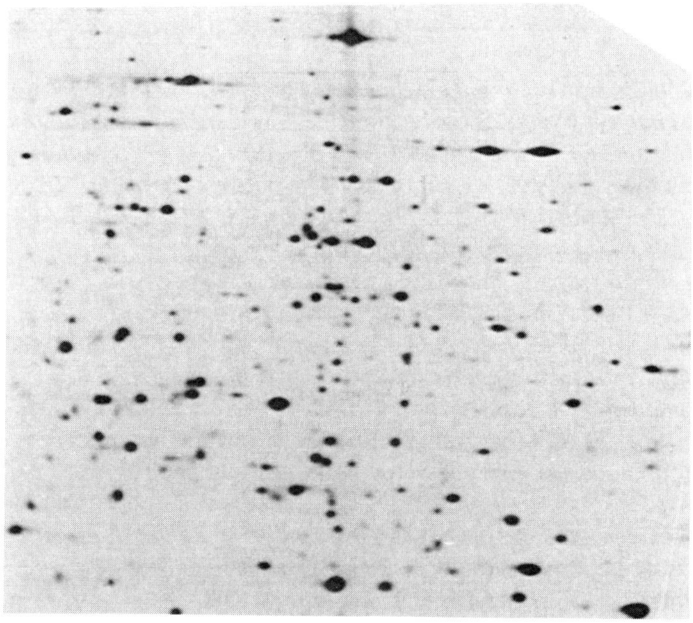

(a)

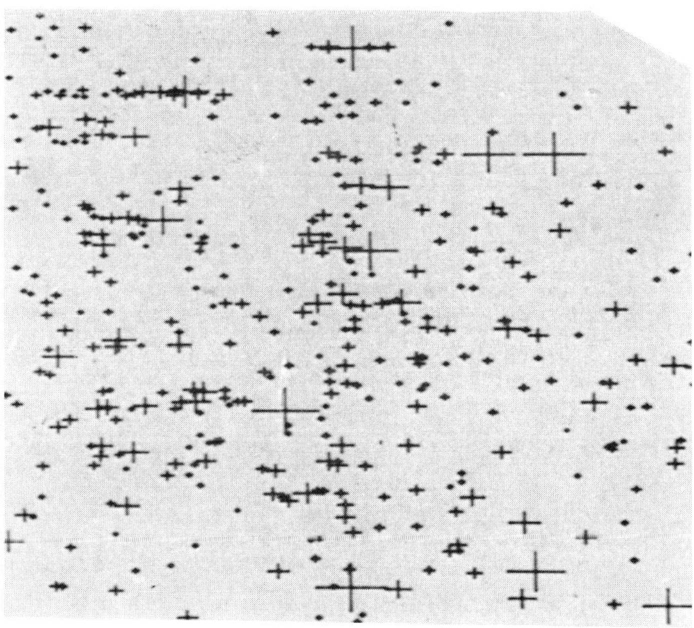

(b)

**Fig. 2.4** Computer analysis of a two-dimensional gel pattern. (a) Portion of a fluorogram of a two-dimensional gel. (b) Computer representation of the spots detected and their integrated densities; the area of the spot is proportional to the measured integrated density. (From Garrels, 1979.)

densitometers, and the intensity can only be expressed as one of 64 levels of grey, but the operator can select a limited region of the fluorogram for study, so the resolution can be adequate. Restricting the area of gel to be analysed, or predetermining the spots to be studied allows considerable simplification of systems. In the method of Garrison and Johnson (1982) the digitized density data from a rotating drum densitometer are displayed on a computer graphics terminal and the operator selects the spots to be analysed and defines their boundaries. In the situation where only a few spots are of interest, the density data need not be digitized and the integrated density can be estimated graphically from the scanning traces in perpendicular directions (Quitschke and Schechter, 1982).

These procedures assume that the polypeptides 'of interest' are known at the start of the analysis; obviously one is most often interested in polypeptides which change their positions or intensities (relative to the majority of proteins) between two related samples. A direct method for identifying such polypeptides uses the optical method developed by astronomers for comparing star maps (Spragg *et al.*, 1983). Two light projectors combine the images of two gels (or photographic negatives) on an X–Y digitizer connected to a microcomputer. The lights of the two projectors flash out of phase with each other; if the gels have identical dimensions or are normalized, common polypeptides are not affected by this alternation of images, whereas polypeptides which differ between the gels either in intensity or position appear to flash on and off, or to jump from side to side. The co-ordinates and intensities of these variable polypeptides can then be digitized for fuller analysis.

## 2.5 Applications of two-dimensional gel electrophoresis

Two-dimensional gel electrophoresis can be applied to any problem which might be tackled by SDS-PAGE but where this one-dimensional technique is likely to have inadequate resolution. Clearly, the scope is very wide: a small number of illustrative applications will be mentioned here.

One powerful use of the technique is to characterize differences in polypeptide composition between related samples – either intrinsic differences between different species, strains, varieties etc., or differences induced by hormones or pharmacological treatments. An excellent example of this is given by the cytochrome *P*-450 apoproteins of mammalian liver microsomal membranes. These exist as a large number of isoenzymes, generally very similar in size and overall properties, so that they are poorly resolved by SDS-PAGE. The cytochrome *P*-450 complement of an animal is a function of its species, sex, age, hormonal status and its past exposure to any of a large number of specific inducers. The successful development by Vlasuk and Walz (1980) of a method for resolving cytochrome *P*-450s on two-dimensional gels makes it possible to see these differences directly. Thus differences between rat strains (Vlasuk and Walz, 1982; Vlasuk *et al.*, 1982) and between rat and hamster proteins (Walz *et al.*, 1983) have been catalogued, as well as the effects on the animals of a wide variety of pretreatments. At the limit, this method provides a very refined form of genetic analysis, since a two-dimensional gel profile of proteins is a very complete representation of the phenotype; thus six homologous cytochrome *P*-450 polypeptides encoded at two genetic loci have been resolved (Rampersaud and Walz, 1983).

Induced changes in mRNA composition, as well as in protein composition can be detected, using *in vitro* translation of mRNA. Thus thyroid hormone-induced changes in gene expression in the liver can be monitored at a pretranslational level (Seelig *et al.*, 1981). Comparison of mature proteins and of *in vitro* translation produced by two-dimensional gel electrophoresis can identify post-translational modifications which are not accompanied by any significant change in $M_r$, since most post-translational changes are associated with some change in charge properties. Many precursor sequences are strongly basic, and this allows easy identification both of processed and unprocessed forms, such as albumin and proalbumin (Vlasuk *et al.*, 1980). Similarly, cytoplasmically synthesized precursors of major mitochondrial proteins can be recognized by comparing translation products obtained from total lymphoid cell mRNA with the polypeptide pattern of lymphoid cells (Anderson, 1981), since the precursors are all distinctly more basic. Conversely, the exact correspondence in two-dimensional gels of specific phenobarbital-inducible cytochrome *P*-450 polypeptides, and translation products of liver mRNA derived from phenobarbital-treated rats, establishes that there are no significant post-translational modifications in this case (Walz *et al.*, 1982).

As well as noting fine differences between closely related preparations, the method can highlight common features of quite diverse source materials. Thus Kaderbhai and Austen (1984) have shown a remarkable identity between the major acidic polypeptides in microsomal membranes from rat liver, dog liver and dog pancreas (see Fig. 2.5).

Topological information can also be obtained by analysis of two-dimensional gels of samples labelled with a membrane-impermeant label. Thus cell surface markers on various lymphocyte cell types can be distinguished directly (Takeda *et al.*, 1983). In another exploitation of labelling methods, the polypeptides in liver microsomes which are covalently modified by the reactive intermediates generated during the metabolism of polycyclic hydrocarbons have been resolved (Kaderbhai *et al.*, 1981) and have been shown to be the set of polypeptides containing reactive thiol groups (Kaderbhai and Freedman, 1984).

As shown in Chapter 1, the introduction of SDS-PAGE permitted a great expansion of work on membrane proteins from about 1970 onwards, allowing quite complex mixtures of rather intractable proteins to be resolved and

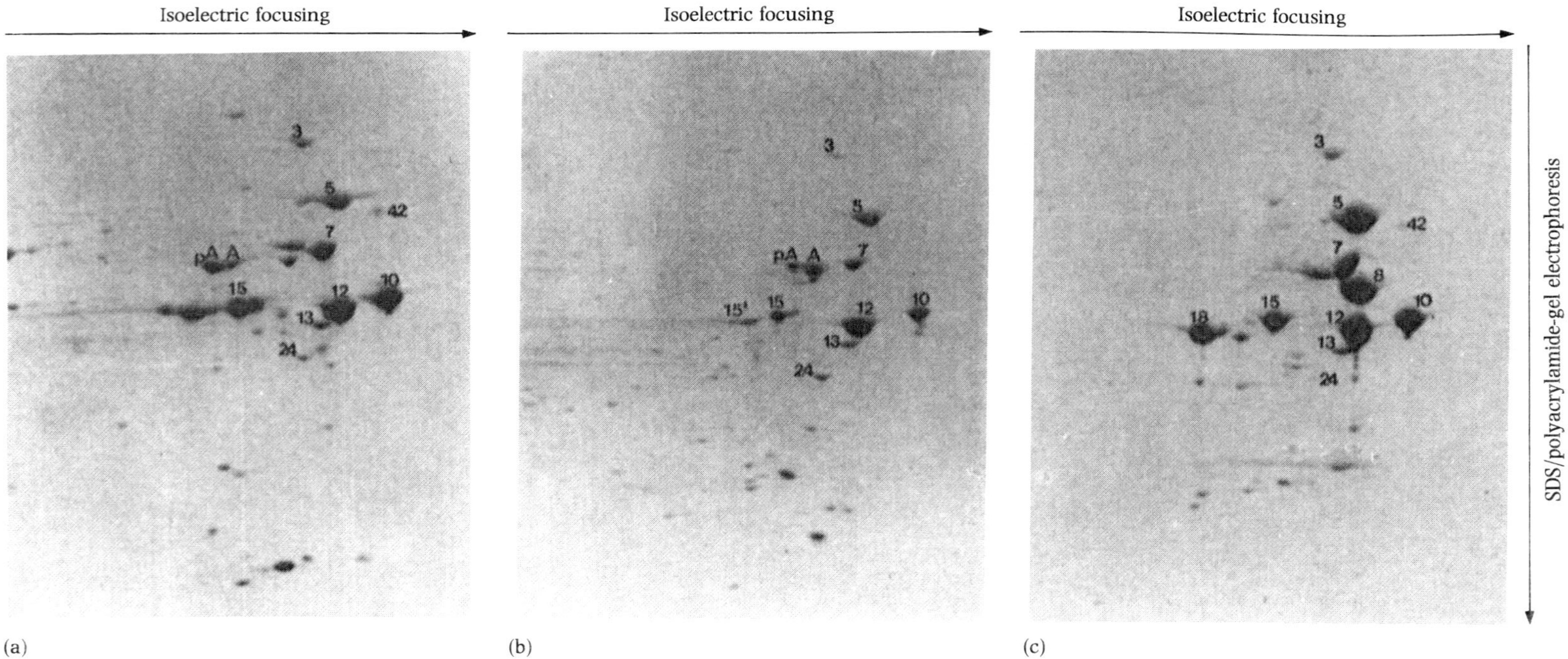


**Fig. 2.5** Two-dimensional electrophoretic analysis of microsomal polypeptides from various sources. (a) Rat liver; (b) dog liver; (c) dog pancreas. pA and A are proalbumin and albumin respectively; spot 18 is amylase. The corresponding spots appearing to be identical in all three sources are numbered. (From Kaderbhai and Austen, 1984.) (Reprinted by permission from Biochemical Journal, Vol. 217, pp. 145–57, © 1984, The Biochemical Society, London.)

characterized. The 100-fold increase in resolution available in two-dimensional electrophoresis will produce a further expansion of the scope of membrane protein studies and will allow even minor components of very complex samples to be studied. The potential for the study of proteins in biological membranes is very great.

## 2.6 References


Agard, D. A., Steinberg, R. A. and Stroud, R. M. (1981) *Anal. Biochem.*, **111**, 275–68.

Ames, G. F. L. and Nikaido, K. (1976) *Biochemistry*, **15**, 616–23.

An der Lan, B. and Chrambach, A. (1982) in *Gel Electrophoresis of Proteins: a Practical Approach* (eds B. D. Hames and D. Rickwood), IRL Press, Oxford, pp. 157–87.

Anderson, L. (1981) *Proc. Natl. Acad. Sci. USA*, **78**, 2407–11.

Anderson, N. G. and Anderson, N. L. (1978a) *Anal. Biochem.*, **85**, 331–40.

Anderson, N. L. and Anderson, N. G. (1978b) *Anal. Biochem.*, **85**, 341–54.

Anderson, N. L. and Hickman, B. J. (1979) *Anal. Biochem.*, **93**, 312–20.

Bhakdi, S., Knufermann, H. and Wallach, D. F. H. (1974) *Biochim. Biophys. Acta*, **345**, 448–57.

Bjellqvist, B. and Ek, K. (1982) LKB application note, No. 321.

Booz, M. L. and Travis, R. L. (1981) *Phytochemistry*, **20**, 1773–9.

Boschetti, A., Sauton-Heiniger, E. and Clemetson, K. J. (1981) in *Membrane Proteins – a Laboratory Manual* (eds A. Azzi, U. Brodbeck and P. Zahler), Springer Verlag, Berlin, pp. 7–12.

Bossinger, J., Miller, M. J., Vo, K. P., Geduschek, E. P. and Xuong, N. H. (1979) *J. Biol. Chem.*, **254**, 7986–98.

Breithaupt, T. B., Nystrom, I. E., Hodges, D. H. and Babitch, J. A. (1978) *Anal. Biochem.*, **84**, 579–82.

Burghes, A. H. M., Dunn, M. J., Statham, H. E. and Dubowitz, V. (1982) *Electrophoresis*, **3**, 185–96.

Cantrell, S. J., Babitch, J. A. and Torres, S. (1981) *Anal. Biochem.*, **116**, 168–73.

Capel, M., Redman, B. and Bourque, D. P. (1979) *Anal. Biochem.*, **97**, 210–20.

Choo, K. H., Cotton, R. G. H. and Danks, D. M. (1980) *Anal. Biochem.*, **103**, 33–8.

Cleveland, D. W., Fischer, S. G., Kirschner, M. W. and Laemmli, U. K. (1977) *J. Biol. Chem.*, **252**, 1102–6.

Dunn, M. J. and Burghes, A. H. M. (1983a) *Electrophoresis*, **4**, 97–116.

Dunn, M. J. and Burghes, A. H. M. (1983b) *Electrophoresis*, **4**, 173–89.

Ek, K., Bjellqvist, B. and Bishop, R. (1983) LKB application note, No. 322.

Finger, J. M. and Choo, K. H. (1981) *Biochem. J.*, **193**, 371–4.

Gaal, O., Medgyesi, G. A. and Vereczkey, L. (1980) *Electrophoresis in the Separation of Biological Macromolecules*, John Wiley, Chichester.

Garrels, J. I. (1979) *J. Biol. Chem.*, **254**, 7961–77.

Garrels, J. I. (1980) *Trends Biochem. Sci.*, **5**, 281–3.

Garrison, J. C. and Johnson, M. L. (1982) *J. Biol. Chem.*, **257**, 13144–9.

Goldman, R. C., Trus, B. L. and Leive, L. (1983) *Eur. J Biochem.*, **131**, 473–80.

Guengerich, F. P. (1979) *Biochim. Biophys. Acta*, **577**, 132–41.

Hare, D. L., Stimpson, D. I. and Cann, J. R. (1978) *Arch. Biochem. Biophys.*, **187**, 274–5.
Hjelmeland, L. M. and Chrambach, A. (1981) *Electrophoresis*, **2**, 1–11.
Hjelmeland, L. M., Nebert, D. W. and Chrambach, A. (1979) *Anal. Biochem.*, **95**, 201–8.
Hjelmeland, L. M., Nebert, D. W. and Osborne, J. C. (1983) *Anal. Biochem.*, **130**, 78–82.
Hunter, L. (1978) *Anal. Biochem.*, **89**, 279–83.
Horst, M. N., Mahaboob, S., Basha, M., Baumback, G. A., Mansfield, E. H. and Roberts, R. M. (1980) *Anal. Biochem.*, **102**, 399–408.
Kaderbhai, M. A., Bradshaw, T. K. and Freedman, R. B. (1981) *Chem. Biol. Interact.*, **36**, 211–27.
Kaderbhai, M. A. and Austen, B. M. (1984) *Biochem. J.*, **217**, 145–57.
Kaderbhai, M. A. and Freedman, R. B. (1980) *Biochim. Biophys. Acta*, **601**, 11–21.
Kaderbhai, M. A. and Freedman, R. B. unpublished work.
Kaltschmidt, E. and Wittmann, H. G. (1970) *Anal. Biochem.*, **36**, 401–12.
Kronenberg, L. M. (1979) *Anal. Biochem.*, **93**, 189–95.
Kurland, C. G. (1977) *Annu. Rev. Biochem.*, **46**, 173–200.
McConkey, E. H. (1979) *Anal. Biochem.*, **96**, 39–44.
Mariash, C. N., Seelig, S. and Oppenheimer, J. H. (1982) *Anal. Biochem.*, **121**, 388–94.
Mills, E. N. C. and Freedman, R. B. (1983) *Biochim. Biophys. Acta*, **734**, 160–7.
Mills, E. N. C., Lambert, N. and Freedman, R. B. (1983) *Biochem. J.*, **213**, 245–8.
Neukirchen, R. O., Schlosshauer, B., Baars, S., Jäckle, H. and Schwarz, U. (1982) *J. Biol. Chem.*, **257**, 15229–34.
Nguyen, N. Y. and Chrambach, A. (1977) *Anal. Biochem.*, **82**, 226–35.
O'Farrell, P. H. (1975) *J. Biol. Chem.* **250**, 4007–21.
O'Farrell, P. Z., Goodman, H. M. and O'Farrell, P. H. (1977) *Cell*, **12**, 1133–42.
Perdew, G. H., Schaup, H. W. and Selivonchick, D. P. (1983) *Anal. Biochem.*, **135**, 453–5.
Privalsky, M. L. and Penhoet, E. E. (1978) *Proc. Natl. Acad. Sci., USA*, **72**, 3625–9.
Quitschke, W. and Schechter, N. (1982) *Anal. Biochem.*, **124**, 231–8.
Rampersaud, A. and Walz, F. G. (1983) *Proc. Natl. Acad. Sci. USA*, **80**, 6542–6.
Righetti, P. G. and Macelloni, C. (1982) *J. Biochem. Biophys. Meth.*, **6**, 1–15.
Righetti, P. G., Tudor, G. and Gianazza, E. (1982) *J. Biochem. Biophys. Meth.*, **6**, 219–27.
Rubin, R. W. and Leonardi, C. L. (1983) *Methods Enzymol.*, **96**, 184–92.
Rubin, R. W. and Milikowski, C. (1978) *Biochim. Biophys. Acta*, 100–10.
Seelig, S., Liaw, C., Towle, H. C. and Oppenheimer, J. H. (1981) *Proc. Natl. Acad. Sci. USA*, **78**, 4733–7.
Siegenthaler, P. A. and Nguyen, T. D. (1983) *Biochim. Biophys. Acta*, **722**, 226–33.
Sinclair, J. and Rickwood, D. (1982) in *Gel Electrophoresis of Proteins: a Practical Approach* (eds B. D. Hames and D. Rickwood), IRL Press, Oxford.
Singer, B. S., Morrissett, H. and Gold, L. (1978) *Anal. Biochem.*, **85**, 224–9.
Spragg, S. P., Jones, M. I. and Hill, B. J. (1983) *Anal. Biochem.*, **129**, 255–68.
Stephenson, G., Marzuki, S. and Linnane, A. W. (1980) *Biochim. Biophys. Acta*, **609**, 329–41.
Takeda, A., Waldron, J. A., Ruddle, N. H. and Cone, R. E. (1983) *Exptl. Cell Res.*, **148**, 83–93.
Tuszynski, G. P., Buck, C. A. and Warren, L. (1979) *Anal. Biochem.*, **93**, 329–38.
Vesterberg, O. (1969) *Acta Chem Scand*, **23**, 2653–66.
Vesterberg, O., Hansen, L. and Sjösten, A. (1977) *Biochim. Biophys. Acta*, **491**, 160–6.

Vlasuk, G. P., Ghrayeb, J. and Walz, F. G. (1980) *Biochem. Biophys. Res. Commun.*, **94**, 366–72.

Vlasuk, G. P., Ryan, D. E., Thomas, P. E., Levin, W. and Walz, F. G. (1982) *Biochemistry*, **21**, 6288–92.

Vlasuk, G. P. and Walz, F. G. (1980) *Anal. Biochem.*, **105**, 112–20.

Vlasuk, G. P. and Walz, F. G. (1982) *Arch. Biochem. Biophys.* **214**, 248–59.

Voris, B. P. and Young, D. A. (1980) *Anal. Biochem.*, **104**, 478–84.

Walz, F. G., Vlasuk, G. P., Omiecinski, C. J., Bresnick, E., Thomas, P. E., Ryan, D. E. and Levin, W. (1982) *J. Biol. Chem.*, **257**, 4023–6.

Walz, F. G., Vlasuk, G. P. and Steggles, A. W. (1983) *Biochemistry*, **22**, 1547–56.

Wilson, D. L., Hall, M. E., Stone, G. C. and Rubin, R. W. (1977) *Anal. Biochem.*, **83**, 33–44.

Wittmann, H. G., Stoffler, G., Hindennach, I., Kurland, C. G., Randall, L., Birge, E. A., Nomura, M., Kaltschmidt, E., Mizushima, S., Traut, R. R. and Bickle, T. A. (1971) *Mol. Gen. Genet.*, **111**, 327–33.

# 3 Solubilization and purification of membrane proteins

YUTAKA YANAGITA and YASUO KAGAWA



## 3.1 Introduction

Recently, there has been an enormous increase in the number of studies on the solubilization and purification of membrane proteins. One of the reasons for this phenomenon is the increase in researchers' interest in problems related to biological membranes such as energy transduction, cell-to-cell recognition and secretion. Another reason is the very rapid development of techniques including new materials and apparatus, for purifying proteins. Availability of these 'secret weapons' from commercial sources may also have encouraged researchers to attempt purification of new membrane proteins. So at the present, purification of membrane proteins is not much more difficult than purification of soluble proteins.

Considering this situation, the purposes of this chapter are (1) to present a survey of and propose some strategies for the purification of membrane proteins, (2) to introduce recently developed techniques and (3) to show possible problems that may be encountered during practical purifications, rather than to make a comprehensive description of a few particular cases as examples.

Since it would be very difficult to comprehensively cover the extensive literature on various aspects of membrane proteins, the articles cited in this review have been confined mostly to those related to the purification of energy-transducing systems and membrane receptors. This choice was made on the basis of our field of interest, and if the readers need other examples, a number of

Techniques for the Analysis of Membrane Proteins. Edited by C. I. Ragan & R. J. Cherry. Published in 1986 by Chapman and Hall, 11 New Fetter Lane, London EC4P 4EE

good monographs exist and these are listed at the top of the reference list as 'General references'.

## 3.2 Preparations before the start of purification

As a prerequisite for success in purifying a new protein, it is important to obtain the maximum information possible about the protein, including the molecular weight, isoelectric point, hydrophobicity etc. In this section we suggest several experiments which make a purification study easier.

### 3.2.1 AMPLIFICATION OF PROTEIN IN THE STARTING MATERIAL

The amount of some groups of membrane proteins strongly depend on the growth conditions of the source organism. Bacterial transport proteins are enriched in the presence of the substrates for these proteins. The composition of bacterial respiratory chains vary according to the oxygen pressure (Poole, 1983; Sone *et al.*, 1983). In a mammalian system, there are also inducible membrane proteins such as the cytochrome *P*-450s (Gotoh *et al.*, 1983). The first thing to be done is to establish the growth conditions for the source organism. If sufficient genetic information on the protein to be purified is available, as in the case of some bacterial proteins, molecular cloning of the protein can be considered as a method of amplification (Kanazawa *et al.*, 1979). The use of strong expression vectors such as pKK223-3 containing human or thermophilic ATPase genes is recommended (Kagawa *et al.*, 1986).

### 3.2.2 IDENTIFICATION OF THE PROTEIN BEFORE PURIFICATION

Requisite to the purification of a particular protein is the identification of that protein in the starting material. The usual method of identification is labelling a protein with radioactivity, fluorescence etc. (Von Dippe and Levy, 1983). After the labelling, detection procedures such as two-dimensional electrophoresis are employed (O'Farrell *et al.*, 1977). If the label introduced is sufficiently specific, purification will become easier. For instance, the lactose carrier was purified from *Escherichia coli* by labelling the carrier with a substrate analogue (Kaback, 1983; Newman *et al.*, 1981). However, application of this procedure is limited to the cases in which the proteins to be purified are labelled with externally added reagents. If a deletion mutant of the protein can be obtained or if the protein is inducible, a novel procedure of identification without external labels can be applied (MacPherson *et al.*, 1981, 1983). This procedure consists of four steps. First, the starting organism of the wild type (or in the induced condition) and the corresponding deletion mutant of the same organism (or in the non-induced condition) are grown separately in media containing different radioisotopes, for example, $^{3}H$ and $^{14}C$. Second, the membrane preparations obtained from these two sources are mixed together and electrophoresed in the same SDS–polyacrylamide gel. Third, the gel is sliced and the radioactivities of these two different isotopes in the gel slices are counted. Finally, the relative radioactivity (wild/mutant) is calculated, and the slice which shows maximum relative radioactivity is identified as the protein to be purified.

## 3.3 Methods of purification

In this section, the procedures after the preparation of starting material (membrane fraction, fragmental or vesicular) are discussed.

### 3.3.1 SOLUBILIZATION

The solubilization of a membrane protein is usually the first step for its purification. Solubilizing agents are used for two purposes: first, they are used to remove unwanted proteins from a particular membrane system; second, they are used to selectively solubilize the desired protein. To achieve this 'differential solubilization', solubilizing agents must be chosen properly. In this subsection, the authors first discuss the properties of detergents and then describe those of other solubilizing agents.

(a) *Detergents*

Detergents are useful for solubilization of integral membrane proteins. They can be classified into four categories according to the charge of their hydrophilic groups: non-ionic, anionic, cationic and zwitterionic. Non-ionic detergents have the advantage that they permit fractionation of the solubilized membrane proteins on ion-exchange columns. Since membrane phospholipids are usually negatively charged, cationic detergents are difficult to remove after solubilization. Anionic detergents are effective in solubilization by weakening both hydrophobic and ionic interactions in the biomembranes. This effect is expected when salts are added to the non-ionic detergents. The physical parameters that are used to characterize detergents include the critical micellar concentration (cmc), the hydrophilic–lipophilic balance (HLB), the Kraft point (Tc) and the cloud point (Davies, 1961). The first two parameters and the molecular shape of a detergent are the most important characteristics. Detergents with an HLB of 12 or higher show solubilizing effects on membrane proteins (Kagawa, 1972). Since phospholipids are amphiphilic, detergent molecules do not simply surround hydrophobic phospholipid micelles but form mixed micelles with individual phospholipid molecules (Helenius and Simons, 1975). Some detailed reviews on solubilization are those of Shinoda *et al.* (1963), Lichtenberg *et al.* (1983) and Tanford and Reynolds (1976).

Solubilization of membrane proteins *per se* is not difficult if we use a strong detergent such as sodium dodecyl sulphate. Very few membrane proteins such as thermophilic ATPase (Yoshida *et al.*, 1975) and bacteriorhodopsin (Huang *et al.*, 1981) can have their higher structure and function restored after complete solubilization with sodium dodecyl sulphate. However, many soluble oligomeric enzymes have been renatured after the removal of sodium dodecyl sulphate with Dowex AG-1 X-2 (Weber and Kuter, 1971). The removal of this agent from membrane proteins causes precipitation even in the presence of urea and mild detergents. Thus, we must find a detergent which will solubilize membrane proteins without impairing their renaturating ability. However, this is difficult since relatively few detergents are well-characterized (Helenius and Simons, 1975; Tanford and Reynolds, 1976), and fewer studies have been carried out to compare the same detergent from different commercial sources (Ashani and Catravas, 1980; Chang and Bock, 1980). The number of studies on the solubilization of a particular membrane system by different kinds of detergents (Gurtabay *et al.*, 1979; Loizaga *et al.*, 1979; Navarette and Serrano, 1983; Newby *et al.*, 1982; Rivnay and Metzgar, 1982 and Womack *et al.*, 1983) are also few. In most cases, choosing the proper detergent is an empirical process rather than a systematic one (Racker, 1983).

In the practical applications of solubilization, two groups of factors must be determined carefully to assure the reproducibility of experiments. The first one is the pH and ionic strength of the buffer used for solubilization. Many of the anionic detergents sometimes form precipitates at acidic pH. This type of problem can also occur with non-ionic detergents because they sometimes form precipitates in buffers of very high ionic strength (several molar concentration of salts). The second factor is the ratio between the detergent and protein (detergent–protein ratio). If the concentration of the protein to be solubilized is not so different from that of the detergent, the concentration of free detergent after solubilization is highly dependent on the detergent–protein ratio in each experiment. As an aid to attempting the solubilization of a new protein, properties of several commonly used detergents are shown in Table 3.1.

Many membrane proteins have been solubilized by a mixture of detergents. A mixture of cholate and deoxycholate was used to solubilize membrane proteins of the thermophilic bacterium PS3 (Hirata, 1979; Hirata *et al.*, 1977; Sone and Yanagita, 1982; Sone *et al.*, 1979). In these examples, the detergent mixture solubilized more proteins than cholate used alone (Sone *et al.*, 1975). Selective solubilization can be effected by selecting the proper ratio of detergents. Other researchers have used a mixture of an anionic detergent plus a non-ionic detergent (Azari and Wiseman, 1982; Watanabe *et al.*, 1979; West *et al.*, 1979) or a mixture of a zwitterionic detergent plus a non-ionic detergent (Hjelmeland *et al.*, 1979).

In recent years, the use of octylglucoside ($\beta$-D-octylglucopyranoside) has increased, especially for studies on the transport systems of *Escherichia coli* inner membrane (Henderson *et al.*, 1983; Kaback, 1983; Newman *et al.*,

**Table 3.1** Properties of commonly used detergents

| *Detergent* | *Triton X-100* | *Tween 20 or 80* | *Octylglucoside* | *Sodium N-lauroyl sarcosinate* | *Sodium cholate* | *Sodium deoxycholate* |
|---|---|---|---|---|---|---|
| Type | A | A | A | A | B | B |
| Ionic properties | Non-ionic | Non-ionic | Non-ionic | Anionic | Anionic | Anionic |
| HLB | 13.5 | 16.7(Tw20)<br>15.0(Tw80) | 10.6 | 31.3 | 18 | 16 |
| Micellar weight | 90 000 | | | | 900–1800 | 1700–4200 |
| cmc (mM)[a] | [b] | [c] | 25 | | 13–15 | 4–6 |
| Recommended pH in use | | (Practically no limitation) | | >6 | >7 | >7 |
| Ion-exchange chromatography | Yes | Yes | Yes | No | (Yes)[d] | (Yes)[d] |
| Removal by dialysis | No | No | Yes | No | Yes | Yes |
| Remarks | [e] | | | | | |

[a] In water (pH > 8 for sodium cholate and sodium deoxycholate).
[b] 0.01–0.001%(w/v).
[c] 60 mg/litre for Tween 20 and 13 mg/litre for Tween 80.
[d] When used in low concentration (for sodium deoxycholate, see Cottingham and Ragan, 1980).
[e] Strong absorption of light in the ultraviolet region.

1981; Tsuchiya *et al.*, 1982). Octylglucoside is non-ionic and colourless in the ultraviolet region. It also has low molecular and micellar weight (Table 3.1) and can be easily removed by dialysis. Most importantly, the proteins solubilized by octylglucoside can be reconstituted into artificial membrane vesicles without special procedures to remove this detergent (Henderson *et al.*, 1983; Kaback, 1983; Newman *et al.*, 1981; Racker *et al.*, 1979; Tsuchiya *et al.*, 1982; Yanagita *et al.*, 1983). However, the high cost of this detergent is a disadvantage, but increased demand for octylglucoside should decrease its price. Similar types of detergents, the alkyl-*N*-gluconamides, are also being used by a few groups (Hanatani *et al.*, 1983; Hildreth, 1982).

Some detergents must be purified before use (Scheule and Gaffney, 1981). Sodium cholate and sodium deoxycholate develop a yellow colour when dissolved in water due to the presence of bilirubin. To remove this contaminant, cholic acid or deoxycholic acid should be purchased instead of their salts, and this should be recrystallized (three times) from ethanol (Kagawa and Racker, 1971). More serious contaminants are found in non-ionic detergents such as Triton X-100 and Brij 35 (Ashani and Catravas, 1980; Chang and Bock, 1980). These detergents contain oxidizing contaminants which react with sulphydryl groups of proteins and also with carbonyl compounds. The amounts of these contaminants vary among different sources or different lots of the detergent and increase during storage in aqueous solution. These contaminants can be removed with two procedures. Ashani and Catravas (1980) treated an aqueous solution of detergent with either $NaHSO_4$ or $SnCl_2$ followed by extraction into $ClCH_2CH_2Cl$ or $CH_2Cl_2$. Chang and Bock (1980) first reduced the detergent with sodium borohydride in ethanol, followed by silica gel chromatography using a chloroform solution and passage through a mixed-bed ion-exchange resin in aqueous solution.

(b) *Other methods of solubilization*

In the section above, we introduced many kinds of detergents and described their properties. However, since it is difficult to remove detergents from protein preparations (see Section 3.3.5), it is best to purify membrane proteins without using detergents. Though the examples cited in this section may be specific successful applications of purification techniques, they can be helpful for developing new strategies for purification.

Proteases were used for protein solubilization in early work (Omura and Takesue, 1970a,b). Proteases cleaved water-soluble moieties of membrane proteins from water-insoluble membrane-buried moieties, and the water-soluble moieties were purified further as ordinary soluble proteins. Of course, the proteins solubilized with proteases have lost some amino acids, and the difference in amino acid composition between the solubilized and membrane-bound forms of the particular protein must be checked (Imai and Sato, 1974). With this technique, two more things should be noted: the proteins solubilized with proteases can no longer be reconstituted into membrane vesicles, and the application of proteases is limited to anchored membrane proteins such as cytochrome $b_5$ (Omura and Takesue, 1970a) or NADH–cytochrome *c* reductase (Omura and Takesue, 1970b).

Some kinds of monovalent cations such as $SCN^-$, $I^-$ or $ClO_4^-$ can solubilize hydrophobic proteins. They have large ionic radii and increase the entropy of the aqueous solution by destroying the structure of water around them, and thus are called 'chaotropic' anions (Hamaguchi and Geiduschek, 1962). The detailed mechanism of solubilization is described elsewhere (Wyman, 1964; Sawyer and Puckridge, 1973). Hatefi's group systematically used chaotropic anions to purify electron transport complexes from bovine heart mitochondria or to solubilize proteins from bacterial plasma membranes (Davis and Hatefi, 1971; Hanstein *et al.*, 1971). Usually, the proteins solubilized by chaotropic anions are collected in the supernatant of a centrifugation and later dialysed to remove the agent. Koval and Murray (1983) have also made a systematic study on chaotropic anions. Chaotropic anions can also be used for pre-extraction of contaminating proteins (Sone and Yanagita, 1982; Sone *et al.*, 1979). Two things are important when using chaotropic anions. First, chaotropic anions may often inactivate proteins because they weaken both ionic interactions and hydrophobic interactions. Second, since they are used in rather high concentration (one to several molar), this makes it difficult to apply further purification procedures such as column chromatography without introducing a different procedure of solubilization.

Organic solvents are sometimes used to solubilize very hydrophobic proteins called proteolipids (Capony *et al.*, 1983; Célis, 1980; Sone *et al.*, 1978). Though they are proteins, proteolipids behave like lipids, and they can be purified by high-performance liquid chromatography using a silica column (Capony *et al.*, 1983). By using organic solvents for phase partition, several peripheral membrane proteins (Böhme *et al.*, 1980) or the water-soluble moiety of a membrane protein complex (Yoshida *et al.*, 1975) are collected in the aqueous phase. Reconstitution of the proteolipid solubilized by organic solvents has also been reported (Célis, 1980).

NADH–cytochrome $b_5$ reductase was solubilized by suspending membrane fractions in 1 M potassium phosphate buffer (Dey *et al.*, 1981). High ionic strength may be a good condition for solubilizing peripheral membrane proteins, because ionic interaction between these proteins and membrane lipids are weakened.

Cytochrome *b* was isolated from electron-transport complex III of yeast mitochondria by treating complex III with 3,4,5,6-tetrahydrophthalic anhydride (T'sai and Palmer, 1982). This reagent modified some of the lysyl residues of cytochrome *b* to increase its solubility in sodium cholate solution. This modification can be reversed by dialysing the purified cytochrome *b* in a solution of sodium taurocholate.

(c) *Confirmation of solubilization*

There is a lot of difference between solubilized protein molecules and dispersed membrane fragments. The properties of the former are as follows:

(1) definite size and shape in the variety of solvents;
(2) usually a molecular weight of less than 1 000 000;
(3) little difference in specific gravity from soluble proteins because of low lipid content.

To avoid any ambiguity in the stage of solubilization, every 'solubilized' fraction must be examined for the size distribution and content of the particles in that fraction. An electron micrograph of a solubilized fraction is the best source of information upon these points, but gel filtration and ultrafiltration can also work well.

In the following paragraphs we discuss preparation of solubilized fractions by centrifugation. Centrifugation is the most common method used to prepare membrane fractions and to extract solubilized proteins. For example, solubilized proteins are usually obtained after centrifugation of a detergent-treated membrane at 100 000 × g for 1 h. However, membrane fragments as large as 30 nm diameter may remain in the supernatant depending on the condition described below. In this section we discuss the reproducibility of the membrane protein preparations from a theoretical background.

There are two methods of centrifugation, namely density gradient centrifugation and differential centrifugation. The principle of separation in the former is the difference in buoyant density, and that in the latter is the difference in speed of sedimentation. The experiments carried out with gradient centrifugation are fairly easy to reproduce if the density of a particular membrane fraction is given. The major remaining parameters are the temperature and the time required to reach equilibrium. On the contrary, many more parameters are needed to establish the exact experimental condition for differential centrifugation. These are discussed in the following paragraphs.

In many articles the condition of differential centrifugation is denoted solely by the centrifugal force ($G$) calculated according to Equation 3.1,

$$G = 1.10 \times 10^{-5} \times R \times (\text{rpm})^2 (\text{gravity}) \qquad (3.1)$$

where $R$ (cm) is the average distance between the sample and the axis of the rotor. However, the distance between the bottom of a centrifuge tube and the surface of the solution in the tube varies according to the type of the rotor and the amount of solution in the tube. If this distance is small, particles near the surface will rapidly precipitate. Considering these conditions, a factor named performance index (PI) is defined according to Equation 3.2 to evaluate the ease of sedimentation in using a particular rotor.

$$\text{PI} = (\text{rpm})^2 / (\ln R_{max} - \ln R_{min}) \qquad (3.2)$$

The time required for precipitation depends also on the medium in which the membrane fraction is dispersed. Assuming that the membrane fraction consists of globular particles, the sedimentation coefficient ($s$) is calculated according to Equation 3.3, where $\eta$ is the viscosity of the medium (poise), $d$ is the diameter of the particle (cm), $\sigma$ and $\rho$ are the density of the particle and the medium (g cm$^{-3}$).

$$s=(\sigma-\rho)d^2/18\eta \tag{3.3}$$

Considering the performance index (Equation 3.2), the time required for precipitation ($T_p$) is calculated according to Equation 3.4,

$$T_p=60/(4\pi^2 s\mathrm{PI})=(27.4\eta)/[\mathrm{PI}d^2(\sigma-\rho)] \tag{3.4}$$

where $T_p$ is given in minutes. Viscous media such as sucrose and detergent solution at low temperature increase $T_p$. $T_p$ also increases if the particles have an extended shape instead of a globular one.

In some papers the description of the duration of centrifugation is quite ambiguous about the period of acceleration. Since the time required for differential centrifugation is much shorter than that for density gradient centrifugation, the time required for acceleration becomes a critical factor for better reproducibility especially on using different types of centrifuge or rotor from those used in the original experiment. Rapid acceleration and braking minimize this kind of problem, and they are required for treating unstable proteins. The Beckmann L8 is an example of a centrifuge suitable for rapid preparation.

### 3.3.2 COLUMN CHROMATOGRAPHY IN THE PRESENCE OF DETERGENTS

In the presence of detergents column chromatography can be used for purification of membrane proteins. In purification of a new membrane protein, the following conditions should be considered.

First, the capacity and selectivity of column materials may be decreased in the presence of detergents. Since protein molecules are surrounded by detergent molecules, the interactions between the proteins and the column materials may be weaker than in the absence of detergent.

Second, in the case of ion-exchange or hydroxyapatite chromatography, the concentration of detergent in a particular elution buffer must be determined precisely, because membrane proteins may be adsorbed on to column materials with unwanted interactions such as hydrophobic interaction.

At the moment, high-performance liquid chromatography (HPLC) is being used for the rapid and small-scale purification of membrane proteins. Hirata *et al.* (1984) have purified the alanine carrier from a thermophilic bacterium using a Bio-Gel HPHT column (a hydroxyapatite column for HPLC) in the presence of 1.0% (w/v) Triton X-100.

### 3.3.3 RECENTLY DEVELOPED SPECIAL TECHNIQUES

Besides the usual chromatographic procedures described in the preceding section (3.3.2), many new techniques have been developed in recent years. In this section affinity chromatography and some specialized techniques are introduced.

Affinity chromatography which has been previously used for soluble proteins is now being applied to membrane proteins (Oppenheimer and Czech, 1983; Serrano and Rivas, 1982; Sigel *et al.*, 1983). Ligands specific for a particular membrane protein are attached to a column matrix to prepare an affinity column. The ligands include substrate analogues, the substrate itself, antibodies (either polyclonal or monoclonal), or hydrophobic molecules (hydrophobic affinity chromatography). After the solution containing the protein to be purified is applied to the affinity column, the column is exposed to mild washing conditions to remove other unwanted proteins. Finally, the protein to be purified is eluted from the column upon addition of substances which weaken that protein–ligand interaction. For the purification of membrane proteins, special techniques have been developed for each step of affinity chromatography. In the following three paragraphs, some examples of applications are described.

Purification of some components of electron-transport systems have been markedly facilitated by using affinity columns containing cytochrome *c*. Besides cytochrome oxidases or cytochrome $bc_1$ complexes (Azzi *et al.*, 1981; Bill *et al.*, 1982; De Vrij *et al.*, 1983; Gennis *et al.*, 1982; Ozawa *et al.*, 1980, 1982; Weiss and Juchs, 1978; Weiss and Kolb, 1979), photosynthetic reaction centres (Brudig *et al.*, 1983) have been purified with these columns. The proteins adsorbed on the cytochrome *c*-affinity column were eluted by increasing the ion concentration, by increasing the detergent concentration, or by adding cytochrome *c* (Gennis *et al.*, 1982). Two different protein complexes, namely, cytochrome oxidase and cytochrome $bc_1$ complex, were purified sequentially from the same column (Bill *et al.*, 1982; Weiss and Kolb, 1979).

Antibodies are good ligands for affinity chromatography because of their strict specificity, especially in the case of monoclonal antibodies. Yoshida-Momoi and Lennon (1982) purified the acetylcholine receptor from human muscle by using a monoclonal antibody as the ligand in an affinity column. They also used an $\alpha$-toxin affinity column for purification of that protein. An affinity column with polyclonal antibody has also been applied successfully (Tsuchida and Sato, 1983). Dextran-bound insulin was also used to purify insulin receptors with a simple gel filtration column equilibrated with a detergent (Fujita-Yamaguchi *et al.*, 1983).

Some hydrophobic groups and molecules act as specific ligands for affinity chromatography. Besides the usual hydrophobic affinity column (Carson and

Konigsberg, 1981; Guengerich and Martin, 1980), a novel affinity column coupled with deoxycholate was developed to purify integral membrane proteins (Cresswell, 1979). By using this column, the hydrophobic proton-channel moiety ($F_0$) was purified from the $H^+$-translocating ATPase complex ($F_0F_1$) of *Escherichia coli* (Schneider and Altendorf, 1982). Detergent-solubilized $F_0F_1$ was adsorbed on the column, and then it was dissociated into $F_0$ and $F_1$ (soluble ATPase moiety) on the column by the addition of buffer with urea but without detergent. After washing the column, $F_0$ was released from the column upon the addition of detergent.

Two more techniques other than affinity chromatography will be discussed in this section. Anderson and Fisher (1978) applied specific adsorption of protein on alumina gel to the purification of mitochondrial transhydrogenase. The adsorbed protein was extracted from the gel by adding dilute solution of detergent. This procedure was also used to concentrate the purified protein. Koepsell *et al.* (1983) utilized selective incorporation of protein into phospholipid vesicles for the purification of the sodium, D-glucose co-transport protein. In their experiment, two polypeptides were selectively incorporated into the vesicles out of more than ten polypeptides.

### 3.3.4 PURIFICATION WITHOUT SOLUBILIZATION

So far we have introduced the methods of purifying solubilized membrane proteins, but some membrane proteins lose their activity immediately after solubilization. Moreover, there are some cases in which the protein of interest is the main component of a membrane system and needs no further purification after preparation of the membrane system.

Purification of ($Na^+ + K^+$)-translocating ATPase and $Ca^{2+}$-translocating ATPase usually means purification of membrane fragments or membrane vesicles (Jørgensen, 1974, 1982). A lot of the studies on these ATPases were carried out using these preparations. Membrane systems are usually prepared using techniques of differential centrifugation, gradient centrifugation and differential solubilization (Peterson and Hokin, 1980; Yoda and Yoda, 1981). In most cases these procedures are more rapid and thus are suitable for purifying unstable proteins. However, since the preparations obtained from these procedures are not uniformly dispersed solutions, experimental conditions must be established precisely. Examination of the prepared protein by electron microscopy is also recommended for determining the average particle size and particle diversity (see Section 3.3.1.c).

### 3.3.5 CONCENTRATION AND STORAGE OF PURIFIED PROTEINS

Concentration and preservation of a purified membrane protein are important in the final stage of purification. In order to concentrate proteins, detergents etc. must be removed by one of four procedures: namely dialysis, ultrafiltration, salt precipitation and adsorption on to resin beads (see also Rivas *et al.*, 1982). However, the application of the first two methods is limited for two reasons. First, they are time-consuming (taking from several hours to several days) and are not suitable for unstable proteins. Second, they can be applied only to the detergents making small micelles (less than 10 000 daltons), because the separation of proteins and detergent micelles is based on the difference of their molecular weights. Bile acid derivatives and alkylglucosides can easily be removed, but most of type A detergents (Helenius and Simons, 1975; Tanford and Reynolds, 1976) can hardly be removed by dialysis or ultrafiltration.

In contrast to those two methods, salt precipitation is rapid and can be applied to a wide variety of membrane proteins. However, in the presence of high concentrations of salts, an oily phase rich in detergent is separated from the water phase, and proteins may be incorporated in the former phase. In order to overcome this problem, the following measures are used. The addition of sodium cholate to the solution containing type A detergents such as Triton X-100 and sodium *N*-lauroylsarcosinate (Sone *et al.*, 1975; Sone and Yanagita, 1982) forms mixed micelles with type A detergents, which elevates the cmc of the entire solution. Recently, Hirata found that addition of 0.5% cholate was sufficient for successful precipitation when the concentration of Triton X-100 was less than 0.25% (unpublished). Another method is salt precipitation at low temperature. As much as 0.23 mg $ml^{-1}$ of ammonium sulphate does not cause separation of Triton X-100 from the aqueous phase if the temperature of the solution is kept at 0° C.

Detergents can be removed by adding hydrophobic adsorbants such as polystyrene resin beads to the solutions (Cheetham, 1979; Holloway, 1973), but some hydrophobic integral proteins may be adsorbed.

The purified membrane proteins can be stored like their soluble counterparts but some proteins such as cytochrome oxidases need exchange of detergent for stabilization (Yamanaka *et al.*, 1981).

### 3.3.6 EXAMINATION OF PURITY

There are many methods for examining purity and homogeneity of the purified proteins, but few are convenient during purification. Gel electrophoresis and electrofocusing are examples. Gel filtration of the purified membrane protein in a detergent solution is also good for the assessment of purity. However, before using these techniques, the following two facts may be considered.

First, some hydrophobic proteins may be resolved into subunits or protomers depending on the detergent used to solubilize the proteins (Ikai,

1980). Thus, some detergents may cause apparent heterogeneity of the purified proteins.

Second, the techniques used for the examination of purity cannot always be applied directly for the estimation of molecular properties of the purified protein such as molecular weight and molecular shape, because observed parameters are those of the protein–detergent complex but not those of the protein itself, and the amount of detergent bound to the protein depends on its hydrophobicity and solvent conditions (Tracy and Chan, 1979).

## 3.4 Possible problems during purification

Although many kinds of membrane proteins have been purified, the techniques used to purify them cannot always be applied directly to purify a new membrane protein. Several problems are listed below and the numbers of sections related to them are indicated in parentheses.

(1) Contaminants in detergents (3.3.1.a).
(2) Limitations in chromatography conditions by the solubilizing agents chosen (3.3.1.a and 3.3.1b).
(3) Contamination by membrane fragments of solubilized fraction (3.3.1.c).
(4) Non-specific adsorption of proteins to column materials (3.3.2).
(5) Decrease in the capacity and specificity of column materials by detergents (3.3.2).
(6) Need for reconstitution to assay activity (3.2.2, 3.3.1a and 3.3.5).

In some cases, reconstitution of proteins into liposomes is inevitable in order to assay their activity (Hirata *et al.*, 1977), because some substrate–protein complexes are often unstable, and a proper labelling agent is not always available. If the protein of interest does not show its activity in a detergent solution and does not have any characteristics such as specific light absorption, reconstitution of the protein in the fractions may be tried. A detailed survey of reconstitution techniques is given in Chapter 4.

## 3.5 Conclusion

The most important step in the purification of a particular membrane protein is its solubilization without causing denaturation. Usually, a membrane protein is solubilized with a suitable detergent. The detergent-solubilized membrane protein can be purified similarly to soluble proteins, if there is sufficient knowledge of the limitations of the use of that detergent. A number of new techniques such as affinity chromatography are now being developed for rapid purification of membrane proteins existing in small quantities.

## References

1. General references
*Biomembranes* (since 1971), Plenum Press, New York.
*Current Topics in Bioenergetics* (since 1966), Academic Press, New York.
*Current Topics in Membranes and Transport* (since 1970), Academic Press, New York.
*The Enzymes of Biological Membranes* (1976), (ed A. Martonosi), Plenum Press, New York.
*Methods in Membrane Biology* (since 1974), Plenum Press, New York.
*Methods in Enzymology* (since 1955), (Vols. 10, 23, 31, 32, 52–56, 81, 88, 96, and 97), Academic Press, New York.

Of these listed above, the following review is an example of the latest information.
Nelson, D. R. and Robinson, N. C. (1983) *Methods Enzymol.*, **97**, 571–618.

2. Articles cited in the text
Anderson, W. M. and Fisher, R. R. (1978) *Arch. Biochem. Biophys.*, **187**, 180–90.
Ashani, Y. and Catravas, G. N. (1980) *Anal. Biochem.*, **109**, 55–62.
Azari, M. R. and Wiseman, A. (1982) *Anal. Biochem.*, **122**, 129–38.
Azzi, A., Brodbeck, U. and Zahler, P. (eds) (1981) *Membrane Proteins. A Laboratory Manual*, Springer-Verlag, Berlin.
Bill, K., Broger, C. and Azzi, A. (1982) *Biochim. Biophys. Acta*, **679**, 28–34.
Böhme, H., Brütsch, S., Weithmann, G. and Böger, P. (1980) *Biochim. Biophys. Acta*, **590**, 248–60.
Brudig, G. W., Worland, S. T. and Sauer, K. (1983) *Proc. Natl. Acad. Sci. USA*, **80**, 683–6.
Capony, J.-P., Rinaldi, M. L., Guilleux, F. and Demaille, J. G. (1983) *Biochim. Biophys. Acta*, **728**, 83–91.
Carson, S. D. and Konigsberg, W. H. (1981) *Anal. Biochem.*, **116**, 398–401.
Célis, H. (1980) *Biochem. Biophys. Res. Commun.*, **92**, 26–31.
Chang, H. W. and Bock, E. (1980) *Anal. Biochem.*, **104**, 112–17.
Cheetham, P. S. J. (1979) *Anal. Biochem.*, **92**, 26–31.
Cottingham, I. R. and Ragan, C. I. (1980) *Biochem. J.*, **192**, 9–18.
Cresswell, P. (1979) *J. Biol. Chem.*, **254**, 414–19.
Davies, J. T. (1961) in *Interfacial Phenomena*, Academic Press, New York.
Davis, K. A. and Hatefi, Y. (1971) *Biochemistry*, **10**, 2509–16.
De Vrij, W., Azzi, A. and Konigs, W. N. (1983) *Eur. J. Biochem.*, **131**, 97–103.
Dey, A. C., Rahal, S., Rimsey, R. L. and Senciall, I. R. (1981) *Anal. Biochem.*, **110**, 373–9.
Fujita-Yamaguchi, Y., Choi, S., Sakamoto, Y. and Itakura, K. (1983) *J. Biol. Chem.*, **258**, 5045–9.
Gennis, R. B., Casey, R. P., Azzi, A. and Ludwig, B. (1982) *Eur. J. Biochem.*, **125**, 189–95.
Gotoh, O., Tagashira, Y., Iizuka, T. and Fujii-kuriyama, Y. (1983) *J. Biochem. (Tokyo)*, **93**, 807–17.
Guengerich, F. P. and Martin, M. V. (1980) *Arch. Biochem. Biophys.*, **205**, 365–79.
Gurtabay, I. G., Azagra, E., Gutiérrez-Arranz, A., Milicua, J. C. G. and Goñi, F. M. (1979) *Biochem. Soc. Trans.*, **7**, 150–2.
Hamaguchi, K. and Geiduscheck, E. P. (1962) *J. Amer. Chem. Soc.*, **84**, 1329.

Hanatani, M., Nishifuji, K., Futai, M. and Tsuchiya, T. (1983) *Seikagaku*, **55**, 651 (in Japanese).
Hanstein, W. G., Davis, K. A. and Hatefi, Y. (1971) *Arch. Biochem. Biophys.* **147**, 534–44.
Helenius, A. and Simons, K. (1975) *Biochim. Biophys. Acta*, **415**, 29–79.
Henderson, P. J. F., Kagawa, Y. and Hirata, H. (1983) *Biochim. Biophys. Acta*, **732**, 204–9.
Hildreth, J. E. K. (1982) *Biochem. J.*, **207**, 363–6.
Hirata, H. (1979) *Methods Enzymol.*, **56**, 430–5.
Hirata, H., Sone, N., Yoshida, M. and Kagawa, Y. (1977) *J. Supramol. Struct.*, **6**, 77–84.
Hirata, H., Kambe, T. and Kagawa, Y. (1984) *J. Biol. Chem.*, **259**(17) 10653–6.
Hjelmeland, L. M., Nebert, D. W. and Chrambach, A. (1979) *Anal. Biochem.*, **95**, 201–8.
Holloway, P. W. (1973) *Anal. Biochem.*, **53**, 304–8.
Huang, K.-S., Bayley, H., Liao, M. J., London, E. and Khorana, H. G. (1981) *J. Biol. Chem.*, **256**, 3802–9.
Ikai, A. (1980) *J. Biochem. (Tokyo)*, **88**, 1349–57.
Imai, Y. and Sato, R. (1974) *J. Biochem. (Tokyo)*, **75**, 689–98.
Jørgensen, P. L. (1974) *Meths Enzymol.*, **32**, 277–90.
Jørgensen, P. L. (1982) *Biochim. Biophys. Acta*, **694**, 27–68.
Kaback, H. R. (1983) *J. Membr. Biol.*, **76**, 95–112.
Kagawa, Y. (1972) *Biochim. Biophys. Acta*, **265**, 297–338.
Kagawa, Y., Hirata, H., Ohta, S., Ishizuka, M. and Karube, Y. (1986) Ion Transport through Membranes (eds K. Yagi and B. Pullman) Academic Press, New York, in press.
Kagawa, Y. and Racker, E. (1971) *J. Biol. Chem.*, **246**, 5477–87.
Kanazawa, H., Miki, T., Tamura, F., Yura, T. and Futai, M. (1979) *Proc. Natl. Acad. Sci. USA*, **76**, 1126–30.
Koepsell, H., Menuhr, H., Ducis, I. and Wissmüller, T. F. (1983) *J. Biol. Chem.*, **258**, 1888–94.
Koval, S. F. and Murray, K. G. E. (1983) *Can. J. Microbiol.*, **29**, 146–50.
Lichtenberg, D., Robson, R. J. and Dennis, E. A. (1983) *Biochim. Biophys. Acta*, **737**, 285–304.
Loizaga, B., Gurtubay, I. G., Macarulla, J. M., Goñi, F. M. and Gómez, J. C. (1979) *Biochem. Soc. Trans.*, **7**, 148–50.
MacPherson, A. J. S., Jones-Mortimer, M. C. and Henderson, P. J. F. (1981) *Biochem. J.*, **196**, 269–83.
MacPherson, A. J. S., Jones-Mortimer, M. C., Horne, P. and Henderson, P. J. F. (1983) *J. Biol. Chem.*, **258**, 4390–6.
Navarette, R. and Serrano, R. (1983) *Biochim. Biophys. Acta*, **728**, 403–8.
Newby, A. C., Chrambach, A. and Bailyes, E. M. (1982) *Tech. Life Sci. Sect. Biochem.*, **B409**, 22 pp.
Newman, M. J., Foster, D. L., Wilson, T. H. and Kaback, H. R. (1981) *J. Biol. Chem.*, **256**, 11804–8.
O'Farrell, P. Z., Goodman, H. M. and O'Farrell, P. H. (1977) *Cell*, **12**, 1133–42.
Omura, T. and Takesue, S. (1970a) *J. Biochem. (Tokyo)*, **67**, 249–57.
Omura, T. and Takesue, S. (1970b) *J. Biochem. (Tokyo)*, **67**, 267–76.
Oppenheimer, C. L. and Czech, M. P. (1983) *J. Biol. Chem.*, **258**, 8539–42.
Ozawa, T., Suzuki, H. and Tanaka, M. (1980) *Proc. Natl. Acad. Sci. USA*, **77**, 928–30.

Ozawa, T., Tanaka, M. and Wakabayashi, T. (1982) *Proc. Natl. Acad. Sci. USA,* **79,** 7175–9.
Peterson, G. L. and Hokin, L. E. (1980) *Biochem. J.,* **192,** 107–18.
Poole, R. K. (1983) *Biochim. Biophys. Acta,* **726,** 205–43.
Racker, E. (1983) *Fed. Proc.,* **42,** 2899–2909.
Racker, E., Violand, B., O'Neal, S., Alfonzo, M. and Telford, J. (1979) *Arch. Biochem. Biophys.,* **198,** 470–7.
Rivas, E., Pasdeloup, N. and le Maire, M. (1982) *Anal. Biochem.,* **123,** 194–200.
Rivnay, B. and Metzger, H. (1982) *J. Biol. Chem.,* **257,** 12800–8.
Sawyer, W. H. and Puckridge, J. (1973) *J. Biol. Chem.,* **248,** 8429–33.
Scheule, R. K. and Gaffney, B. J. (1981) *Anal. Biochem.,* **117,** 61–6.
Schneider, E. and Altendorf, K. (1982) *Eur. J. Biochem.,* **126,** 149–53.
Serrano, A. and Rivas, J. (1982) *Anal. Biochem.,* **126,** 109–15.
Shinoda, K., Nakagawa, T., Tamamushi, B. and Isemura, T. (1963) *Colloidal Surfactants,* Academic Press, New York.
Sigel, E., Stephenson, F. A., Mamalaki, C. and Barnard, E. A. (1983) *J. Biol. Chem.,* **258,** 6965–71.
Sone, N., Kagawa, Y. and Orii, Y. (1983) *J. Biochem. (Tokyo),* **93,** 1329–36.
Sone, N., Ohyama, T. and Kagawa, Y. (1979) *FEBS Lett.,* **106,** 39–42.
Sone, N. and Yanagita, Y. (1982) *Biochim. Biophys. Acta,* **682,** 216–26.
Sone, N., Yoshida, M., Hirata, H. and Kagawa, Y. (1975) *J. Biol. Chem.,* **250,** 7917–23.
Sone, N., Yoshida, M., Hirata, H. and Kagawa, Y. (1978) *Proc. Natl. Acad. Sci. USA,* **75,** 4219–23.
Tanford, C. and Reynolds, J. A. (1976) *Biochim. Biophys. Acta,* **457,** 133–70.
Tracy, R. P. and Chan, S. H. P. (1979) *Biochim. Biophys. Acta,* **576,** 109–17.
T'sai, A.-L. and Palmer, G. (1982) *Biochim. Biophys. Acta,* **681,** 484–95.
Tsuchida, S. and Sato, K. (1983) *Biochim. Biophys. Acta,* **756,** 341–8.
Tsuchiya, T., Misawa, A., Miyake, Y., Yamasaki, K. and Niiya, S. (1982) *FEBS Lett.,* **142,** 231–4.
Von Dippe, P. and Levy, D. (1983) *J. Biol. Chem.,* **258,** 8896–901.
Watanabe, H., Kamita, Y., Nakamura, T. and Yamanaka, T. (1979) *Biochim. Biophys. Acta,* **547,** 70–8.
Weber, K. and Kuter, J. (1971) *J. Biol. Chem.,* **246,** 4504–9.
Weiss, H. and Juchs, B. (1978) *Eur. J. Biochem.,* **88,** 17–28.
Weiss, H. and Kolb, H. J. (1979) *Eur. J. Biochem.,* **99,** 139–49.
West, S. B., Huang, M.-T., Miwa, G. T. and Lu, A. Y. H. (1979) *Arch. Biochem. Biophys.,* **193,** 42–50.
Womack, M. D., Kendall, D. A. and MacDonald, R. C. (1983) *Biochim. Biophys. Acta,* **733,** 210–15.
Wyman, J. (1964) *Adv. Protein Chem.,* **19,** 224–86.
Yamanaka, T., Kamita, Y. and Fukumori, Y. (1981) *J. Biochem. (Tokyo),* **89,** 265–73.
Yanagita, Y., Sone, N. and Kagawa, Y. (1983) *Biochem. Biophys. Res. Commun.,* **113,** 575–80.
Yoda, A. and Yoda, Y. (1981) *Anal. Biochem.,* **110,** 82–8.
Yoshida, M., Sone, N., Hirata, H. and Kagawa, Y. (1975) *J. Biol. Chem.,* **250,** 7910–16.
Yoshida-Momoi, M. and Lennon, V. A. (1982) *J. Biol. Chem.,* **257,** 12757–64.

# 4

# Reconstitution of membrane proteins into vesicular membranes

HAJIME HIRATA



## 4.1 Introduction

Membrane-associated functions such as energy transduction, transport of nutrients, recognition and transmission of external signals, and cell–cell interactions are catalysed by an appropriate specific protein(s) embedded in the membrane. These membrane proteins have been identified, solubilized and purified from membranes by various procedures as described in the preceding chapters. Their activities can be divided into two essential categories; one is non-vectorial catalytic activity such as electron transport, ATP hydrolysis and binding of substrates, and the other is vectorial activity such as translocation of ions and substrates across the membrane. The former is analysed with the respective proteins under conditions similar to those applied to soluble proteins when appropriate lipids or detergents are used. However, the latter requires a strictly enclosed membrane structure that partitions the aqueous internal space from its surroundings. To meet this requirement for isolated membrane proteins, they are introduced into lipid bilayer vesicles, liposomes, to form proteoliposomes. For example, a purified $H^+$-translocating ATPase from membranes of thermophilic bacteria, which catalyses ATP synthesis in oxidative phosphorylation, manifests its ATP hydrolysing activity when a small amount of phospholipids is added to form non-vesicular protein–lipid complexes. However, an ATP-dependent translocation of $H^+$ can be observed

Techniques for the Analysis of Membrane Proteins. Edited by C. I. Ragan & R. J. Cherry. Published in 1986 by Chapman and Hall, 11 New Fetter Lane, London EC4P 4EE

only after reconstituting it into proteoliposomes. Thus the reconstitution of membrane proteins into proteoliposomes is essential for the study of vectorial membrane function. Moreover, other membrane proteins, which have no catalytic activity other than mediation of the transport of nutrients, have been purified by following the activity facilitating the translocation of specific nutrient molecules into the reconstituted proteoliposomes. In these cases, the reconstitution of proteoliposomes is critical not only for the study of the transport mechanism but also for the assay during the purification process.

The methods of proteoliposome reconstitution are diverse and the reader must choose carefully, depending on his purpose. This chapter will describe various procedures of reconstitution, their advantages and disadvantages, and characteristics of the prepared proteoliposomes. However, the reader may also consult recently published reviews such as the articles of Racker (1979), Montal *et al.* (1981), Eytan (1982), Houslay and Stanley (1982), Kagawa *et al.* (1982), Darszon (1983) and Banerjee and Datta (1983).

## 4.2 Materials

Proteoliposomes consist of membrane proteins and phospholipids. The membrane proteins to be reconstituted are not necessarily in the most purified state, but detergents and/or other agents used for their solubilization and purification should be removed as far as possible, unless they can be removed during formation of proteoliposomes. In particular the type A detergents (see Chapter 3), even at low concentration, will inhibit the formation of proteoliposomes or produce leaky proteoliposomes.

Phospholipids to be used are either synthetic or natural. The soybean phospholipids (asolectin), which can be purchased commercially, are the most convenient to start with. They are a mixture of phosphatidylcholine (45%, w/w), phosphatidylethanolamine (15%), phosphatidylinositol (25%) and small amounts of cardiolipin and phosphatidic acid. The commercial product may be used as it is or purified partially by washing with excess acetone to remove neutral lipids (Kagawa and Racker, 1971). Some membrane proteins require specific phospholipids for either their activation or successful reconstitution. For this, the specific phospholipids should be prepared from the original membranes or obtained from commercial sources. However, these requirements sometimes seem to be very ambiguous since they originate not only from the needs of protein function but also from those of the reconstitution technique itself. For example, the alanine carrier from membranes of the thermophilic bacterium PS3 was first successfully reconstituted by the cholate dialysis method (see Section 4.3.1.a) only with phospholipids from its original membranes, and asolectin was the least effective phospholipid for reconstitution (Hirata *et al.*, 1976b). However, it was demonstrated later that asolectin can form active proteoliposomes equally as well as the native phospholipids when the freeze–thaw–sonication method (see Section 4.3.2.b) was applied (Hirata, 1979a,b). Therefore, one should be very cautious when describing the specific phospholipid requirements of particular membrane proteins.

## 4.3 Reconstitution procedures

The methods of proteoliposome reconstitution now available are diverse, but they essentially fall into three major categories with many variations: one method is to solubilize phospholipids together with membrane proteins by using detergents with subsequent removal or dilution of the detergents: a second method is to use mechanical force; and a third method is to insert membrane proteins directly into preformed liposomes. These all have both advantages and disadvantages.

### 4.3.1 SOLUBILIZATION BY DETERGENTS

Phospholipids and membrane proteins are first solubilized together by use of detergents, and then the detergents are removed slowly or diluted eventually to form proteoliposomes. Detergents with a high critical micellar concentration (cmc) and a small micelle size are usually used. Sodium cholate, sodium deoxycholate and octylglucoside have been the most widely and successfully used.

(a) *Detergent–dialysis method*
This method was first successfully used to reconstitute the ATP-$^{32}P_i$ exchange reaction of oxidative phosphorylation using sodium cholate as detergent (Kagawa and Racker, 1971). Since then, many membrane processes have been reconstituted using this method (for reviews and articles, see Racker (1979), Eytan (1982), Kagawa *et al.* (1982), Banerjee and Datta (1983), Darszon (1983)).

In this method, phospholipids and proteins of an appropriate ratio are first mixed with 1–3% sodium cholate (commercial products are usually contaminated with bilirubin and other impurities, so purification before use is strongly recommended, as described in Chapter 3.3.1.) in an appropriately buffered medium and sonicated until the suspension becomes translucent. The sonication period is usually less than 5 min with a probe-type sonicator. The sonicated suspension is then dialysed against 200- to 500-fold of an appropriate buffer for 20 h or more at an appropriate temperature. The dialysing buffer may be changed once, if necessary, 2–4 h after the start of dialysis. There are various points to be optimized empirically: addition of deoxycholate (0.5–1%) improves the formation of tight proteoliposomes in some instances (Knowles and Racker, 1975; Sone *et al.*, 1975). The presence of the anti-oxidant dithiothreitol during the sonication and dialysis may be required to avoid the auto-oxidation of the phospholipids. Also the dialysing buffer (pH, ionic strength, etc.), the dialysing temperature and the dialysing period should be optimized in each case. From the author's experience, too rapid removal of cholate results in the formation of less active proteoliposomes. The simple way to decelerate the dialysing rate is to increase the internal pressure of the dialysis bag by infusing air.

The resulting proteoliposomes are used for the assay as they are or fractionated by discontinuous Ficoll or sucrose gradient centrifugation to obtain protein-rich proteoliposomes (see Section 4.4.3). There are some disadvantages in this procedure: the prolonged exposure to the detergent may be harmful to the protein and the dialysing time of 20 h or more for complete reconstitution may be troublesome when a prompt answer is required such as during a purification process.

Recently, octylglucoside ($\beta$-D-octylglucopyranoside), a non-ionic and dialysable detergent, has been increasingly utilized instead of sodium cholate, not only for solubilization of membrane proteins but also for reconstitution of proteoliposomes (Baron and Thompson, 1975). Its characteristics are described in the preceding chapter (3.3.1).

(b) *Detergent–dilution method*

The detergent–dilution method was first described by Racker *et al.* (1975). In this method, the detergent–phospholipid–protein mixture is prepared as described in Section 4.3.1.a. Then the mixture is simply diluted 20- to 40-fold into the appropriate assay mixture at room temperature. The solution seems clear at first, but it becomes turbid after a few minutes of incubation. The proteoliposomes thus formed are collected and washed to remove the excess detergent by centrifugation (140 000 ***g***, 30 min). Serrano *et al.* (1976) reported that the optimal concentration of sodium cholate was in the range 0.5–0.8% for the reconstitution of the oligomycin-sensitive ATPase complex; however, 2% sodium cholate plus 1% sodium deoxycholate was the most effective when the alanine carrier of thermophilic bacterium PS3 was reconstituted with the PS3 phospholipids (Hirata, unpublished).

The major advantages of this procedure are its rapidity and its reproducibility. The exposure of the protein to the detergents is limited to only a few minutes, and the total procedure can be accomplished in an hour including centrifugation. Therefore, this procedure is quite useful for assays of multiple protein fractions after chromatographic separation and particularly for fragile proteins. However, some membrane proteins, such as bacteriorhodopsin, are not effectively reconstituted by the cholate–dilution method (Racker, 1979).

More recently, octylglucoside instead of sodium cholate has been successfully used with this method for the reconstitution and purification of the sugar transporter of *E. coli* (Newman *et al.*, 1981; Kaback, 1983; Henderson *et al.*, 1983).

### 4.3.2 MECHANICAL TREATMENTS

Phospholipids tend to form vesicles (liposomes) simply by exposing them to mechanical agitation, shaking or sonication in an aqueous medium. When membrane proteins are present during the mechanical treatments, proteoliposomes are eventually formed.

(a) *Sonication method*
In this method, dried phospholipids are placed in a test tube to which the detergent-free proteins in an appropriate aqueous medium are added. The mixture is then sonicated with either a probe-type sonicator or a bath-type sonicator. With the former instrument, the tip of the probe is immersed directly into the mixture, and its power output is much stronger than the latter; however, it is rather hard to control and produces local heat which denatures the proteins easily. In contrast, the bath-type sonicator yields a smaller sonication effect since the sonic wave is transduced through the glass wall of the test tube. Therefore, it takes a rather longer period of sonication (sometimes 30 min or more of sonication is required). At any event, the experimental conditions should be optimized in individual cases depending on sample volume, size and thickness of the glassware, temperature and power output of the instrument. Among them, the duration of sonication is the most critical point to be determined with each protein to be reconstituted.

The major advantages of this method are that it is rapid and requires no detergents. However, the serious disadvantages are that it is very difficult to control the instrument to obtain reproducible results and cannot be applied to proteins that are sensitive to sonic oscillation.

(b) *Freeze–thaw–sonication method*
The pioneering work of the freeze–thaw–sonication method was first described by Kasahara and Hinkle (1976) for the reconstitution of glucose carrier from red blood cells into proteoliposomes.

In this method, the liposomes are first prepared by sonicating phospholipids suspended in an appropriate aqueous medium until the suspension becomes translucent. To the resulting liposome suspension, the proteins to be reconstituted are added. The mixture is then frozen quickly by immersing the test tube with gentle shaking in liquid nitrogen or a dry ice–acetone bath. (If these coolants are not available, an ethanol bath kept in a deep freezer (−80° C) is just sufficient and economical.) After a few minutes (5–10 ) of immersion, the test tube is taken out and the mixture is thawed at room temperature. A rapid thawing with warm water is not recommended. The thawed mixture, which becomes turbid and viscous, is then sonicated briefly (10–120 s) and this freeze–thaw–sonication procedure is repeated several times, if necessary. The freezing and thawing induces not only the incorporation of proteins in the liposomes but also the fusion of liposomes (Pick, 1981), which results in the formation of relatively large size proteoliposomes.

The main advantages of this method are that it requires no detergents, and the exposure of proteins to sonic waves is limited to only a few seconds. The deleterious effect of detergents and sonication inevitably accompanying other methods are thus minimized. Therefore, this procedure may be the most moderate for proteins and yields the most reproducible results.

### 4.3.3 INCORPORATION AND FUSION METHODS

Some membrane proteins are incorporated directly into preformed liposomes with or without the aid of dilute detergents. The detergents used in this method are either lysolecithin or sodium cholate.

Liposomes are prepared first from phospholipids, including 10% lysolecithin, by sonication, to which the membrane proteins to be reconstituted are added. The mixture is incubated at 0° C for several hours. The proteins are inserted unidirectionally into the lipid bilayer of liposomes in a spontaneous manner. A very low concentration of sodium cholate (0.1%) can be substituted for lysolecithin. However, this detergent-facilitated incorporation procedure has some serious disadvantages: unnaturally high concentrations of lysolecithin and prolonged periods of incubation are required, which may affect the properties of lipid bilayer and proteins.

The incorporation of membrane proteins into liposomes is also attainable in the absence of detergent when acidic phospholipids such as cardiolipin, phosphatidylinositol or phosphatidylserine are supplemented. The mixture of phosphatidylcholine, phosphatidylethanolamine and the acidic phospholipids (10–30% of the total phospholipids) in an appropriate aqueous medium is subjected to sonication to form liposomes. The resulting liposomes are incubated with membrane proteins at room temperature for 30 min in the presence of $Mg^{2+}$. The insertion of membrane proteins into the lipid bilayer of liposomes occurs spontaneously and rapidly (completed in a few minutes). The absence of $Mg^{2+}$ and the presence of EDTA or lower temperature decelerate the proteoliposome formation, and detergents such as lysolecithin or sodium cholate inhibit the incorporation process even at very low concentrations. The optimal concentration of acidic phospholipids must be determined in each case.

The successful application of the incorporation procedures so far known is limited to only a small number of examples (Eytan *et al.*, 1976; Racker, 1979); however, the mode of insertion of proteins into the lipid bilayer is thought to be unidirectional, and the orientation of the proteins in the lipid bilayer will be asymmetric. Furthermore the incorporation process may serve as a model system of membrane biogenesis or protein secretion.

In addition, the liposomes containing acidic phospholipids are known to fuse with each other when $Ca^{2+}$ is present (Papahadjopoulos *et al.*, 1974). Using this $Ca^{2+}$-induced fusion technique, two different proteoliposome preparations were fused to form proteoliposomes with mixed functions (Miller and Racker, 1976).

### 4.3.4 SPECIAL TECHNIQUES

There are several techniques available for the reconstitution of proteoliposomes other than those thus far described. These, including the reconstitution of large proteoliposomes (macroproteoliposomes) and highly stable proteoliposomes, have been developed recently for specific purposes. The proteoliposomes prepared by the methods described above are small in size, ranging from 30 to 100 nm in diameter. With these microproteoliposomes, accurate kinetic studies are usually difficult unless very sensitive probes and instruments are available, and in particular, direct electrophysiological measurements with microelectrodes are impossible. To overcome these difficulties, several methods for reconstituting macroproteoliposomes have been developed.

(a) *Macroproteoliposomes on the surface of Sephasorb*

This is a modified detergent–dialysis method (Rögner *et al.*, 1979). The detergent–protein–phospholipid mixture (see Section 4.3.1.a) is dialysed in the presence of Sephasorb (Pharmacia, HP superfine, 10–23 $\mu$m). After dialysis for 18–20 h, macroproteoliposomes with diameters of approximately 1 $\mu$m are formed on the dimpled hydrophobic surface of the Sephasorb particle.

(b) *Macroproteoliposomes from protein–lipid complexes in organic solvent*

In this method, an apolar solvent, ether or hexane, is used to extract protein–lipid complexes from preformed proteoliposomes in the presence of $Ca^{2+}$. The extracted protein–lipid complexes are flushed with argon or nitrogen to evaporate the solvent, and subsequently, the complexes are swollen with an appropriate aqueous medium. Macroproteoliposomes with an average diameter of 20–30 $\mu$m are spontaneously formed after 48–72 h incubation without agitation (Darszon *et al.*, 1980). These macroproteoliposomes can be utilized directly for electrophysiological measurements with microelectrodes or patch clumping. The limitation of this method is that only those proteins tolerant to the organic solvent can be reconstituted.

(c) *Highly stable proteoliposomes*

The macroproteoliposomes prepared by the methods described above are fascinating since they can be easily observed and directly impaled by a microelectrode under a light microscope; however, they are usually heterogeneous in size and protein content and fragile to osmotic or mechanical shock. So one needs patience and luck to look for a stable, protein containing proteoliposome. In this context, the recently developed technique using polymerizable phospholipids looks promising. Ringsdorf's group have described the preparation of highly stable liposomes with lipids bearing diacetylene, butadiene or vinyl moieties as polymerizable groups (Hub *et al.*, 1980). These lipids stabilize intensively the lipid bilayer of liposomes upon polymerization by UV irradiation at 254 nm. To these stable liposomes, the ATP synthetase complex has been successfully reconstituted (Wagner *et al.*, 1981). Using this technique together with those described above, it may be possible to prepare highly stable macroproteoliposomes.

## 4.4 Characterization of proteoliposomes

The procedures described above have been successfully applied to reconstitute a variety of membrane proteins into proteoliposomes. However, several factors controlling the characteristics of the proteoliposomes should be determined for the analysis of the reconstituted membrane proteins.

### 4.4.1 ORIENTATION OF PROTEINS IN PROTEOLIPOSOMES

Proteins in native biological membranes are integrated in the lipid bilayer asymmetrically, which in turn enables the membrane-related functions to be vectorial. However, these membrane proteins, in general, distribute randomly in artificial proteoliposomes except in minor cases: purple membranes of *Halobacterium halobium* reconstituted by the sonication procedure (Section 4.3.2.a) or cytochrome oxidase and oligomycin-sensitive ATPase by the incorporation procedure (Section 4.3.3) have been reported to be oriented unidirectionally (Eytan *et al.*, 1976; Racker and Hinkle, 1974). Although the control of protein orientation in proteoliposomes is difficult, the vectorial functions can be achieved by adding the respective substrate anisotropically or by imposing an ion gradient with the aid of specific ionophores.

Several ways of determining the orientation of the proteins in the lipid bilayer have been: electron spin resonance study (Kornblatt *et al.*, 1974, 1975), immunological interactions (Curman *et al.*, 1980), binding of ligands (Wu *et al.*, 1981), chemical labelling with membrane-impermeable agents (Eytan and Broza, 1978), proteolytic digestion (Huang *et al.*, 1980), and so on. However, these techniques require some information on the structure of the proteins, including their subunit composition, primary structure and subunit–subunit interactions.

### 4.4.2 SIZE OF PROTEOLIPOSOMES

The size of proteoliposomes is one of the most important parameters for estimating the functional activities of the reconstituted proteins. Since the proteoliposomes prepared by a variety of methods described in this chapter are usually small and heterogeneous ranging from 20 to 100 nm in diameter, a direct observation by a light microscope is impossible, and an electron microscopic observation yields only particular and specific information on an individual proteoliposome. Therefore, the measurement of the entrapped volume of the proteoliposomes, instead of an exact dimension of internal space, is much more useful and meaningful. For this purpose, the estimation of the amount of marker entrapped in the proteoliposomes during their formation is the most popular. The markers used in these studies should be membrane-impermeable, physiologically inert, and not deleterious to the membranes. Carbohydrates, nucleic acids or proteins labelled with radioisotopes have been successfully used as the marker (Szoka and Papahadjopoulos, 1978; Olson *et al.*, 1979). The procedures include the formation of proteoliposomes in the presence of the marker followed by removal of the non-trapped external markers by centrifugation or gel filtration, and subsequently, the radioactivity entrapped in the proteoliposomes is measured. Therefore, the markers which easily diffuse out from the proteoliposomes or firmly attach to the external surface of the proteoliposomes will lead to under- or over-estimation, respectively.

The method of Oku *et al.* (1982) utilizes a chelating derivative of fluorescein, calcein (2′,7′-{[bis(carboxymethyl)amino]methyl}fluorescein), as a marker. This dye is quenched by cations such as cobalt or copper. The proteoliposomes are prepared in the presence of calcein, and then the external calcein is quenched by the addition of $CoCl_2$. The resulting fluorescence is mainly due to the dye inside the proteoliposomes. Subsequently, the proteoliposomes are disrupted by adding a small amount of Triton X-100, and the residual fluorescence is subtracted as a base. This method is very rapid, simple and accurate since the removal of the external marker after proteoliposome formation is not necessary.

### 4.4.3 DISTRIBUTION OF PROTEINS AMONG PROTEOLIPOSOMES

The distribution of membrane proteins among the proteoliposomes is another important factor for a quantitative analysis of membrane function. If the proteins are distributed unevenly among the proteoliposomes prepared, the estimations and calculations cannot be done on the basis of the total internal volume of the preparation.

The different populations of the proteoliposomes can be fractionated by either discontinuous Ficoll or sucrose gradient centrifugation (Carroll and

Racker, 1977) or gel filtration on a Sepharose 2B column (Goldin, 1977). The density of proteins distributed in an individual proteoliposome from each fraction is then estimated by a freeze–fracture electron microscopic observation. Mitochondrial cytochrome oxidase, reconstituted by either the cholate–dialysis method or the direct incorporation method, is distributed unevenly among the proteoliposomes and less than 10% of the total proteoliposomes contain active enzyme (Carroll and Racker, 1977). On the contrary, some other proteins, such as ($Na^{+}K^{+}$)-ATPase, distribute evenly among the proteoliposomes (Goldin, 1977; Skriver *et al.*, 1980).

## 4.5 Reconstitution of energy-transducing proteoliposomes

It may not be too exaggerated to state that the development of techniques of proteoliposome reconstitution has been mainly driven by the need to study the mechanisms of oxidative phosphorylation and photophosphorylation in mitochondria, chloroplasts and bacteria. Mitchell's chemiosmotic hypothesis, which is now fundamentally accepted by researchers in this field, states that the components of the electron-transport chain are anisotropically oriented in the membrane so that the energy liberated by oxidoreduction is utilized to translocate protons across the membrane which results in the formation of transmembranous electrochemical potential difference of protons ($\Delta\bar{\mu}H^{+}$). The $\Delta\bar{\mu}H^{+}$ thus established drives protons through an anisotropic proton-translocating ATPase ($H^{+}$-ATPase) and this chemiosmotic energy is utilized to generate ATP from ADP and $P_i$. The $H^{+}$-ATPase is a proton pump which reversibly generates $\Delta\bar{\mu}H^{+}$ upon hydrolysis of ATP into ADP and $P_i$.

The hypothesis has been verified from various points of view (for reviews and articles, see those of Boyer *et al.* (1977), Mitchell (1979), Skulachev (1981)); however, the main difficulty in interpreting $\Delta\bar{\mu}H^{+}$ generation in natural membranes is due to their complicated enzymic and ion-transporting characteristics. Therefore, a simplified system of reconstituted proteoliposomes consisting of a purified $\Delta\bar{\mu}H^{+}$ generator and phospholipids is obviously suitable in this context.

In the following section typical experiments carried out in our laboratory using reconstituted proteoliposomes capable of energy transduction are described.

### 4.5.1 $H^{+}$-ATPase FROM THERMOPHILIC BACTERIA

$H^{+}$-ATPase was first extracted from mitochondria (Kagawa and Racker, 1966) and called oligomycin-sensitive ATPase. It is composed of a soluble ATPase portion ($F_1$) and a membrane-integrated portion ($F_0$). The soluble

ATPase itself is insensitive to energy-transfer inhibitors, such as *N,N'*-dicyclohexylcarbodiimide (DCCD) and oligomycin; however, it becomes sensitive when attached to the $F_0$ part. Mitochondrial $H^+$-ATPase, although not in a purified state, was first reconstituted into proteoliposomes by the cholate–dialysis method (see Section 4.3.1.a) and the resulting proteoliposomes were shown to catalyse both the ATP–$^{32}P_i$ exchange reaction and proton translocation driven by ATP hydrolysis (Kagawa and Racker, 1971).

It is now known that $H^+$-ATPase distributes universally in membranes of almost all kinds of cell. Furthermore, ATPases from various sources including both eukaryotes and prokaryotes are essentially similar not only in their functional aspects but also in their structural aspects. The $F_1$ portion of $H^+$-ATPase has a molecular weight of $3.5 \times 10^5$–$4.0 \times 10^5$ and consists of five kinds of subunit ($\alpha$, $\beta$, $\gamma$, $\delta$ and $\varepsilon$). The stoichiometry of these subunits is 3:3:1:1:1. The $F_0$ portion of $H^+$-ATPase consists of at least two kinds of extremely hydrophobic subunit: one binds $F_1$ and the other binds DCCD. The latter is the smallest subunit (an apparent molecular weight of some $8 \times 10^3$–$10 \times 10^3$) in $H^+$-ATPase and is called DCCD-binding protein. DCCD-binding proteins from various sources have been isolated and characterized, and a remarkable homology in their primary structures has been demonstrated (Sebald *et al.*, 1982).

$H^+$-ATPase from the thermophilic bacterium PS3 ($TF_0$:$F_1$) has been isolated and characterized (Sone *et al.*, 1975). The enzyme is essentially similar to those found in mitochondria and other bacteria except that it is highly stable against various physicochemical treatments. This peculiar property enabled us to characterize its subunit structure and function on a molecular basis. Namely, the five subunits of $TF_1$ have been isolated and reassembled into an active form (Yoshida *et al.*, 1977). The role of each subunit has been determined (Yoshida *et al.*, 1977; Kagawa, 1982). Image reconstruction studies on the electron micrographs of the two-dimensional crystals of $TF_1$ in the presence of heavy metals or various ligands showed a pseudohexagonal structure with a pseudo 6-fold and 2-fold symmetry (Wakabayashi *et al.*, 1977). On the other hand, the $TF_0$ portion of $TF_0$:$F_1$ has been found to contain three kinds of subunit; they are called band 4, 6 and 8 according to the SDS-polyacrylamide gel electrophoresis pattern. Band 4 protein can be removed from $TF_0$ without loss of $TF_0$ activity, so that its existence as a subunit of $F_0$ is still uncertain. Band 6 and 8 proteins have been shown to be the $F_1$ binding and DCCD-binding protein, respectively (Sone *et al.* (1978). The subunit composition of $TF_0$ is not known because its exact molecular weight is difficult to determine. However, it is supposed that $TF_0$ contains 4–6 DCCD-binding proteins since it loses activity when one-third of the all DCCD-binding proteins are modified (Sone *et al.*, 1979a,b; Sebald *et al.*, 1979).

Using this thermostable $H^+$-ATPase ($TF_0$:$F_1$), various types of experiment have been performed after reconstituting it into proteoliposomes.

(a) *ATP formation by reconstituted proteoliposomes*
Proteoliposomes were reconstituted by the cholate–dialysis method from purified $TF_0:F_1$ and phospholipids of the thermophilic bacterium PS3. These proteoliposomes were able to synthesize ATP from ADP and $P_i$ with energy from an electrochemical proton gradient formed by acid/base transition and/or $K^+$ diffusion mediated by valinomycin.

The proteoliposomes were first incubated in malonate buffer (pH 5.5) in the presence of valinomycin, and then rapidly transferred to glycylglycine buffer (pH 8.4) containing 150 mM KCl. This procedure instantaneously induced both a transmembrane pH gradient and $K^+$ diffusion potential. Although the reaction was transient (completed in about 40 s), the initial $\Delta\bar{\mu}H^+$ was calculated to be above 300 mV (interior acidic and positive), which in turn drove ATP synthesis. The amount of ATP synthesized was about 100 nmol/mg of $TF_0:F_1$, and this value corresponds to 55 turnovers of the enzyme. Either a proton concentration difference (ΔpH) or a membrane potential was effective for driving ATP synthesis, and the minimal $\Delta\bar{\mu}H^+$ required was 180–200 mV (Sone *et al.*, 1977a). This experiment clearly demonstrated that $H^+$-ATPase can transduce the potential energy of $H^+$ to the chemical-bond energy of ATP, and eliminated the possible involvement of unknown factor(s) other than the $H^+$-ATPase. Similar results were obtained with proteoliposomes containing $H^+$-ATPase from several sources (Pick and Racker, 1979).

Energization of the $TF_0:F_1$ to form ATP is also possible in combination with other proton translocators from foreign sources such as bacteriorhodopsin. The bacteriorhodopsin is the chromoprotein embedded in the cytoplasmic membrane of *Halobacterium halobium*, which transforms light energy into $\Delta\bar{\mu}H^+$. Proteoliposomes containing both $TF_0:F_1$ and bacteriorhodopsin were prepared by the detergent–dialysis method. When these proteoliposomes were illuminated by a projector lamp, $\Delta\bar{\mu}H^+$ of 180–190 mV (inside positive and acidic) was established and concomitantly ATP formation from ADP and $P_i$ was observed. The ATP formation was completely dependent on light and inhibited by uncoupler and DCCD (Sone *et al.*, 1977b).

The experiments described above demonstrate that $H^+$-ATPase is the enzyme capable of ATP formation in response to $\Delta\bar{\mu}H^+$; however, the time resolution is poor and the energy components are somewhat complex. An alternative method is to drive ATP synthesis by an external electric field. This has been successfully applied to chloroplasts (Witt *et al.*, 1976; Schlodder and Witt, 1981; Vinkler and Korenstein, 1982; Vinkler *et al.*, 1982), mitochondria (Hamamoto *et al.*, 1982) and submitochondrial particles (Teissie *et al.*, 1981). It can be applied to $H^+$-ATPase-containing proteoliposomes when sufficient membrane potential (above 200 mV) is attained. For this purpose, the external electric field strength or the diameter of the reconstituted proteoliposomes must be increased. Owing to the production of Joule heat by the current, even at low electrolyte concentrations, the electric field strength cannot be raised above 1000 V/cm. Therefore, the diameter of the proteoliposomes must be increased to the order of micrometres. Large proteoliposomes (macroproteoliposomes) containing $TF_0:F_1$ were thus prepared as described in Section 4.3.4.a: a detergent–$TF_0:F_1$–phospholipid mixture (see Section 4.3.1.a) was dialysed in the presence of Sephasorb at 30° C for 18 h. Macroproteoliposomes with diameters of approximately 1 μm formed on the Sephasorb particles were separated from microproteoliposomes by centrifugation. These macroproteoliposomes were placed between two plate electrodes (5 $cm^2$ area, 0.1 mm thick) in the presence of $^{32}P_i$ and ADP, and rectangular electric pulses of 200 V, 20 ms duration with a rise time of 1 μs were applied to the electrodes. The electric field generated between the electrodes introduced a membrane potential which in turn drove ATP formation. The amount of ATP formed was dependent on the number of electric pulses applied and decreased by adding either uncoupler or DCCD (Rögner *et al.*, 1979).

(b) *Active $H^+$ translocation by $H^+$-ATPase*

$H^+$-ATPase translocates $H^+$ from the $F_1$ side to the $F_0$ side on hydrolysis of ATP and establishes $\Delta\bar{\mu}H^+$. This $H^+$ translocation can be measured with reconstituted proteoliposomes containing $H^+$-ATPase by the use of fluorescent dyes.

In proteoliposomes reconstituted by the detergent–dialysis method, most of the $F_1$ is spontaneously arrayed on the outer surface of the liposomes. Therefore, ATP added to these proteoliposomes is hydrolysed by the $F_1$ and an active $H^+$ translocation into the liposome lumen occurs. The $H^+$ translocation results in the formation of $\Delta\bar{\mu}H^+$, which is composed of a membrane potential ($\Delta\psi$) and a pH gradient (ΔpH). The $\Delta\psi$ is measured with negatively charged dyes (8-anilinonaphthalene-1-sulphonate (ANS), cyanine dye, etc.), which are attracted to the membrane, and their limited intramolecular rotation causes fluorescence enhancement (ANS) or absorption change (cyanine dye) (Bashford and Smith, 1979). The ΔpH is measured as the distribution of a membrane-permeable weak base (9-aminoacridine or quinacrine), which is distributed both inside and outside the liposomes according to the Henderson-Hasselbach equation. 9-Aminoacridine is accumulated in the liposomes in response to pH (inside acidic) and shows self-quenching of its fluorescence (Schuldiner *et al.*, 1972).

Using these techniques, $\Delta\bar{\mu}H^+$ formation depending on ATP hydrolysis has been estimated with $TF_0:F_1$-containing proteoliposomes (Sone *et al.*, 1977c). The addition of ATP to the proteoliposomes induced both enhancement of ANS fluorescence and quenching of 9-aminoacridine fluorescence. These were reversed by the addition of uncoupler or DCCD. From these results, more than 200 mV of $\Delta\bar{\mu}H^+$ (inside positive and acidic) was established at the steady state (Table 4.1). With this value, the $H^+$/ATP ratio was calculated to be 2.

**Table 4.1** $\Delta\psi$, ΔpH and $\Delta\bar{\mu}H^+$ formed in $H^+$-ATPase proteoliposomes following the addition of ATP

| *Phospholipids* | *Medium* | $\Delta F/F$ (%) | $\Delta\Psi$ (mV) | $Q/1-Q$ | ΔpH | $\Delta\bar{\mu}H^+$ (mV) |
|---|---|---|---|---|---|---|
| PS3 phospholipids | Tris | 102 | 145 | 0 | 0 | 145 |
| | Tricine | 48 | 70 | 0.67 | 2.9 | 253 |
| | Tricine + $NO_3^-$ | 0 | 0 | 2.45 | 3.5 | 221 |
| Asolectin | Tris | 141 | 140 | 0 | 0 | 140 |
| | Tricine | 74 | 74 | 0.29 | 2.6 | 238 |
| | Tricine + $NO_3^-$ | 0 | 0 | 0.58 | 2.9 | 183 |

$H^+$-ATPase (0.25 mg) was reconstituted into liposomes with either PS3 phospholipids or asolectin, and the relative increases in the medium were determined. The relative increase in the fluorescence of ANS ($\Delta F/F$), membrane potential ($\Delta\Psi$), the fraction of the total 9-aminoacridine fluorescence quenched by the addition of ATP (Q), pH difference across the liposome membrane (ΔpH), and the electrochemical potential difference across the membrane ($\Delta\bar{\mu}H^+$) are as described (Sone *et al.*, 1977c).

### (c) *Passive $H^+$ permeation through the $H^+$ channel and reconstitution of $H^+$-ATPase*

Submitochondrial particles or thylakoid membrane vesicles become permeable to $H^+$ after removal of $F_1$, and this $H^+$ permeability is inhibited by DCCD, oligomycin and tributyltin chloride. These indicate that the $F_0$ portion of the $H^+$-ATPase is embedded in the membrane and forms an $H^+$ channel through which $H^+$ permeates passively down its electrochemical gradient, and those reagents described above react with one of the $F_0$ subunits resulting in blockage of $H^+$ permeation.

The passive $H^+$ permeation through $F_0$ can be measured with a pH electrode or with a $\Delta\bar{\mu}H^+$ probe (ANS, 9-aminoacridine, or others) as described in the previous section. The $H^+$-driving force is supplied artificially by a valinomycin-mediated $K^+$ diffusion potential and/or acid–base transition.

Studies on reconstituted $TF_0$ proteoliposomes revealed that the rate of $H^+$ conduction was proportional to the imposed membrane potential and the amount of $TF_0$ incorporated (Okamoto *et al.*, 1977). $H^+$ conduction was inhibited by DCCD and the inhibition followed pseudo-first-order kinetics. Dependence of $H^+$ conductivity on $TF_0$ concentration fitted a Michaelis–Menten equation showing a $V_{max}$ of 31.3 μg-ions $min^{-1}$ $mg^{-1}$ of $TF_0$ and $K_m$ of 0.095 μg-ions $l^{-1}$ (Sone *et al.*, 1981). This fact confirms that the ion actually translocated through $TF_0$ is not $OH^-$ but $H^+$ (or $H_3O^+$). Modification of $TF_0$ with iodine resulted in loss of affinity towards $H^+$, which indicates that a single tyrosyl residue in the DCCD-binding subunit of $TF_0$ plays an essential role as an $H^+$ donor in the rate-limiting step of $H^+$ permeation through $TF_0$.

The passive $H^+$ permeation by $TF_0$-incorporated proteoliposomes was also blocked by adding $TF_1$. The resulting proteoliposomes restored ATP-dependent generation of $\Delta\bar{\mu}H^+$. The blockage of $H^+$ permeation and the restoration of $\Delta\bar{\mu}H^+$ generation were parallel and a 1:1 stoichiometry of $TF_1$ and $TF_0$ was found (Sone *et al.*, 1981). Furthermore, attempts were made to reconstitute $H^+$-ATPase from purified subunits of $TF_1$ (Yoshida *et al.*, 1977). The complete assembly of $\alpha+\beta+\gamma+\delta+\varepsilon$ blocked the $H^+$ permeation of $TF_0$-incorporated proteoliposomes (Fig. 4.1) and showed an ATP dependent $\Delta\bar{\mu}H^+$ generation determined by ANS fluorescence. Among various combinations of subunits tested, $\gamma+\delta+\varepsilon$ was the minimum combination required for stopping the $H^+$ permeation. Although the resulting proteoliposomes failed to show $H^+$-ATPase activity, further addition of $\alpha$ and $\beta$ subunits yielded ATP-dependent $\Delta\bar{\mu}H^+$ generation.

From these results and others (for reviews, see Kagawa *et al.*, 1979; Kagawa, 1982), it is suggested that subunits $\gamma$, $\delta$ and $\varepsilon$ form an $H^+$ gate complex which effectively controls $H^+$ flow through the $H^+$ channel ($TF_0$) in the membrane, and that subunits $\alpha$ and $\beta$ in combination with subunit $\gamma$ form an ATP-driven $H^+$ pump (Fig. 4.2).

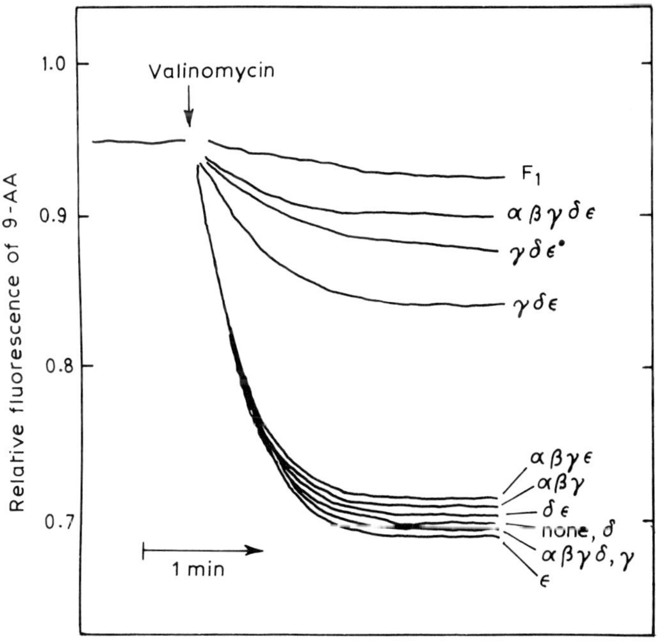


**Fig. 4.1** Reconstitution of the $H^+$ gate on $TF_0$–liposomes. Influx of $H^+$ into $TF_0$–proteoliposomes was assayed by 9-aminoacridine (9-AA). The following amounts of subunits were added to 50 $\mu l$ of $TF_0$–proteoliposome suspensions which contained 17.5 $\mu g$ of $TF_0$, as indicated: $\alpha$, 31 $\mu g$; $\beta$, 45 $\mu g$; $\gamma$, 15 $\mu g$; $\delta$, 11 $\mu g$; $\varepsilon$, 11 $\mu g$. In the experiment designated $\gamma\delta\varepsilon$*, double concentrations of the three subunits were used (Yoshida *et al.*, 1977).

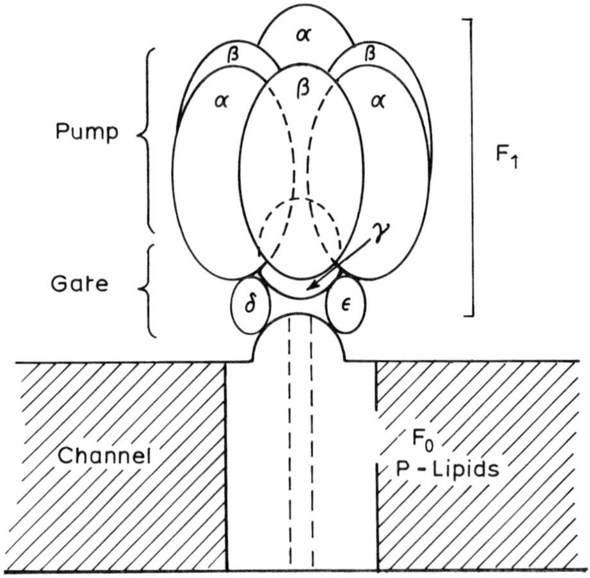


**Fig. 4.2** A tentative model of $H^+$-ATPase (Yoshida *et al.*, 1977).

## 4.5.2 ALANINE CARRIER FROM THERMOPHILIC BACTERIA

The secondary active transport of nutrients across the bacterial membrane is catalysed by a specific carrier protein(s) embedded in the membrane in a chemiosmotic manner as envisaged by Mitchell (for reviews, see Eddy, 1978; Harold, 1978; Rosen and Kashket, 1978; Mitchell, 1979; West, 1980). In order to study the mechanism at the molecular basis, the isolation of the carrier protein(s) from the membranes is an essential step. However, since the carrier protein(s) is expected to possess no catalytic activity other than mediation of the transport of nutrients, identification and purification of the protein can only be performed by following the activity facilitating the translocation of specific nutrient molecules into the reconstituted proteoliposomes unless a specific agent or substrate analogue, which covalently binds to the active centre of the carrier protein(s), is available. Here, a typical experiment on isolation of alanine carrier protein from membranes of thermophilic bacterium PS3 is introduced.

Membrane vesicles of thermophilic bacterium PS3 are capable of active transport of alanine in response to $\Delta\bar{\mu}H^+$ generated by either respiration or $K^+$ diffusion mediated by valinomycin (Hirata *et al.*, 1976a). From these membrane vesicles, alanine carrier activity has been solubilized by ionic detergents and reconstituted into proteoliposomes (Hirata *et al.*, 1976b). The detergent-soluble fraction was subjected to purification procedures including ion-exchange column chromatography and high-performance liquid chromatography with a hydroxylapatite column (Hirata *et al.*, 1984). At appropriate steps, proteins were reconstituted into proteoliposomes loaded with $K^+$ salt according to the freeze–thaw–sonication method (see Section 4.3.2.b). Partially purified soybean phospholipids, 40 mg in 1 ml of 0.25 M potassium phosphate (pH 8), were sonicated five times for 3 min at 1 min intervals under argon at 0° C with a probe-type sonifier (Branson, model 200). To the resulting liposome suspension, samples of the preparations obtained at each purification step (0.1–0.5 mg of protein) were added. The mixture was then sonicated for 1 min, frozen quickly in a dry ice–ethanol bath and thawed at room temperature, followed by a brief sonication (10 times of 0.5 s pulse). This freeze–thaw–sonication procedure was repeated three times. The resulting proteoliposomes were centrifuged at 140 000 ***g*** for 30 min, and suspended in 0.2 M sucrose containing 10 mM $MgSO_4$. Then the activity of the carrier protein(s) was determined as protein(s) catalysing the accumulation of radioactive alanine in response to a membrane potential created by $K^+$ diffusion via valinomycin. Addition of valinomycin to the reconstituted proteoliposomes in a potassium-free medium induced efflux of $K^+$ down the concentration gradient with concurrent generation of a membrane potential (interior negative). In the course of $K^+$ efflux, radioactive alanine accumulated in the proteoliposomes.

Using these methods, the alanine carrier protein has been purified as a single polypeptide with an apparent $M_r$ of 42 500 estimated by SDS–polyacrylamide gel electrophoresis (Hirata *et al.*, 1984).

## 4.6 Concluding remarks

A variety of membrane functions have been successfully reconstituted into artificial membrane systems from respective membrane proteins and phospholipids. These approaches have allowed us not only to define the minimum components participating in particular membrane functions but also to elucidate the detailed mechanism of membrane function.

As described in this chapter, reconstitution procedures are diverse and each has both advantages and disadvantages. However, there is no established general strategy for reconstitution, and it is rather simplistic to apply one of them to reconstitution of new membrane proteins. Care should be taken not to be misled by various artifacts which inevitably accompany reconstitution procedures.

## 4.7 References

Banerjee, R. K. and Datta, A. G. (1983) *Mol. Cell. Biochem.*, **50**, 3–15.
Baron, C. and Thompson, T. E. (1975) *Biochim. Biophys. Acta*, **382**, 276–85.

Bashford, C. L. and Smith, J. C. (1979) *Methods Enzymol.*, **55**, 569–86.
Boyer, P. D., Chance, B., Ernster, L., Mitchell, P., Racker, E. and Slater, E. C. (1977) *Annu. Rev. Biochem.*, **46**, 955–1026.
Caroll, R. C. and Racker, E. (1977) *J. Biol. Chem.*, **252**, 6981–90.
Curman, B., Klareskog, L. and Peterson, P. A. (1980) *J. Biol. Chem.*, **255**, 7820–6.
Darszon, A. (1983) *J. Bioenerg. Biomemb.*, **15**, 321–34.
Darszon, A., Vandenberg, C. A., Ellisman, M. H., Spitzer, N. and Montal, M. (1980) *Proc. Natl. Acad. Sci. USA*, **77**, 239–43.
Eddy, A. A. (1978) *Curr. Top. Membr. Transp.*, **10**, 279–360.
Eytan, G. D. (1982) *Biochim. Biophys. Acta*, **694**, 185–202.
Eytan, G. D. and Broza, R. (1978) *J. Biol. Chem.*, **253**, 3196–202.
Eytan, G. D., Matheson, M. J. and Racker, E. (1976) *J. Biol. Chem.*, **251**, 6831–7.
Goldin, S. M. (1977) *J. Biol. Chem.*, **252**, 4631–7.
Hamamoto, T., Ohno, K. and Kagawa, Y. (1981) *J. Biochem. (Tokyo)* **91**, 1759–66.
Harold, F. M. (1978) in *The Bacteria* (eds L. N. Ornston and J. R. Sokatch), New York, Vol. VI, pp. 463–521. Academic Press.
Henderson, P. J. F., Kagawa, Y. and Hirata, H. (1983) *Biochim. Biophys. Acta*, **732**, 204–9.
Hirata, H. (1979a) in *Function and Molecular Aspects of Biomembrane Transport* (eds E. Quagliariello, F. Palmieri, S. Papa and M. Klingenberg), Elsevier/North-Holland Biomedical Press, Amsterdam, New York and Oxford, pp. 505–12.
Hirata, H. (1979b) *Methods Enzymol.*, **56**, 430–5.
Hirata, H., Kambe, T. and Kagawa, Y. (1984) *J. Biol. Chem.*, **259**, 10653–6.
Hirata, H., Sone, N., Yoshida, M. and Kagawa, Y. (1976a) *J. Biochem. (Tokyo)*, **79**, 1157–66.
Hirata, H., Sone, N., Yoshida, M. and Kagawa, Y. (1976b) *Biochem. Biophys. Res. Commun.*, **69**, 665–71.
Houslay, M. D. and Stanley, K. K. (1982) in *Dynamics of Biological Membranes*, John Wiley and Sons, Chichester, New York, Brisbane, Toronto and Singapore.
Huang, K.-S., Bayley, H. and Khorana, H. G. (1980) *Proc. Natl. Acad. Sci. USA*, **77**, 323–7.
Hub, H.-H., Hupfer, B., Koch, H. and Ringsdorf, H. (1980) *Angew. Chem. Int. Ed. Engl.*, **19**, 938–40.
Kaback, H. R. (1983) *J. Membr. Biol.*, **76**, 95–112.
Kagawa, Y. (1982) in *Membranes and Transport* (ed N. Martonosi), Plenum Publishing Corp., Amsterdam, Vol. 1, pp. 439–46.
Kagawa, Y., Ide, C., Hamamoto, T., Rögner, M. and Sone, N. (1982) in *Membrane Reconstitution* (eds G. Poste and G. L. Nicolson), Elsevier Biomedical Press, Amsterdam, pp. 137–60.
Kagawa, Y. and Racker, E. (1966) *J. Biol. Chem.*, **241**, 2475–82.
Kagawa, Y. and Racker, E. (1971) *J. Biol. Chem.*, **246**, 5477–87.
Kagawa, Y., Sone, N., Hirata, H. and Yoshida, M. (1979) *J. Bioenerg. Biomembr.*, **11**, 39–78.
Kasahara, M. and Hinkle, P. C. (1976) *Proc. Natl. Acad. Sci. USA*, **73**, 396–400.
Knowles, A. F. and Racker, E. (1975) *J. Biol. Chem.*, **250**, 3538–44.
Kornblatt, J. A., Chen, W. L., Hsia, J. C. and Williams, G. R. (1974) *Proc. Can. Fed. Biol. Soc.*, **17**, 279.

Kornblatt, J. A., Chen, W. L., Hsia, J. C. and Williams, G. R. (1975) *Can. J. Biochem.*, **53**, 364–70.
Miller, C. and Racker, E. (1976) *J. Membr. Biol.*, **30**, 271–82.
Mitchell, P. (1979) *Science*, **206**, 1148–59.
Montal, M., Darszon, A. and Schindler, H. (1981) *Q. Rev. Biophys.*, **14**, 1–79.
Newman, M. J., Foster, D. L., Wilson, T. H. and Kaback, H. R. (1981) *J. Biol. Chem.*, **256**, 11804–8.
Okamoto, H., Sone, N., Hirata, H., Yoshida, M. and Kagawa, Y. (1977) *J. Biol. Chem.*, **252**, 2160–5.
Oku, N., Kendall, D. A. and MacDonald, R. C. (1982) *Biochim. Biophys. Acta*, **691**, 332–40.
Olson, F., Hunt, C. A., Szoka, F. C., Vail, W. J. and Papahadjopoulos, D. (1979) *Biochim. Biophys. Acta*, **557**, 9–23.
Papahadjopoulos, D., Poste, G., Schaeffer, D. E. and Vail, W. J. (1974) *Biochim. Biophys. Acta*, **325**, 10–28.
Pick, U. (1981) *Arch. Biochem. Biophys.*, **212**, 186–94.
Pick, U. and Racker, E. (1979) *J. Biol. Chem.*, **254**, 2793–9.
Racker, E. (1979) *Methods Enzymol.*, **55**, 699–711.
Racker, E., Chien, T.-F. and Kandrach, A. (1975) *FEBS Lett.* **57**, 14–18.
Racker, E. and Hinkle, P. C. (1974) *J. Membr. Biol.*, **17**, 181–8.
Rögner, M., Ohno, K., Hamamoto, T., Sone, N. and Kagawa, Y. (1979) *Biochem. Biophys. Res. Commun.*, **91**, 362–7.
Rosen, B. P. and Kashket, E. R. (1978) in *Bacterial Transport* (ed B. P. Rosen), Dekker, New York, pp. 559–620.
Schlodder, E. and Witt, H. T. (1981) *Biochim. Biophys. Acta*, **635**, 571–84.
Schuldiner, S., Rottenberg, H. and Avron, M. (1972) *Eur. J. Biochem.*, **25**, 64–70.
Sebald, W., Friedl, P., Schairer, H. U. and Hoppe, J. (1982) *Ann. N.Y. Acad. Sci.*, **402**, 28–44.
Sebald, W., Graf, T. and Lukins, H. B. (1979) *Eur. J. Biochem.*, **93**, 587–99.
Serrano, R., Kanner, B. I. and Racker, E. (1976) *J. Biol. Chem.*, **251**, 2453–61.
Skulachev, V. P. (1981) in *Chemiosmotic Proton Circuits in Biological Membranes* (eds V. P. Skulachev and P. C. Hinkle), Addison-Wesley, Reading, MA, pp. 3–46.
Skriver, E., Maunsbach, A. B. and Jørgensen, P. L. (1980) *J. Cell Biol.*, **86**, 746–54.
Sone, N., Hamamoto, T. and Kagawa, Y. (1981) *J. Biol. Chem.*, **256**, 2873–7.
Sone, N., Ikeba, K. and Kagawa, Y. (1979a) *FEBS Lett.*, **97**, 61–4.
Sone, N., Takeuchi, Y., Yoshida, M. and Ohno, K. (1977b) *J. Biochem. (Tokyo)*, **82**, 1751–8.
Sone, N., Yoshida, M., Hirata, H. and Kagawa, Y. (1975) *J. Biol. Chem.*, **250**, 7917–23.
Sone, N., Yoshida, M., Hirata, H. and Kagawa, Y. (1977a) *J. Biol. Chem.*, **252**, 2956–60.
Sone, N., Yoshida, M., Hirata, H. and Kagawa, Y. (1977c) *J. Biochem. (Tokyo)*, **81**, 519–28.
Sone, N., Yoshida, M., Hirata, H. and Kagawa, Y. (1978) *Proc. Natl. Acad. Sci. USA*, **75**, 4219–23.
Sone, N., Yoshida, M., Hirata, H. and Kagawa, Y. (1979b) *J. Biochem. (Tokyo)*, **85**, 503–9.
Szoka, F. C. and Papahadjopoulos, D. (1978) *Proc. Natl. Acad. Sci. USA*, **75**, 4194–8.

Teissie, J., Knox, B. E., Tsong, T. Y. and Wehrle, J. (1981) *Proc. Natl. Acad. Sci. USA*, **78**, 7473–7.

Vinkler, C. and Korenstein, R. (1982) *Proc. Natl. Acad. Sci. USA*, **79**, 3183–7.

Vinkler, C., Korenstein, R. and Farkas, D. L. (1982) *FEBS Lett.*, **145**, 235–40.

Wagner, N., Dose, K., Koch, H. and Ringsdorf, H. (1981) *FEBS Lett.*, **132**, 313–18.

Wakabayashi, T., Kubota, M., Yoshida, M. and Kagawa, Y. (1977) *J. Mol. Biol.*, **117**, 515–19.

West, I. C. (1980) *Biochim. Biophys. Acta*, **604**, 91–126.

Witt, H. T., Schlodder, E. and Graber, P. (1976) *FEBS Lett.*, **69**, 272–6.

Wu, W. C. S., Moore, H. P. and Raftery, M. A. (1981) *Proc. Natl. Acad. Sci. USA*, **78**, 775–9.

Yoshida, M., Sone, N., Hirata, H. and Kagawa, Y. (1977) *J. Biol. Chem.*, **252**, 3480–5.

# 5

# Functional reconstitution of membrane proteins in planar lipid bilayer membranes

MAURICIO MONTAL



## Abbreviations used:

A – electron acceptor
AChR – acetylcholine receptor
ACh – acetylcholine
BLM black lipid membrane
BTX – bathracotoxin
D – electron donor
$M_r$ – molecular weight
PC – phosphatidylcholine
PE – phosphatidylethanolamine
PS – phosphatidylserine
RCs – reaction centres
STX – saxitoxin
TTX – tetrodotoxin

## 5.1 The question

Understanding the mechanism of action of membrane proteins depends upon the convergence of knowledge about the structure of the protein with a

Techniques for the Analysis of Membrane Proteins. Edited by C. I. Ragan & R. J. Cherry.
Published in 1986 by Chapman and Hall, 11 New Fetter Lane, London EC4P 4EE

detailed characterization of its function at the molecular level. Hybridoma technology, recombinant DNA technology and electrical recordings from reconstituted membrane systems are three new approaches to the study of membranes. These technologies are rapidly expanding our ability to study membrane function at the molecular level. In this chapter, the latter approach is briefly discussed.

Progress in the biochemical characterization of membrane systems has led to the association of complex biological phenomena with identifiable membrane proteins. Many membrane proteins have been purified to homogeneity and chemically characterized in terms of physicochemical properties, subunit compositions, amino acid compositions and even amino acid sequence. Examples of these advances are *Halobacterium halobium* bacteriorhodopsin (see Khorana *et al.*, 1979; Ovchinnikov *et al.*, 1979), the visual pigment rhodopsin (Ovchinnikov *et al.*, 1982; Hargrave *et al.*, 1983), the nicotinic acetylcholine receptor (AChR) from *Torpedo californica* electric organs (Changeux, 1981; Noda *et al.*, 1982, 1983; Claudio *et al.*, 1983; Devilliers-Thiery *et al.*, 1983), cytochrome *c* oxidase (for a review see Wikström *et al.*, 1983) and others. This structural information is leading to specific proposals about the arrangements of the polypeptide chains in the membrane lipid bilayer and opening up suggestions about the correlation between the structure of the protein and its function at the molecular level.

The question now arises: How can one assay the function of the purified membrane protein in a membrane system? Such questions are prompted by the fact that membrane proteins are purified in the presence of detergents and, consequently, their transmembrane function cannot be assayed in the micellar solution. Given this limitation, one is led to develop techniques that would allow the reassembly of the membrane protein in a lipid bilayer membrane suitable for the assay of the specific function with high sensitivity. During the past decade, several technical developments have occurred in this endeavour. This chapter presents a brief description of the salient technical approaches and provides a few selected examples of the membrane proteins reconstituted accordingly. Comprehensive reviews of the field appeared in 1981 and 1986 (Montal *et al.*, 1981; Miller, 1986). Other reviews cover more specific aspects of the field (Goldin *et al.*, 1983; Latorre and Alvarez, 1981; Miller, 1983; Schindler, 1982a; Latorre and Miller, 1983).

## 5.2 A definition

Several recent reports have described the incorporation of membrane fragments or crude membrane extracts into lipid bilayers and referred to this approach as the reconstitution of a membrane. It is important to emphasize the distinction between membrane reconstitution and the incorporation of membrane material into a lipid bilayer.

We explicitly refer to '*reconstitution*' of membrane proteins into lipid bilayers when a specific membrane protein is solubilized from its native membrane, purified from other membrane constituents, recombined with specific phospholipids and then reassembled into a well-defined planar lipid bilayer membrane that exhibits functional activity. In the biochemical tradition, this involves the resolution of a complex membrane into its molecular components and the reconstitution of its components into a functional unit. In this context, we refer to the '*incorporation*' of membrane proteins into lipid bilayers when crude membrane extracts or membrane fragments are '*associated*' with or '*transformed*' into lipid bilayers. In the biochemical tradition, this implies that the native membrane is not dissociated into its components and, therefore, a genuine reconstitution of the system from the components cannot, by necessity, be achieved. This distinction is timely since reports of putative reconstitutions are misleading in the fact that they do not constitute a genuine resolution and reconstitution of the membrane but the incorporation or transformation of native membranes into lipid bilayers. Clear examples of these contributions are the assays in lipid bilayers of functional *Torpedo* AChRs (Schindler and Quast, 1980; Schindler *et al.*, 1983) or of *Torpedo* $Cl^-$ channels (Miller, 1982a; Tank *et al.*, 1982) and of voltage-dependent $Na^+$ channels from rat brain membrane vesicles (Krueger *et al.*, 1983; French *et al.*, 1984).

## 5.3 Approaches to the reassembly of membrane proteins in planar lipid bilayers

Planar lipid bilayers bounded on both sides by aqueous media can be assembled by two distinct conceptual modalities that determine, in turn, the use of two different experimental procedures for their formation. For both techniques, immersion of an electrode in each aqueous compartment allows electrical recordings across the bilayer membrane.

### 5.3.1 BLACK LIPID FILMS – BLM (MUELLER *et al.*, 1962; MUELLER AND RUDIN, 1969)

The principle involves the extension and drainage of a mixture of lipid in an organic solvent over a hydrophobic aperture until the two monolayers at the water–organic solvent interface coalesce. This forms a focus of bilayer that eventually propagates to cover the entire area of the aperture. The process is formally a phase transition and the resultant metastable structure consists of a bilayer of lipid in hydrocarbon solvent (Haydon, 1970).

### 5.3.2 PLANAR LIPID BILAYERS FROM MONOLAYERS

These membranes are formed by adjoining the hydrocarbon tails of two lipid monolayers, initially at the air/water interface, through an aperture in a hydrophobic septum that separates two aqueous phases. The resultant structure is that of a lipid bilayer which is virtually solvent-free (Montal and Mueller, 1972; cf. Montal *et al.*, 1981). Fig. 5.1 is a schematic idealization of the process of bilayer formation: an aperture in a Teflon septum separates two lipid monolayers at the air/water interface. The water levels are successively raised above the aperture within which the two monolayers combine to form a stable planar lipid bilayer.

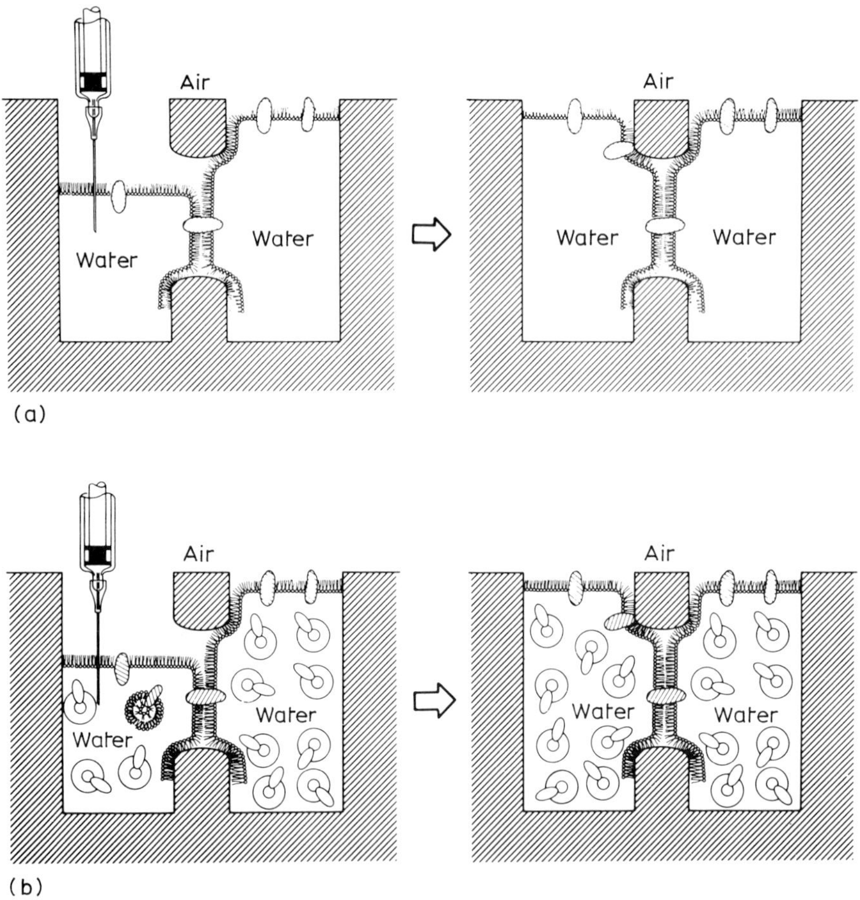


**Fig. 5.1** Schematic (idealized) representation of the process of bilayer formation from monolayers at the air/water interface in the aperture of a Teflon partition that separates two aqueous compartments. The water level in one of the compartments is illustrated covering the entire aperture while the second is raised only partially in the left-hand panel to indicate the sequence of steps in the process. The water level is raised by injecting the aqueous medium from the syringe until the bilayer covers the entire aperture (right-hand panel). (a) Monolayer self-assembly from protein–lipid complexes in hexane. (Modified from Montal, 1974a.) (b) Monolayer self-assembly from vesicle suspensions. (Modified from Schindler, 1982a,b.)

It is evident from this scheme that an obligatory intermediate step in the process of bilayer formation is the generation of monolayers at the air/water interface. From this fact, it emerges that there are two main pathways to achieve the self-assembly of a lipid–protein monolayer: either from lipid–protein complexes in organic solvents (Fig. 5.1a; Montal, 1974a) or from vesicle suspensions (Fig. 5.1b; Schindler, 1979, 1980, 1982a,b). Lipids are amphiphilic molecules that spontaneously orient themselves at air/water interfaces to maximize the hydrophobic and hydrophilic interactions and thus minimize their configurational free energy. Monolayers form by spreading the lipid at the interface from a solution in an evaporatable solvent (Fig. 5.1a). The alternative pathway of monolayer assembly proceeds from the aqueous phase (Fig. 5.1b): a suspension of lipid vesicles spontaneously break at the interface leading to the eventual formation of a monomolecular film which remains in equilibrium with the vesicle suspension (Schindler, 1979, 1980, 1982a,b; Schindler *et al.*, 1980; see also Verger and Pattus, 1976; Pattus *et al.*, 1978a,b).

Recently, a novel application of the principle of bilayer formation from monolayers has been developed (Wilmsen *et al.*, 1982; Coronado and Latorre, 1983; Schuerholz and Schindler, 1983; Suarez-Isla *et al.*, 1983). This is schematically illustrated in Fig. 5.2. It consists of the formation of a planar bilayer in the tip of a patch pipette, a pipette similar to those used in patch clamp electrophysiological recordings (see Hamill *et al.*, 1981). A monolayer is initially formed either from a solution of lipids in an organic solvent or from a suspension of lipid vesicles in an aqueous medium. The patch pipette is introduced through the air/water interface (step 1 in Fig. 5.2). Thereafter, the pipette is removed from the solution. The lipid head groups of the monolayers are attached to the pipette while the hydrocarbon tails contact air. In step 3, the pipette is reimmersed into the solution. This leads to the apposition of the hydrocarbon tails of the attached monolayer to those of the original monolayer thereby forming a bilayer. The complete bilayer separates two aqueous compartments when it is submerged in the solution. This method offers the possibility of obtaining single-channel recordings with higher signal-to-noise ratio and faster time resolution. One of the disadvantages of this technique is that the aqueous compartment contained within the patch pipette is not readily accessible to manipulations.

### 5.3.3 FUSION OF PREFORMED VESICLES WITH PREFORMED BILAYERS

Another valuable strategy for the incorporation of membrane proteins into planar bilayers is that of membrane fusion (Drachev *et al.*, 1974; Miller and Racker, 1976; Miller *et al.*, 1976; Cohen *et al.*, 1980, 1982; Krueger *et al.*, 1983; Miller, 1982a,b). In this modality, the preformed lipid bilayer can be assembled either as a black film or as a bilayer formed from monolayers either

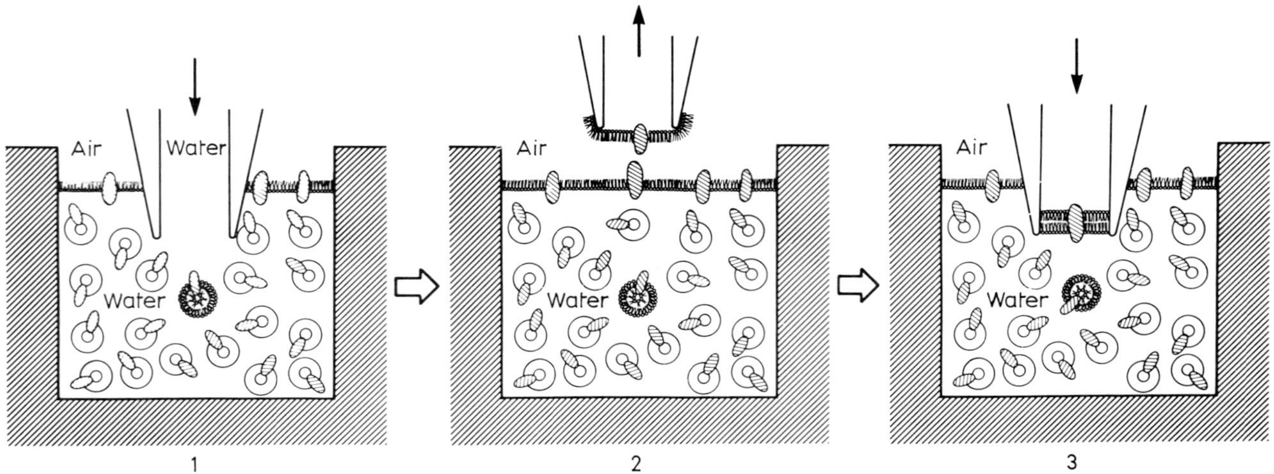


**Fig. 5.2** Schematic (idealized) representation of the process of bilayer formation from monolayers at the air/water interface at the tip of patch pipettes. In step 1, monolayers are self-assembled from a lipid solution in hexane (not shown) or from reconstituted vesicles (one vesicle is shown in more detail). A patch pipette is introduced into the solution. Next, the pipette is removed from the solution (step 2) and immediately reimmersed into the solution (step 3). A bilayer is formed at the tip of the pipette by the apposition of the hydrocarbon tails of the attached and the original monolayer. (From Suarez-Isla *et al.*, 1983.)

on Teflon septi or on patch pipettes (see, for example, Coronado and Latorre, 1983). To incorporate the protein into the preformed bilayer, a membrane extract or a vesicle suspension is introduced into the aqueous phase.

### 5.3.4 LIPOSOME-PATCH METHOD

A recent technical development has allowed the recording of electrical parameters directly from reconstituted vesicles. In this technique, the solubilized membrane protein is incorporated into phospholipid vesicles; the vesicles are subject to a freeze–thaw cycle (Kasahara and Hinkle, 1977) in order to produce large vesicles ($\sim 1\ \mu m$). Thereafter, a patch pipette is brought into the proximity of the reconstituted vesicle; gentle suction of the vesicles into the pipette produces a high-resistance seal (of the order of 1 GΩ) of a patch of membrane in the pipette. This patch is then isolated from the reconstituted vesicle and used to obtain electrical recordings either by measuring membrane current at a constant applied voltage (voltage clamp) or by measuring membrane voltage at constant applied current (current clamp). This method was used to record single $Cl^-$ channels from a crude preparation of *Torpedo californica* electric organ membranes (Tank *et al.*, 1982) as well as single channels from reconstituted vesicles containing purified *Torpedo* AChR (Suarez-Isla *et al.*, 1983; Tank *et al.*, 1983).

We now turn to describe the results obtained by the application of the methods described to specific membrane proteins.

## 5.4 The nicotinic acetylcholine receptor from *Torpedo californica*

The nicotinic AChR from *Torpedo californica* electric organ is by far the best characterized membrane channel protein of nervous tissue (see Anholt *et al.*, 1985; McNamee and Ochoa, 1982). The AChR molecule has an $M_r$ of approx. 270 000 and is composed of five membrane glycoproteins with a subunit stoichiometry of $\alpha_2\beta\gamma\delta$ (Reynolds and Karlin, 1978; Lindstrom *et al.*, 1979; Raftery *et al.*, 1980; Noda *et al.*, 1982, 1983; Claudio *et al.*, 1983; Devilliers-Thiery *et al.*, 1983). Recently, the entire amino acid sequence of all the subunits was elucidated (Noda *et al.*, 1982, 1983; Claudio *et al.*, 1983; Devilliers-Thiery *et al.*, 1983) and specific models for the arrangement of the polypeptide chains across the membrane bilayer are being proposed (Noda *et al.*, 1982, 1983; Claudio *et al.*, 1983; Devilliers-Thiery *et al.*, 1983; Guy, 1984). The biochemical information which has given insights into the structure of the AChR molecule together with a wealth of electrophysiological and pharmacological information regarding the function of the AChR make it an attractive candidate to use to address questions aimed at establishing correlations between the structure of the molecule and its function. The reconstitution studies would allow the biochemical manipulation of the receptor in detergent solution and the subsequent assessment of functional activity after reassembly of the AChR into a model membrane. Several important questions have been addressed in the reconstituted system. Is the ion channel a component of the purified AChR? Which subunits contribute to the channel structure? What molecular mechanisms underlie the conformational transitions which take place after the binding of an agonist? How is the function of the AChR channel affected by chemical modifications of the protein or by alterations in the lipid environment in which the receptor is embedded?

The AChR was solubilized from electrical organ membranes with a cholate–lipid mixture (Epstein and Racker, 1978). The solubilized protein was purified by affinity chromatography techniques and reconstituted into lipid vesicles by detergent dialysis (Huganir *et al.*, 1979; Changeux *et al.*, 1979; Lindstrom *et al.*, 1980; Anholt *et al.*, 1980, 1981, 1982; Huganir and Racker, 1982). The reconstituted vesicles have been extensively characterized in terms of their structure as well as their permeability response to cholinergic agonists. The reconstituted purified AChR showed two prominent features of the postsynaptic membrane, namely activation and desensitization induced by cholinergic agonists (Anholt *et al.*, 1980, 1981, 1982; Walker *et al.*, 1982; McNamee *et al.*, 1984; Popot *et al.*, 1981; Epstein and Racker, 1978; Huganir *et al.*, 1979; Huganir and Racker, 1982; Wu *et al.*, 1981; Hess *et al.*, 1982, 1983; Lüdi *et al.*, 1983).

Several strategies have been used in order to obtain single-channel recordings from a purified AChR: (1) The reconstituted vesicles were transformed into monolayers at an air/water interface which subsequently were joined together into a planar lipid bilayer (Nelson *et al.*, 1980; Labarca *et al.*, 1981, 1982, 1983a,b, 1984; Montal *et al.*, 1984; Schindler, 1982b; Schindler, *et al.*, 1983; see also Schindler and Quast, 1980); (2) The reconstituted vesicles were subjected to a freeze–thaw cycle and then a high-resistance seal was formed by suction of the vesicle into the tip of the pipette (Suarez-Isla *et al.*, 1983; Tank *et al.*, 1983); (3) The AChR was reconstituted into vesicles composed of a single specific phospholipid and thereafter fused with a preformed lipid bilayer composed of the same phospholipid (Boheim *et al.*, 1981).

We now turn to describe in more detail the results obtained with each one of the strategies outlined above.

The AChR reconstituted vesicles were spontaneously transformed into monolayers at the air/water interface and subsequently a lipid bilayer was formed by joining two such monolayers (Schindler, 1979, 1980, 1982b; Schindler and Quast, 1980; Nelson *et al.*, 1980; Labarca *et al.*, 1981, 1982, 1983, 1984a; Montal *et al.*, 1984). The bilayers were formed both in the orifice of a Teflon septum (Nelson *et al.*, 1980; Labarca *et al.*, 1981, 1982, 1983a, 1984; Montal *et al.*, 1984) and in the tip of a patch pipette (Suarez-Isla *et al.*, 1983).

### 5.4.1 MACROSCOPIC PHENOMENA

The reconstituted planar bilayer responded to the addition of cholinergic agonist with a transient increase in membrane conductance which relaxed to a steady-state value. The agonist-activated membrane conductance increased with agonist concentration and was inhibited by cholinergic antagonists (Nelson *et al.*, 1980; Labarca *et al.*, 1984b). The increased membrane conductance was accompanied by an increase in the membrane noise. At high current resolution this increasing membrane noise could be resolved into discrete transitions between clearly identifiable current levels. This is shown in Fig. 5.3 which is a composite illustration of membrane current recordings at constant applied voltage obtained under a variety of conditions. Fig. 5.3A illustrates the macroscopic conductance, showing the transient increase in membrane conductance following the addition of the agonist and the spontaneous relaxation to a steady value which is associated with the phenomenon of desensitization (Katz and Thesleff, 1957). Fig. 5.3B illustrates a single-channel current record, activated by suberyldicholine, an analogue of acetylcholine (Ach), at a concentration of 1 $\mu$M. This recording reveals a characteristic pattern of occurrence of single-channel activity at high agonist concentrations: paroxysms of activity were followed by quiescent periods during which little or no activity was observed. This pattern of channel activity was more pronounced as the agonist concentration increased, a behaviour that is reminiscent of that of single AChR channel currents in muscle at desensitizing concentrations of ACh (Sakmann *et al.*, 1980). This phenomenon seems to result from the transit of the AChR through its active conformation, from which it can open several times before desensitizing (Sakmann *et al.*, 1980; Labarca *et al.*, 1983a,b; Montal *et al.*, 1984). Fig. 5.3C illustrates a single-channel recording at faster time resolution; notice that the groups of openings are composed of many single-channel openings; each opening is frequently interrupted by brief closings.

From single-channel recordings, as illustrated in Fig. 5.3C, information concerning the single-channel conductance and the channel open and closed times is derived (see Labarca *et al.*, 1984; Montal *et al.*, 1984). The single-channel conductance, that is the amplitude of the current transitions between the current level associated with the closed state of the channel and the level associated with the open state of the channel, was independent of the applied voltage and of the nature of the agonist used to activate the channel. The single-channel current increased with sodium concentration and showed saturation. The AChR channel is cation-selective with the apparent ion selectivity sequence: $NH_4^+ > Cs^+ > Rb^+ \geq Na^+$, $Cl^-$, $F^-$, $SO_4^{2-}$.

The lifetime of the open state of the channel depended on the affinity of the agonist, but not on its concentration, and was longest for suberyldicholine, shortest for carbamoylcholine and intermediate for ACh, in line with single

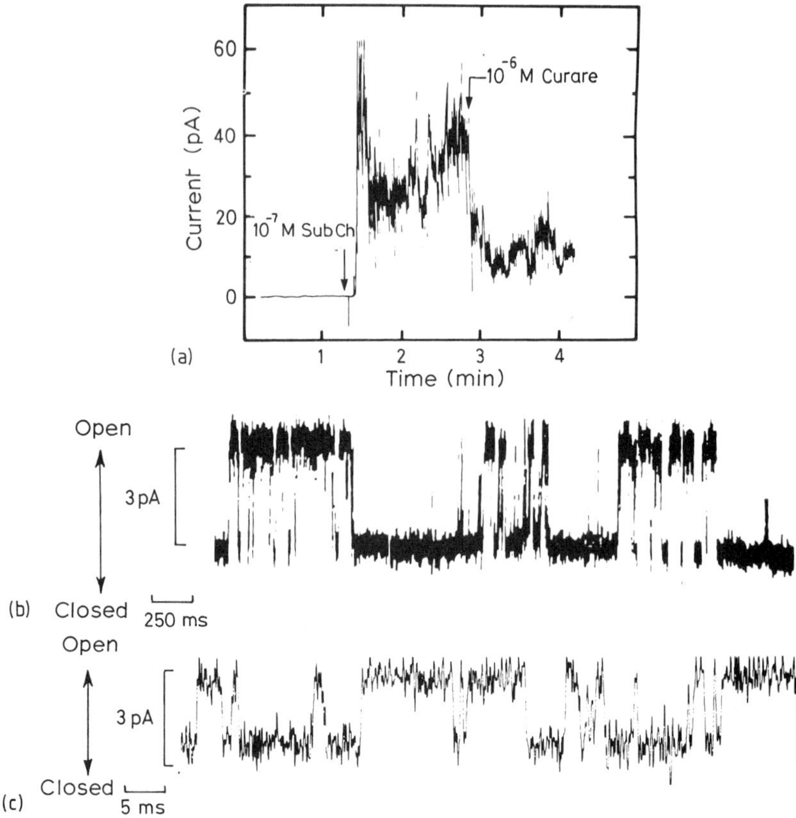


**Fig. 5.3** Electrical recordings from reconstituted AChRs in planar lipid bilayers. (a) Time course of the agonist-activated conductance increase in planar lipid bilayers containing purified AChR. Bilayers were formed in 0.5 M NaCl/5 mM $CaCl_2$/2.5 mM Hepes, pH 7.4. A voltage step of + 50 mV was applied across the membrane and held constant throughout the record. After the response activated by 0.1 $\mu$M suberyldicholine (SubCh) had reached a plateau, 1 $\mu$M curare was added to the cis chamber. Note that curare depresses both the membrane conductance and the current noise. (Modified from Labarca *et al.*, 1984.) (b) Single AChR channel currents activated by desensitizing concentrations of agonist (1 $\mu$M suberyldicholine) at $V = 100$ mV. Other conditions as for a. (Modified from Labarca *et al.*, 1983a.) (c) Single AChR channel currents activated by agonist (0.1 $\mu$M suberyldicholine) at $V = 100$ mV. (Modified from Labarca *et al.*, 1984.)

AChR channels recorded *in situ* (Labarca *et al.*, 1981, 1982, 1983, 1984; Montal *et al.*, 1984; Neher and Sakmann, 1976a,b; Jackson *et al.*, 1982). A detailed analysis of the channel lifetimes revealed an interesting feature of the AChR channel: no fit was obtained for the open state lifetime distributions by a single exponential but a good fit was obtained with the sum of two exponentials (Labarca *et al.*, 1983, 1984; Montal *et al.*, 1984; Tank *et al.*, 1983). A double-exponential distribution of open times has been observed for the AChR in muscle cells (Colquhoun and Sakmann, 1981; Jackson *et al.*,

1983). Two exponentials could arise from two independent populations of AChR, from a single population of AChR displaying two open states, or from more complicated kinetic schemes (Labarca *et al.*, 1983a). An analysis based on conditional probability demonstrated that the results could be attributed most simply to the presence of one population of AChR having two distinct open states (Labarca *et al.*, 1983a). Jackson *et al.* (1983b) analysed the correlation between fast and slow opening events and postulated two open states for the AChR in muscle cells. Further analysis is required to understand the origin of the two open states.

In an independent study, Schindler and collaborators have reconstituted the purified AChR from *Torpedo marmorata* into planar lipid bilayers and reported similar results to those described for *Torpedo californica* (Schindler and Quast, 1980; Schindler, 1982b; Schindler *et al.*, 1983). They emphasized the exquisite dependence of the channel properties on the surface properties of the monolayers as well as on the lipid composition of the reconstituted vesicles and of the monolayers themselves. Two particular requirements for the function of the AChR channel in the reconstituted bilayer were found: (1) the presence of cholesterol in the membrane which seemed to stabilize the channel; (2) a cohesive pressure in the monolayer of about 30–32 dynes/cm; this relatively low cohesiveness of the membrane appeared to reduce the aggregation of AChRs in the bilayer. Their extensive analysis established that there is a correlation between the monolayer properties and the AChR channel properties in the planar bilayer (Schindler *et al.*, 1983).

The channel properties of the purified AChR were studied in reconstituted bilayers formed at the tip of patch pipettes. The bilayers were formed either by gentle suction of the vesicles into the pipette (Tank *et al.*, 1983; Suarez-Isla *et al.*, 1983) or by the formation of a reconstituted bilayer from a monolayer at the tip of the pipette (Suarez-Isla *et al.*, 1983). Virtually all the properties hitherto described for the purified AChR channel reconstituted into planar bilayers in Teflon apertures have been recorded using bilayers in the tip of patch pipettes. These new developments, however, have improved the sensitivity and time resolution of the single-channel recordings from the purified receptor and, therefore, are providing new detailed information on the kinetics of channel gating. It is anticipated that a more extensive characterization of the channel properties with these techniques will provide new insights into the details of the operation of the AChR channel.

The purified AChR from *Torpedo marmorata* was reconstituted into planar lipid bilayers using the fusion approach (Boheim *et al.*, 1981). The method involves the incorporation of the purified protein into vesicles composed of a pure mixed-chain lipid [1,3-stearoyl-3-myristoylglycero-2-phosphocholine (1,3-SMPC)]. Thereafter, the reconstituted vesicles are fused with a preformed planar bilayer containing the same lipid, 1,3-SMPC, by incubating the vesicles with a preformed bilayer a few degrees below the phase transition temperature of the lipid. Single-channel currents were activated by cholinergic agonists and displayed similar properties to those previously described for *Torpedo californica* AChR. The reconstitution technique was tested by comparing the incorporation of AChR into the bilayer at different extents of resolution: receptor-enriched membrane fragments, before and after removal of non-receptor peripheral proteins, mixtures of purified receptor monomer and dimer, and the isolated receptor monomer, Virtually identical single-channel recordings were obtained from the preparations at the three distinct levels of purification. The results suggest that the AChR monomer is the active species. Such functional equivalence between monomers and dimers [dimers being two monomers linked covalently by a disulphide bridge between their $\delta$ subunits (Hamilton *et al.*, 1979; Chang and Bock, 1977)] had been established for purified AChR from *Torpedo californica* in reconstituted vesicles (Anholt *et al.*, 1980). One limitation of this reconstitution strategy is its requirement for the specific lipid to achieve the fusion between the reconstituted vesicle and the planar bilayer. This condition severely hampers the application of the reconstitution approach to the study of the effects of lipids on channel function.

The results summarized for the AChR indicate that a variety of procedures have led to the reconstitution of the purified AChR into planar lipid bilayers. It is evident that for this receptor the full cycle of solubilization, purification and reconstitution can be achieved without impairment of channel function. This initial characterization of the properties of the purified AChR channel established the data base upon which to compare the effects introduced by modifications of the protein or alterations introduced by specific phospholipids. It is anticipated that the use of the reconstituted membrane will get us one step closer towards the identification of the functional role of specific receptor subunits.

In this section the leading reconstitution strategy was that of forming a planar lipid bilayer from monolayers derived from the reconstituted vesicles containing the purified protein. This strategy has been applied to many other proteins, among them: porin (Schindler and Rosenbusch, 1978, 1981; Schindler *et al.*, 1980); the proton channel of the proton ATPase (Schindler and Nelson, 1982); bacteriorhodopsin (Bamberg *et al.*, 1981); the reaction centres from photosynthetic bacteria (Gopher *et al.*, 1983, 1985; Blatt *et al.*, 1983) and cytochrome *c* oxidase from thermophilic bacteria (Hamamoto and Montal, 1984, Hamamoto *et al.*, 1985).

## 5.5 Reaction centres from photosynthetic bacteria

The primary event in photosynthesis occurs in membrane-bound chlorophyll–protein complexes known as reaction centres (RCs). The RC of *Rhodopseudomonas sphaeroides* has been extensively characterized (Okamura *et al.*, 1982). It is an integral membrane protein, of $M_r \sim 80\,000$, composed of three polypeptide subunits (L, M and H, standing for light, medium and heavy according to their molecular weight). The following prosthetic groups are attached to the subunits: four bacteriochlorophyll molecules, two bacteriopheophytin molecules, two ubiquinone molecules and one iron molecule. Absorption of a photon by the RC results in the transfer of an electron from a donor, D (a specialized bacteriochlorophyll dimer) to an acceptor, A (a ubiquinone magnetically coupled to $Fe^{2+}$). The secondary reactants of the electron-transfer chain are a cytochrome *c* on the donor side and a secondary ubiquinone on the acceptor side. The primary charge separation is stabilized by secondary electron transfer from the reduced bacteriopheophytin to the primary ubiquinone, $Q_A$ and subsequently from reduced $Q_A$ to a secondary ubiquinone, $Q_B$. The oxidized bacteriochlorophyll dimer ($D^+$) can be re-reduced by a back-reaction from either reduced $Q_A$ or reduced $Q_B$. In the presence of '*c*'-type cytochromes, $D^+$ is re-reduced in the submillisecond time range and thereby the two back-reactions are hindered. The kinetics of these different steps in the photochemical reaction have been well characterized in detergent-solubilized RCs (see Okamura *et al.*, 1982).

The current view is that the RC protein spans the bacterial plasma membrane from one surfacc to the other. The binding sites for cytochrome *c* 2 and quinone are considered to be located on opposite sides of the membrane (Prince *et al.*, 1975; Valkirs and Feher, 1982). Therefore, the primary function of the RC is to drive electrons to one side of the membrane, leaving a net positive charge on the opposite side, thereby generating a transmembrane electric current and potential (see Mitchell, 1968).

A major motivation for the study of RCs in planar lipid bilayers is precisely to study the transmembrane events associated with the photochemical reactions which cannot be studied in detergent-solubilized RCs. The electrogenic nature of the primary photoinduced charge separation, which has been inferred from a variety of indirect measurements (see Jackson and Dutton, 1973; Chance and Baltscheffsky, 1975), can be recorded directly in a reconstituted lipid bilayer. In addition, the system offers the possibility of clarifying the kinetics of the various steps in the electron-transfer chain and of investigating the coupling between redox potentials and membrane potentials. The functional role of the L, M and H subunits which thus far has remained elusive can, in principle, be evaluated by using the reconstitution approach.

The direct demonstration of photosynthetic transduction of light energy into transmembrane electric currents by intact RCs reconstituted in planar lipid bilayers has been demonstrated by several groups (Schönfeld *et al.*, 1979; Packham *et al.*, 1980, 1982; Gopher *et al.*, 1983, 1985; Blatt *et al.*, 1983; Appell *et al.*, 1983).

Two methods were used to reconstitute RCs into planar lipid bilayers. The first, and most studied, involves the extraction of an RC–lipid complex in hydrocarbon solvent (either hexane or octane). Planar lipid bilayers were formed by apposing two RC–lipid monolayers assembled from the RC–lipid complex in hexane (Schönfeld *et al.*, 1979, 1980). The RC–lipid complex in octane was used to prepare BLMs or planar membranes with a thickness ≥500 Å (Packham *et al.*, 1980, 1982; Appell *et al.*, 1983). In the second method, RCs were incorporated into liposomes which were subsequently transformed at the air/water interface into monolayers. Bilayers were then formed by apposing two such monolayers (Schindler, 1979, 1980; Gopher *et al.*, 1983; Blatt *et al.*, 1983).

We now consider in more detail the preparation derived from RC–lipid complex extracted into alkanes. The technique consists of three basic steps (see Montal, 1974a; Darszon *et al.*, 1978; Schönfeld *et al.*, 1980; Kendall-Tobias *et al.*, 1981; Packham *et al.*, 1980; Appell *et al.*, 1983): (1) partial removal of detergent from purified RCs; (2) recombination of RCs with the desired phospholipids in the aqueous phase by sonication; (3) extraction of the lipoprotein vesicles into the organic solvent promoted by charge neutralization of the complex with divalent cations, $Mg^{2+}$ (Schönfeld *et al.*, 1980) or $Ca^{2+}$ (Packham *et al.*, 1980; Appell *et al.*, 1983). The organic solvents were hexane (Schönfeld *et al.*, 1980) or octane (Packham *et al.*, 1980, 1982; Appell *et al.*, 1983), but not other more polar organic solvents (Schönfeld *et al.*, 1980). This extraction protocol resulted in the transfer of RCs as a protein–lipid complex into the organic solvent. In the visible and infrared regions the absorbance spectrum of the organic solvent extract was similar to that of RCs in aqueous detergent solution. The light-induced absorbance changes were only about 5% of the expected values; however, the photoactivity was almost fully restored by adding exogenous ubiquinone. This indicated that the quinones were extracted by the organic solvent, resulting initially in loss of activity (Schönfeld *et al.*, 1980; Kendall-Tobias *et al.*, 1981; Packham *et al.*, 1980). EPR (electron paramagnetic resonance) spectroscopy was also used to evaluate the RC–lipid complex in hexane. Active RCs displayed two light-induced EPR signals: a narrow one at $g=2.0026$, associated with oxidized $D^+$, and a broad signal at $g=1.8$, associated with reduced $A^-$ (Feher and Okamura, 1978). The light-induced EPR absorbance changes and the low-temperature kinetics of the back-reaction in the dark were found to be similar to those determined for RCs in the chromatophore membrane or when purified. In terms of the composition of the RC–lipid complex, the molar ratios of phospholipid–RC, detergent–RC, $H_2O$–phospholipid and RC–$Mg^{2+}$ were 4000–6000, 400, 40, and ~2, respectively (Schönfeld *et al.*, 1980). The RCs in the phospholipid–octane solution were also shown to retain photochemical activity (Packham *et al.*, 1980). The kinetics of the flash-induced oxidation of D and its dark reduction were studied. In the absence of supplementary quinone, the oxidation of D was recorded followed by the subsequent reduction in the dark. The half-time of the dark reaction was ~75 ms, which was similar to the half-time of return of an electron to $D^+$ from the primary endogenous quinone, $Q_A$, measured in detergent-solubilized RCs. It was concluded, therefore, that the secondary quinone, $Q_B$, was lost in the preparative procedures. Addition of ubiquinone-10 to the RC–phospholipid–octane solution increased the amount of D that could be oxidized in the millisecond time range following the flash and prolonged the time course of reduction in the dark, now to a half-time of 165 ms. This retardation suggests that the added ubiquinone reconstituted the electron-transfer reaction between the primary and secondary quinones (Packham *et al.*, 1980).

### 5.5.1 THE PHOTORESPONSES OF RCs IN THE RECONSTITUTED LIPID BILAYER

(a) *Steady state illumination*

The activity of the RCs in the bilayer was investigated by measuring the photoelectric responses when the RCs were illuminated. Electrodes were immersed into the solutions in the two compartments to measure light-induced open-circuit photovoltages between them, or currents flowing when the circuit was completed, short-circuit photocurrents. RCs are randomly oriented in the lipid bilayer. Consequently, in the absence of exogenous secondary reactants, no photoresponse was observed (Schönfeld *et al.*, 1979; Packham *et al.*, 1980, 1982; Appell *et al.*, 1983). However, when cytochrome *c* was added to one compartment, only those RCs whose donors faced the compartment containing cytochrome *c* can perform the secondary electron transfer. Thus, a functional orientation of the RCs can be achieved. These resulted in a photoresponse as shown in Fig. 5.4A, where the top panel shows the short-circuit photocurrent and the bottom the open-circuit photovoltage in response to a 20-second illumination of the reconstituted RCs in a planar lipid bilayer that separates the secondary donor (cytochrome *c*) from the secondary electron acceptor (quinone). The photoresponse can be divided into a transient, which occurred when the light is turned on and off, and a steady-state current and voltage. The initial current transient at the onset of illumination and the steady-state current (about 10 pA) indicated different electron-transfer processes. A simple model based on the transfer of charges across the membrane which explains the salient features of the photoresponse was presented (Schönfeld *et al.*, 1979). The wavelength dependence of this photoresponse matched the absorption spectrum of the RCs and, therefore, unequivocally identified the RCs as the molecular species responsible for the photoelectric effect (Schönfeld *et al.*, 1979). Planar membranes derived from the RC–phospholipid–octane solution also displayed no photoelectric responses unless ferrocytochrome *c* was added to one side of the membrane (Packham *et al.*, 1980, 1982; Appell *et al.*, 1983). Addition of ubiquinone-10 to the RC–phospholipid–octane solution prior to bilayer formation increased the amplitude of the short-circuit photocurrent. This effect is in agreement

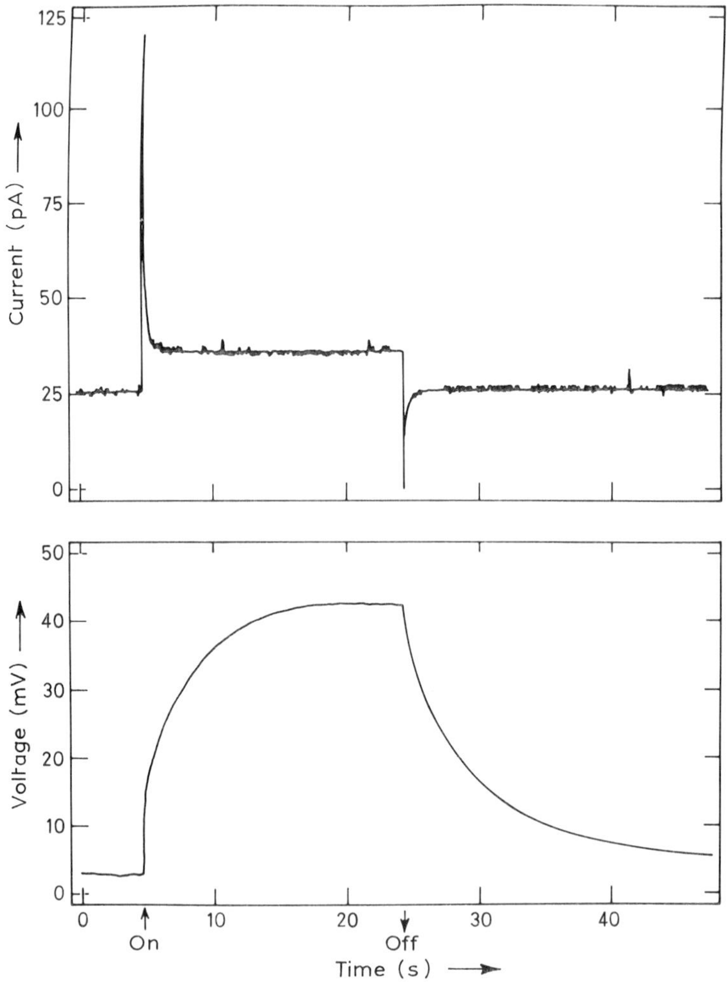


**Fig. 5.4** Photoresponses of reconstituted RCs in planar lipid bilayers. (a) Short-circuit photocurrent (top) and open-circuit photovoltage (bottom) obtained when a secondary donor (cytochrome *c*) is added to one compartment and a secondary acceptor (quinone) to the other. The arrows labelled 'on' and 'off' mark the onset and cessation of illumination. The photoresponse consists of a transient which occurs when the light is turned on and off and a steady-state current and voltage. (Modified from Schönfeld *et al.*, 1979 and Feher *et al.*, 1980.)

with the photochemical results in which the excess ubiquinone increased the amount of photo-oxidizable D by increasing the number of RCs which contain a full complement of ubiquinone. The action spectrum of the peak current response generated during a steady-state illumination was in agreement with the absorption spectrum of the RC–phospholipid–octane solution (Packham *et al.*, 1980; see also Appell *et al.*, 1983).

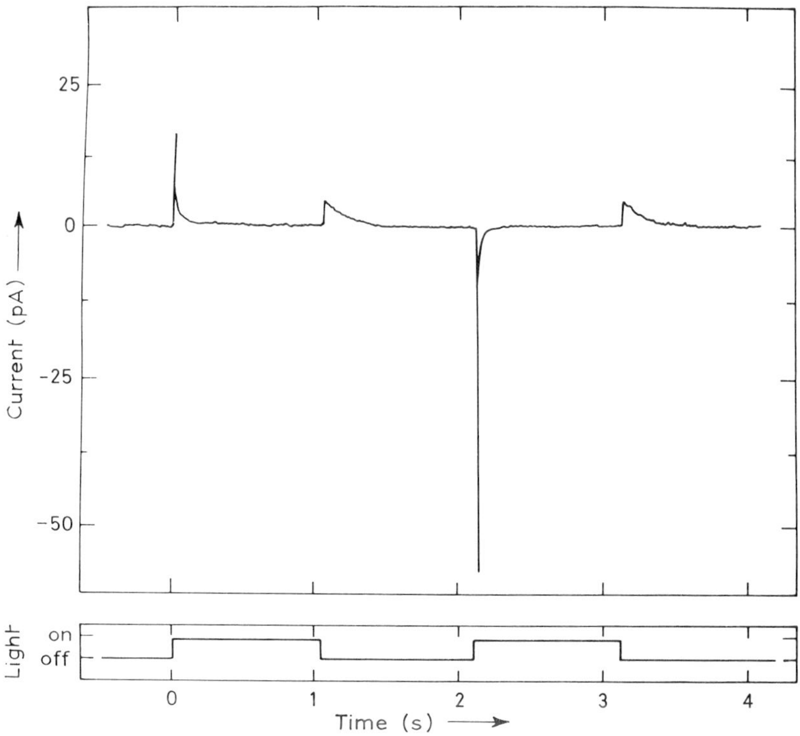


**Fig. 5.4** (b) Photocurrent response to a periodic pulse sequence (as shown) in the presence of reduced cytochrome *c* in one compartment. During the first pulse, the RCs are functionally oriented by cytochrome *c* reducing $D^+$ in the RC population in contact with it. During the second pulse only those RCs that did not react with cytochrome *c* are active. The time constant of the charge recombination event can be measured at the cessation of the second light pulse. (Modified from Feher *et al.*, 1981; see Gopher *et al.*, 1985.)

(b) *Pulsed light conditions*

Using a series of single flashes spaced 25 ms apart it was shown that the first flash produced a small response while subsequent flashes elicited much larger current transients. However, if the membrane was supplemented with ferricyanide on the side opposite to that where ferrocytochrome *c* was added, the light-induced photocurrents were of the same amplitude as the first one (Packham *et al.*, 1980). These results indicated that the symmetrical planar membrane contained two RC populations of opposite orientation: on the first turnover, both populations of RCs were activated but their opposing current transients effectively cancelled. However, by the second turnover, only the RC population with an orientation accessible to reduced cytochrome *c* was returned to a functional state while the other was still in the photochemically inactive oxidized state. Addition of the membrane impermeant potassium ferricyanide to the side opposite to that containing reduced cytochrome *c* chemically oxidized only the RC population inaccessible to reduced cytochrome *c* and thereby allowed a net current transient following the first turnover (Packham *et al.*, 1980, 1982). Thus, functional orientation of RCs in planar bilayers can be obtained either by the selective reduction of D with reduced cytochrome *c* or by the selective oxidation of D with ferricyanide. Under these conditions one can study a single population of functionally oriented RCs in the bilayer and attempt to correlate the electrical response with the known electron-transfer reactions.

The electron transfer from the primary donor to the primary quinone acceptor, $Q_A$, was studied by Packham *et al.* (1982), under the conditions of selective oxidation with ferricyanide of one RC population. The photocurrent response showed a transient peak that relaxed to zero current followed by a transient reverse current of opposite polarity detected at the termination of the light pulse. The dark reverse current had a relaxation half-time of 60 ms, a time which is in agreement with that determined for the back-reaction from reduced $Q_A$ to $D^+$ in octane-solubilized RCs (Packham, 1980; Kendall-Tobias *et al.*, 1981) and in the chromatophore membrane (Blankenship and Parson, 1979). Thus, the decay time constant of this transient is a measure of the intrinsic rate of this electron-transfer reaction. The time course of this dark reaction was also obtained from the recovery kinetics of photocurrent activity between two light pulses (Packham *et al.*, 1982).

Electron transfer from D to $Q_A$ was also studied in functionally oriented RCs where the strategy of selective reduction of one population of RCs with cytochrome *c* was used: Fig. 5.4B shows the photocurrent from reconstituted RCs in lipid bilayers illuminated with the indicated periodic pulse sequence. In the first pulse, both RC populations underwent one electron transfer in opposite directions, an event which was electrically silent. The reduction by cytochrome *c* of the $D^+$ population in contact with it gave rise to a photocurrent indicating, therefore, that the reduction of $D^+$ by cytochrome *c* is electrogenic, as was suggested before (Jackson and Dutton, 1973; Packham *et al.*, 1982; Tiede *et al.*, 1982; Blatt *et al.*, 1983). The dark reverse current at the cessation of the first pulse has a decay time constant of about 100 ms*. This time constant is in agreement with that for the charge recombination between the reduced $Q_A$ and $D^+$ (Gopher *et al.*, 1983, 1985; Blatt *et al.*, 1983). The RC population reduced by cytochrome *c* cannot undergo the charge recombination at the cessation of the pulse or additional charge separations due to the lack of secondary acceptor and it is, therefore, practically inactivated. In the other population, that which was not reduced by cytochrome *c*, D remained oxidized as long as the illumination continued and

*A decay time constant of about 100 ms corresponds to a half-time of about 60 ms as reported by Packham *et al.* (1980, 1982).

photocurrents due to charge recombination were detected upon the cessation of the illumination. The photocurrents generated by the second pulse were, effectively, single turnover events due to electron transfer in those RCs that were not reduced by cytochrome *c*. This double-pulse experiment permits the recording of the light-driven charge separation from $D^+$ to $Q_A$ at the onset of the light pulse and the kinetics of the charge recombination event at the cessation of the pulse (Feher *et al.*, 1980, 1981; Gopher *et al.*, 1983, 1985; Blatt *et al.*, 1983).

Electron transfer from $D^+$ to the secondary quinone, $Q_B$, was studied using the recovery kinetics of the photocurrent activity between light pulses (Packham *et al.*, 1982). Addition of ubiquinone-10 led to an increase in the peak current presumably due to a partial reconstitution of electron transfer from $Q_A$ to $Q_B$. However, the recovery of the peak current would not fit with a single exponential function: a fraction of the peak current (about 30%) was recovered with a 60 ms half-time, a time characteristic for RCs containing only $Q_A$. The residual current recovered with a half-time of 1 s, a time characteristic for the charge recombination from reduced $Q_B$ to $D^+$. Electron transfer from $Q_A$ to $Q_B$ was electrically silent as determined either in RCs devoid of $Q_B$ or using *o*-phenanthroline, a known blocker of electron transfer between $Q_A$ and $Q_B$ (Packham *et al.*, 1980, 1982; Tiede *et al.*, 1982; Blatt *et al.*, 1983; Appell *et al.*, 1983).

Thus, electron-transfer reactions from cytochrome *c* to the primary donor, from $D^+$ to $Q_A$ and to $Q_B$ as well as the back-reactions from $Q_A^-$ to $D^+$ and from $Q_B^-$ to $D^+$, have been measured. Among these electron-transfer events, the cytochrome *c* to $D^+$ and the $D^+$ to $Q_A^-$ emerge as electrogenic steps whereas the electron transfer between $Q_A^-$ and $Q_B$ appears to be electrically silent. Similar conclusions were obtained by studying the photovoltages from RCs incorporated into interfacial lipid layers (Tiede *et al.*, 1982; Blatt *et al.*, 1983). Appell *et al.* (1983) have provided a detailed analysis of the kinetics of the reconstituted membrane.

One of the salient features of the planar lipid bilayer is its inherent accessibility for the application of voltages across the membrane and for the investigation of the effect of transmembrane voltages on the electron-transfer reactions proceeding in the reconstituted RCs. As yet, the evidence is solely based on the effect of diffusion potentials generated across the chromatophore membrane on the intensity and the kinetics of delayed light emission (Fleischman, 1971; Crofts *et al.*, 1972) or exogenous voltage-sensitive dyes (Chance and Baltscheffsky, 1975). Gopher *et al.* (1983, 1985) used a double-pulse strategy (Fig. 5.4B) to study the rate of charge recombination in functionally oriented reconstituted RCs in which the primary acceptor quinone was either ubiquinone or anthraquinone. The experimental design was based on the report of Gunner *et al.* (1982) that the recombination rate increased 10-fold for every 60 mV decrease in quinone redox potential. Gopher *et al.* (1983, 1985) showed that, when $Q_A$ was the natural ubiquinone-10, the rate of charge recombination was unaffected by applied voltages up to $\pm 125$ mV. However, when $Q_A$ was anthraquinone then an e-fold change in the apparent rate of charge recombination was measured for an applied voltage of 175 mV. The experiment showed that when the potential was negative in the compartment facing the quinone-binding site in the RC, the recombination process was accelerated, whereas when the potential in this compartment was inverted, the rate was decelerated. Thus, the interplay between the photoinduced transmembrane electric current and transmembrane applied voltages has now been directly demonstrated.

In summary, RCs reconstituted in planar lipid bilayers generate photoelectric signals that arise from the transfer of electrons within the RCs as well as from secondary donors to secondary acceptors located on opposite sides of the membrane and, therefore, provide a direct demonstration for the transduction of light energy into transmembrane electric current.

In this section, the main strategy for the assembly of the membrane protein in planar lipid bilayers was that of extracting lipid–protein complexes into organic solvents. This approach has also been used for other membrane proteins: cytochrome *c* oxidase (Montal, 1974a; Chien and Mueller, 1976); bacteriorhodopsin (Hwang *et al.*, 1977; Bamberg *et al.*, 1981); the voltage-dependent anion channel of mitochondria (Schein *et al.*, 1976; Colombini, 1979).

## 5.6 The voltage-dependent sodium channel

The processes which underlie the well-known phenomenon of nerve electrical excitability are associated with the changes in membrane ionic conductance in response to changes in membrane potential. These membrane conductances are produced by distinct, ion-selective channels formed by integral membrane proteins that span the bilayer to provide the permeability pathway for cation translocation. There is a considerable body of evidence in support of the view that these channels are somehow opened or closed, i.e. gated, by the membrane potential. The control of the gating mechanism by the membrane potential exhibits conspicuous characteristics which were first recorded and extensively analysed in the squid axon (Hodgkin and Huxley, 1952) and have been observed in other cells irrespective of the ionic selectivity of the different channels. The $Na^+$ channel of electrically excitable cells and tissues has been studied in great electrophysiological and biophysical detail. The advent of the patch clamp recording technique has allowed measurements of single $Na^+$ channel currents in a variety of excitable cells (Sigworth and Neher, 1980; Horn *et al.*, 1981; Fukushima, 1981; Quandt and Narahashi, 1982; Huang *et al.*, 1982). In contrast, the biochemical identification and characterization of the molecular entities responsible for the voltage-sensitive conductance changes have been lagging behind. In the last few years, several groups reported techniques for the identification, isolation and purification of the protein components of the $Na^+$ channel. The availability of specific neurotoxins that interact with the channel protein has allowed their use as molecular tools to identify the channels electrophysiologically, as markers during the purification of the channel and as specific functional–pharmacological assays in reconstitution studies. Three distinct neurotoxin-binding sites on the $Na^+$ channel have been characterized (for reviews see Catterall, 1980, and Narahashi, 1974). Site 1 binds saxitoxin (STX) and tetrodotoxin (TTX) which block $Na^+$ flux. Site 2 binds the lipid-soluble alkaloid toxins bathracotoxin (BTX) and veratridine, which shift the voltage-dependence of $Na^+$ activation to more negative membrane potentials and block inactivation. The combination of these two effects causes the $Na^+$ channel to be persistently activated and, therefore, these toxins can be used as operationally effective agonists for the $Na^+$ channel. The action of BTX is inhibited by TTX in a non-competitive fashion (Alburquerque and Daly, 1976; Khodorov, 1978). Site 3 binds the polypeptide scorpion toxins as well as the sea-anemone toxin, which leads to an inhibition of the inactivation of the $Na^+$ channel (see Catterall, 1980).

The neurotoxin-binding components of the $Na^+$ channel have been solubilized and purified from electric eel electroplax (Agnew *et al.*, 1978; Miller *et al.*, 1983; Nakayama *et al.*, 1982), rat sarcolemma (Barchi *et al.*, 1980; Weigele and Barchi, 1982; Barchi, 1983) and rat brain (Hartshorne and Catterall, 1981, 1984; Hartshorne *et al.*, 1982). The $Na^+$ channel protein has been solubilized from membranes with a variety of non-ionic detergents and has been partially purified by ion-exchange chromatography and affinity chromatography leading to purifications of up to about 1000-fold. Analysis of the polypeptide composition of the partially purified preparations from eel electroplax and rat brain identified a large glycoprotein of $M_r$ about 250 000–270 000, designated as the α-subunit of the $Na^+$ channel. In preparations from rat brain and rat skeletal muscle sarcolemma, two smaller peptides were identified as components of the purified $Na^+$ channel: for rat brain, a $\beta_1$ polypeptide of $M_r$ 39 000 and a $\beta_2$ component of $M_r$ about 37 000 which is linked to the α subunit via a disulphide bond (Hartshorne *et al.*, 1982; Hartshorne and Catterall, 1984).

Recently, $Na^+$ channel preparations from rat muscle sarcolemma (Weigele and Barchi, 1982; Tanaka *et al.*, 1983; Barchi and Tanaka, 1984) and rat brain (Talvenheimo *et al.*, 1982; Catterall *et al.*, 1984) were reconstituted into phospholipid vesicles. The solubilized membranes in the non-ionic detergent were supplemented with phospholipids. The detergent–lipid mixtures were used to preserve the labile toxin-binding activity of the $Na^+$ channel, as initially demonstrated by Agnew *et al.* (1978). These reports must be distinguished from previous claims of recovery of functional activity from crude membrane extracts. These include reports on the 'reconstitution' of $Na^+$ channel function determined as the recovery of veratridine-stimulated $Na^+$ uptake into vesicles containing membrane fragments from the lobster walking nerve (Villegas *et al.*, 1970) or from cholate extracts of brain membranes (Malysheva *et al.*, 1980). Goldin *et al.* (1980) showed a veratridine-stimulated transport of $Cs^+$ into PC vesicles containing cholate-solubilized brain membranes. Tamkun and Catterall (1981) demonstrated neurotoxin binding at all three toxin receptor sites in cholate-solubilized $Na^+$ channels from rat brain synaptosomes incorporated into PC vesicles which also exhibited neurotoxin-activated $Na^+$ uptake. All these studies demonstrate that the activity of the $Na^+$ channel can be preserved during solubilization of crude membrane extracts and after the incorporation into lipid vesicles by detergent removal.

Weigele and Barchi (1982) reconstituted the purified preparation from rat muscle into PC vesicles and showed that BTX increased the uptake of radioactive $Na^+$ into the vesicles and that STX specifically and quantitatively blocked the BTX-stimulated $Na^+$ influx. The concentration dependence for a BTX-stimulated flux response showed an apparent $K_d$ of 1.5 $\mu$M. Veratridine also stimulated the uptake of $Na^+$ with an apparent $K_d$ of 35 $\mu$M, and was also blocked by STX. Tanaka *et al.* (1983) further characterized the cation selectivity of the reconstituted vesicles. Quenched flow kinetic techniques were used to attempt a time resolution of the kinetics of cation translocation. At 181° C, the half-times for the uptake for cation equilibration were 350 ms, 2.5 s and 10 s for $K^+$, $Rb^+$ and $Cs^+$, respectively. The equilibrations of $Na^+$ occurred within the time resolution of the technique and an upper limit of 50 ms was estimated. These results led to the following cation selectivity sequence for the BTX-activated channel: $Na^+ > K^+ > Rb^+ > Cs^+$. The BTX-stimulated uptake was blocked by STX with a $K_i$ of 5 nM at 18° C. These results indicate that the reconstituted STX-binding protein from rat skeletal muscle sarcolemma exhibits characteristics attributed to a functional $Na^+$ channel: i.e. selective cation translocation activated by BTX and veratridine and blocked by STX at concentrations comparable to those required to block the $Na^+$ channel *in situ*.

The $Na^+$ channel preparation from rat brain was also reconstituted into PC vesicles. The reconstituted vesicles exhibited a TTX-sensitive and veratridine-stimulated $Na^+$ uptake. $K_i$ for the TTX inhibition was 15 nM, a comparable concentration to that required in native membranes. In addition, the veratridine-activated $Na^+$ channels expressed cation selectivity with the apparent sequence of $Na^+ > Rb^+ > Cs^+$ (Talvenheimo *et al.*, 1984). These results show that the purified $Na^+$ channel from rat brain composed of $\alpha\beta_1\beta_2$ subunits contains the components necessary for the neurotoxin-activated and selective-ion permeability (Talvenheimo *et al.*, 1984).

The ultimate criterion for reconstitution is to mimic in detail the native $Na^+$

channel. This requires measuring the function of single $Na^+$ channels, and the determination of the voltage-dependent activation and inactivation of the channel. The conductance of single $Na^+$ channels has recently been measured in small patches of a variety of cells (Sigworth and Neher, 1980; Patlak and Horn, 1982; Fukushima, 1981; Quandt and Narahashi, 1982). Therefore, the $Na^+$ channel has the advantage of a sound biophysical background and the availability of neurotoxins that provide both a specific assay and an end point in the reconstitution attempts. A first step in the reconstitution of the $Na^+$ channel protein was taken with the demonstration that $Na^+$ channels can be transferred from crude brain membrane vesicles into preformed black lipid membranes by the fusion technique (Krueger *et al.*, 1983; French *et al.*, 1984). We now turn to describe these results in more detail to illustrate the fusion approach.

The strategy of membrane fusion basically consists of making membrane vesicles interact with preformed planar bilayers in the presence of a negatively charged lipid, calcium ions and osmotic conditions which promote the swelling of the vesicles. Fusion is believed to occur by the osmotic swelling of the vesicles in contact with the planar membrane, with subsequent rupture of vesicular and planar membranes in the region of contact (Drachev *et al.*, 1974, 1981; Miller and Racker, 1976; Miller *et al.*, 1976; Miller, 1978; Zimmerberg *et al.*, 1980; Cohen *et al.*, 1980, 1982).

The procedure used by Krueger *et al.* (1983) and French *et al.* (1984) was as follows: BLMs were formed from a decane solution of bovine brain phosphatidylethanolamine (PE) and phosphatidylserine (PS), at a weight ratio of 5:2. The ionic composition of the aqueous medium was 0.5 M NaCl, 0.15 mM $CaCl_2$, 0.1 mM $MgCl_2$ at pH. 7.0. The membrane vesicles were prepared from rat brain homogenates and suspended in 0.4 M sucrose. This fraction was enriched in STX-binding sites about 5-fold relative to the crude homogenate. This preparation was introduced into one of the aqueous compartments of the bilayer chamber, the *cis* chamber, and 0.6 $\mu$M BTX was present in the opposite compartment, the *trans* side. Without further manipulations, a stepwise increase in membrane conductance was recorded. Under these conditions, BTX induced a steady-state macroscopic current that could be virtually blocked by addition of 0.3 $\mu$M STX, to the *cis* side. Half-maximal block of the conductance occurred at about 5 nM STX (at an applied voltage of +30 mV) a value which is similar to the $K_i$ for block of macroscopic $Na^+$ current in nerves by STX and to the $K_d$ for binding of radiolabelled STX. All the channels incorporated in the bilayer appear to be oriented with the STX-blocking sites facing the *cis* compartment: addition of STX to the *trans* side was ineffective. Discrete current transitions due to the opening and closing of single BTX-activated $Na^+$ channels were observed when very low protein concentrations were used (0.2 $\mu$g/ml) and incubation periods of up to 30 min. The single-channel conductance in 0.5 M NaCl was 30 pS, a value consistent with data obtained by patch recordings from cells considering the saturating conductance–activity relationship (Sigworth and Neher, 1980; Patlak and Horn, 1983; Hille, 1975). The BTX-activated channel is cation-selective, as determined from reversal potential measurements of the single-channel currents in a NaCl gradient. The transference number for $Na^+$ under the conditions tested was 0.93 and the estimated permeability ratio of $K^+/Na^+$ was 0.06. The BTX-activated channel was voltage-dependent: the probability of the channel being found in the closed state increased as the transmembrane potential was more negative than $-60$ mV (in the direction of hyperpolarization). From the conductance voltage relationship a value for the voltage at which the channel was half-open of $-93$ mV was obtained (Krueger *et al.*, 1983; French *et al.*, 1984). The neurotoxin sensitivity, ion selectivity and voltage-dependent gating of the incorporated channel are consistent with the notion that the incorporated entities are associated with a functional $Na^+$ channel.

The studies of Krueger *et al.* (1983) and French *et al.* (1984) have already provided new information on the $Na^+$ channel: blocking by STX of the incorporated $Na^+$ channels was voltage-dependent, with hyperpolarizing potentials promoting the block. This phenomenon together with several other properties of the incorporated $Na^+$ channel are illustrated in Fig. 5.5 (taken from French *et al.*, 1984). The upper trace illustrates the membrane currents in the control preparation starting at an applied voltage of $+60$ mV (depolarizing potential). After several seconds, the potential was turned off for a few seconds and then switched to a hyperpolarizing potential, $-60$ mV. This record illustrates that the BTX-activated channels were preferentially open with very brief closing events (transitions towards the zero-applied voltage current level). These transitions were too fast to be resolved at the bandwidth of these recordings but at higher resolution, discrete current steps with a single-channel conductance of 30 pS in 0.5 M NaCl were measured. These rapid fluctuations were presumably due to the voltage-dependent gating of the BTX-activated channel. It is worth noting that in the studied range of $-60$ mV to $+60$ mV, the channels were open almost all of the time (more than 98%). This feature allowed the study of the voltage-dependence of the block exerted by STX in the steady-state. This is shown in the lower trace obtained in the presence of 1 nM STX where a number of closing events of rather long duration are evident. These long closed periods were promoted by increasing the STX concentration, and therefore, are presumed to be caused by STX block. This record also shows the voltage-dependence of the STX block, since at $+60$ mV the channels are preferentially open while at $-60$ mV, and after several seconds, long closed periods occur. A detailed analysis of the voltage-dependence of the STX block demonstrated that the equilibrium dissociation constants computed from association (blocking) and dissociation (unblocking) rate constants for the STX block agreed with those determined from a steady-

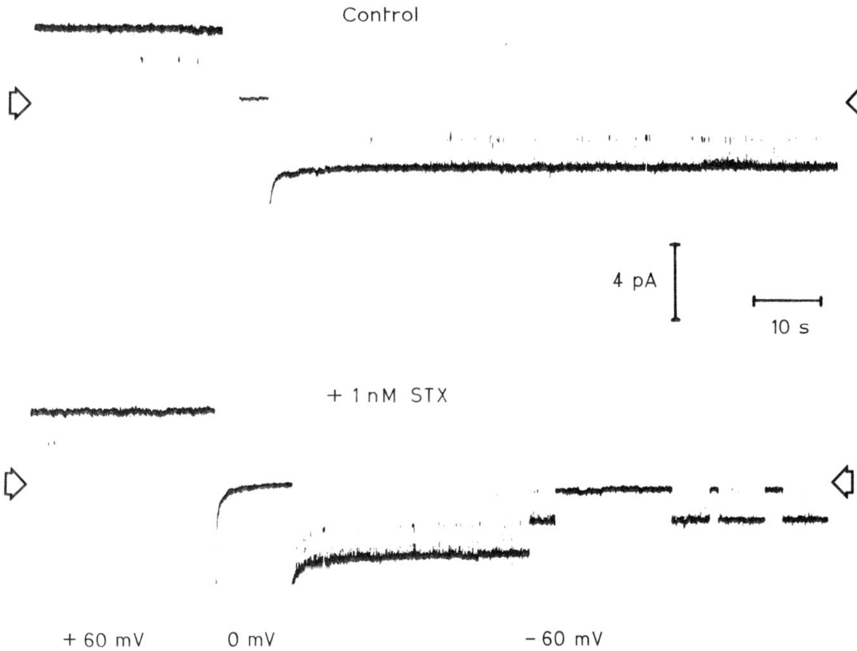


**Fig. 5.5** Current fluctuations from a membrane containing two BTX-activated $Na^+$ channels. Zero current levels are shown by the arrows. Top trace: control record. The channels remained open almost all of the time, with brief closings towards the zero-current level. These closings were presumably due to the voltage-dependent gating of the channels. Lower trace: record taken after addition of 1 nM STX to the *cis* side. The long closing steps, including an extended period during which neither channel was conducting, were most likely due to the block of individual channels resulting from the binding of STX. Records were filtered (low pass) at 60 Hz on playback. (Reproduced with permission of R. French and B. Krueger, and the Rockefeller University Press, from French *et al.*, 1984.)

state fractional reduction in the current (French *et al.*, 1984). The results are consistent with a model in which the divalent cation toxin senses about 30–40% of the transmembrane voltage (French *et al.*, 1984). This report illustrates how the model system can be used to learn about the properties of the $Na^+$ channel which are not readily accessible in the native membrane.

These experiments demonstrated that the fusion technique can be successfully used to incorporate functional $Na^+$ channels from crude membrane vesicles into the planar bilayer. Recently, this approach was applied to the reconstituted vesicles containing the purified and well-characterized brain $Na^+$ channel (Hartshorne *et al.*, 1984, 1985): Fig. 5.6 illustrates single $Na^+$ channel currents activated by 2 μM BTX and recorded at a potential of −70 mV (electrophysiological convention). A section of the record is shown below at faster time resolution. The single channel

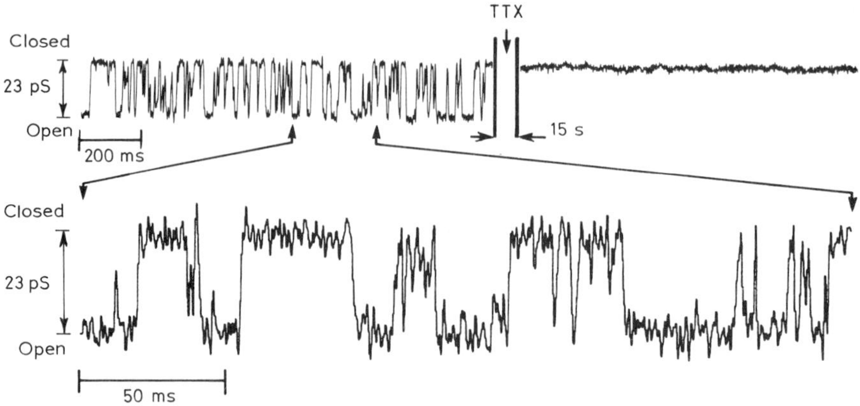


**Fig. 5.6** Single sodium channel currents from purified brain sodium channels reconstituted in planar lipid bilayers. Sodium channel currents were activated by 2 $\mu$M BTX and recorded at a potential of $-70$ mV (electrophysiological convention). A section of the record (filtered at 500 Hz) is shown below at faster time resolution. The single-channel conductance was 23 pS in a 0.5 M/0.2 M NaCl gradient. Channel openings were abolished by the addition of 3 $\mu$M TTX. (Reproduced with permission of The Society for Neuroscience.)

conductance is 23 pS in a 0.5 M/0.2 M NaCl gradient. Channel openings are abolished by the addition of 3 $\mu$M TTX. The $K_i$ for TTX inhibition is about $10^{-8}$ M at $-60$ mV while at $+60$ mV it is about $10^{-7}$ M. $Na^+$ channel gating is voltage-dependent in the presence of BTX with negative potentials increasing the probability of channel closure. The channels are closed 50% of the time at $-98$ mV $\pm 14$ mV. Thus, purified $Na^+$ channels in planar lipid bilayers display the known single channel conductance, neurotoxin sensitivity and voltage dependence associated with native $Na^+$ channels. The kinetic parameters of channel function as derived from single-channel recordings in combination with biochemical studies of the $Na^+$ channel protein will provide further understanding of the molecular basis underlying channel function and will provide a system for the study of the structure/function correlations at the level of the $Na^+$ channel protein.

The technique of fusion just described for the rat brain $Na^+$ channel has also been used to incorporate other channels: the $Cl^-$ channel from *Torpedo* electroplax was incorporated into both BLMs (White and Miller, 1979; Miller, 1982a) and into phospholipid vesicles made from crude cholate extracts of *Torpedo* electroplax plasma membranes (Tank *et al.*, 1982); the $K^+$ channel of sarcoplasmic reticulum (Labarca *et al.*, 1980; Miller, 1982b; Coronado and Miller, 1979; Coronado *et al.*, 1980; Bell and Miller, 1984); the $K^+$ channel from lobster axon membranes (Coronado *et al.*, 1984); a $Ca^{2+}$-activated $K^+$ channel from rat muscle and rabbit muscle (Latorre *et al.*, 1982; Vergara *et al.*, 1984); a voltage-dependent anion channel (VDAC) from rat liver mitochondria (Cohen *et al.*, 1980); and bacteriorhodopsin (see Bamberg *et al.*, 1981).

## 5.7 Conclusion and perspectives

This chapter has illustrated a variety of techniques used to reconstitute well-defined membrane proteins into planar lipid bilayers. Three membrane proteins have been discussed in more detail where each one of them exemplified a particular reassembly strategy. The AChR from *Torpedo californica* was reconstituted by transforming reconstituted vesicles into a planar lipid bilayer; the photosynthetic RCs were reconstituted into planar bilayers by first extracting them into organic solvents in the form of a lipid–protein complex; and the voltage-dependent $Na^+$ channel from rat brain was incorporated into preformed planar lipid bilayers by 'fusing' reconstituted vesicles with the preformed membrane. These three examples illustrate the current availability of various techniques which should be tested for any particular protein of interest: while one method may be suitable for a given membrane protein, it may not be so for another.

The techniques here described have come a long way: from a novel venue to merge biochemical and biophysical knowledge on membrane proteins to an established technology with a fairly substantiated theoretical framework. These advances validate the notion that the molecular physiology of signal transduction and energy transduction can be investigated at the level of the membrane protein itself. It is anticipated that the planar lipid bilayer in combination with the biochemical/structural characterization of membrane proteins will bring us one step further towards the understanding of the molecular basis of membrane protein function.

## Acknowledgements

I acknowledge with gratitude the collaboration of P. Labarca, J. Lindstrom and Myrta Montal on the acetylcholine receptor project, of M. Schönfeld, A. Gopher, Y. Blatt and G. Feher on the reaction centre project, and R. Hartshorne, B. Keller and W. Catterall on the sodium channel project. I am indebted to R. French and B. Krueger for allowing me to include their results, to R. French, B. Krueger and H. Schindler for allowing me to read work in progress, and to Marcia Chadwick for her invaluable assistance. This work was supported by grants from the National Institutes of Health (EY-02084) and The Department of the Army Medical Research (17-82-C221).

## 5.8 References

Agnew, W., Levinson, S. R., Brabson, J. S. and Raftery, M. A. (1978) *Proc. Natl. Acad. Sci. USA*, **75**, 2606–10.

Alburquerque, E. X. and Daly, J. (1976) in *The Specificity and Action of Animal, Bacterial and Plant Toxins. Receptors and Recognition Series B* (ed P. Cuatrecasas), Chapman and Hall, London, Vol. 1, pp. 297–338.

Anholt, R., Fredkin, D. R., Deerinck, T., Ellisman, M., Montal, M. and Lindstrom, J. (1982) *J. Biol. Chem.*, **257**, 7122–34.

Anholt, R., Lindstrom, J. and Montal, M. (1980) *Eur. J. Biochem.*, **109**, 481–7.

Anholt, R., Lindstrom, J. and Montal, M. (1981) *J. Biol. Chem.*, **256**, 4377–87.

Anholt, R., Lindstrom, J. and Montal, M. (1985) in *The Enzymes of Biological Membranes* (ed A. Martonosi), 2nd edn., Plenum Press, New York pp. 335–401.

Anholt, R., Montal, M. and Lindstrom, J. (1983) *Peptide Protein Rev.*, **1**, 95–137.

Appell, H.-J., Snozzi, M. and Bachofen, R. (1983) *Biochim. Biophys. Acta*, **724**, 258–77.

Bamberg, E., Dencher, N. A., Fahr, A. and Heyn, M. P. (1981) *Proc. Natl. Acad. Sci. USA*, **78**, 7502–6.

Barchi, R. L. (1983) *J. Neurochem.*, **40**, 1377–85.

Barchi, R. L., Cohen, S. and Murphy, L. (1980) *Proc. Natl. Acad. Sci. USA*, **77**, 1306–10.

Barchi, R. L. and Tanaka, J. C. (1984) *Biophys. J.*, **45**, 35–7.

Bell, J. E. and Miller, C. (1984) *Biophys. J.*, **45**, 279–287.

Blankenship, R. E. and Parson, W. W. (1979) in *Photosynthesis in Relation to Model Systems* (ed J. Barber), Elsevier/North Holland Biomedical Press, Amsterdam, pp. 71–114.

Blatt, Y., Gopher, A., Montal, M. and Feher, G. (1983) *Biophys. J.*, **41**, 121a.

Boheim, G., Hanke, W., Barrantes, F. J., Eibl, H., Sakmann, B., Fels, G. and Maelicke, A. (1981) *Proc. Natl. Acad. Sci. USA*, **78**, 3586–90.

Catterall, W. A. (1980) *Annu. Rev. Pharmacol. Toxicol.*, **20**, 15–43.

Chance, B. and Baltscheffsky, M. (1975) *Biomembranes*, **7**, 33–50.

Chang, H. W. and Bock, E. (1977) *Biochemistry*, **18**, 155–63.

Changeux, J.-P. (1981) *Harvey Lect.*, **75**, 85–254.

Changeux, J.-P., Heidmann, T., Popot, J.-L. and Sobel, A. (1979) *FEBS Lett*, **105**, 181–7.

Chien, T. F. and Mueller, P. (1976) *Fed. Proc.*, **35**, 1599.

Claudio, T., Ballivet, M., Patrick, J. and Heinemann, S. (1983) *Proc. Natl. Acad. Sci. USA*, **80**, 111–15.

Cohen, F. S., Akabas, M. and Finkelstein, A. (1982) *Science*, **217**, 458–60.

Cohen, F. S., Zimmerberg, J. and Finkelstein, A. (1980) *J. Gen. Physiol.*, **77**, 251–70.

Colombini, M. (1979) *Nature (London)*, **279**, 643–5.

Colquhoun, D. and Sakmann, B. (1981) *Nature (London)*, **294**, 464–6.

Coronado, R. and Latorre, R. (1983) *Biophys. J.*, **43**, 231–6.

Coronado, R., Latorre, R. and Mautner, H. G. (1984) *Biophys. J.*, **45**, 289–299.

Coronado, R. and Miller, C. (1979) *Nature (London)*, **280**, 807–10.

Coronado, R., Rosenberg, R. and Miller, C. (1980) *J. Gen. Physiol.*, **76**, 425–46.

Crofts, A. R., Jackson, J. B., Evans, E. H. and Cogdell, R. J. (1972) *Proc. Int. Congr., 2nd, Photosynth.*, 873–902.

Darszon, A., Philipp, M., Zarco, J. and Montal, M. (1978) *J. Membr. Biol.*, **43**, 71–90.

Devilliers-Thiery, A., Giraudat, J., Bentaboulet, M. and Changeux, J.-P. (1983) *Proc. Natl. Acad. Sci. USA*, **80**, 2067–71.

Drachev, L. A., Jesaitis, A. A., Koulen, A. D., Kondrashin, H. A., Liberman, E. H., Hayrecek, I. B., Ostrounov, S. A., Semenov, A. Y. and Skulachev, V. P. (1974) *Nature (London)*, **249**, 321–3.

Drachev, L. A., Semenov, A. Y., Skulachev, V. P., Smirnova, I. A., Chamorovsky, S. K., Kononenko, A. A., Rubin, A. B. and Uspenskaya, N. Y. (1981) *Eur. J. Biochem.*, **117**, 483–9.

Epstein, M. and Racker, E. (1978) *J. Biol. Chem.*, **253**, 6660–2.
Feher, G. and Okamura, M. Y. (1978) in *The Photosynthetic Bacteria* (eds W. R. Sistrom and R. K. Clayton), Plenum Press, New York, pp. 349–86.
Feher, G., Schönfeld, M. and Montal, M. (1980) *1st Eur. Bioenergetics Conf.*, Urbino, Italy, June 29–July 5, pp. 419–20.
Feher, G., Schönfeld, M. and Montal, M. (1981) *VII Int. Biophys. Congr. and III Pan-American Biochem. Congr.*, Mexico City, Mexico, Aug. 23–28, 117 pp.
Fleischman, D. E. (1971) *Photochem. Photobiol.*, **14**, 277–86.
French, R. J., Worley, J. F. and Krueger, B. K. (1984) *Biophys. J.*, **45**, 301–10.
Fukushima, Y. (1981) *Proc. Natl. Acad. Sci. USA*, **78**, 1274–7.
Goldin, S. M., Moczydlowski, E. G. and Papazian, D. M. (1983) *Annu. Rev. Neurosci.*, **6**, 419–46.
Goldin, S. M., Rhoden, V. and Hess, E. J. (1980) *Proc. Natl. Acad. Sci. USA*, **77**, 6884–8.
Gopher, A., Blatt, Y., Okamura, M. Y., Feher, G. and Montal, M. (1983) *Biophys. J.*, **41**, 121a.
Gopher, A., Blatt, Y., Schonfeld, M., Okamura, M. Y., Feher, G. and Montal, M. (1985) *Biophys. J.*, **48**, 311–20.
Gunner, M. R., Tiede, D. M., Prince, R. C. and Dutton, P. L. (1982) in *Function of Quinones in Energy Conserving Systems* (ed B. L. Trumpower), Academic Press, New York, pp. 265–9.
Guy, H. R. (1984) *Biophys. J.*, **45**, 249–61.
Hamamoto, T. and Montal, M. (1984) *Biophys. J.*, **45**: Abstr. 38a, 126.
Hamamoto, T., Carrasco, N., Matsushita, K., Kaback, H. R. and Montal, M. (1985) *Proc. Natl. Acad. Sci. USA*, **82**, 2570–3.
Hamill, O., Marty, A., Neher, E., Sakmann, B. and Sigworth, F. J. (1981) *Pfluegers Arch.*, **391**, 85–100.
Hamilton, S. L., McLaughlin, M. and Karlin, A. (1979) *Biochemistry*, **18**, 155–63.
Hargrave, P. A., McDowell, J. H., Curtis, D. R., Wang, J. K., Juszczak, E., Fong, S. L., Mohana Rao, J. K. and Argos, P. (1983) *Biophys. Struct. Mech.*, **9**, 235–44.
Hartshorne, R. P. and Catterall, W. A. (1981) *Proc. Natl. Acad. Sci. USA*, **78**, 4620–4.
Hartshorne, R. P. and Catterall, W. A. (1984) *J. Biol. Chem.*, **259**, 1667–75.
Hartshorne, R., Keller, B., Talvenheimo, J., Catterall, W. and Montal, M. (1984) *Soc. Neurosci.*, Abstr. 10, Part 2, p. 864.
Hartshorne, R. P., Keller, B. U., Talvenheimo, J. A., Catterall, W. A. and Montal, M. (1985) *Proc. Natl. Acad. Sci. USA*, **82**, 240–4.
Hartshorne, R. P., Messner, D. J., Coppersmith, J. C. and Catterall, W. A. (1982) *J. Biol. Chem.*, **257**, 13888–91.
Haydon, D. A. (1970) *Membr. Ion Transp.*, **1**, 64–92.
Hess, G. P., Cash, D. J. and Aoshima, H. (1983) *Annu. Rev. Biophys. Bioengineer.*, **12**, 443–73.
Hess, G. P., Pasquale, E. B., Walker, J. W. and McNamee, M. G. (1982) *Proc. Natl. Acad. Sci. USA*, **79**, 963–7.
Hille, B. (1975) *J. Gen. Physiol.*, **66**, 535–60.
Hodgkin, A. L. and Huxley, A. F. (1952) *J. Physiol. (London)*, **116**, 449–506; **117**, 500–44.
Horn, R., Patlak, J. and Stevens, C. F. (1981) *Nature (London)*, **291**, 426–7.
Huang, L.-Y., M., Moran, N. and Ehrenstein, G. (1982) *Proc. Natl. Acad. Sci. USA*, **79**, 2082–5.

Huganir, R. L. and Racker, E. (1982) *J. Biol. Chem.*, **257**, 9372–8.
Huganir, R. L., Schell, M. A. and Racker, E. (1979) *FEBS Lett.*, **108**, 155–60.
Hwang, S.-B., Korenbrot, J. I. and Stoeckenius, W. (1977) *J. Membr. Biol.*, **36**, 115–58.
Jackson, J. B. and Dutton, P. L. (1973) *Biochim. Biophys. Acta*, **325**, 102–13.
Jackson, M. B., Lecar, H., Askanas, V. and Engel, W. K. (1982) *J. Neurosci.*, **2**, 1465–73.
Jackson, M. B., Wong, B. S., Morris, C. E., Lecar, H. and Christian, C. N. (1983) *Biophys. J.*, **42**, 109–14.
Kasahara, M. and Hinkle, P. C. (1977) *J. Biol. Chem.*, **252**, 7384–90.
Katz, B. and Thesleff, S. (1957) *J. Physiol. (London)*, **138**, 63–80.
Kendall-Tobias, M. W., Celis, H., Celis, S. A. and Crofts, A. R. (1981) *Biochim. Biophys. Acta*, **635**, 585–601.
Khodorov, B. I. (1978) in *Membrane Transport Processes* (ed D. C. Tosteson, Y. A. Ovchinnikov and R. Latorre), Raven Press, New York, Vol. 2, 153–74.
Khorana, H. G., Gerber, G. C., Herlihy, W. C., Gray, C. P., Anderegg, R. J., Nihei, K. and Biemann, K. (1979) *Proc. Natl. Acad. Sci. USA*, **76**, 5046–50.
Krueger, B. K., Worley, J. F. and French, R. J. (1983) *Nature (London)*, **303**, 172–5.
Labarca, P., Coronado, R. and Miller, C. (1980) *J. Gen. Physiol.*, **76**, 397–424.
Labarca, P., Fredkin, D. R., Lindstrom, J. and Montal, M. (1983) *Biophys. J.*, **41**, 135a.
Labarca, P., Lindstrom, J. and Montal, M. (1981) *Abstr. VII Int. Biophys. Congr. and III Pan-American Biochem. Congr.*, Mexico City, Mexico, 258 pp.
Labarca, P., Lindstrom, J. and Montal, M. (1982) *Biophys. J.*, **37**, 170a.
Labarca, P., Lindstrom, J. and Montal, M. (1984a) *J. Neurosci.*, **4**, 502–7.
Labarca, P., Lindstrom, J. and Montal, M. (1984b) *J. Gen. Physiol.*, **83**, 473–96.
Latorre, R. and Alvarez, O. (1981) *Physiol. Rev.*, **61**, 77–150.
Latorre, R. and Miller, C. (1983) *J. Membr. Biol.*, **71**, 11–30.
Latorre, R., Vergara, C. and Hidalgo, C. (1982) *Proc. Natl. Acad. Sci. USA*, **77**, 7484–6.
Lindstrom, J., Anholt, R., Einarson, B., Engel, A., Osame, M. and Montal, M. (1980) *J. Biol. Chem.*, **255**, 8340–50.
Lindstrom, J., Merlie, J. and Yogeeswaran, G. (1979) *Biochemistry*, **18**, 4465–70.
Lüdi, H., Oetliker, H., Brodbeck, U., Ott, P., Schwendimann, B. and Fulpius, B. W. (1983) *J. Membr. Biol.*, **74**, 75–84.
Malysheva, M. K., Lishko, U. K. and Chagovetz, A. M. (1980) *Biochim, Biophys. Acta*, **602**, 70–7.
McNamee, M. G. and Ochoa, E. L. M. (1982) *Neuroscience*, **7**, 2305–19.
McNamee, M. G., Richardson, C. A. and Walker, J. W. (1984) *Biophys. J.*, **45**, 18–20.
Miller, C. (1978) *J. Membr. Biol.*, **40**, 1–23.
Miller, C. (1982a) *Philos. Trans. R. Soc. London*, **299**, 401–11.
Miller, C. (1982b) *J. Gen. Physiol.*, **38**, 869–91.
Miller, C. (1983) in *Current Methods in Cellular Neurobiology* (eds J. Barker and J. F. McKelvey), Wiley, New York (in press).
Miller, C. (1986) *Ion Channel Reconstitution*, Plenum Press, New York.
Miller, C., Arvan, P., Telford, J. N. and Racker, E. (1976) *J. Membr. Biol.*, **30**, 271–82.
Miller, C. and Racker, E. (1976) *J. Membr. Biol.*, **30**, 283–300.
Miller, J. A., Agnew, W. S. and Levinson, S. R. (1983) *Biochemistry*, **22**, 462–70.
Mitchell, P. (1968) *Chemiosmotic coupling and energy transduction*, Glynn Research, Bodmin.
Montal, M. (1974a) in *Perspectives in Membrane Biology*, Academic Press, New York, pp. 591–622.

Montal, M. (1974b) *Methods Enzymol.*, **32**, 545–56.
Montal, M., Darszon, A. and Schindler, H. (1981) *Q. Rev. Biophys.*, **14**, 1–79.
Montal, M., Labarca, P., Fredkin, D. R., Suarez-Isla, B. A. and Lindstrom, J. (1984) *Biophys. J.*, **45**, 165–74.
Montal, M. and Mueller, P. (1972) *Proc. Natl. Acad. Sci. USA*, **69**, 3561–6.
Mueller, P. and Rudin, D. O. (1969) in *Current Topics in Bioenergetics* (ed D. R. Sanadi) Academic Press, New York, Vol. 3, pp. 157–249.
Mueller, P., Rudin, D. O., Tien, H. Ti. and Wescott, W. C. (1962) *Nature (London)*, **194**, 979–81.
Nakayama, H., Witby, R. and Raftery. M. A. (1982) *Proc. Natl. Acad. Sci. USA*, **79**, 7575–9.
Narahashi, T. (1974) *Physiol. Rev.*, **54**, 813–89.
Neher, E. and Sakmann, B. (1976a) *Nature (London)*, **260**, 799–802.
Neher, E. and Sakmann, B. (1976b) *J. Physiol.*, **258**, 705–29.
Nelson, N., Anholt, R., Lindstrom, J. and Montal, M. (1980) *Proc. Natl. Acad. Sci. USA*, **77**, 3057–61.
Noda, M., Takahashi, H., Tanabe, T., Toyosato, M., Fusutani, Y., Hirose, T., Asai, M., Inayama, S., Miyata, T. and Numa, S. (1982) *Nature (London)*, **299**, 793–7.
Noda, M., Takahashi, H., Tanabe, T., Toyosato, M., Kikyotani, S., Hirose, T., Asai, M., Takashima, H., Inayama, S., Miyata, T. and Numa, S. (1983) *Nature (London)*, **301**, 251–5.
Okamura, M. Y., Feher, G. and Nelson, N. (1982) in *Photosynthesis: Energy Conversion by Plants and Bacteria*, Academic Press, New York, Vol. 1, pp. 195–272.
Ovchinnikov, Yu. A., Abdulaev, N. G., Feigina, M. Yu., Artamonov, I. D., Zolotarev, A. S., Kostina, M. B., Bogachuck, A. S., Moroshnikov, A. I., Martinov, V. I. and Kudelin, A. B. (1982) *Bioorg. Khim.*, **8**, 1011–14.
Ovchinnikov, Yu. A., Abdulaev, N. G., Feigina, M. Yu., Kisllev, A. V. and Lobanov, N. A. (1979) *FEBS Lett.*, **100**, 219–24.
Packham, N. K., Dutton, P. L. and Mueller, P. (1982) *Biophys. J.*, **37**, 465–73.
Packham, N. K., Packham, C., Mueller, P., Tiede, D. M. and Dutton, P. L. (1980) *FEBS Lett.*, **110**, 101–6.
Patlak, J. and Horn, R. (1982) *J. Gen. Physiol.*, **79**, 333–51.
Pattus, F., Desmuelle, P. and Verger, R. (1978a) *Biochim. Biophys. Acta*, **507**, 62–70.
Pattus, F., Piorant, M. C. L., Lardunski, C. J., Desmuelle, P. and Verger, R. (1978b) *Biochim. Biophys. Acta*, **507**, 71–82.
Popot, J.-L., Cartaud, J. and Changeux, J.-P. (1981) *Eur. J. Biochem.*, **118**, 203–14.
Prince, R. C., Baccarini-Melandri, A., Hauska, G. A., Melandri, B. A. and Crofts, A. R. (1975) *Biochim. Biophys. Acta*, **387**, 212–27.
Quandt, F. N. and Narahashi, T. (1982) *Proc. Natl. Acad. Sci. USA*, **79**, 6732–6.
Raftery, M. A., Hunkapiller, M. W., Strader, C. D. and Hood, L. E. (1980) *Science*, **208**, 1454–7.
Reynolds, J. A. and Karlin, A. (1978) *Biochemistry*, **17**, 2035–8.
Sakmann, B., Patlak, J. and Neher, E. (1980) *Nature (London)*, **286**, 71–3.
Schein, S. J., Colombini, M. and Finkelstein, A. (1976) *J. Membr. Biol.*, **30**, 99–120.
Schindler, H. (1979) *Biochim. Biophys. Acta*, **555**, 316–36.
Schindler, H. (1980) *FEBS Lett.*, **122**, 77–9.
Schindler, H. (1982a) in *Transport in Biomembranes: Model Systems and Reconstitution* (eds R. Antolini, A Gliozi and A Gorio), Raven Press, New York, 75–85.

Schindler, H. (1982b) *Neurosci. Res. Program Bull.*, **20**, 295–301.
Schindler, H. and Nelson, N. (1982) *Biochemistry*, **21**, 5787–94.
Schindler, H. and Quast, U. (1980) *Proc. Natl. Acad. Sci. USA*, **77**, 3052–6.
Schindler, H. and Rosenbusch, J. (1978) *Proc. Natl. Acad. Sci. USA*, **75**, 3751–5.
Schindler, H. and Rosenbusch, J. P. (1981) *Proc. Natl. Acad. Sci. USA*, **78**, 2302–6.
Schindler, H., Rosenbusch, J. P. and Quast, U. (1980) *Neurochem. Int.*, **2**, 291–8.
Schindler, H., Spillecke, F. and Neumann, E. (1984) *Proc. Natl. Acad. Sci. USA*, **81**, 6222–6.
Schönfeld, M., Montal, M. and Feher, G. (1979) *Proc. Natl. Acad. Sci. USA*, **76**, 6351–5.
Schönfeld, M., Montal, M. and Feher, G. (1980) *Biochemistry*, **19**, 1535–42.
Schuerholz, T. and Schindler, H. (1983) *FEBS Lett.*, **152**, 187–90.
Sigworth, F. J. and Neher, E. (1980) *Nature (London)*, **287**, 447–9.
Suarez-Isla, B. A., Wan, K., Lindstrom, J. and Montal, M. (1983) *Biochemistry*, **22**, 2319–23.
Talvenheimo, J. A., Tamkun, M. M. and Catterall, W. A. (1982) *J. Biol. Chem.*, **257**, 11868–71.
Talvenheimo, J. A., Tamkun, M. M., Messner, D. J., Hartshorne, R. P., Sharkey, R. M. and Catterall, W. A. (1984) *Biophys. J.*, **45**, 37–40.
Tamkun, M. M. and Catterall, W. A. (1981) *J. Biol. Chem.*, **256**, 11457–63.
Tanaka, J. C., Eccleston, J. F. and Barchi, R. L. (1983) *J. Biol. Chem.*, **258**, 7519, 7526.
Tank, D. W., Huganir, R. L., Greengard, P. and Webb, W. W. (1983) *Proc. Natl. Acad. Sci. USA*, **80**, 5129–33.
Tank, D. W., Miller, C. and Webb, W. W. (1982) *Proc. Natl. Acad. Sci. USA*, **79**, 7749–53.
Tiede, D. M., Mueller, P. and Dutton, P. L. (1982) *Biochim. Biophys. Acta*, **681**, 191–201.
Valkirs, G. E. and Feher, G. (1982) *J. Cell. Biol.*, **95**, 179–88.
Vergara, C., Moczydlowski, E. and Latorre, R. (1984) *Biophys. J.*, **45**, 73–6.
Verger, R. and Pattus, F. (1976) *Chem. Phys. Lipids*, **16**, 285–91.
Villegas, R., Villegas, G. M., Barnola, F. V. and Racker, E. (1970) *Adv. Cytopharmacol.*, **3**, 373–85.
Walker, J. W., Takiyasu, K. and McNamee, M. G. (1982) *Biochemistry*, **21**, 5384–9.
Weigele, J. B. and Barchi, R. L. (1982) *Proc. Natl. Acad. Sci. USA*, **79**, 3651–5.
White, M. M. and Miller, C. (1979) *J. Biol. Chem.*, **254**, 10161–6.
Wikström, M., Saraste, M. and Penttilä, T. (1985) *The Enzymes of Biological Membranes* (ed A. Martonosi), 2nd edn, Plenum Press, New York (in press).
Wilmsen, U., Methfessel, C., Hanke, W. and Boheim, G. (1983) in *Physical Chemistry of Transmembrane Ion Motions* (ed G. Spach), Elsevier Science Publishers B.V., Amsterdam, pp. 479–85.
Wu, W. C.-S., Moore, H.-P. H. and Raftery, M. A. (1981) *Proc. Natl. Acad. Sci. USA*, **78**, 775–9.
Zimmerberg, J., Cohen, F. S. and Finkelstein, A. (1980) *J. Gen. Physiol.*, **75**, 241–50.

# 6

# Topography of membrane proteins – determination of regions exposed to the aqueous phase

PAUL A. HARGRAVE



## 6.1 Introduction

Integral membrane proteins are firmly associated with the lipid bilayer and have a portion of their mass exposed at one or both aqueous membrane surfaces. It is an operational hypothesis that membrane protein topography is related to function, and that a knowledge of the mode of association of the protein in the membrane will lead us to understand its function.

A variety of probes have been employed in order to determine what portions of membrane proteins are exposed to an aqueous environment. High-molecular-weight non-penetrating probes include proteolytic enzymes, the enzymes lactoperoxidase and transglutaminase and non-enzyme binding proteins such as antibodies and lectins. Low-molecular-weight hydrophilic surface probes have been selected for their chemical reactivity with protein side chains and their inability to penetrate or traverse the lipid bilayer. Each method has its own inherent strengths and weaknesses. No single technique is sufficient to give an adequate picture of the aqueous-exposed surface of a membrane protein. However, once several complementary methods have been properly applied, a consistent picture of the protein topography emerges.

Success in application of topographic probes is highly dependent upon characteristics of the membrane preparation. The membrane system to be studied must have well-defined criteria for evaluation of purity and integrity.

Techniques for the Analysis of Membrane Proteins. Edited by C. I. Ragan & R. J. Cherry. Published in 1986 by Chapman and Hall, 11 New Fetter Lane, London EC4P 4EE

The membranes must not undergo substantial inversion or suffer degradative changes as a result of isolation or due to modification with the topographic probes. If pure membranes can be prepared in a reproducible manner and maintained with defined sidedness, then valid results should be obtained from topographic studies. This chapter discusses the application of several such probes for hydrophilic surfaces of membrane proteins.

## 6.2 Transglutaminase

### 6.2.1 USE OF TRANSGLUTAMINASE TO LABEL MEMBRANE PROTEINS

Transglutaminase is an enzyme which can be used to covalently attach fluorescent or radioactive amines to glutamine residues in protein substrates (Scheme 6.1). The enzyme transglutaminase is found in many tissues in the

Scheme 6.1

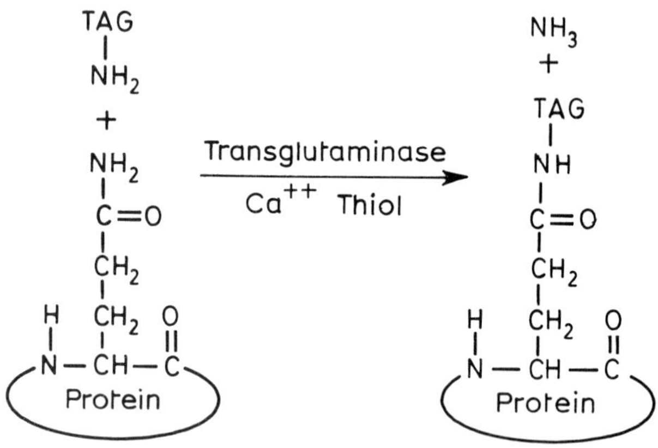


body and normally carries out reactions between glutamines and lysine ε-amino groups in which both reactants are components of proteins (Folk, 1983). This leads to cross-linked proteins such as those found in the fibrin clot. In the presence of high concentrations of a suitable low-molecular-weight amine substrate, the enzyme substitutes the amine for the amide group of a protein-bound glutamine residue.

### 6.2.2 TRANSGLUTAMINASE LABELS 'HIGHLY EXPOSED' GLUTAMINES

One cannot predict whether a protein will be a substrate for transglutaminase on the basis of the protein's glutamine content. Only a few of the glutamines in a protein may become labelled by transglutaminase, and many proteins are not labelled at all (Iwanij, 1977). There is a high correlation between proteins which can become phosphorylated and proteolytically modified and those which are substrates for transglutaminase. Thus, it is reasonable to suppose that a glutamine must be present in a 'highly exposed' surface region of a protein in order to be accessible to transglutaminase.

### 6.2.3 TRANSGLUTAMINASE LABELS SOME MEMBRANE PROTEINS

Few membrane proteins have been found to be labelled by transglutaminase in intact cells (Iwanij, 1977). A protein in sarcoplasmic reticulum membranes is a substrate as are several proteins in erythrocyte 'ghost' membranes (Dutton and Singer, 1975; Iwanij, 1977). Some of the proteins in the membrane viruses influenza virus and bacteriophage PM2 become labelled (Brewer and Singer, 1974; Iwanij, 1977). The best studied system seems to be that of rhodopsin in retinal disc membranes. Pober *et al.* (1978) reported incorporation of 1 mol of either [$^{14}$C]putrescine or dansylcadaverine into a protease-sensitive site on rhodopsin. Prior limited proteolysis of the membranes by thermolysin eliminated the transglutaminase site. When dansylcadaverine-modified rhodopsin was subjected to proteolysis by thermolysin, proteolysis was greatly inhibited. Long-term digestion of the modified rhodopsin by subtilisin was successful in cleaving rhodopsin and removing the fluorescent label. The location of the dansylcadaverine tag has been reported to be glutamine-237 in the rhodopsin sequence (Ovchinnikov, 1982) which is consistent with known sites of proteolysis. A recent study of reductively methylated disc membranes employed a different source of transglutaminase and found that as much as 3.5 mol of putrescine could be incorporated into four sites in rhodopsin: glutamine-236, glutamine-237, glutamine-238 and glutamine-344 (McDowell *et al.*, 1986; Fig. 6.1). A consideration of parameters of the transglutaminase reaction explains many of the differences observed.

### 6.2.4 HOW TO MODIFY MEMBRANE PROTEINS USING TRANSGLUTAMINASE

The traditional source of transglutaminase has been the guinea pig liver (Connellan *et al.*, 1971). This starting material can be expensive, is often not readily available, and the preparation procedure is not always trouble-free. Transglutaminase has recently been purified from outdated human erythrocytes by a simple and straightforward procedure (Brenner and Wold, 1978). It is not necessary to purify the enzyme to homogeneity; it can be used successfully following two ion-exchange steps followed by gel filtration. The enzyme may be frozen in liquid nitrogen and stored at −80° C (J. H. McDowell, personal communication). The action of the erythrocyte transglutaminase appears to be the same as that of the liver enzyme when compared using 20 soluble protein substrates and red cell membranes (Brenner and Wold, 1978). A convenient way to follow activity of the enzyme during purification is the

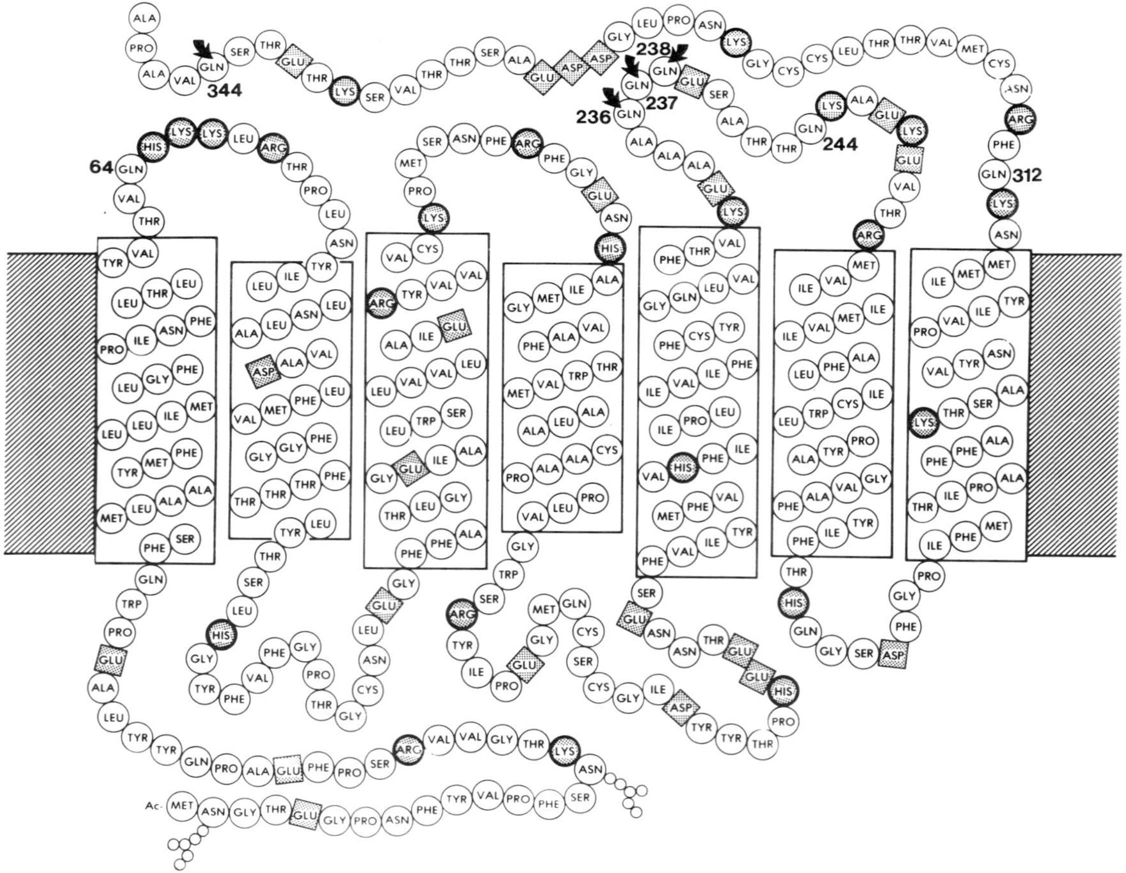


**Fig. 6.1** A model for the organization of the vertebrate rhodopsin polypeptide chain with respect to the disc membrane lipid bilayer. The low-polarity portion of the lipid bilayer is indicated by striped areas at each side of the drawing. The *N*-terminal Ac-Met is at the bottom of the drawing; this surface of rhodopsin is exposed to the aqueous interior of the disc membrane. The *C*-terminal Ala is at the top of the drawing; this surface of rhodopsin is the external aqueous-exposed surface of the disc membrane. Glutamines-236, -237,

filter paper assay of Lorand *et al.* (1972) using $\beta$-lactoglobulin as the protein substrate and [$^{14}$C]putrescine as the amine. At least 5 mM amine is required and even then the reaction is slow. For completion of the reaction with membranes a time course should be performed over a 24–28 h period, although maximum incorporation may occur after about 4–8 h (Pober *et al.*, 1978). A radioactive amine should be employed. Accurate stoichiometry of the reaction is obtained by using a radioactive amine of known specific radioactivity, purifying the modified protein and measuring the amount of protein by amino acid analysis. Determination of incorporation of radioactivity into individual membrane proteins is best obtained from gel slices following SDS–polyacrylamide gel electrophoresis.

Cross-linking of protein substrates by transglutaminase seems to be a competing reaction in all reported studies to date and leads to heterogeneity of products and uncertain stoichiometry. Cross-linked aggregates are observed even with 10–20 mM amine (Pober *et al.*, 1978). Prior reductive methylation of membrane amino groups with formaldehyde and sodium cyanoborohydride (Jentoft and Dearborn, 1979) eliminates the protein cross-linking side reaction (McDowell *et al.*, 1986), since secondary and tertiary amines are not substrates for the enzyme. Use of diamines such as putrescine as an amine substrate can promote cross-linking reactions (Schrode and Folk, 1978) but this does not appear to be a serious problem in practice compared to a failure to block lysine amino groups.

Extensive studies have been performed to determine the amine substrate specificity of transglutaminase (Lorand *et al.*, 1979; Folk, 1983). An amine with a hydrophobic group attached to an alkylamine side chain of about 7 Å (such as dansylcadaverine) is the best substrate. However, affinity for the amine substrate is not the only important consideration. When there were three adjacent sites which could be modified in rhodopsin, only 1 mol of the rather large dansylcadaverine could be incorporated (Pober *et al.*, 1978). In contrast the sterically smaller amine putrescine could modify each of the three adjacent glutamines (McDowell *et al.*, 1986). There are few systematic data available on the effect of the chemical nature of the amine and site specificity of incorporation, but the previous discussion suggests that amines of different sizes should be compared. There may not only be differences in extent and site of incorporation with different amines, but differences in properties of the modified membrane protein may also be observed. For example, rhodopsin containing 1 mol of dansylcadaverine is a poor substrate for proteolysis by thermolysin (Pober *et al.*, 1978), yet proteolysis is unaffected after incorporation of 2–3 mol of putrescine (McDowell *et al.*, 1986). Some amines may be chosen for reasons other than their reactivity. For example, transglutaminase-catalysed incorporation of the fluorescent dansyl probe into rhodopsin has been employed for distance measurements by energy-transfer spectroscopy (Pober *et al.*, 1978).

## 6.3 Iodination

### 6.3.1 SURFACE-SPECIFIC IODINATION OF MEMBRANE PROTEINS

Both chemical and enzymatic iodination methods have been employed with the intent to vectorially iodinate membrane proteins in intact cells or membrane preparations. Oxidation of $I^-$ with chloramine-T, which has proven so useful for iodinating soluble proteins (Greenwood *et al.*, 1963), has been applied to the surface labelling of viruses, bacteria and eukaryotic cells (Frost, 1977; Munford and Gotschlich, 1977). However, even under carefully controlled conditions, the use of such a powerful oxidizing agent (which requires strong reducing conditions to terminate the reaction) greatly detracts from the general utility of this method for labelling membranes. Several other iodination methods have been developed. The most widely used method is lactoperoxidase-catalysed iodination (Morrison, 1980).

### 6.3.2 LACTOPEROXIDASE-LABELLING OF MEMBRANE PROTEINS

Lactoperoxidase is a $M_r \cong 80\,000$ enzyme which utilizes $I^-$ and $H_2O_2$ in catalysing the iodination of proteins. Tyrosine, and to some extent histidine, residues may be iodinated. Under optimum conditions the halogenation reaction takes place on the enzyme active site. Therefore the reaction should be specific for only surface-exposed amino acids. Evidence in support of this comes from selectivity in iodination of tyrosine-74 in cytochrome *c* (Osheroff *et al.*, 1977). Lactoperoxidase has been used to demonstrate the surface exposure of proteins in erythrocyte membranes (Phillips and Morrison, 1971), photoreceptor membranes (Fung and Hubbell, 1978b), bacteriophage PM2 (Brewer and Singer, 1974) as well as numerous other systems.

### 6.3.3 EXPERIMENTAL PARAMETERS OF LACTOPEROXIDASE LABELLING

The lactoperoxidase reaction may be initiated by adding $H_2O_2$ to a membrane or cell suspension containing $^{125}I^-$. The reaction will be vectorial as long as negligible amounts of diffusible iodinating agents such as $I_2$ are generated. $I_2$ can dissolve in the lipid bilayer and iodinate unsaturated lipids (Mersel *et al.*, 1976; Schlager, 1979), or diffuse through the membrane and iodinate proteins in the cell or membrane interior. In order to minimize formation of $I_2$, the amounts of $I^-$ and $H_2O_2$ added must be carefully controlled. Morrison (1980) recommends keeping the concentration of $I^-$ to $\leq 10^{-6}$ M, which may be achieved by use of an iodide-sensitive electrode and an autotitrator. Maintenance of excess protein substrate at relatively high concentrations also minimizes the diffusible $I_2$ concentration. Membranes should be iodinated only to a limited extent (a few percent of iodinatable sites), rather than exhaustively. This will favour better vectorial labelling by both limiting $I_2$ formation and avoiding possible membrane damage due to excessive protein modification.

The lactoperoxidase reaction may incorporate up to 1 mol of iodine into protein for each mol of $H_2O_2$ consumed (if lipid labelling is kept low). Thus a carefully regulated supply of $H_2O_2$ is desirable. Many investigators employ the convenient *in situ* generation of $H_2O_2$ using the glucose–glucose oxidase system (Hubbard and Cohn, 1972). Morrison (1980) recommends either continuous deliveryof $H_2O_2$ or addition of multiple small aliquots. This avoids addition of other proteins and allows both measurement and control of the amount of $H_2O_2$ added. Excess $H_2O_2$ ($>10^{-4}$ M) must be avoided since it inhibits the lactoperoxidase and can oxidize proteins and unsaturated fatty acids. Many cells contain enzymes which destroy $H_2O_2$, and if these are present in the preparation under study, additional peroxide will need to be added. Some cells contain peroxidases which can generate $I_2$ internally from $I^-$. Artifactual labelling from this latter source can be detected by performing a control experiment to determine whether iodination occurs when lactoperoxidase has been omitted from the reaction mixture.

Following iodination it is desirable to reisolate the cells (viruses, organelles, etc.) in such a way as to separate them from any which have been damaged. Appropriate tests for biological integrity or function should be performed (exclusion of Trypan Blue, maintenance of enzyme activity, retention of viral infectivity, etc.). Incorporation of radioactivity into particular proteins is then conveniently measured following separation by SDS-gel electrophoresis (Fig. 6.2). If the membranes are washed to remove excess $I^-$ following labelling, radioactivity at the gel front may be indicative of labelling of lipids. Lipid labelling may be investigated by $CHCl_3$/MeOH extraction (Bligh and Dyer, 1959) and, if present, indicates that conditions may be too vigorous for reliable surface labelling of proteins. Conditions should then be moderated until lipid labelling is negligible.

Surface-specific labelling may be achieved if the above precautions are taken. It is possible, of course, that proteins at a membrane surface may lack amino acid side chains capable of being iodinated. In such a case it is possible to chemically modify lysine or cysteine side chains in order to make it possible to perform surface iodination (Fung and Hubbell, 1978b).

### 6.3.4 IODINATION USING CHLOROGLYCOLURIL

The iodinating reagent 1,3,4,6-tetrachloro-3α,6α-diphenylglycoluril (Iodogen, Pierce Chemical Company, Rockford, Illinois) achieves high levels of iodination from $I^-$ (Markwell and Fox, 1978). To perform the reaction, membranes and $^{125}I^-$ are placed in a glass reaction vessel whose walls are coated with the reagent. The reaction is conveniently terminated simply by

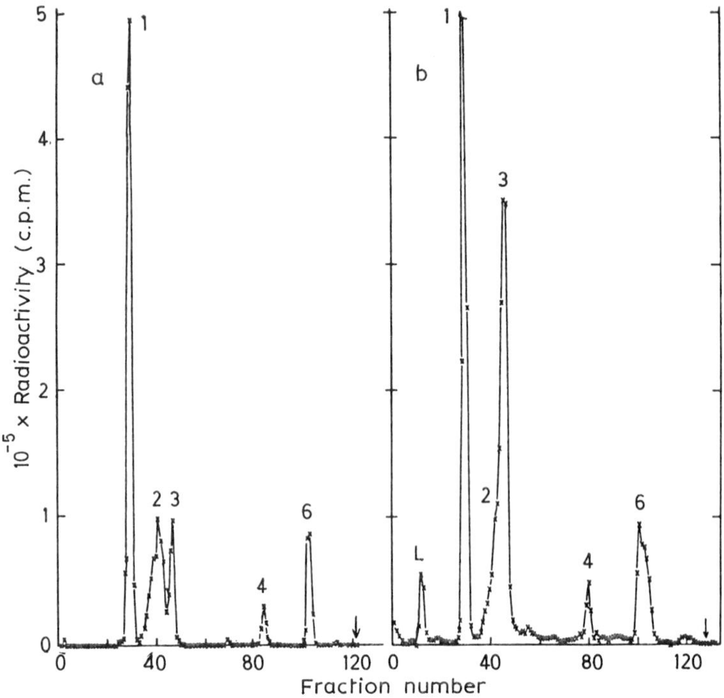


**Fig. 6.2** SDS–polyacrylamide gel electrophoresis of bacteriophage PM2 treated with $^{125}I^-$ and lactoperoxidase. (a) treatment of intact PM2; (b) treatment of PM2 disrupted by freeze–thaw. Numbers refer to membrane protein. L indicates self-labelled lactoperoxidase. (Reprinted with permission from Brewer and Singer, 1974.)

removing the sample from the reaction vessel and washing the membranes in buffer containing non-radioactive NaI. When the chloroglycoluril method was compared with lactoperoxidase iodination of red cell membranes and viruses, the same vectorial labelling was achieved but with increased specific radioactivity (Markwell and Fox, 1978). The conditions of reaction are important. Optimal surface labelling occurs at 4–10 μg of chloroglycoluril/100 μg of protein and 5 μM iodide. Non-vectorial labelling will occur at high iodide concentrations. It should be noted that the oxidation reaction is not catalytic and will lead to a burst of the iodinating species rather than a constant production. The method has been usefully applied, e.g., to label bacteria (Sullivan and Williams, 1982), platelets (Tuszynski *et al.*, 1983) and thyrocytes (Islam *et al.*, 1983). This reaction is not temperature dependent, unlike an enzyme-catalysed reaction, and may be conveniently performed at 4° C in order to enhance stability of cells or membranes. It is also unaffected by high salt concentrations which would be needed for halobacteria, or the presence of azide, cyanide or fluoride – conditions which are not compatible with the lactoperoxidase reaction. Experience with the method is limited, however, and proper controls would be particularly important in its use.

### 6.3.5 *N*-CHLOROBENZENESULPHONAMIDE

Protein iodination has been reported using a solid state reagent in which the oxidant *N*-chlorobenzenesulphonamide is covalently attached to non-porous polystyrene beads (Markwell, 1982). The reaction achieves almost quantitative incorporation of radioiodide and is technically easy to perform. Evidence has been presented that the active iodinating species is non-diffusible (possibly *N*-iodobenzenesulphonamide formed by substitution of an iodine atom for the chlorine) (Markwell, 1982). This suggests that the reagent should be useful for vectorial labelling, and experience to date is encouraging (M. A. Markwell, personal communication). The beads are too large to be engulfed by cells (2.8 mm diameter) and are non-porous – features which would be useful in labelling of cells.

## 6.4 Limited proteolysis of membrane proteins

### 6.4.1 INTRODUCTION

Limited proteolysis of membranes has proven to be perhaps the most useful tool to date for investigating the structure and topography of membrane proteins. Proteolytic methods have served to identify regions of proteins which are exposed at the membrane external surface or which are exposed at the membrane internal surface. Digestion of membrane proteins by proteases has yielded information concerning structure–function relationships and provided large protein fragments for structural analysis.

### 6.4.2 LIMITED PROTEOLYSIS

The use of proteases to study the topography of membrane proteins benefits from an extensive literature on the use of proteases to study the structure, function and topography of soluble globular proteins (Mihalyi, 1978a,b). It had been observed that exposure of native proteins to small amounts of proteases often led to formation of a small number of well-defined products rather than to extensive proteolytic digestion. In such cases it was frequently observed that several proteases of different specificities would act on a protein to produce similar proteolytic fragments. This led to the conclusion that the three-dimensional structure of the protein substrate must limit access of the proteases to certain highly susceptible regions. In the case of the digestion of ribonuclease A by subtilisin to form a two-fragment complex (Richards and Vithayathil, 1959), the protease-susceptible region was found to be an exposed surface loop. The protease-susceptible region in the immunoglobulin IgG molecule is an exposed segment of the polypeptide chain connecting two tightly folded domains (Zubay, 1983). Although such detailed structural information is not yet available for intrinsic membrane proteins, we can anticipate that their aqueous surfaces will contain all of the basic elements of protein secondary structure and will be susceptible to proteases for identical structural reasons. Depending upon the extent to which the protein is embedded in the membrane, a smaller fraction of the protein surface will be exposed to an aqueous environment, thus leading to potentially fewer available sites for proteolysis than for a non-membrane protein of comparable size.

### 6.4.3 WHAT KINDS OF RESULTS CAN BE EXPECTED FROM THE USE OF PROTEOLYTIC ENZYMES AS PROBES OF TOPOGRAPHY?

The results obtained will be as varied as the proteases employed and as the membrane protein substrates investigated. Some proteins may be relatively resistant to digestion by a protease. For example, trypsin fails to digest Band 3 in intact erythrocytes (Steck *et al.*, 1976), and cleaves only a single bond in rhodopsin in disc membranes (Molday and Molday, 1979; Hargrave *et al.*, 1980). In contrast, subtilisin and papain cleave bonds in at least three surface regions of the rhodopsin molecule, and produce a membrane-bound complex of three proteolytic fragments (Albert and Litman, 1978; Dratz *et al.*, 1979). There is considerable variation in how much of a membrane-bound protein becomes solubilized as a result of protease digestion. Substantial portions of histocompatability antigens (Nathenson *et al.*, 1981) and cytochrome $b_5$ (Ozols *et al.*, 1976) become solubilized. Approximately 41 K is released from the ~93 K erythrocyte Band 3 when the cytoplasmic surface of 'ghost' membranes is digested (Steck *et al.*, 1976). In contrast, the action of several proteases on both membrane surface regions of the purple membrane of *Halobacterium halobium* solubilizes less than 10% of the mass of bacteriorhodopsin (Walker *et al.*, 1979).

### 6.4.4 WHAT PROTEASE SHOULD BE USED?

The nature of the proteolytic cleavages obtained is best understood by progressing from use of the most specific to the least specific proteases. Several proteases of differing specificity should be examined as topographic probes for the system of interest. Proteases such as trypsin and *Staphylococcus aureus* protease which have a restricted range of specificity for the polypeptide bonds which they can hydrolyse may fail to cleave some proteins or may cleave at only one or two bonds. Proteases such as papain and subtilisin have a much broader range of bonds susceptible to cleavage and may cleave at more sites on the protein as well as at different bonds within the same general site. The mixture of bacterial enzymes, Pronase, has been employed for an exhaustive search for surface exposed sites (Bender *et al.*, 1971; Walker *et al.*, 1979).

### 6.4.5 CONDITIONS FOR THE REACTION

The experimental conditions must ensure integrity of the membrane and activity of the proteolytic enzyme. Parameters include pH, ionic strength, buffer type and concentration, temperature, presence of anti-oxidants and metal ions, etc. Knowledge of the behaviour of the particular membrane system will determine optimal conditions for the membrane; a slightly hypo-osmotic solution can keep the membranes swollen and disaggregated yet not promote membrane lysis. Membrane lipid oxidation can be minimized by bubbling buffers with argon, including EDTA to chelate trace heavy metals, and ensuring that membrane preparations contain reasonable amounts of anti-oxidants (Farnsworth and Dratz, 1976). Buffer type and pH can affect protease activity and, in some cases, protease specificity. For example, *Staphylococcus aureus* protease cleaves only glutamyl bonds in 50 mM ammonium bicarbonate buffer (pH 7.8) or ammonium acetate buffer (pH 4.0). In 50 mM phosphate buffer it will cleave bonds on the C-terminal side of either aspartate or glutamate (Drapeau, 1976). Findlay *et al.* (1981) used 2% (w/w) *S. aureus* protease to digest photoreceptor membranes in 67 mM sodium phosphate (pH 7.0) for 3 h at 30° C. Two bonds were cleaved releasing a small peptide from rhodopsin and producing two large membrane-bound peptides. Two other sets of conditions yielded less satisfactory results. When the same amount of enzyme was used with 50 mM ammonium bicarbonate (pH 7.8) or with 5 mM Hepes (pH 7.0), far less cleavage resulted owing to effects on either the enzyme or the membrane or both (P. A. Hargrave and H. Kühn, unpublished results). Some effects of conditions of digestion are illustrated by another experiment in which thermolysin was employed. The same extent of digestion of a membrane protein was to be achieved per unit time under three sets of conditions: (1) pH 6.5, 20° C, (2) pH 5.1, 0° C and (3) pH 8.5, 0° C (Kühn *et al.*, 1982). This was accomplished by altering the weight ratio of enzyme from 1:400, to 1:100 and 1:11, respectively, in the three different experiments in order to compensate for the differences in enzyme activity under the different conditions.

In order to easily reproduce the results of experiments, it may be necessary to use enzyme from the same manufacturer, and even from the same lot number. If trypsin is used, it should be treated with tosylphenylchloromethylketone (Carpenter, 1967) in order to inactivate small and variable amounts of chymotryptic activity. Most enzymes can be prepared in quantity and stored frozen in aliquots so that experimental conditions can be easily duplicated. *S. aureus* protease may be stored frozen in distilled water (Drapeau, 1976). Thermolysin may be frozen in $Ca^{2+}$-containing buffers (Matsubara, 1970). Subtilisin, trypsin and chymotrypsin are conveniently stored in 1 mM HCl.

Once the reaction has been initiated, there must be a reliable method for terminating it. The effectiveness of the termination reaction must be verified by adding inactivated enzyme to an aliquot of the membrane and carrying this sample through the analytical procedure in parallel with treated and untreated membranes (see Section 6.4.6). Phenylmethanesulphonyl fluoride (PMSF) is often used as a general protease inhibitor (Bender *et al.*, 1971). It is effective on such serine proteases as subtilisin and chymotrypsin but relatively less effective when used against trypsin (Fahrney and Gold, 1963). Enzymes which have been inactivated by PMSF and which are to be treated with SDS should be kept inactivated by rapid boiling, since the PMSF modification is reversible upon heating. The toxic organophosphorous compound diisopropylfluorophosphate (DIFP) inactivates trypsin. However, commercial samples of DIFP available in recent years have been found to cause aggregation of rhodopsin when added to photoreceptor membranes (P. Hargrave, unpublished results). When large amounts of trypsin are used, addition of excess trypsin inhibitor (e.g. pancreatic trypsin inhibitor) is not always sufficient to stop all tryptic digestion as monitored by SDS-gel electrophoresis (see Chapter 1). Even when membranes are washed in order to remove most of the trypsin–trypsin inhibitor complex, the residual membrane-bound trypsin may be able to transiently digest the membrane protein in SDS before becoming denatured itself. Postreaction inhibition of proteases is the major technical problem in the use of this technique. The action of proteases on proteins in SDS forms the basis for a successful method of peptide mapping of proteins by SDS-gel electrophoresis (Cleveland *et al.*, 1977).

Proteases which need metal ions for activity are more easily inactivated. For example, thermolysin, a $Ca^{2+}$-requiring enzyme, may be inactivated by adding EDTA.

Other proteases may present other possible problems, particularly if used in large quantities such as 5% or more by weight of membrane protein. Papain, a sulphydryl active-site enzyme, must be activated from an inactive precursor by addition of a reducing agent such as cysteine. Following reaction it is usually inactivated by addition of a reagent which reacts with sulphydryl groups (e.g. iodoacetic acid). If the membrane proteins are to be examined by SDS-gel electrophoresis, the reaction mixture is then dissolved in SDS containing a reducing agent, $\beta$-mercaptoethanol or dithiothreitol. There is the potential that if traces of papain had not been activated from its precursor, it could be activated at this point. Proper controls could assess whether this is in fact a problem.

### 6.4.6 METHODS FOR THE IDENTIFICATION OF THE PRODUCTS OF PROTEOLYSIS

In order to be able to obtain information about protein topography following limited proteolysis of a membrane, it is necessary to be able to identify the products, soluble or membrane-bound, and account for all portions of the membrane protein molecule. Soluble products are recovered following termination of the digestion by centrifugation and removal of the supernatant. Whatever tests are applied to the supernatant must also be applied to a control sample which contains all of the same components; i.e. the protease is first inactivated and then added to the membrane sample. In order to test for the release of small peptides, samples of digestion supernatant may be examined by thin-layer chromatography (Aromatorio *et al.*, 1983) or paper electrophoresis (Bennett, 1967). Such techniques were used to identify an octapeptide and a tetrapeptide produced by the action of thermolysin on rhodopsin in photoreceptor membranes (Hargrave and Fong, 1977). In such experiments it is necessary to estimate the number of nanomoles of product expected so that a large enough sample of membranes is used. A known amount of an authentic peptide should be submitted to the same procedures in parallel.

Analysis for soluble peptides of size greater than 20 amino acids is conveniently performed by gel electrophoresis. The SDS–urea polyacrylamide system of Swank and Munkres (1971) is well designed for separation of peptides in the $M_r$ range 2000 to 10 000.

The most common method for identification of the membrane-bound products of limited proteolysis of membrane proteins is polyacrylamide gel electrophoresis in SDS (see Chapter 1). In order to identify the products of proteolysis it is most common to fix the proteins in the gel with acetic acid, and to stain to visualize the products with Coomassie Blue. From such a procedure one can conclude that a protein (with a particular apparent molecular weight) has been converted by proteolysis to a product or products, whose apparent molecular weights are determined by comparison with selected standards. Thus human erythrocyte Band 3 (about 90 K) is converted by chymotrypsin to membrane-bound 55 K and 38 K fragments (Steck *et al.*, 1976). Bovine rhodopsin (about 37 K) is digested by thermolysin to yield two membrane-bound fragments of about 26 K and 12.5 K (Pober and Stryer, 1975). Often the molecular weights do not add up, or apparent molecular weights obtained from one gel system do not closely match values obtained with a different gel system. Explanations for such electrophoretic behaviour of polypeptides are discussed in Chapter 1.

One should not rely solely on gel staining in order to identify products of proteolysis. Small peptides, heterogeneous peptide products or acidic weakly staining peptides may be missed. Various other labels are useful for product identification. Intrinsic markers such as oligosaccharides may serve as labels to identify proteolytic fragments. Chemical methods or the use of lectins (Molday and Molday, 1979) can be employed for identification of such oligosaccharide-containing fragments. Alternatively, a natural prosthetic group may be covalently attached, e.g, by reduction of the Schiff base linkage of retinal to bacteriorhodopsin (Bayley *et al.*, 1981) or vertebrate rhodopsin (Wang *et al.*, 1980). Radioactive labels may be introduced, often biosynthetically. Both carbohydrate and amino acid (Basinger and Hoffman, 1982) precursors may be useful.

Chemical modification of already synthesized membrane proteins may be usefully employed. General labelling of the polypeptide chain by methylation has been reported (Kiehn and Holland, 1968; Moore, 1975) but this method appears to have been seldom used. Reductive methylation of lysyl side chains [with [$^{3}$H] or [$^{14}$C]formaldehyde (Jentoft and Dearborn, 1979)] locates any product which contains lysyl residues. Specific chemical characteristics of the membrane protein may be exploited; the different reaction of two rhodopsin cysteines with different sulphydryl reagents (Sale *et al.*, 1977) or the reaction of erythrocyte Band 3 with stilbene isothiocyanates (Cabantchik *et al.*, 1978). If the membrane protein is phosphorylated by a kinase, $\gamma$-[$^{32}$P]ATP can be used to specifically label phosphorylation sites. Fluorescent rather than radioactive tags may be introduced, either chemically (Gabriel, 1971) or enzymically (Pober *et al.*, 1978).

### 6.4.7 TIME COURSE OF THE DIGESTION

It can be instructive to follow the time course of a digestion in order to identify intermediates in a reaction and to determine precursor–product relationships. Such studies showed that there was a rapid proteolytic attack on rhodopsin's C-terminal region prior to converting it to the two-fragment membrane-bound complex, F1–F2 (Hargrave and Fong, 1977; Fung and Hubbell, 1978a). Quantitative scanning of the SDS-gels and plotting of the results showed that the digestion proceeds from rhodopsin ($M_r = 37\,500$) to an intermediate of $M_r = 34\,000$, which is then converted to two membrane-bound fragments ($M_r = 27\,000$ and $M_r = 12\,500$). Data from this experiment are shown in Fig. 6.3.

### 6.4.8 CHARACTERIZATION OF THE FRAGMENTS

The fragments from limited proteolysis of membrane proteins form convenient defined portions of the molecule for functional and structural studies. Some fragments, such as the polar haem-containing chymotryptic moiety of cytochrome $b_5$, are readily isolated and studied by conventional techniques of the protein chemist (Ozols *et al.*, 1976). Other fragments, such as the two chymotryptic membrane-bound portions of bacteriorhodopsin, have been

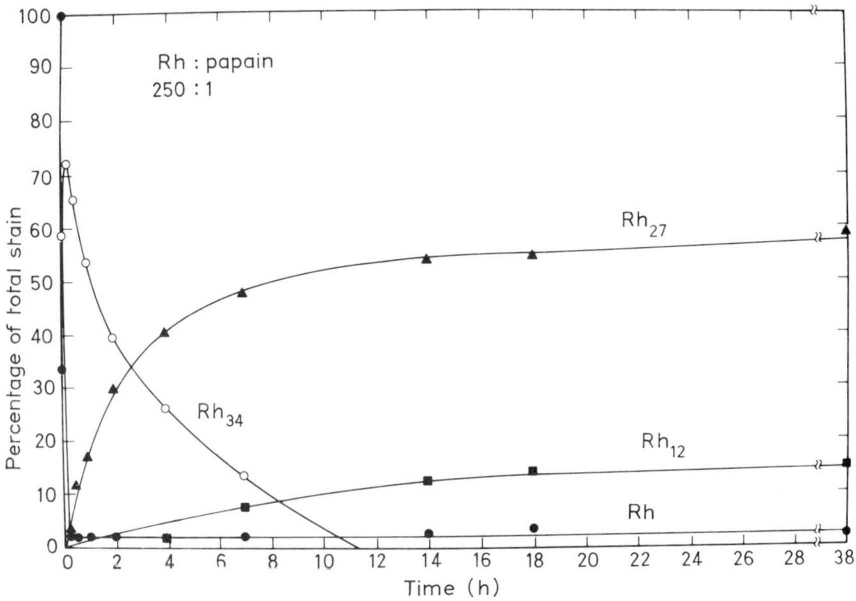


**Fig. 6.3** Kinetics of papain digestion of membrane-bound rhodopsin. The percentage of total Coomassie Blue stain in the gel bands representing rhodopsin and its proteolysis products is plotted vs time of proteolysis. The amount of stain in each band was measured by densitometric scanning. Rhodopsin ($Rh_{37.5}$) was converted to an intermediate ($Rh_{34}$) which formed two smaller membrane bound products ($Rh_{27}$, $Rh_{12}$). (Reprinted with permission from Fung and Hubbell, 1978a).

structurally characterized (Khorana *et al.*, 1979) with great difficulty. The development of new techniques has been required in order to structurally characterize the non-polar fragments of membrane proteins. Such techniques are described in Chapter 9.

## 6.5 Vectorial chemical modification of aqueous-exposed surfaces of membrane proteins

### 6.5.1 GENERAL CONSIDERATIONS

Chemical labelling procedures are widely employed in order to detect those proteins or portions of proteins which are exposed at membrane aqueous surfaces. Some of the early studies have been reviewed (Carraway, 1975). The labelling techniques derive from the extensive literature on the use of chemical modification to study the structure and function of proteins (Cohen, 1970; Means and Feeney, 1971). For vectorial labelling, however, water-soluble charged reagents are used which are shown or assumed not to traverse the lipid bilayer and to react only with amino acid side chains of proteins on the membrane external surface. By this method it is possible to introduce tags which may be chromophoric, fluorescent, radioactive, detectable by ESR, or visible by electron microscopy. The presence of an amino acid side chain of appropriate specificity on a protein surface does not guarantee that it will react with a particular reagent. Each group will have its own unique microenvironment with amino acids surrounding it which will alter the dielectric constant, solvation and charge environment, affecting the group's p*K* and reactivity. Each reagent used will react differently due to steric factors, pH and its mechanism of reaction. Such factors will make some groups super-reactive compared to the free amino acid in solution, and other groups essentially unreactive.

The labelling procedure must be mild enough so that the membrane integrity is not disrupted. It should be possible to reisolate the intact membrane at the end of the procedure. The label should in fact be impenetrable, and there should be methods available to assess whether membrane penetration has occurred. If red blood cells are modified, does haemoglobin become modified, or does one of the polypeptides known to be associated only with the inside surface? A control experiment should verify that, in disrupted membranes, known interior proteins can in fact be modified by the reagent. The reagent should not partition into the membrane or adsorb to it non-covalently.

Chemical modification reagents show a variety of specificities ranging from ability to modify only a single type of amino acid side chain, to potential to modify all of them. We will consider a few of the most widely used and generally useful of these reagents.

### 6.5.2 IMIDOESTERS

Although the imidoester ethyl acetimidate (EAI) is lipid-soluble and membrane-permeable, the reagent made from a charged alcohol, isethionyl acetimidate (IAI, $CH_3\text{–}\overset{NH_2^+}{\overset{\|}{C}}\text{–}OCH_2CH_2\text{–}SO_3^-$) is polar and lipid-insoluble. Whiteley and Berg (1974) synthesized IAI for use as a non-penetrating label for red cell membranes and reported it to be a vectorial reagent. Both EAI and IAI form the same positively charged product upon reaction with ε-amino groups of lysyl residues. Reaction of intact cells with the non-penetrating [$^3$H] IAI was followed by reaction with the penetrating reagent [$^{14}$C] EAI in order to quantify groups present on each surface (Whiteley and Berg, 1974). Depending upon reaction conditions, some penetration of IAI into the red cell was measurable (e.g. at pH 8 and 37° C in phosphate buffer) as monitored by modification of haemoglobin. Some cross-linking of proteins also occurred because of a side reaction when EAI is used at pH less than 10 (Browne and Kent, 1975).

The kinetics and mechanism of amidination are intimately dependent on the pH of the medium (Hand and Jencks, 1962; Browne and Kent, 1975). If the pH of the medium is too low for a given reagent it may not be possible to obtain quantitative conversion of all amino groups accessible to amidines by a single treatment with an excess of imidoester (Browne and Kent, 1975). The optimal conditions for quantitative labelling are different for different imidoesters, a factor which needs to be considered when contemplating the use of these reagents as probes.

Nemes *et al.* (1980) used IAI and newly synthesized choline acetimidate (CAI) and a sulphone imidoester (SAI), which react well at pH 7.5, in a careful study of imidoester modification of rhodopsin in disc membranes. They loaded disc membranes with [$^3$H]lysine and monitored its modification during the reaction of imidoesters with rhodopsin. The use of lysine-loaded membranes served as a monitor for membrane integrity and extent of penetration of reagent – a control which should be more generally used in vectorial-labelling studies.

Nemes *et al.* (1980) performed labelling with the new penetrating reagent SAI that is highly reactive at pH 7.5 and compared its labelling with the 'non-penetrating' reagents IAI and CAI. Disc membranes are leaky above pH 8 so reagents that react well below this pH are essential. There was substantial membrane penetration of the anionic IAI as well as the cationic CAI at pH 7.5 used for the labelling studies. Measurement of the extent of penetration of the reagents allowed the authors to correct their data and also suggested that IAI may not be as generally useful as a vectorial label as had been previously believed. However, the usefulness of IAI (and CAI) as a vectorial label was not conclusively evaluated in this study because some of the membrane penetration to the disc interior appeared to be due to the presence of a leaky subfraction of membranes in the preparations used.

Nemes *et al.* (1980) concluded that the surface lysine residues were equally divided between the inside and outside surface of the disc membrane. No further studies of lysine distribution in rhodopsin have been performed, but the preponderance of data from other types of experiments leads to a model in which nine lysines are located at the external aqueous surface and only one at the internal aqueous surface (Dratz and Hargrave, 1983). This suggests that the less polar penetrating imidoester is able to modify lysines in regions of the hydrophilic external surface which are inaccessible to the charged reagent IAI.

### 6.5.3 DIAZOSULPHANILIC ACID

In contrast to the imidoesters which are specific for reaction with amino groups, diazonium salts can react with the side chains of lysine, cysteine, tyrosine and histidine residues. A great deal of experience has been gained in the use of diazosulphanilic acid and its derivatives. These reagents include diazo[$^{35}$S]sulphanilic acid (Berg, 1969), diazo[$^{125}$I]iodosulphanilic acid (Smith *et al.*, 1979) and diazo[$^{125}$I]di-iodosulphanilic acid (Luthra *et al.*, 1978).

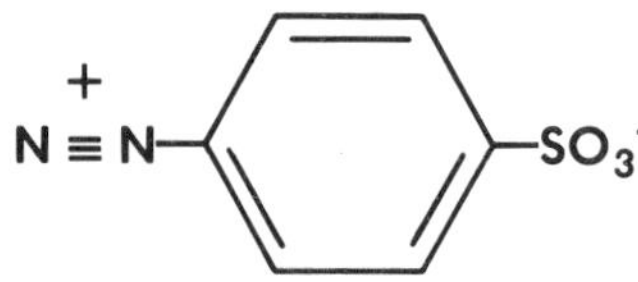

In early studies in which low specific radioactivity reagent was employed and exhaustive labelling was attempted, there was membrane damage which led to cell lysis (Berg, 1969; Carraway, 1975). However, with the use of higher specific radioactivity [$^{35}$S]sulphanilic acid (Berg and Hirsch, 1975) and the iodinated reagents, this problem could be avoided.

Introduction of iodine to the sulphanilic acid makes the reagent less water-soluble. The diazodi-iodosulphanilic acid becomes non-covalently associated with the membrane and is removed only following multiple washings with albumin-containing buffer (Sears *et al.*, 1971). This strongly suggests that the reagent preferentially partitions into the membrane and that special care should be taken to look for transmembrane labelling when a new system is investigated. On the other hand, the application of diazodi-iodosulphanilic acid labelling to red cells labels known external proteins and does not appear to label known internal membrane components (Luthra *et al.*, 1978). A truly penetrating reagent such as *N*-ethylmaleimide will label *all* proteins in the membrane (Carraway and Shin, 1972).

Specific reaction conditions appear to be particularly important in controlling the relative reactivity of membrane proteins toward diazodi-iodosulphanilic acid. Both pH and temperature of reaction alter the extent of labelling of many of the red cell membrane proteins by the reagent (Luthra *et al.*, 1978).

### 6.5.4 PHOTOCHEMICAL LABELLING OF SURFACE PROTEINS

In contrast to the relatively restricted specificity of reaction shown by most protein chemical modification reagents, nitrenes and carbenes have the potential to be extremely non-specific in their reaction with amino acid side chains. This would suggest that membrane proteins which might not have the necessary surface amino acids available for reaction with group-specific reagents might be detected by labelling with such reagents which have a broad spectrum of reactivity. Although nitrenes are capable of inserting into C–H bonds (Knowles, 1972) and under exhaustive labelling conditions appear to do so (Matheson *et al.*, 1977), the nitrenes do, in fact, show considerable selectivity for nucleophiles (Staros, 1980). We will restrict our comments to nitrenes since no carbene precursor has been described for hydrophilic surface labelling.

(a) *NAP-Taurine*

The best studied reagent for membrane protein photochemical labelling is NAP-taurine. This reagent labels all of the red cell membrane exterior surface proteins shown to be present by other techniques, as well as several additional minor proteins (Staros and Richards, 1974). The reagent was found to be impermeable to the red cell membrane at 0° C (Staros *et al.*, 1975). The ability of NAP-taurine to penetrate the red cell membrane at higher temperatures (Staros *et al.*, 1975) is due to its specific interaction with the anion-transporting protein, Band 3, for which it serves as a photoaffinity label (Knauf *et al.*, 1978).

One problem with use of the reagent NAP-taurine concerns the variety of reactions which it, and other nitrene precursors, exhibit following photolysis. The insertion into C–H or N–H bonds on a putative sub-millisecond time scale appears to be a minor reaction compared to hydrogen abstraction, nucleophilic attack, rearrangement or addition. Hydrogen abstraction appears to occur; the labelling of the cytoplasmic surface of red cell membranes was accompanied by what appears to be free radical polymerization of haemoglobin (Staros *et al.*, 1975). Nitrene rearrangements appear to occur. Mas *et al.* (1980) found that prephotolysis of the reagent followed by mixing with their membranes in the dark gave substantial incorporation of the reagent over a half-hour period, suggesting that labelling was occurring due to reaction of protein with a long-lived nitrene rearrangement product. For some purposes these side reactions may not be serious. The reaction may be quenched with mercaptans after several minutes of photolysis in order to minimize non-nitrene labelling, if desired. Even then it has been observed that label is lost from peptides during their purification, indicating that unstable products have been formed which are not the result of a nitrene-insertion reaction. This precludes making strictly quantitative statements regarding the labelling of peptides in a membrane protein, but comparative and semiquantitative comparisons are still possible.

A serious drawback of NAP-taurine for protein aqueous-surface labelling is that the reagent is amphipathic and partitions into the membrane. Following reaction the reagent and its products must be removed by extensive washing or by repurification of the membrane. This strongly suggests that labelling may not be restricted totally to the aqueous-exposed region of the membrane proteins but could penetrate to a depth of several angstroms in the lipid bilayer. Some of the minor red cell membrane proteins labelled by Staros and Richards (1974) may, in fact, be deeply penetrating from the interior surface rather than exposed to the external surface (Richards and Brunner, 1980).

(b) *A naphthyldisulphonyl azide*
A reagent which is designed to be less penetrating than NAP-taurine is the fluorescent compound 3-azido-2,7-naphthalene disulphonate, or ANDS (Dockter, 1979; Moreland and Dockter, 1980). This nitrene precursor has been found to label all of the red cell membrane proteins which are generally available to other chemical probes. It has recently been prepared in radioactive form (Dockter and Koseki, 1983). Although experience with it is somewhat limited, ANDS may be a useful alternative to NAP-taurine.

(c) *General considerations*
For non-penetrating photochemical labelling, ANDS may be the reagent of choice. It is commercially available (Pierce Chemical Co., Rockford, Illinois, USA) although not in radioactive form. NAP-taurine may be conveniently synthesized from high specific radioactivity [$^{35}$S]- or [$^{3}$H]-taurine. The findings with NAP-taurine concerning non-nitrene labelling and lability of the reaction product to some experimental conditions may also apply to products of any photochemical reagent.

Finally, photolysis of any membrane system in the presence or absence of a nitrene precursor can be damaging to the protein. Control experiments should be performed in the absence of the reagent. Use of inert atmosphere and reduced temperatures for the reaction is desirable.

## Acknowledgements

The author would like to thank Drs Edward Dratz and J. Hugh McDowell for helpful comments. During the preparation of this manuscript, the author was supported in part by grant EY6225 from the US National Institutes of Health, a Jules and Doris Stein Professorship from Research to Prevent Blindness, Inc., and by an unrestricted departmental grant from Research to Prevent Blindness, Inc.

## 6.6 References

Albert, A. D. and Litman, B. J. (1978) *Biochemistry*, **17**, 3893–900.
Aromatorio, D. K., Parker, J. and Brown, W. E. (1983) *Methods Enzymol.*, **91**, 384–91.
Basinger, S. F. and Hoffman, R. T. (1982) *Methods Enzymol.*, **81**, 772–82.
Bayley, H., Huang, K.-S., Radhakrishran, R., Ross, A. H., Takagaki, Y. and Khorana, H. G. (1981) *Proc. Natl. Acad. Sci. USA*, **78**, 2225–9.
Bender, W. W., Garan, H. and Berg, H. C. (1971) *J. Mol. Biol.*, **58**, 783–97.
Bennett, J. C. (1967) *Methods Enzymol.*, **11**, 330–9.
Berg, H. C. (1969) *Biochim. Biophys. Acta*, **183**, 65–78.
Berg, H. C. and Hirsh, D. (1975) *Anal. Biochem.*, **66**, 629–31.
Bligh, E. G. and Dyer, W. J. (1959) *Can. J. Biochem. Physiol.*, **37**, 911–17.

Brenner, S. C. and Wold, F. (1978) *Biochim. Biophys. Acta,* **522,** 74–83.
Brewer, G. J. and Singer, S. J. (1974) *Biochemistry,* **13,** 3580–8.
Browne, D. T. and Kent, S. B. H. (1975) *Biochem. Biophys. Res. Commun.,* **67,** 126–32.
Cabantchik, Z. I., Knauf, P. A. and Rothstein, A. (1978) *Biochim. Biophys. Acta,* **515,** 239–302.
Carpenter, F. H. (1967) *Methods Enzymol.,* **11,** 237.
Carraway, K. L. (1975) *Biochim. Biophys. Acta,* **415,** 379–410.
Carraway, K. L. and Shin, B. C. (1972) *J. Biol. Chem.,* **247,** 2102–8.
Cleveland, D. W., Fischer, S. G., Kirschner, M. W. and Laemmli, V. K. (1977) *J. Biol. Chem.,* **252,** 1102–6.
Cohen, L. A. (1970) *The Enzymes,* Vol. 1, 3rd edn. (ed P. D. Boyer), Academic Press, New York, pp. 147–211.
Connellan, J. M., Chung, S. I., Whetzel, N. K., Bradley, L. M. and Folk, J. E. (1971) *J. Biol. Chem.,* **246,** 1093–8.
Dockter, M. E. (1979) *J. Biol. Chem.,* **254,** 2161–4.
Dockter, M. E. and Koseki, T. (1983) *Biochemistry,* **22,** 3954–61.
Drapeau, G. R. (1976) *Methods Enzymol.,* **45,** 469–75.
Dratz, E. A. and Hargrave, P. A. (1983) *Trends Biochem. Sci.,* **8,** 128–31.
Dratz, E. A., Miljanich, G. P., Nemes, P. P., Gaw, J. E. and Schwartz, S. (1979) *Photochem. Photobiol.,* **29,** 661–70.
Dutton, A. and Singer, S. J. (1975) *Proc. Natl. Acad. Sci. USA,* **72,** 2568–71.
Fahrney, D. E. and Gold, A. M. (1963) *J. Amer. Chem. Soc.,* **85,** 997–1000.
Farnsworth, C. C. and Dratz, E. A. (1976) *Biochim. Biophys. Acta,* **443,** 556–70.
Findlay, J. B. C., Brett, M. and Pappin, D. J. C. (1981) *Nature (London),* **293,** 314–16.
Folk, J. E. (1983) *Adv. Enzymol.,* **54,** 1–56.
Frost, E. H. E. (1977) *J. Gen. Virol.,* **35,** 181–5.
Fung, B. K.-K. and Hubbell, W. L. (1978a) *Biochemistry,* **17,** 4396–402.
Fung, B. K.-K. and Hubbell, W. L. (1978b) *Biochemistry,* **17,** 4403–10.
Gabriel, O. (1971) *Methods Enzymol.,* **22,** 578–604.
Greenwood, F. C., Hunter, W. M. and Glover, J. S. (1963) *Biochem. J.,* **89,** 114–23.
Hand, E. S. and Jencks, W. P. (1962) *J. Amer. Chem. Soc.,* **84,** 3505–14.
Hargrave, P. A. and Fong, S.-L. (1977) *J. Supramol. Struct.,* **6,** 559–70.
Hargrave, P. A., Fong, S.-L., McDowell, J. H., Mas, M. T., Curtis, D. R., Wang, J. K., Juszczak, E. and Smith, D. P. (1980) *Neurochem. Int.,* **1,** 231–44.
Hargrave, P. A., McDowell, J. H., Curtis, D. R., Wang, J. K., Juszczak, E., Fong, S.-L., Rao, J. K. M. and Argos, P. (1983) *Biophys. Struct. Mech.,* **9,** 235–44.
Hubbard, A. L. and Cohn, Z. A. (1972) *J. Cell Biol.,* **55,** 390–405.
Islam, M. N., Briones-Urbina, R., Bako, G. and Farid, N. R. (1983) *Anal. Biochem.,* **130,** 260–5.
Iwanij, V. (1977) *Eur. J. Biochem.,* **80,** 359–68.
Jentoft, N. and Dearborn, D. G. (1979) *J. Biol. Chem.,* **254,** 4359–65.
Khorana, H. G., Gerber, G. E., Herlihy, W. C., Gray, C. P., Anderegg, R. J., Nihel, K. and Biemann, K. (1979) *Proc. Natl. Acad. Sci. USA,* **76,** 5046–50.
Kiehn, E. D. and Holland, J. J. (1968) *Proc. Natl. Acad. Sci. USA,* **61,** 1370–7.
Knauf, P. A., Brenner, W., McCulloch, L. and Rothstein, A. (1978) *J. Gen. Physiol.,* **72,** 631–49.
Knowles, J. R. (1972) *Acc. Chem. Res.,* **5,** 155–60.

Kühn, H., Mommertz, O. and Hargrave, P. A. (1982) *Biochim. Biophys. Acta,* **679,** 95–100.

Lorand, L., Campbell-Wilkes, L. K. and Cooperstein, L. (1972) *Anal. Biochem.,* **50,** 623–31.

Lorand, L., Parameswaran, K. N., Stenberg, P., Tong, Y. S., Velasco, P. I., Jönsson, N. A., Mikiver, L. and Moses, P. (1979) *Biochemistry,* **18,** 1756–65.

Luthra, M. G., Friedman, J. M. and Sears, D. A. (1978) *J. Biol. Chem.,* **253,** 5647–53.

Markwell, M. A. K. (1982) *Anal. Biochem.,* **125,** 427–32.

Markwell, M. A. K. and Fox, C. F. (1978) *Biochemistry,* **17,** 4807–17.

Mas, M. T., Wang, J. K. and Hargrave, P. A. (1980) *Biochemistry,* **19,** 684–92.

Matheson, R. R., Van Wart, H. E., Burgess, A. W., Weinstein, L. I. and Scheraga, H. A. (1977) *Biochemistry,* **16,** 396–403.

Matsubara, H. (1970) *Methods Enzymol.,* **19,** 642–51.

McDowell, J. H., Ubel, A. and Hargrave, P. A. (1986) *Arch. Biochem. Biophys.* (in press).

Means, G. E. and Feeney, R. E. (1971) *Chemical Modification of Proteins,* Holden-Day, San Francisco.

Mersel, M., Benenson, A. and Doljanski, F. (1976) *Biochem. Biophys. Res. Commun.,* **70,** 1166–71.

Mihalyi, E. (1978a) *Application of Proteolytic Enzymes to Protein Structure Studies,* Vol. 1, 2nd edn., CRC Press, West Palm Beach, FL.

Mihalyi, E. (1978b) *Application of Proteolytic Enzymes to Protein Structure Studies,* Vol. 2, 2nd edn., CRC Press, West Palm Beach, FL.

Molday, R. S. and Molday, L. L. (1979) *J. Biol. Chem.,* **254,** 4653–60.

Moore, G. (1975) *Can. J. Biochem.,* **53,** 328–37.

Moreland, R. B. and Dockter, M. E. (1980) *Anal. Biochem.,* **103,** 26–32.

Morrison, M. (1980) *Methods Enzymol.,* **70,** 214–20.

Munford, R. S. and Gotschlich, E. C. (1977) *J. Bacteriol.,* **130,** 775–80.

Nathenson, S. G., Uehara, H., Ewenstein, B. M., Kindt, T. J. and Coligan, J. E. (1981) *Annu. Rev. Biochem.,* **50,** 1025–52.

Nemes, P. P., Miljamich, G. P., White, D. L. and Dratz, E. A. (1980) *Biochemistry,* **19,** 2067–74.

Osheroff, N., Feinberg, B. A., Margoliash, E. and Morrison, M. (1977) *J. Biol. Chem.,* **252,** 7743–51.

Ovchinnikov, Y. A. (1982) *FEBS Lett.,* **148,** 179–91.

Ozols, J., Gerard, C. and Nobrega, F. G. (1976) *J. Biol. Chem.,* **251,** 6767–74.

Phillips, D. R. and Morrison, M. (1971) *Biochemistry,* **10,** 1766–71.

Pober, J. S., Iwanij, V., Reich, E. and Stryer, L. (1978) *Biochemistry,* **17,** 2163–9.

Pober, J. S. and Stryer, L. (1975) *J. Mol. Biol.,* **95,** 477–81.

Richards, F. M. and Brunner, J. (1980) *Ann. N.Y. Acad. Sci.,* **346,** 144–64.

Richards, F. M. and Vithayathil, P. J. (1959) *J. Biol. Chem.,* **234,** 1459–65.

Sale, G. J., Towner, P. and Akhtar, M. (1977) *Biochemistry,* **16,** 5641–9.

Schlager, S. I. (1979) *J. Immunol.,* **123,** 2108–13.

Schrode, J. and Folk, J. E. (1978) *J. Biol. Chem.,* **253,** 4837–40.

Sears, D. A., Reed, C. F. and Helmkamp, R. W. (1971) *Biochim. Biophys. Acta,* **233,** 716–19.

Smith, W. P., Tai, P.-C. and Davis, B. D. (1979) *Biochemistry,* **18,** 198–202.

Staros, J. V. (1980) *Trends Biochem. Sci.,* **5,** 320–2.

Staros, J. V. and Richards, F. M. (1974) *Biochemistry*, **13**, 2720–6.
Staros, J. V., Richards, F. M. and Haley, B. E. (1975) *J. Biol. Chem.*, **250**, 8174–8.
Steck, T. L., Ramos, B. and Strapazon, E. (1976) *Biochemistry*, **15**, 1154–61.
Sullivan, K. H. and Williams, R. (1982) *Anal. Biochem.*, **120**, 254–8.
Swank, R. T. and Munkres, K. D. (1971) *Anal. Biochem.*, **39**, 462–77.
Tuszynski, G. P., Knight, L. C., Kornecki, E. and Srivastava, S. (1983) *Anal. Biochem.*, **130**, 166–70.
Walker, J. E., Carne, A. F. and Schmitt, H. W. (1979) *Nature (London)*, **278**, 653–4.
Wang, J. K., McDowell, J. H. and Hargrave, P. A. (1980) *Biochemistry*, **19**, 5111–17.
Whiteley, N. M. and Berg, H. C. (1974) *J. Mol. Biol.*, **87**, 541–61.
Zubay, G. (1983) *Biochemistry*, Addison-Wesley Co., Reading, MA, p. 127.

# 7

# Topology of membrane proteins – determination of regions exposed to the lipid bilayer

ROBERTO BISSON and CESARE MONTECUCCO



## Abbreviations

RBC – red blood cell
SDS-PAGE – sodium dodecyl sulphate–polyacrylamide gel electrophoresis
s-Ig – surface immunoglobulins
HLA – histocompatibility or transplantation antigens
CD – circular dichroism
HPLC – high-performance liquid chromatography
TLC – thin-layer chromatography
PLEP – phospholipid exchange protein

## 7.1 Introduction

The fluid mosaic model and the chemiosmotic hypothesis have stimulated a great deal of research on biological membranes (Singer and Nicholson, 1972; Mitchell, 1966). As a result our knowledge of their structure and function is reaching the molecular level.

Protein embedded partially or totally in the lipid bilayer, classified as integral (Singer, 1974), can be divided into structurally and functionally different classes.

Techniques for the Analysis of Membrane Proteins. Edited by C. I. Ragan & R. J. Cherry.
Published in 1986 by Chapman and Hall, 11 New Fetter Lane, London EC4P 4EE

Transport proteins are transmembrane with two hydrophilic domains connected by a hydrophobic sector, which interacts with the hydrocarbon portion of lipids. Their polypeptide chain(s) must span the membrane more than once to account for the large bulk of their hydrophobic domains. To this group of proteins belong bacteriorhodopsin, the red cell anion transporter, the $Ca^{2+}$- and the $(Na^{+}+K^{+})$-ATPases and the mitochondrial complexes (Henderson and Unwin, 1975; Rothstein, 1982; Hobbs and Albers, 1980; Jorgensen, 1980; Azzi, 1980; Capaldi, 1982).

The hydrophobic domain of other integral proteins, not involved in transport functions, appear to be simpler, being formed by one or two polypeptide chains not necessarily transversing the membrane. Examples of such proteins are cytochrome $b_5$, glycophorin, the sucrase–isomaltase complex, the histocompatibility antigens and surface immunoglobulins (Strittmatter and Dailey, 1982; Marchesi *et al.*, 1976; Brunner *et al.*, 1979; Strominger *et al.*, 1980; Rogers *et al.*, 1980). Fatty acids may be covalently linked to the protein to act as a membrane anchor and/or to strengthen and stabilize the insertion of a polypeptide chain in the membrane (Schlesinger and Magee, 1982; Hantke and Braun, 1973; Kamiya *et al.*, 1979; Rose *et al.*, 1984).

In addition to membrane proteins classified as peripheral (Singer, 1974), there are now several examples of proteins such as certain toxins (van Heyningen, 1977; Luria and Suit, 1982) and complement (Esser, 1982), which exist in a water-soluble form and yet are able to penetrate the hydrophobic membrane interior.

Membrane proteins may be composed of more than one subunit and may interact specifically with other proteins both in and out of the lipid phase (Depierre and Ernster, 1977). It has also been well established that the activity of membrane-embedded enzymes is lipid-dependent. In certain cases a strong selectivity for a defined class of lipids has been demonstrated (Sandermann, 1978).

Integral proteins are difficult to crystallize. Only recently have three-dimensional X-ray-diffraction-grade crystals of *Escherichia coli* porin and of *Rhodopseudomonas viridis* reaction centres been obtained (Garavito *et al.*, 1983; Michel, 1982). The tendency of membrane proteins is to form two-dimensional crystalline arrays but only in the case of bacteriorhodopsin has the resolution been sufficient to show details of the secondary structure, which appears to be formed by seven α-helices spanning the membrane (Henderson and Unwin, 1975). Circular dichroism studies indicate that the membrane domain of other integral proteins have high α-helical content (Lenard and Singer, 1966; Spiess *et al.*, 1982). The adoption of helical conformations in the membrane environment is also suggested by some theoretical considerations on peptide folding in a hydrophobic phase (Kennedy, 1978; Henderson, 1981). Such conformations minimize the exposure of the polar groups of the peptide backbone since they are all involved in intrachain hydrogen-bonding (see Fig. 7.1a). For the same reasons $\beta$-turns in the hydrocarbon region of the membrane are unlikely even though they cannot be excluded *a priori* (Rose *et al.*, 1983). Maximum hydrogen-bonding can also be obtained in a $\beta$-barrel structure, which has been found in the hydrophobic interior of some soluble proteins (Richardson, 1981; Mavridis *et al.*, 1982). As shown in Fig. 7.1(b) the lateral residues point alternately in and out of the barrel so that even charged stretches can span the membrane with their polar residues hidden within the barrel. For the *E. coli* outer membrane protein porin CD spectroscopy indicates a high content of $\beta$ structure (Rosenbusch, 1974).

Unlike their hydrophilic domains, the hydrophobic sector of integral proteins is difficult to study; it contains few reactive amino acid residues, it is not degraded by proteases *in situ*, it generally does not contain the enzyme active site and it is poorly antigenic. Moreover the difficulty of handling hydrophobic peptides has made the determination of their amino acid

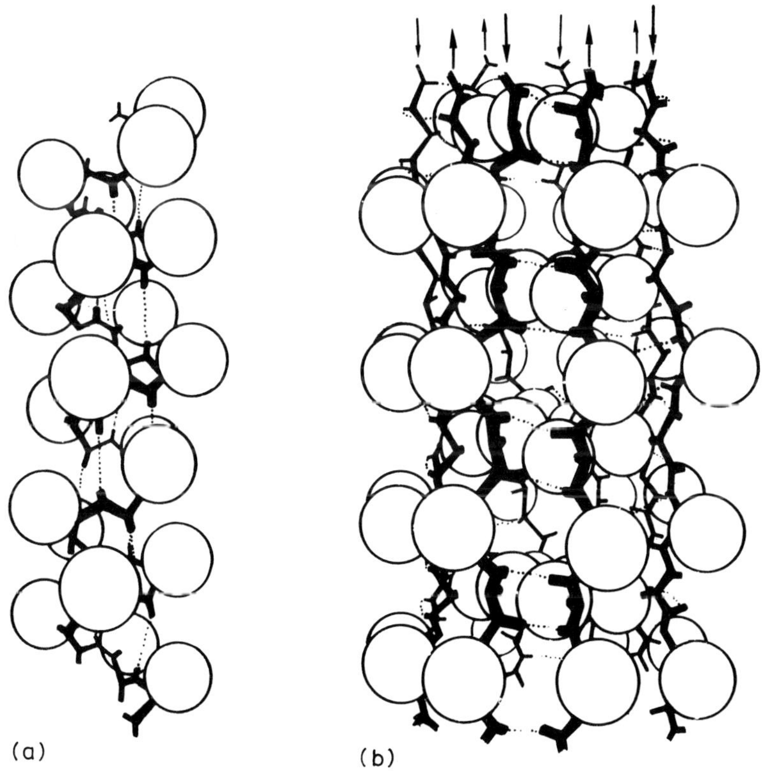


**Fig. 7.1** Two of the possible configurations which could be adopted by an integral protein in the hydrophobic membrane phase: (a) $\alpha$-helix; (b) $\beta$-barrel. Lateral residues are indicated by circles. Hydrogen-bonds are indicated by dashed lines.

sequences very difficult. Nevertheless several primary structures of integral membrane proteins are now known, mainly from knowledge of their nucleic acid sequences (see Chapter 9). Most of these show stretches of hydrophobic residues which could well be intercalated in the lipid bilayer, even though this does not necessarily imply an involvement in the hydrophobic domain.

New approaches have been developed to identify those polypeptide chains exposed to lipids and to gain some information on their spatial arrangement and their mode of interaction with phospholipids. The present review will discuss at length the use of photoreactive probes, which offers the highest potential, and briefly that of group-specific reagents and cross-linkers, which have been reviewed by others (Peters and Richards, 1977; Das and Fox, 1979; Sigrist and Zahler, 1982; Ji, 1983).

## 7.2 Hydrophobic labelling reagents

### 7.2.1 WHAT WE WOULD LIKE TO HAVE

(a) *Labelling restricted to the hydrophobic domain*
This can be obtained with the use of highly hydrophobic molecules, which will partition efficiently into the membrane or more safely by anchoring the reagent to a lipid molecule.

(b) *High reactivity*
A hydrophobic labelling reagent should be highly reactive and unselective towards the different residues exposed to the hydrocarbon phase. It should be able to insert into single C–H bonds in order to label proteins having a high content of unreactive aliphatic residues. It should have a very short lifetime in order to allow the study of dynamic phenomena.

(c) *Absence of chemical reactivity*
The probe should be unable to react before it partitions in the membrane so as to avoid the labelling of the water-exposed protein surface.

(d) *Stability of the probe*
It should be stable under storage conditions in the radioactive form so as to allow one to perform many experiments before the formation of by-products, which could have labelling properties different from those of the parent compound.

(e) *Sensitive detection of the labelled products*
In this respect highly radioactive chemicals should be preferred over, for example, fluorescent reagents.

(f) *Stability of the labelled derivative*
The product of labelling should be stable to all the conditions used in the identification of the modified protein residues.

### 7.2.2 WHAT WE HAVE

At the present time the chemical reagents that best meet these requirements are carbene and nitrene precursors (Knowles, 1972; Bayley and Knowles, 1977; Chowdry and Westheimer, 1979; Staros, 1980). These molecules can be made very hydrophobic and highly radioactive and in the dark they are stable under most conditions used in biochemical research. Carbenes and nitrenes can be photogenerated *in situ* by illumination. These are electron-deficient unstable intermediates which can react with neighbouring molecules forming covalent derivatives or rearrange intramolecularly as shown in Fig. 7.2. The relative rates of the different routes of reaction depend on the group directly linked to the carbon or nitrogen atom and on the reactivity of neighbouring molecules. Needless to say, only the first of these two parameters can be manipulated by properly designed synthetic schemes.

The use of hydrophobic photogenerated probes has been the subject of a symposium and of some reviews (Tometsko and Richards, 1980; Brunner, 1981; Robson *et al.*, 1982; Bayley, 1982, 1983). This chapter will focus mainly on the biologically relevant information obtained by hydrophobic photolabelling and will discuss critically the use of the various probes which have been introduced.

## 7.3 Lipophilic photoreactive probes

### 7.3.1 GENERAL PROPERTIES

Table 7.1 lists the reagents used in membrane studies together with some of their properties. Most of them are aromatic compounds because these are more stable than the corresponding aliphatics; they also have a lower tendency to rearrange and a lower chemical reactivity. Rearrangement should be avoided since it generally decreases the yield of cross-linked product. In fact, the photogenerated rearranged intermediate may be chemically stable or show group-specific reactivity (see Fig. 7.2).

It has been shown for arylnitrenes that reactivity varies with the aromatic system, increasing from 1-azidoanthracene to 1-azidonaphthalene to azidobenzene (Reiser *et al.*, 1966, 1968). Moreover the presence of a substituent on the aromatic ring affects the arylnitrene lifetime in the sense that it is decreased by electron-withdrawing groups and increased by electron-donor groups. For example in a soft polystyrene matrix the presence of the weakly electron-withdrawing acetoxy group is sufficient to lower the

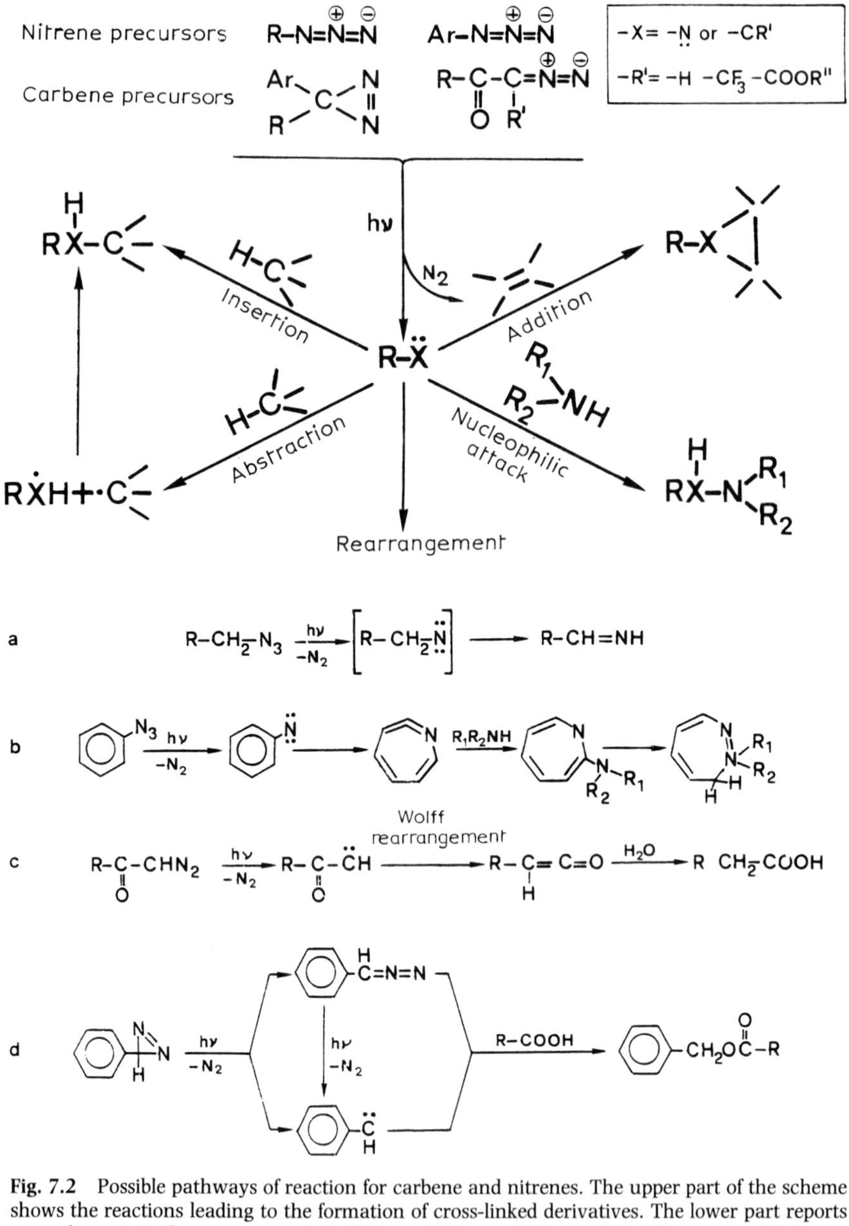


**Fig. 7.2** Possible pathways of reaction for carbene and nitrenes. The upper part of the scheme shows the reactions leading to the formation of cross-linked derivatives. The lower part reports some characterized rearrangements (a: Kyba and Abramovitch, 1980; b: Chapman and Le Roux, 1978; Nielsen and Buchardt, 1982; c: Chowdhry and Westheimer, 1979; d: Smith and Knowles, 1975). They may or may not lead to cross-linking; in any case the rearranged intermediate will be more selective than the parent carbene or nitrene.

nitrenobenzene lifetime from 0.1 to 0.02 s (Reiser and Leyshon, 1970). It is expected that the powerful nitro group will further decrease this value. In addition, it should be emphasized that the lifetime in a biological membrane will be much shorter because of the presence of reactive amino acid residues, lipid double bonds and water. This aspect is exemplified by the finding that the lifetime of 1-nitrenoanthracene is around 1 s in an unreactive polystyrene matrix whereas it is between 3 and 10 $\mu$s in an organic glass at 77 K (Reiser *et al.*, 1966; Reiser and Leyshon, 1970). Moreover in a biological membrane the diffusion of the probe makes possible the reactions of nitrene recombination and nitrene attack upon azide which lead to self-quenching of the reagent. This appears to be a major pathway of decay even at concentrations as low as $10^{-5}$ M (Reiser *et al.*, 1968).

From the above considerations it was estimated that the lifetime of a nitrenonitrobenzene in a lipid phase is in the region of $10^{-4}$–$10^{-5}$ s, provided it cannot recombine (Reiser, personal communication). If recombination is the main reaction, as probably occurs in biological membranes because of the high concentrations always used (millimolar up to molar in the lipid phase) the lifetime surely is much shorter. Similar data are not available for arylcarbenes but these general considerations are expected to hold also in this case.

It should be made clear that a short lifetime does not necessarily mean high reactivity and high yields of cross-linking because both the precursor and the photogenerated intermediate may undergo internal rearrangement and because of self-quenching.

It is a general rule that increased reactivity results in decreased selectivity. However, even the highly reactive aryltrifluoromethylcarbene, when photogenerated in methanol, shows a 95% insertion into O–H bonds with respect to C–H bonds (despite the amount of C–H being three times larger than that of O–H) (Brunner *et al.*, 1980). The goal of a low selectivity is far from being reached by any of the available reagents.

Another relevant aspect for consideration is the type of radioactive isotope used and its specific radioactivity, which is related to the rate of self-decomposition of the radiochemical (Evans, 1976). Compounds of high specific radioactivity are desirable because they can be used in trace amounts so causing minimal perturbation of the biological system.

As reported in Table 7.1 the various probes show different absorption characteristics. In general terms those which can be photoactivated with non-protein-damaging UV or visible radiations should be preferred.

### 7.3.2 LABELLING PROCEDURES

The probe is added from an ethanolic stock solution to the membrane suspension (purged with $N_2$ for at least 30 min) with stirring (final ethanol concentration lower than 1%). Incubations of 30–60 min are generally

**Table 7.1** Lipophilic photoreactivatable reagents

| Structure | Name and symbol | $\lambda$(nm)/$\varepsilon_{mM}$ | Specific radioactivity (Ci/mol) | References |
|---|---|---|---|---|
| | | *Nitrene precursors* | | |
| 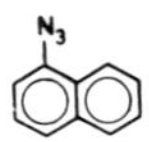 | 1-[$^{3}$H]Azidonaphthalene NA | 300 | 2.5 | Klip and Gitler (1974) |
| 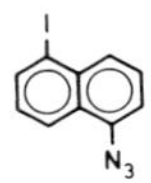 | 5-[$^{125}$I]Iodonaphthyl azide INA | 310/21400 | 15 000 | Bercovici and Gitler (1978) |
| 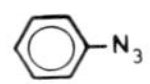 | [$^{3}$H] [$^{14}$C] Phenyl azide PA | 250/10200 | 97<br>2.2 | Bayley and Knowles (1978) |
| 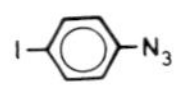 | 1-Azido-4-[$^{125}$I]iodobenzene IPA | 258 | 10 | Wells and Findlay (1979) |
|  | [illegible] | [illegible] | [illegible] | Carlotti and Schatz (1979) |

| | | | | |
|---|---|---|---|---|
|  | PySA | | | (1979) |
| 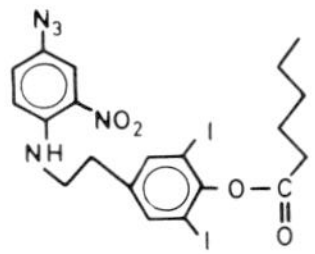 | [$^{125}$I]Hexanoyldi-iodo-*N*-(4-azido-2-nitrophenyl)-tyramine HNT | 460/5800 | 200 000 | Hebdon *et al.* (1980) |
| | *Carbene precursors* | | | |
| 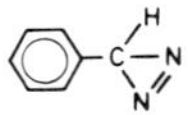 | 3-Phenyl-3H-diazirine PD | 362/299 | | Smith and Knowles (1975) |
| 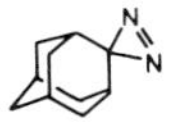 | 1-[$^3$H]Spiro(adamantane-4,4′-diazirine) AD | 372/245 | 850 | Bayley and Knowles (1980) |
| 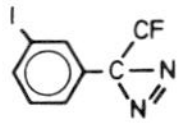 | 3-[$^{125}$I]Trifluoromethyl-3-(*m*-iodophenyl)diazirine TID | 353/266 | 10 000 | Brunner and Semenza (1981) |

sufficient to allow equilibration. Photolysis is carried out at wavelengths longer than 300 nm. The irradiation time depends on the probe, on the power of the light source, its distance from the sample and on the optical density and turbidity of the membrane suspension. It is therefore advisable to define for any experiment the particular set of conditions leading to optimal covalent incorporation of the probe into the membrane components. Times ranging from seconds to half an hour have been found to be sufficient in most experiments. When present, enzymatic activity must be checked in order to test for irradiation damage. Unbound photoreacted probe is washed out with an albumin-containing buffer. The labelled species are then determined by SDS-PAGE, gel chromatography in the presence of detergents, HPLC, TLC or other suitable methods.

### 7.3.3 THE RED CELL MEMBRANE AS A MODEL

To date most studies employing lipophilic photoreactive reagents have been concerned more to test their responses than to obtain new information on membrane structure. Among the available model systems, the red cell membrane has been most widely used. Fig. 7.3 shows a pictorial view of the red cell membrane with its three best characterized proteins. Spectrin, the main component of the cytoskeleton, accounts for about 30% of the

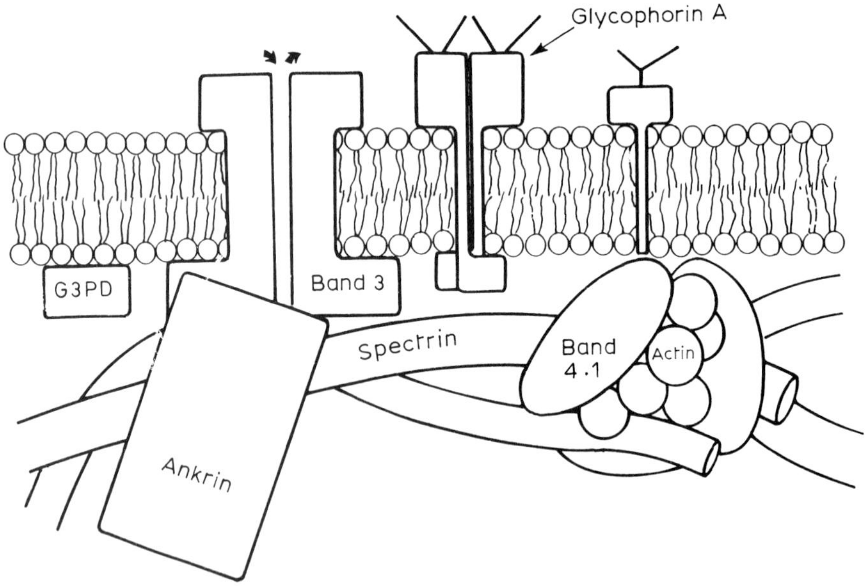


**Fig. 7.3** A pictorial view of the red cell membrane showing some of the best characterized protein components; G3PD is glyceraldehyde-3-phosphate dehydrogenase.

membrane protein; it is a heterodimer ($M_r$ over 200 000) and it is known to be peripheral to the lipid bilayer (Marchesi, 1979). Band 3, the anion-transport system, has an $M_r$ of 90 000–110 000 and it is present in around $10^6$ copies per cell (Cabantchik *et al.*, 1978). Its polypeptide chain transverses the membrane several times and hence a major portion of the protein is embedded in the membrane. Glycophorin ($M_r$ 36 000) spans the lipid bilayer once and most of its mass is outside the membrane (Marchesi *et al.*, 1976).

Only on this membrane have both carbene and nitrene probes been used under comparable (even though not identical) conditions. The results of labelling the red blood cell membrane with different reagents, reported in Table 7.2, can be compared on the basis of the above mentioned structure and localization of some of its protein components. Therefore one would expect no labelling of the extrinsic spectrin, but intense labelling of band 3 and a lower extent of glycophorin modification. The result nearest to this prediction was obtained with TID, while both AD and INA gave significant labelling of spectrin, which was still more highly labelled by PA; glycophorin reacted more heavily than band 3 by INA (Bercovici and Gitler, 1978; Kahane and Gitler, 1978; Gitler and Bercovici, 1980; Wells and Findlay, 1979; Bayley and Knowles, 1980; Brunner and Semenza, 1981). Another interesting aspect is represented by the amount of radioactivity incorporated into lipids with respect to proteins. Thus labelling of the inert hydrocarbon part of lipids is directly indicative of high reactivity, which in turn should result, at least in

**Table 7.2** Red blood cell membrane labelling with small lipophilic reagents*

| *Probe* | | *P*† | *Yield of protein labelling (%)* | *Lipid over protein labelling* | *Percentage of labelling on:* | | |
|---|---|---|---|---|---|---|---|
| | | | | | *Spectrin* | *Band 3* | *Glycophorin* |
| PA | 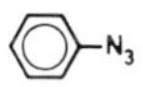 | 240 | 5 | — | 23 | | 61 |
| INA | 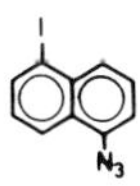 | 163 000 | 18 | 2 | 4 | 15 | 70 |
| AD | 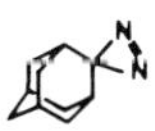 | 1 750 | 20 | 4 | 5 | 37 | 34 |
| TID | 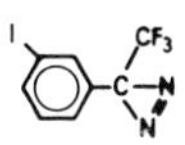 | 39 000 | 2.7 | 16 | 0.4 | 48 | 7.4 |

*Data are taken or estimated from: Wells and Findlay (1979a); Gitler and Bercovici (1980); Bayley and Knowles (1980); Brunner and Semenza (1981).
†Partition coefficient defined as $P$=(ligand bound/milligram of lipid)/(free ligand/microlitre external solution).

principle, in the labelling of the aliphatic lateral residues of proteins. In this respect TID again appears to give a better result.

It is noteworthy that the patterns of labelling both in terms of topographical distribution of the label and of the yield of cross-linking are not strictly correlated with the hydrophobicity of the reagent.

Control experiments can easily be performed to test the possibility that the label is bound to the hydrophilic domain(s) of an integral protein. Membrane-impermeant thiols, such as glutathione, can effectively scavenge nitrenes and carbenes present in the aqueous medium (Staros *et al.*, 1978; Takagaki *et al.*, 1980). For TID, AD and INA no reduction of their incorporation in the membrane was found while glutathione reduced fivefold the membrane labelling with PA. For AD there was a 50% decrease in the amount of glycophorin labelling. An additional control is the lack of labelling of soluble proteins able to bind to the membrane surface, i.e. antibodies, lectins, hormones etc. (Prochaska *et al.*, 1980; Bayley and Knowles, 1980). However, none of these treatments can exclude the possibility of labelling the hydrophilic domain due to absorption of the probe into hydrophobic pockets. Also the possibility of alterations of the membrane due to protein disulphide reduction or ligand-induced redistribution of receptors should be considered. Another potential source of membrane damage during photolabelling experiments is irradiation. These artifacts can be to some extent evaluated if the membrane possesses measurable enzyme activities.

A further complication in studying biological membranes may be the difficulty of resolving its protein components. With SDS-PAGE, the most widely used method for the analysis of the pattern of labelling of red cell membrane, a co-migration of unidentified (yet surely present) integral proteins with the few identified components is possible. These proteins (i.e. the $Ca^{2+}$ and $Na^{+}+K^{+}$-ATPases), even though present in very low quantity, may incorporate a significant amount of label because of their large hydrophobic domains and because of the possible presence of exceptionally reactive lateral residues exposed to lipids (see Section 7.4.5).

In the case of the abundantly available glycophorin, we have a membrane protein whose sequence and fragmentation patterns are known and therefore the label distribution along the polypeptide chain can be determined. Fig. 7.4 shows the sites of cleavage with trypsin and CNBr. Tryptic fragment 62–97 contains a 22-residue hydrophobic sequence, which is believed to be intercalated in the lipid bilayer (Marchesi *et al.*, 1976). The percentage of radioactive label associated with this fragment was: 96% for TID, 80–90% for INA and for IPA and 70–85% for AD (Brunner and Semenza, 1981; Bayley and Knowles, 1980; Kahane and Gitler, 1978; Bercovici and Gitler, 1978; Wells and Findlay, 1979). Residual radioactivity could not be attributed to any of the other tryptic peptides except for some 5% of INA and of IPA, which were bound to the hydrophilic 1–31 fragment. The analysis of CNBr fragments

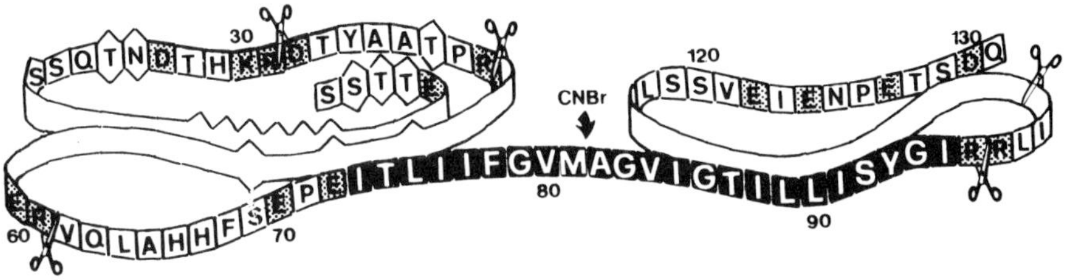


**Fig. 7.4** Patterns of fragmentation of glycophorin with trypsin (scissor) and CNBr. The largest tryptic fragment, containing the hydrophobic stretch 73–95 is shown in black. ⬡ glycosilated and ▩ charged residues.

shows in the case of AD that more than one residue was labelled. Further information can be obtained only by sequencing and determination of the modified site(s). In Section 7.4.5 the single available analysis with small photoreactive probes will be discussed.

### 7.3.4 ISOLATED MULTISUBUNIT MEMBRANE PROTEINS

Labelling of membrane proteins *in situ* is always to be preferred. However, the analysis of the pattern of labelling of a protein may be very difficult when present in low amounts with respect to the other membrane components and when it cannot be recovered with reasonable yield. This problem can be partially overcome by increasing the amount of probe introduced into the membrane at the expense of increasing its non-specific perturbing effect and its self-quenching. An alternative approach is to label the protein in an active reconstituted lipid–protein complex (for reconstitution procedures see Chapter 4).

A multisubunit protein, which has been studied with different photoreactive probes, is the $(Na^+ + K^+)$-ATPase (Hobbs and Albers, 1980; Jorgensen, 1980). This plasma membrane enzyme can be resolved into two different subunits: a larger polypeptide ($M_r$ 90 000–100 000), which is phosphorylated during the enzymic cycle, and a sialoglycoprotein subunit ($M_r$ 45 000–50 000) of unknown function. In the isolated enzyme INA and AD label only the large subunit (Karlish *et al.*, 1977; Farley *et al.*, 1980) indicating that it is present in the lipid bilayer and this has supported the widely accepted view that the glycoprotein subunit was external to the membrane. However, it should be made clear that a negative result in any kind of labelling experiment (with any chemical or biological reagent) does not tell anything about the location of an unlabelled peptide. In fact in the case of hydrophobic photolabelling the location could be external to the membrane or alternatively it could be within the membrane and shielded by the labelled peptides. The unlabelled protein could also be composed of amino acid residues less reactive than those of the labelled polypeptides, which effectively trap the photogenerated reagent before it can react with the other membrane components. There is also the possibility that the unlabelled peptide interacts with lipid so strongly as to be virtually inaccessible to the exogenous probe (Thelestam *et al.*, 1983). One additional possibility is related to the aggregation state of the enzyme, which may be different in its isolated, reconstituted and membrane states. Indeed returning to the $(Na^+ + K^+)$-ATPase the sialoglycoprotein subunit did become labelled when four different photoreactive lipids were used to probe the reconstituted enzyme, isolated from two unrelated sources (Montecucco *et al.*, 1981). The same result was obtained when HNT was used in the intact microsomal membrane (Girardet *et al.*, 1983). Thus both subunits of the $(Na^+ + K^+)$-ATPase contribute to form its lipid–protein boundary. Low concentrations of INA and AD appear to label different regions of the larger polypeptide chain and also give different patterns of labelling for the $Na^+$ and $K^+$ forms of the enzyme (Farley *et al.*, 1980; Jorgensen *et al.*, 1982). Whatever the reason for these apparently conflicting results, these studies indicate that different conformational states of a protein can be detected by this method opening up a new possibility for the use of photoreactive probes.

### 7.3.5 NATIVE MEMBRANE

Small photoreactive probes easily insert into complex biological membranes and have been used to label the integral proteins of a basophilic leukaemia cell line, of intestinal microvilli, of influenza virus and of lymphocytes (Holowka *et al.*, 1981; Noren and Sjostrom, 1980; Goldman *et al.*, 1979; Owen *et al.*, 1980).

The latter study, which introduced the use of HNT, reports data on the transplantation antigens, on surface immunoglobulins and on Thy-1 antigen. These proteins were isolated from the photolabelled membranes by specific immunoprecipitation, a method which allows the recovery of very low amounts of antigenic proteins. Several studies have shown that the transplantation antigens are transmembrane (Strominger *et al.*, 1980), while at the time the localization of s-Ig and of Thy-1 was controversial (Williams *et al.*, 1978; Feinstein *et al.*, 1979). Much evidence was presented which indicated that labelling was restricted to the hydrophobic domain. Only the heavy chain of surface immunoglobulins was labelled indicating that this is at least partially inserted into the membrane. Also Thy-1 was labelled indicating its integral nature. This is particularly interesting in the light of its sequence, now available (Campbell *et al.*, 1981), showing no evident hydrophobic peptides which could be intercalated in the membrane. This example stresses that the absence of long hydrophobic stretches in a primary sequence cannot be diagnostic of its not being an integral protein.

### 7.3.6 ADDITIONAL COMMENTS

Several aspects of the use of simple lipophilic photoactivatable probes deserve further discussion. Among these reagents, INA has been most widely used. Its preparation is relatively simple and its specific radioactivity can be high (Bercovici and Gitler, 1978). Both its $^{3}H$ and $^{125}I$ forms decompose very rapidly on storage and hence must be used within a few days of synthesis (Gitler and Bercovici, 1980). Compared to other photoreactive probes its reactivity appears to be rather low. As a result certain membrane-embedded peptides may not be labelled because of their high content of unreactive aliphatic residues (Hoppe *et al.*, 1983c). Another nitrene-generating probe is HNT, whose synthesis is relatively simple yielding a highly radioactive compound (Hebdon *et al.*, 1980). No data are available for its shelf life. It is very hydrophobic and it can be expected to be much more reactive than INA because of the phenyl instead of naphthyl aromatic group and also because of the presence of the electron-withdrawing nitro group. Moreover the nitro substituent shifts the absorption spectrum into the visible with a high molar absorption, lowering the possibility of irradiation damage to the membrane. The preparations of AD and of TID are not accessible to most biochemists (Bayley and Knowles, 1980; Brunner and Semenza, 1981). The introduction of radioactivity into AD is cumbersome and the specific radioactivity is not high. TID can instead be rendered radioactive at the last step at a high specific value.* Both are rather stable if stored at $-20°$ C in ethanol. On irradiation a diazirine can be converted not only to a carbene, but also to the corresponding diazo compound (see Fig. 7.2). The rearranged adamantane diazo compound is more reactive than the corresponding TID derivative and hence it could label proteins via the selective diazo intermediate.

## 7.4 Photoreactive lipids

### 7.4.1 WHY?

As discussed above the small probes while possessing several advantages also present a number of drawbacks, which may lead to erroneous information. In general, small hydrophobic molecules have an undefined location in the lipid bilayer and hence they cannot distinguish between the different depths of the membrane. They can be adsorbed on hydrophobic pockets or specific sites present on the hydrophilic protein surface (Ueda *et al.*, 1976; Halsey *et al.*, 1978; Stefanini *et al.*, 1979). They can also be excluded from those hydrophobic protein sites involved in strong interaction with lipids (Thelestam *et al.*, 1983). They can photolyse to intermediates with higher polarity than the parent compound and hence repartition out of the lipid bilayer (Bayley and

*TID is now commercially available from The Radiochemical Centre, Amersham, Bucks., UK.

Knowles, 1978a). Once partitioned into the cell's plasma membrane the probe can be re-extracted if serum proteins are present in the cell culture medium (Steckel *et al.*, 1983).

These limitations can be overcome, at least in principle, by anchoring the photoreporter group at a selected position of one fatty acid chain of a phospholipid (Hubbell and McConnell, 1971; Chakrabarti and Khorana, 1975; Stoffel *et al.*, 1977).

### 7.4.2 SOME PROPERTIES

Table 7.3 shows the formulae of the photoactivatable amphipatic probes which have been used in membrane studies.

At variance from simple lipophilic reagents and from probes XIV and XV which partition readily into membranes, the insertion of photoactive phospholipids may be more difficult. Generally it can be performed by liposome fusion or uptake (Pagano and Weinstein, 1978), by PLEP-mediated exchange (Akeroyd and Wirtz, 1982) and by biosynthetic incorporation of photoreactive fatty acids into the cell phospholipids (Greenberg *et al.*, 1976; Stoffel *et al.*, 1978; Leblanc *et al.*, 1982). In the case of red cell 'ghosts', sarcoplasmic reticulum and acetylcholine-receptor-rich membranes, photoactivatable phospholipids appeared to be able to insert spontaneously into the membrane (Ross *et al.*, 1982; Brunner *et al.*, 1983; Gutweniger and Montecucco, 1984; Giraudat *et al.*, 1985). With isolated and reconstituted lipid–protein complexes, the phospholipid probe is intermixed with the bulk of lipids during the reconstitution procedure (Bisson and Montecucco, 1981).

All the evidence, available only for some of the amphipathic probes of Table 7.3, shows that their labelling is restricted to the hydrophobic membrane phase (Bisson *et al.*, 1979b, 1982; Hu and Wisnieski, 1979; Montecucco *et al.*, 1980; Hoppe *et al.*, 1983b).

Brunner and Richards (1980) have compared the efficiency of lipid-bound arylnitrene and arylcarbene in cross-linking to saturated and unsaturated phosphatidylcholines. While the yields of reaction were similar for the addition to the double bond (4.3% versus 5.8%), the arylcarbene was 6.5 times more effective (18.5% versus 2.8%) than the arylnitrene in reacting with saturated lipids. It is noteworthy that the ability of an arylnitrene to insert into C–H bonds increases by an order of magnitude when anchored to a phospholipid fatty acid chain (0.25% for arylnitrene versus 2.8% for the phospholipid-bound arylnitrene) (Bayley and Knowles, 1978a; Brunner and Richards, 1980). Recently it was found that a lipid-bound nitrophenylnitrene inserts into the C–H bonds of dipalmitoyllecithin with a yield lower than 1%, raising the possibility that the previous results obtained with probe IX and X were at least partially due to radicals generated by S–S bridge fragmentation (Brunner, personal communication).

### 7.4.3 CAN THEY ACT AS MEMBRANE CHEMICAL RULERS?

Brunner and Richards (1980) also tested the potential use of photoactivatable phospholipids to serve as chemical rulers of the depth of penetration of proteins into the lipid bilayer in the lipid–gramicidin A model system. As shown in Fig. 7.5 most evidence indicates that gramicidin A forms a transmembrane ion channel via a $NH_2$ terminal to a $NH_2$ terminal dimer (Urry, 1971). Even though the two phospholipid probes IX and X bear the photoreactive group at the fatty acid methyl terminus, they label predominantly the four tryptophan residues located close to the lipid bilayer surface so implying a folding back of the photoreactive group as shown in Fig. 7.5. This indicates that the tryptophan residues are much more reactive than the other gramicidin A residues toward both the nitrene and carbene intermediates. The experiment suggests that photoreactive phospholipids can be used as depth probes only when amino acid residues of similar reactivity are present at the different levels of the membrane. However, this limitation does not hold for phospholipid analogues, whose photoactivatable group has a restricted mobility, such as probes III and XI. In these cases the reactive group is strictly localized at the membrane surface and hence it can label only those peptide regions intercalated with the polar head groups of phospholipids. Moreover the use of a 'deep' probe such as XII in conjunction with a shallow probe such as probe XI can give useful information about different degrees of penetration of a protein in the membrane. The following two experiments illustrate the potential of this approach.

Succinate ubiquinone reductase, complex II of the inner mitochondrial membrane, is formed by four subunits: $SDH_1$ ($M_r$ 70 000), $SDH_2$ ($M_r$ 27 000), $C_{II-3}$ ($M_r$ 13 500) and $C_{II-4}$ ($M_r$ 7000). Succinate dehydrogenase ($SDH_1$ + $SDH_2$) can be reversibly dissociated from $C_{II-3}$ and $C_{II-4}$ and be maintained in solution without detergents (Trumpower and Katki, 1979). Incubation of the soluble enzyme with liposomes tagged with probes XI and XII led to its insertion in the lipid bilayer. After illumination, $SDH_2$ was labelled to a considerable extent whereas significant labelling of $SDH_1$ only occurred with the shallow probe XI. When bound to $C_{II-3}$ and $C_{II-4}$, as in the inner mitochondrial membrane, $SDH_1$ was not labelled by any of the probes (Girdlestone *et al.*, 1981). The first result, although not related to the physiological situation, suggests that the two subunits penetrate the membrane to a different extent. The two smaller polypeptides appear to serve as membrane anchors for the SDH subunits in such a way as to exclude $SDH_1$ from contact with lipids.

Another example is given by cholera toxin. This protein is composed of three different subunits: $\alpha(M_r$ 21 000), $\beta(M_r$ 11 590) and $\gamma(M_r$ 5395) with the stoichiometry $\alpha\gamma\beta_5$. The $\beta$ subunit binds specifically to ganglioside $G_{M1}$, the specific cellular receptor of cholera toxin. The enzymic $\alpha$ subunit is linked to the

**Table 7.3** Amphipatic photoactivatable reagents

| Code number | Structure | | | Specific radioactivity (Ci/mol) | References |
|---|---|---|---|---|---|
| | 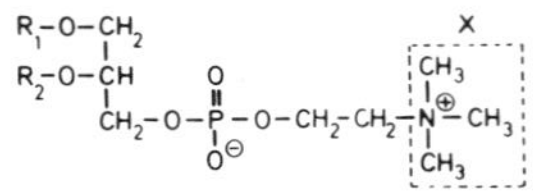 | | | | |
| | $R_1$ | $R_2$ | X | | |
| I | Palmitoyl | 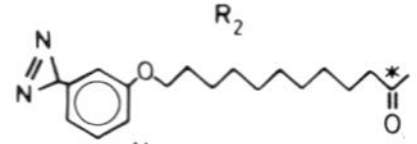 | | 48 | |
| II | Palmitoyl* | 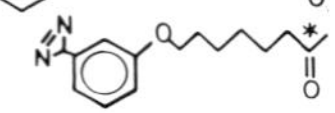 | | 5.6 | Robson *et al.* (1982) |
| III | Myristoyl | 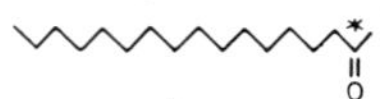 | 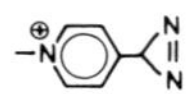 | 5.6 | |
| IV | | 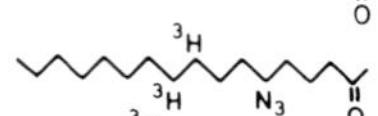 | | | |
| V | Palmitoyl or | 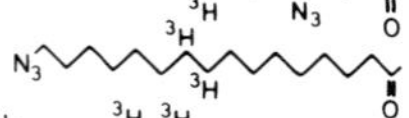 | | 100 | Stoffel *et al.* (1982) |
| VI | Stearoyl | 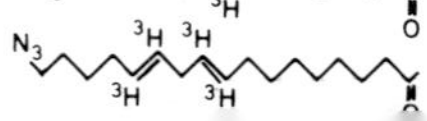 | | | |

| | | | | |
|---|---|---|---|---|
| VIII | Palmitcyl | 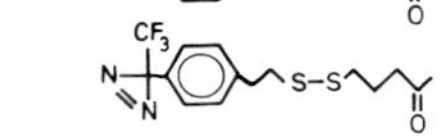 | — | Brunner and Richards (1980) |
| IX | | 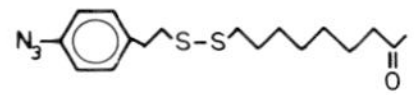 | — | |
| X | | 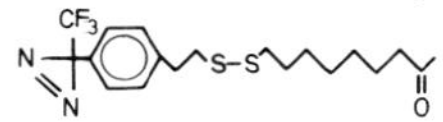 | — | |
| XI | Palmitoyl | 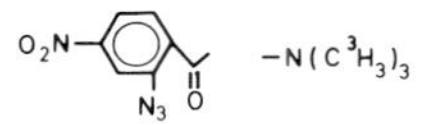 | 3,900 | Bisson and Montecucco (1981) |
| XII | Myristoyl | 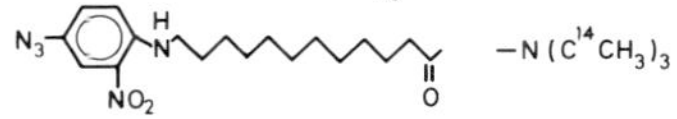 | 177 | |
| XIII | Palmitoyl | 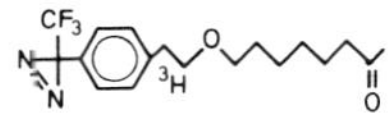 | 3 600 | Brunner *et al.* (1983) |
| XIV | | 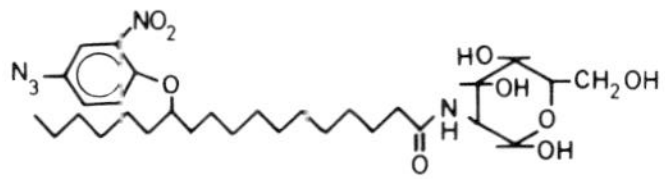 | 50 | Bramhall *et al.* (1978) |
| XV | | 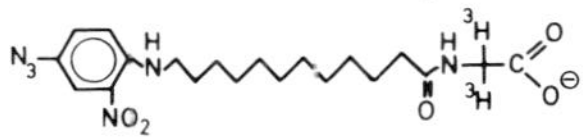 | 23 000 | Montecucco *et al.* (1981a) |

*Shows the location of the $^{14}C$ isotope.

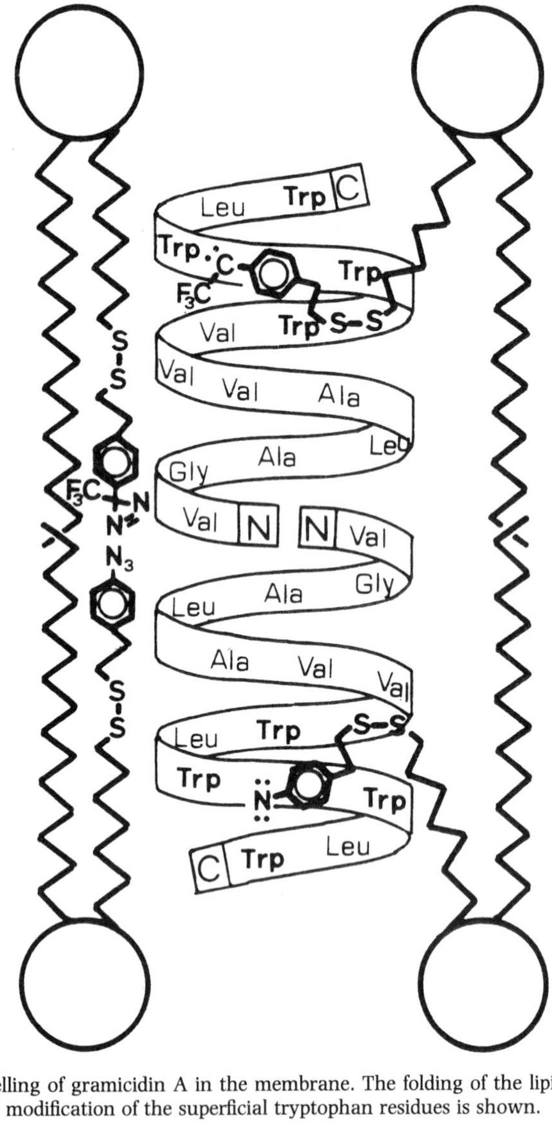


**Fig. 7.5** Labelling of gramicidin A in the membrane. The folding of the lipid fatty acid chain which leads to modification of the superficial tryptophan residues is shown.

cell binding region $\gamma\beta_5$ via a single disulphide bridge (van Heyningen, 1977). After binding to $G_{M1}$-containing liposomes, tagged with photoreactive phospholipids XI and XII or with an azidoglycolipid probe similar to XIV, and illumination, the $\beta$ and $\gamma$ polypeptides were labelled only with the shallow probe XI (Tomasi and Montecucco, 1981; Tomasi *et al.*, 1982). This indicates that the membrane–toxin interaction is mediated by subunits $\beta$ and $\gamma$ and that

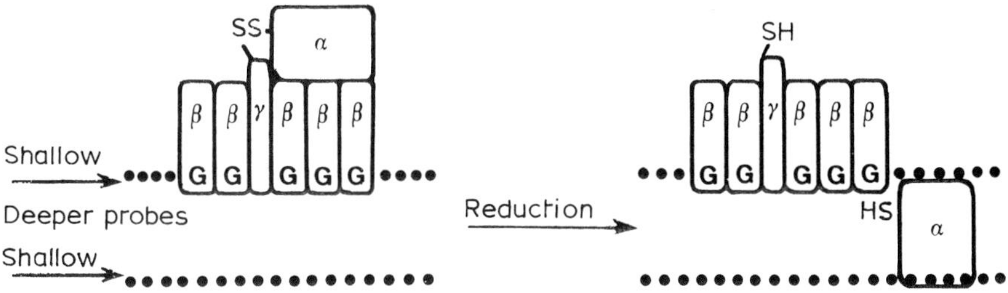


**Fig. 7.6** Schematic view of the interaction of cholera toxin with $G_{MI}$-containing liposomes as deduced by hydrophobic photolabelling (after Tomasi and Montecucco, 1981).

this is restricted to the membrane surface. The enzyme activity of the α subunit is freed after reduction of the α–γ disulphide bridge and is exerted on the cytoplasmic side of the plasma membrane. Hence, somehow, the α subunit has to enter the lipid bilayer. The pattern of labelling under reducing conditions is very different. While again the β and γ subunits were labelled only by the shallow probe, the α subunit was now labelled by both the superficial and the deep probes. This indicates that, after reduction, the α subunit becomes a hydrophobic entity and penetrates deeply into the lipid bilayer as depicted in the scheme of Fig. 7.6. A membrane insertion of the α subunit is also suggested by the use of probe XIV on the Newcastle Disease virus (Wisnieski and Bramhall, 1981). It should be stressed at this point that this kind of topographical information cannot be obtained with small probes. Wisnieski and Bramhall (1981) have also tried to follow the kinetics of penetration of α in the membrane by using a 15 s photolysis period at different time intervals of incubation. Although some aspects need further investigation, this study shows that photoreactive lipids can be used to follow dynamic events in the membrane.

Since in these studies the probes are distributed among the two lipid bilayer halves one cannot distinguish between a half-way or a full penetration of a polypeptide chain in the membrane.

Takagaki *et al.* (1983) have elegantly solved this problem by preparing with PLEP, reconstituted systems where the two halves of the bilayer were probed by $^{3}H$- and $^{14}C$-labelled photoreactive phospholipids respectively. The analysis of the $^{3}H/^{14}C$ distribution along the polypeptide chain provided an unambiguous demonstration that cytochrome $b_5$ transverses the lipid bilayer when present in its non-membrane transferable form.

### 7.4.4 PROBING LIPID–PROTEIN AND PROTEIN–PROTEIN INTERACTION

Photoreactive lipids have also been used to study the lipid–protein boundary of several multisubunit membrane proteins and to probe the arrangement of peptides in the membrane hydrophobic phase.

Mitochondrial respiratory chain components are among the most complex membrane-bound enzymes both in terms of prosthetic groups and their subunit composition. They range from the four subunits of the relatively simple complex II to the over 25 different polypeptides present in complex I (Capaldi, 1982; Trumpower and Katki, 1979; Azzi, 1980; Ragan, 1980). All the inner mitochondrial complexes have been probed in reconstituted active systems with photoreactive lipids (Earley and Ragan, 1981; Girdlestone *et al.*, 1981; Gutweniger *et al.*, 1981; Bisson *et al.*, 1979a; Montecucco *et al.*, 1980). Two examples will be discussed here for their relevance to the present chapter.

Isolated mammalian cytochrome *c* oxidases, the terminal enzyme of the respiratory chain, are composed of 12 different polypeptides. These are listed in Table 7.4. The enzyme isolated from lower eukaryotes such as yeasts has a simpler composition; however, structural homologies are expected to exist among different phyla. This possibility can be tested with photoreactive lipids, by identifying the lipid-exposed subunits. The patterns of labelling of the bovine and yeast enzymes are remarkably similar for the subunits whose sequence homology has been established (Bisson *et al.*, 1979a; Gutweniger *et al.*, 1981). As reported in Table 7.4 the primary sequence, since completed, shows that only those polypeptides containing at least one 18–22-residue-long hydrophobic stretch becomes labelled.

Another complex mitochondrial enzyme is the ATP synthase, which is composed of an as yet undefined number of components. It is available in two preparations differing by the presence or absence of the adenine nucleotide carrier. In this context it is useful to report that, when present, the carrier is the

**Table 7.4** Relative labelling of subunits of beef heart cytochrome *c* oxidase by arylazido phospholipids

| *Polypeptide* | *N-Terminus* | *Residues* | *Number of hydrophobic segments* | *Labelling intensity* |
|---|---|---|---|---|
| I | F-Met | 514 | 7 | ++ |
| II | F-Met | 227 | 2 | + |
| III | (Met)Thr | 261 | >3 | +++ |
| IV | Ser | 147 | 1 | + |
| VIa | Ala | 109 | — | |
| V | Ser | 98 | — | |
| VIb | Ala | 84–5 | * | +++ |
| VIc | Ser | 73 | * | |
| VII | N-AcAla | 85 | — | |
| VIIIa | Ser | 47 | 1 | |
| VIIIb | Ile | 46 | 1 | +++ |
| VIIIc | Phe | 50 | 1 | |

*Unpublished sequence.

most heavily labelled polypeptide (Montecucco *et al.*, 1983a). In carrier-depleted preparations two previously unlabelled peptides become exposed to the radioactive phospholipid probes. The simplest explanation of this result is that, when present, the carrier shields the two peptides from contact with lipids and hence that these two proteins are adjacent to the carrier in the membrane. It thus appears that in certain conditions photoactivatable phospholipids can give information on protein–protein contacts within the membrane and this approach can be useful in following monomer–multimer transitions.

### 7.4.5 PEPTIDE FOLDING IN THE MEMBRANE

Valuable information on the arrangement in the membrane of a labelled polypeptide can be obtained if the modified residues are known. Up to date this has been accomplished only in a few cases.

Subunit II of cytochrome *c* oxidase, which forms the cytochrome *c*-binding site of the enzyme, contains two uninterrupted stretches of hydrophobic residues. The polypeptide was isolated from the enzyme labelled with the shallow probe XI (Bisson *et al.*, 1982). This probe was chosen because the photoreporter group is strictly localized at the head group region of the lipid bilayer; thus it labels the terminal sequences of the peptide(s) spanning the hydrocarbon membrane phase. Sequencing of proteolytic fragment 20–98 (see Fig. 7.7b), which contained all the radioactivity, showed that a cluster of residues at the N-terminus of the first hydrophobic peptide was labelled. No radioactivity was found at the level of the intermediate region of the first hydrophobic peptide. Unfortunately no data are available for the rest of the sequence because of the low degradation yield and of the loss of radioactivity due to washing out from the spinning cup of the sequenator.

A less fortunate case is represented by the labelling of glycophorin with phospholipids I, II and III containing carbene precursors (Ross *et al.*, 1982). The deep probes I and II are expected to label mainly the central hydrophobic region of the membrane while probe III labels only the polypeptide regions present at the membrane surface. Both deep probes, despite their different length, label specifically the same residue, glutamate-70. This cross-linking was not attributed to the photogenerated carbene, but to the long-lived linear diazo isomer formed during irradiation (see Fig. 7.2d) (Ross *et al.*, 1982). The photoactivatable head group of phospholipid probe III appears to label different protein regions. However, the site(s) of modification could not be identified owing to the instability of the cross-linked derivative under the sequencing conditions. As well as in the case of cytochrome *c* oxidase subunit II a CNBr-induced loss of radioactive label was noticed.

A further example of determination of photochemically modified residues is that of the subunit b of *E. coli* ATP synthase, where both a photoreactive phospholipid (probe XI) and simpler reagents (TID and INA) have been used

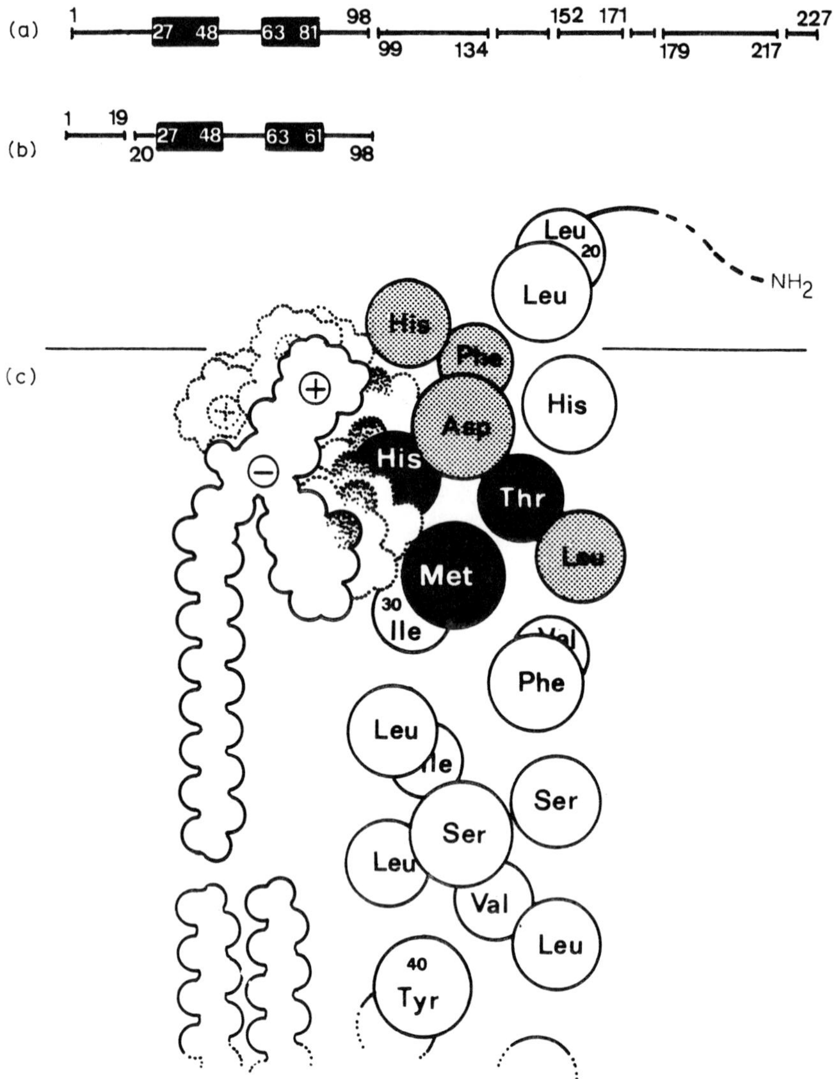


**Fig. 7.7** Fragmentation patterns of (a) cytochrome *c* oxidase subunit II with trypsin and (b) of segment 1–98 with *S. aureus* protease. Black boxes show the positions of the two hydrophobic stretches. (c) Proposed α-helix arrangement in the membrane for the N-terminus of peptide 20–98 based on the sequencing of the probe XI-labelled fragment. Different positions of the photoreactive group, which may explain the labelling of different residues, are depicted. Full circles indicate heavy labelling, dotted circles poor labelling, open circles no labelling (after Bisson *et al.*, 1982).

(Hoppe *et al.*, 1983a,b,c). This enzyme is composed of eight different subunits: five of them belong to a hydrophilic water-exposed domain, while the other three (a,b and c) form a transmembrane proton channel (Sebald and Hoppe, 1981). In a properly reconstituted system (lipid/protein molar ratio over 1000) phospholipid probe XI labelled predominantly subunit b and in minor proportion subunit c, while no labelling of the large hydrophilic domain was found (Montecucco *et al.*, 1983a). In a detergent-solubilized enzyme TID labelled all the three membranous polypeptides and also in minor amount the hydrophilic domain while INA bound preferentially to subunit b (Hoppe *et al.*, 1983a,c). Solid-phase sequencing of this peptide showed that the superficial probe XI modified only two short sequences located at the two ends of the *N*-terminal hydrophobic segment of subunit b. No label was found in the intermediate part of the hydrophobic stretch nor in its hydrophilic portion (Hoppe *et al.*, 1983b). This kind of label distribution suggests that the *N*-terminal part of subunit b transverses the membrane adopting an α-helical configuration (Montecucco *et al.*, 1983b). TID, which can diffuse freely in the hydrocarbon membrane region, labelled also the intermediate section of the subunit b hydrophobic sequence, further supporting its membrane location (Hoppe *et al.*, 1983a). With both probes the most heavily labelled residue was cysteine-21 (while INA modified exclusively this residue) which again points to the very different reactivity of the various amino acid lateral residues (see also Section 7.4.3). Comparison of the labelling of the two terminal regions of the hydrophobic segment of subunit b, the two commonly modified by the different reagents, indicates that the nitroarylnitrene and the aryltrifluoro-acetylcarbene have rather similar reactivities.

Hoppe *et al.* (1983a) also analysed the distribution of TID on subunit c, which has two hydrophobic sequences connected by a hydrophilic segment (Sebald and Hoppe, 1981). TID labelled only the two hydrophobic segments in a native system, while in an SDS-denatured state the intervening hydrophilic sequence was also modified. This result provides good evidence for the proposed hairpin arrangement of subunit c in the membrane (Sebald and Hoppe, 1981).

Several technical problems, which may be encountered during polypeptide isolation, fragmentation and sequencing, deserve further comments. The labelling alters the physicochemical properties of the peptides: it is clearly shown in SDS-PAGE by the shift of the radioactive peak with respect to the corresponding unmodified protein band, owing to the covalent attachment of a lipid molecule. This effect is pronounced only with small peptides and is less evident when simple probes are used. Further possible difficulties are the heterogeneity of the labelled peptides (because of the several possible modified sites) and the low yield of labelling. These problems may render the purification and sequencing of small labelled protein fragments difficult. Moreover most modified peptides are hydrophobic and may be lost when a spinning-cup sequenator is used (in this sense a solid-phase sequenator is preferable). Some of these difficulties can be overcome when large fragments are used.

### 7.4.6 BIOLOGICAL MEMBRANES

Photoreactive phospholipids have been used not only in reconstituted but also in native systems. Viral membrane integral proteins have been identified using probe XIV (Bramhall *et al.*, 1978) and in elegant studies using viral phospholipids obtained by feeding cells, prior to infection, with photoactivatable radioactive fatty acids (Stoffel *et al.*, 1978; Leblanc *et al.*, 1982). This biosynthetic incorporation, which appears to work equally well for all the different photoreactive groups (Greenberg *et al.*, 1976), is likely to be the most reliable and least perturbing method for introducing lipid probes into the cell membranes. Moreover it promises to be a useful way to prepare certain classes of photoactivatable radioactive phospholipids of problematic synthesis.

### 7.4.7 CONCLUSIONS

Although the different classes of amphipathic probes listed in Table 7.3 have been used on the same system rarely, some comparison between their properties and use can be made. The simple lipid probes XIV and XV are easy to prepare from commercially available precursors with a very high specific radioactivity (Bramhall *et al.*, 1978; Montecucco *et al.*, 1981a). However, at present little evidence that their labelling is restricted to the hydrophobic domain is available (Hu and Wisnieski, 1979). Among the various kinds of photoreactive phospholipids the easiest to synthesize are XI and XII (Bisson and Montecucco, 1981). They are made highly radioactive only at the last step in two different isotopic forms ($^{14}C$ and $^{3}H$) allowing the direct comparison of the patterns of labelling at two different levels of the lipid bilayer. As discussed in Sections 7.4.3 and 7.4.5, this is a most informative approach in studying the penetration and folding of an integral protein in the membrane. One problem encountered with these probes is the heavy loss of radioactivity during chemical degradation of peptides and isolation of the labelled fragments. This shortcoming is common to carbene-generating probes (Ross *et al.*, 1982) and may be attributed to the low stability under the conditions used of both the covalent bond formed after photoactivation and/or the ester bonds present in between the photoreacted group and the radioactive isotopes. In this respect it is preferable to use lipid probes such as I, II, IV, V, VI, XIV, and XV where very stable bonds link the radioactive tag to the photoreactive group. Photogenerated adducts appear instead to be rather stable to Edman degradation, although in the case of probe III-labelled glycophorin nearly all the radioactivity was lost in the first cycle. The synthesis of probes I to X and XIII appears difficult, requires many steps and, with the exception of probe XIII, the specific radioactivity is low. The aliphatic nitrene-containing phospholipids IV, V and VI have low molar absorption at around 290 nm, which may result in protein damage (Stoffel *et al.*, 1977). Moreover aliphatic nitrenes rearrange rapidly and hence their yield of labelling is expected to be low (Knowles, 1972). However, this is not the case for reconstituted apolipoprotein (yields of cross-linking as high as 20–30%; Stoffel *et al.*, 1982). Owing to the reduced size of the photoreactive group these probes introduce the least perturbation into a biological system.

There is no single membrane where all the different probes of Table 7.3 have been tested. If one may make a comparison within the limits of the few studies reported at the present time, then the conclusion is that very similar biologically relevant information has been obtained with the different lipid probes (compare for viral membranes: Stoffel *et al.*, 1978 with Capone *et al.*, 1983; for complement-membrane interaction: Podack *et al.*, 1981 with Hu *et al.*, 1981 and Steckel *et al.*, 1983; for gramicidin A: Brunner and Richards, 1980).

## 7.5 Lipophilic group-specific reagents

The analysis of the available sequences of membrane proteins and some hydrophobic photolabelling experiments indicate that the lipid-exposed protein surface is largely devoid of reactive lateral residues. This restricts the possibility of using group-specific reagents, which modify only certain residues, in the labelling of the hydrophobic domain of integral proteins. A list of these is given in reviews by Sigrist and Zahler, 1982 and Ji, 1983. A related topic is that of selective modification of residues present in the interior of the hydrophobic domain of transport proteins such as in the case of DCCD and SITS (Sebald and Hoppe, 1981; Cabantchik *et al.*, 1978). This kind of study may be considered to be an active-site modification, which is beyond the scope of this review.

## 7.6 Next-neighbour relationship within the membrane

Another important aspect of membrane protein topology is that of the proximity relation among different polypeptides of a multisubunit protein and among different proteins in the membrane. This has been explored in one case with photoreactive lipids (see Section 7.4.4). On the basis of earlier work with hydrophilic cross-linkers on protein–DNA complexes and on soluble proteins, hydrophobic cross-linkers can be expected to give relevant information (Peters and Richards, 1977; Das and Fox, 1979; Ji, 1983). However, the usefulness of such reagents may be limited by several additional difficulties. The low reactivity of lipid-exposed protein domains has already been mentioned. For this reason a very poor yield of cross-linking is to be expected with homobifunctional group-specific reagents. Heterobifunctional molecules with a group-specific arm, which reacts via a chemical reaction, and a photoactivatable group on the other arm should give better results (Ji, 1983; Sigrist and Zahler, 1982). However, the photoreactive group may be improperly oriented with respect to the neighbouring polypeptide. Moreover it will cross-link also the lipids and this reaction will be more pronounced with highly reactive photogenerated intermediates. The lipid reaction will decrease the yield of protein cross-linking. With homobifunctional photoactivatable reagents the overall cross-linking efficiency is expected to be low because it is the product of the yields of the two single reactions, which as we have seen cannot be very high (Section 7.3.2). For these reasons it may be necessary to use radioactive reagents because this allows detection of very low amounts of cross-linked products. One great advantage of hetero (chemical and photoreactive) bifunctional probes is that, owing to its short lifetime, the photogenerated intermediate will react before the chemically labelled protein can encounter too many other non-associated membrane proteins giving rise to spurious products.

## 7.7 Conclusions and future developments

Photoreactive reagents are now available which can give information on several aspects of membrane structure such as: (a) determination of the integral nature of a membrane protein; (b) identification of the polypeptide forming the lipid–protein boundary in a multisubunit membrane protein; (c) protein–protein interaction within the membrane; (d) analysis of the folding of polypeptides in the hydrophobic membrane phase; (e) insertion of soluble proteins into the lipid phase.

Several additional developments of these probes can be already foreseen: (a) more reactive and hence less selective photoactivatable groups; (b) synthesis of specific classes of photoactive lipids to probe specific lipid–protein interactions; (c) preparation of rigid photoreactive lipids to be used as faithful chemical rulers of the membrane penetration of proteins.

## Acknowledgements

We thank Drs J. Brunner, B. D. Gomperts, H. Monaco and M. Zoratti for critical reading of the manuscript and Professor G. F. Azzone for support and encouragement. The skilful secretarial assistance of Mrs Franca Mazzari is gratefully acknowledged. We also thank Drs H. Bayley, B. J. Wisnieski and J. Brunner for making available to us manuscripts prior to publication.

## 7.8 References


Akeroyd, R. and Wirtz, K. W. A. (1982) *Membr. Trans.*, **1**, 93–7.

Azzi, A. (1980) *Biochim. Biophys. Acta*, **594**, 231–52.

Bayley, H. (1982) *Membr. Trans.*, **1**, 185–94.

Bayley, H. (1983) *Photogenerated Reagents in Biochemistry and Molecular Biology*, North-Holland/Elsevier, Amsterdam.

Bayley, H. and Knowles, J. R. (1977) *Methods Enzymol.*, **46**, 69–114.

Bayley, H. and Knowles, J. R. (1978a) *Biochemistry*, **17**, 2414–19.

Bayley, H. and Knowles, J. R. (1978b) *Biochemistry*, **17**, 2420–3.

Bayley, H. and Knowles, J. R. (1980) *Biochemistry*, **19**, 3883–92.

Bercovici, T. and Gitler, G. (1978) *Biochemistry*, **17**, 1484–9.

Bisson, R. and Montecucco, C. (1981) *Biochem. J.*, **193**, 757–63.

Bisson, R., Montecucco, C. and Capaldi, R. (1979b) *FEBS Lett.*, **106**, 317–20.

Bisson, R., Montecucco, C., Gutweniger, H. and Azzi, A. (1979a) *J. Biol. Chem.*, **254**, 9962–5.

Bisson, R., Steffens, G. C. M. and Buse, C. (1982) *J. Biol. Chem.*, **257**, 6716–20.

Bramhall, J. S., Shiflett, M. A. and Wisnieski, B. J. (1978) *Biochem. J.*, **177**, 765–8.

Brunner, J. (1981) *Trends Biochem. Sci.*, **6**, 44–6.

Brunner, J. and Richards, F. M. (1980) *J. Biol. Chem.*, **255**, 3319–29.

Brunner, J. and Semenza, G. (1981) *Biochemistry*, **20**, 7174–82.

Brunner, J., Hauser, H., Braun, H., Wilson, K. J., Wacker, H., O'Neill, B. and Semenza, G. (1979) *J. Biol. Chem.*, **254**, 1821–8.

Brunner, J., Senn, H. and Richards, F. M. (1980) *J. Biol. Chem.*, **255**, 3313–18.

Brunner, J., Spiess, M., Aggeler, R., Huber, P. and Semenza, G. (1983) *Biochemistry*, **22**, 3812–20.

Cabantchik, Z. I., Knauf, P. A. and Rothstein, A. (1978) *Biochim. Biophys. Acta*, **515**, 239–302.

Campbell, D. G., Gagnon, J., Reid, K. B. M. and Williams, A. F. (1981) *Biochem. J.*, **195**, 15–30.

Capaldi, R. A. (1982) *Biochim. Biophys. Acta*, **694**, 291–306.

Capone, J., Leblanc, P., Gerber, G. E. and Gosh, H. P. (1983) *J. Biol. Chem.*, **258**, 1395–8.

Cerletti, N. and Schatz, G. (1979) *J. Biol. Chem.*, **254**, 7746–51.

Chapman, O. L. and LeRoux, J. P. (1978) *J. Amer. Chem. Soc.*, **100**, 282–5.

Chakrabarti, P. and Khorana, H. G. (1975) *Biochemistry*, **14**, 5021–33.

Chowdhry, V. and Westheimer, F. H. (1979) *Ann. Rev. Biochem.*, **48**, 293–325.

Das, M. and Fox, C. F. (1979) *Annu. Rev. Biophys. Bioeng.*, **8**, 165–93.

De Pierre, J. W. and Ernster, L. (1977) *Annu. Rev. Biochem.*, **46**, 201–62.

Earley, F. G. P. and Ragan, C. I. (1981) *FEBS Lett.*, **127**, 45–7.

Esser, A. F. (1982) *Biological Membranes*, Academic Press, New York, Vol. 4, pp. 277–325.

Evans, E. A. (1976) *Self-Decomposition of Radiochemicals*, The Radiochemical Centre, Amersham.

Farley, R. A., Goldman, D. W. and Bayley, H. (1980) *J. Biol. Chem.*, **255**, 860–4.

Feinstein, A., Richardson, N. E. and McIlhinney (1979) in *P and S Biomedical Sciences Symposia. Cells of Immunoglobulin Synthesis* (eds B. Pernis and H. J. Vogel) Academic

Press, New York, pp. 165–88.
Garavito, R. M., Jenkins, J., Jansonins, J. M., Karlsson, R. and Rosenbusch, J. F. (1983) *J. Mol. Biol.*, **164**, 313–27.
Girardet, M., Geering, K., Rossier, B. C., Kraehenbuhl, J. P. and Bron, C. (1983) *Biochemistry*, **22**, 2296–300.
Giraudat, J., Montecucco, C., Bisson, R. and Changeaux, J. P. (1985) *Biochemistry*, **24**, 3121–7.
Girdlestone, J., Bisson, R. and Capaldi, R. A. (1981) *Biochemistry*, **20**, 152–6.
Gitler, C. and Bercovici, T. (1980) *Ann. N.Y. Acad. Sci.*, **346**, 199–211.
Goldman, D. W., Pober, J. S., White, J. and Bayley, H. (1979) *Nature (London)*, **280**, 841–3.
Greenberg, G. R., Chakrabarti, P. and Khorana, H. G. (1976) *Proc. Natl. Acad. Sci. USA*, **73**, 86–90.
Gutweniger, H., Bisson, R. and Montecucco, C. (1981a) *J. Biol. Chem.*, **256**, 11132–6.
Gutweniger, H., Bisson, R. and Montecucco, C. (1981b) *Biochim. Biophys. Acta*, **635**, 187–93.
Gutweniger, H. and Montecucco, C. (1984) *Biochem. J.*, **220**, 613–16.
Hantke, K. and Braun, V. (1973) *Eur. J. Biochem.*, **34**, 284–96.
Halsey, M. J., Brown, F. F. and Richards, R. E. (1978) *Molecular Interaction and Activity in Proteins: CIBA Found. Symp.*, **60**, 123–36.
Hebdon, G. M., Knott, J. C. A. and Green, N. M. (1980) *Biochemistry*, **19**, 6216.
Henderson, R. (1981) in *Biophysics of Membranes and Intercellular Communication* (eds M Chabre and P. F. Devaux), North-Holland, Amsterdam, pp. 232–49.
Henderson, R. and Unwin, P. N. T. (1975) *Nature (London)*, **257**, 28–32.
Hobbs, A. S. and Albers, R. W. (1980) *Annu. Rev. Biophys. Bioeng.*, **9**, 259–91.
Holowka, D., Gitler, C., Bercovici, T. and Metzger, H. (1981) *Nature (London)*, **289**, 806–8.
Hoppe, J., Brunner, J. and Jorgensen, B. B. (1983a) in *Structure and Function of Membrane Proteins* (eds F. Palmieri and E. Quagliariello), North-Holland/Elsevier, Amsterdam (in press).
Hoppe, J., Friedl, P. and Jorgensen, B. B. (1983c) *FEBS Lett.*, **160**, 239–42.
Hoppe, J., Montecucco, C. and Friedl, P. (1983b) *J. Biol. Chem.*, **258**, 2882–5.
Hu, V. W., Esser, A. F., Podack, E. R. and Wisnieski, B. J. (1981) *J. Immunol.*, **127**, 380–6.
Hu, V. W., and Wisnieski, B. J. (1979) *Proc. Natl. Acad. Sci. USA*, **76**, 5460–4.
Hubbell, W. L. and McConnell, H. M. (1971) *J. Amer. Chem. Soc.*, **93**, 314–26.
Ji, T. H. (1983) *Methods Enzymol.*, **91**, 580–609.
Jorgensen, P. L. (1980) *Physiol. Rev.*, **60**, 864–917.
Jorgensen, P. L., Karlish, S. J. D. and Gitler, C. (1982) *J. Biol. Chem.*, **257**, 7435–42.
Kahane, I. and Gitler, C. (1978) *Science*, **201**, 351–2.
Kamiya, Y., Sakurai, A., Tamura, S., Takahashi, N., Tsuchiya, E., Abe, K. and Fukui, S. (1979) *Agric. Biol. Chem.*, **43**, 363–9.
Karlish, S. J. D., Jorgensen, P. L. and Gitler, C. (1977) *Nature (London)*, **269**, 715–17.
Kennedy, S. J. (1978) *J. Membr. Biol.*, **42**, 265–79.
Klip, A. and Gitler, C. (1974) *Biochem. Biophys. Res. Commun.*, **60**, 1155–62.
Knowles, J. R. (1972) *Acc. Chem. Res.*, **5**, 155–60.
Kyba, P. E. and Abramovich, R. A. (1980) *J. Amer. Chem. Soc.*, **102**, 735–40.

Leblanc, P., Capone, J. and Gerber, E. G. (1982) *J. Biol. Chem.*, **257**, 14586–9.
Lenard, J. and Singer, S. J. (1966) *Proc. Natl. Acad. Sci. USA*, **56**, 1828–35.
Luria, S. E. and Suit, J. L. (1982) *Membr. Transp.*, **2**, 279–84.
Marchesi, V. T. (1979) *J. Membr. Biol.*, **51**, 101–31.
Marchesi, V. T., Furthmayr, H. and Tomita, M. (1976) *Annu. Rev. Biochem.*, **45**, 667–97.
Mavridis, I. M., Hatada, M. H., Tulinsky, A. and Lebioda, L. (1982) *J. Mol. Biol.*, **162**, 419–44.
Michel, H. (1982) *J. Mol. Biol.*, **158**, 567–72.
Mitchell, P. (1966) *Biol. Rev.*, **41**, 445–502.
Montecucco, C., Bisson, R., Dabbeni-Sala, F., Pitotti, A. and Gutweniger, H. (1980) *J. Biol. Chem.*, **255**, 10040–3.
Montecucco, C., Bisson, R., Gache, C. and Johannsson, A. (1981a) *FEBS Lett.*, **128**, 17–20.
Montecucco, C., Bisson, R., Zaccolin, G. P., Dabbeni-Sala, F. and Galante, Y. (1981b) in *Vectorial Reactions in Electron and Ion Transport in Mitochondria and Bacteria* (eds F. Palmieri, E. Quagliariello and S. Papa), Elsevier/North-Holland Biomedical Press, Amsterdam, pp. 227–30.
Montecucco, C., Dabbeni-Sala, F., Friedl, P. and Galante, Y. M. (1983a) *Eur. J. Biochem.*, **132**, 189–94.
Montecucco, C., Friedl, P. and Hoppe, J. (1983b) in *Structure and Function of Membrane Proteins* (eds F. Palmieri and E. Quagliariello), Elsevier/North-Holland Biomedical Press, Amsterdam (in press).
Nielsen, P. E. and Buchardt, O. (1982) *Photochem. Photobiol.*, **35**, 317–23.
Noren, O. and Sjostrom, H. (1980) *Eur. J. Biochem.*, **104**, 25–31.
Owen, M. J., Knott, J. C. A. and Crumpton, M. J. (1980) *Biochemistry*, **19**, 3092–9.
Pagano, R. E. and Weinstein, J. N. (1978) *Annu. Rev. Biophys. Bioeng.*, **7**, 435–68.
Peters, K. and Richards, F. M. (1977) *Annu. Rev. Biochem.*, **46**, 523–51.
Podack, E. R., Stoffel, W., Esser, A. F. and Muller-Eberhard, H. (1981) *Proc. Natl. Acad. Sci. USA*, **78**, 4544–8.
Prochaska, L. J., Bisson, R. and Capaldi, R. A. (1980) *Biochemistry*, **19**, 3174–9.
Ragan, C. I. (1980) *Subcell. Biochem.*, **7**, 267–307.
Reiser, A. and Leyshon, L. (1970) *J. Amer. Chem. Soc.*, **92**, 7487.
Reiser, A., Terry, G. C. and Willets, F. W. (1966) *Nature (London)*, **211**, 410.
Reiser, A., Willets, F. W., Terry, G. C., Williams, V. and Marley, R. (1968) *Trans. Faraday Soc.*, **64**, 3265–75.
Richardson, J. S. (1981) *Adv. Protein Chem.*, **34**, 167–339.
Robson, R. J., Radhakrishnan, R., Ross, A. H., Tagakaki, Y. and Khorana, H. G. (1982) *Lipid–Protein Interact.*, **2**, 149–92.
Rogers, J., Early, P., Carter, C., Calame, K., Bond, M., Hood, L. and Wall, R. (1980) *Cell*, **20**, 303–12.
Rose, G. D., Young, W. B. and Gierasch, L. M. (1983) *Nature (London)*, **304**, 654–7.
Rose, J. K., Adams, G. A. and Gallione, C. J. (1984) *Proc. Natl. Acad. Sci. USA*, **81**, 2050–4.
Rosenbusch, J. P. (1974) *J. Biol. Chem.*, **249**, 8019–29.
Ross, A. H., Radhakrishnan, R., Robson, R. J. and Khorana, H. G. (1982) *J. Biol. Chem.*, **257**, 4152–61.

Rothstein, A. (1982) *Membr. Transp.* **2**, 435–40.
Sandermann, H. (1978) *Biochim. Biophys. Acta*, **515**, 209–37.
Sator, V., Gonzalez-Ros, J. M., Calvo-Fernandez, P. and Martinez-Carrion, M. (1979) *Biochemistry*, **18**, 1200–6.
Schlesinger, M. J. and Magee, A. I. (1982) *Biophys. J.*, **37**, 126–7.
Sebald, W. and Hoppe, J. (1981) *Curr. Top. Bioenerg.*, **12**, 2–59.
Sigrist, H. and Zahler, P. (1982) *Membr. Transp.*, **1**, 173–84.
Singer, S. J. (1974) *Annu. Rev. Biochem.*, **43**, 805–33.
Singer, S. J. and Nicholson, G. L. (1972) *Science*, **175**, 720–31.
Smith, R. A. G. and Knowles, J. R. (1975) *J. Chem. Soc. Perkin Trans. 2*, 686–94.
Spiess, M., Brunner, J. and Semenza, G. (1982) *J. Biol. Chem.*, **257**, 2370–7.
Staros, J. V. (1980) *Trends Biochem. Sci.*, **5**, 320–2.
Staros, J. V., Bayley, H., Standring, D. N. and Knowles, J. R. (1978) *Biochem. Biophys. Res. Commun.*, **80**, 568–72.
Steckel, E. W., Wellbaum, B. E. and Sodetz, J. M. (1983) *J. Biol. Chem.*, **258**, 4318–24.
Stefanini, S., Finazzi-Agrò, A., Chiancone, E. and Antonini, E. (1979) *FEBS Lett.*, **100**, 296–300.
Stoffel, W., Darr, W. and Salm, K. P. (1977) *Hoppe-Seyler's Z. Physiol. Chem.*, **358**, 453–62.
Stoffel, W., Schreiber, C. and Scheefers, H. (1978) *Hoppe-Seyler's Z. Physiol. Chem.*, **359**, 923–31.
Stoffel, W., Sahn, K. P. and Muller, M. (1982) *Hoppe Seyler's Z. Physiol. Chem.*, **363**, 1–18.
Strittmatter, P. and Dailey, H. A. (1982) *Membr. Transp.*, **1**, 71–82.
Strominger, J. L., Orr, H. T., Parham, P., Ploegh, H. L., Mann, D. L., Bilofsky, H., Saroff, H. A., Wu, T. T. and Kabat, E. A. (1980) *Current Topics in Developmental Biology, Developmental Immunology*, vol. 14, Academic Press, New York.
Takagaki, Y., Gupta, C. M. and Khorana, H. G. (1980) *Biochem. Biophys. Res. Commun.*, **95**, 589–95.
Takagaki, Y., Radhakrishnan, R., Wirtz, K. W. A. and Khorana, H. G. (1983) *J. Biol. Chem.*, **258**, 9136–42.
Thelestam, M., Jolivet-Reynaud, C. and Alouf, J. E. (1983) *Biochem. Biophys. Res. Commun.*, **111**, 444–9.
Tomasi, M. and Montecucco, C. (1981) *J. Biol. Chem.*, **256**, 11177–81.
Tomasi, M., D'Agnolo, G. and Montecucco, C. (1982) *Biochim. Biophys. Acta*, **692**, 339–44.
Tometsko, A. M. and Richards, F. M. (1980) *Ann. N.Y. Acad. Sci.*, **346**, 1–502.
Trumpower, B. L. and Katki, A. G. (1979) in *Membrane Proteins in Energy Transduction*, Dekker, New York, pp. 89–206.
Ueda, I., Kamaya, H. and Eyring, H. (1976) *Proc. Natl. Acad. Sci. USA*, **73**, 481–5.
Urry, D. W. (1971) *Proc. Natl. Acad. Sci. USA*, **68**, 672–6.
Van Heyningen, S. (1977) *Biol. Rev.*, **52**, 509–49.
Wells, E. and Findlay, J. B. C. (1979a) *Biochem. J.*, **179**, 257–64.
Wells, E. and Findlay, J. B. (1979b) *Biochem. J.*, **179**, 265–72.
Williams, P. B., Kubo, R. T. and Grey, H. M. (1978) *J. Immunol.*, **121**, 2435–9.
Wisnieski, B. J. and Bramhall, J. S. (1981) *Nature (London)*, **289**, 319–21.

# 8

# Immunochemical analysis of membrane proteins

MILTON R. J. SALTON



## 8.1 Introduction

Membranes have a deceptively simple appearance when viewed in the electron microscope as profiles of thin sections of cells and organelles or as the inner and outer fracture faces of freeze-fractured membranes. This apparent physical simplicity masks the complexity of the variety of proteins found in the majority

Techniques for the Analysis of Membrane Proteins. Edited by C. I. Ragan & R. J. Cherry. Published in 1986 by Chapman and Hall, 11 New Fetter Lane, London EC4P 4EE

of membrane structures, with of course a few notable exceptions such as myelin and bacterial purple (bacteriorhodopsin) membranes with a very restricted complement of proteins. Resolving the complex arrays of membrane proteins, determining their functions and molecular mechanisms of action *in situ* has been a major challenge to the ingenuity of biochemists and immunochemists for several decades. Peripheral membrane proteins soluble in aqueous media, in contrast to the integral proteins (Singer and Nicolson, 1972), have been much more amenable to biochemical and immunochemical analysis, as for example erythrocyte membrane spectrin (Marchesi and Steers, 1968; Reynolds and Trayer, 1971) and the $F_1$-ATPases that can be released from the membranes by mild treatments and obtained in detergent-free solutions (Pedersen, 1975; Downie *et al.*, 1979). Resolution and purification of the integral proteins as either active enzymes or antigens has been much more of a problem, requiring detergent-dissociation ('solubilization') of the membranes by surface-active agents that do not destroy their biological activities and even proteolytic cleavage of hydrophobic polypeptides. With the variety of receptor, transport, biosynthetic, electron transport and proton-pump functions performed by structurally diverse membrane proteins, it is not surprising that there is no universal detergent that can be used for solubilization and purification for biochemical studies; the same is also true for immunochemical investigations although less extensively documented. A wide variety of non-ionic detergents has been used successfully for biochemical and immunochemical studies and generally have been more suitable than charged, anionic, cationic or zwitterionic detergents for retention of biological activities. Although sodium dodecyl sulphate (SDS) has been used extensively in the SDS–polyacrylamide gel electrophoretic (SDS-PAGE) analysis of the variety of polypeptides in virtually all types of biomembranes and cell surface structures, with rare exceptions the dissociated proteins have been enzymically inactive and may (but not invariably) exhibit reduced reactivity with antibodies.

Serological analysis has been used in the study of cell surface antigens for a very long time and, indeed, provided the major source of our knowledge of the nature of erythrocyte and bacterial cell surfaces for example. Antibody adsorption studies contributed further information on the variety of cell surface and membrane antigens and the cross-reactivities of some of the antigens. Immunochemical studies in combination with subcellular fractionation and/or electron microscopy (e.g. ferritin-labelled antibody, see Oppenheim and Nachbar, 1977) revealed further information on the distribution and topography of membrane antigens. Such procedures have been invaluable in the identification and localization of individual membrane proteins and glycoproteins by immunochemical techniques but they have been largely qualitative. Early attempts to resolve the variety of antigens in membranes (erythrocytes, bacterial envelopes, cytoplasmic membranes) by conventional immunoelectrophoresis and agar-gel diffusion methods were disappointing and generally revealed relatively few, poorly defined immunoprecipitates. Thus Poulik and Bron (1969) found seven or more antigens in the lipid-poor stromal protein fractions of red cell membranes. Considering the multiplicity of biochemical functions performed by bacterial plasma membranes, the small number of antigens resolvable was also incompatible with the known variety of individual enzymes present (Salton, 1974). Demonstration of the real antigenic complexity of the protein components of cell membranes thus had to await the development of the high-resolution, two-dimensional crossed immunoelectrophoresis (CIE) method following Laurell's (1965) dramatic demonstration of electroimmunoprecipitation on electrophoresis of antigens into agarose gels containing antibody. The procedure could also be adapted to the study of detergent-solubilized membrane antigens simply by the addition of non-ionic detergents to the agarose gels containing antibody (Bjerrum, 1977). Both the rocket (one-dimensional) and so-called crossed (two-dimensional) electroimmunochemical techniques have not only revolutionized the analysis of membrane proteins but have also facilitated the resolution of highly complex mixtures of soluble proteins (as antigens) such as those found in serum and cell cytoplasmic fractions (Axelsen *et al.*, 1973). In addition to being so powerful because of its high two-dimensional resolving capability, the methodology is notable for the great versatility of the electroimmunochemical approach and the variety of associated techniques that can be adapted for the study of membrane proteins as emphasized by Bjerrum (1983). Rocket and two-dimensional immunoelectrophoretic methods can be combined with conventional absorption studies to determine the topological arrangement of proteins in membranes (Owen, 1983a); they can be used in radioimmunochemical analysis of membrane antigens (Bjerrum and Gianazza, 1983); they have great utility in monitoring membrane fractionation and protein purification and enzyme identification by zymogram staining; and combined with lectins by affinoelectrophoresis they can be used for the quantification and identification of carbohydrate, glycoproteins and other cell surface glycolipid and lipopolysaccharide antigens (Bjerrum, 1983; Bjerrum and Gianazza, 1983; Owen and Smyth, 1977). Moreover, the potentials for the use of this approach in the analysis of membrane proteins in the normal and pathological states are only just beginning to be realized and it can be anticipated that the application of these techniques to membrane protein analysis will greatly expand our qualitative and quantitative knowledge of the structure and functions of membrane antigens. This chapter will accordingly dwell principally on selected aspects of the application of electroimmunochemical techniques to the study of membrane proteins and in addition briefly refer to immunoadsorbent methods that have already been widely documented (Ruoslahti, 1976; Clausen, 1981; Danielsen, 1983).

## 8.2 Crossed immunoelectrophoretic and rocket analysis of membrane proteins

### 8.2.1 PREPARATION OF TWO-DIMENSIONAL IMMUNOPRECIPITATE REFERENCE PATTERNS OF MEMBRANE PROTEINS

(a) *General considerations*

Analysis of the complexity of membrane proteins as antigens can be achieved by the use of the sensitive, high-resolution technique of two-dimensional immunoelectrophoresis in the presence of detergent; the detergent-solubilized membrane proteins (or complexes) having been separated by electrophoresis in the first dimension are then electrophoresed into the agarose gel containing the reference antibody immunoglobulins in the second dimension thereby giving a two-dimensional display of immunoprecipitate peaks. For membrane proteins the whole immunoelectrophoretic procedure is performed in the presence of a non-ionic detergent such as Triton X-100 in both agarose gel and in the electrophoresis buffers. There are three basic requirements for the preparation and establishment of a crossed immunoelectrophoresis reference pattern of membrane antigens. The first requirement is the generation of polyspecific antiserum (usually in rabbits) by immunization with isolated membrane fractions and collection of a suitably sized pool (about 500 ml) of antiserum. The route and duration of immunization may have to be determined empirically but in general it is advisable to immunize long enough (several months) to obtain adequate antibody responses for the detection of minor and poorly immunogenic membrane antigens. Partial purification of the immunoglobulin fraction is necessary in order to avoid the side effects seen with use of crude antiserum (Bjerrum, 1983). The second basic requirement is the selection of a suitable membrane solubilization procedure, involving the choice of detergent (non-ionic types used most frequently in immunochemical analysis) giving the highest degree of solubilization and minimal denaturation of the proteins (Bjerrum, 1983). As part of this requirement for optimal membrane solubilization, factors such as the pH, ionic strength of buffers, ratios of detergent/protein (often 2:1), addition of EDTA and in some instances inhibitors of membrane proteases, should all be considered. The third basic requirement in the preparation of the reference pattern involves experimental adjustment of the amounts of membrane antigens and reference immunoglobulin (IgG) fractions to optimize the resolution and detection of the array of individual antigens as immunoprecipitates on the immunoplates. It may not always be possible to achieve this under one set of conditions and both high and low antigen loadings and electrophoresis of antigens for short and longer periods may have to be performed for the detection of the maximum number of antigens as immunoprecipitate peaks (Axelsen *et al.*, 1973). To conserve valuable antisera much of the preliminary two-dimensional separation and adjustments of antigen/antibody ratios can be performed on 2 in × 2 in slides and, moreover, the resolution on this size immunoplate is generally perfectly satisfactory for most analytical studies of membrane proteins by CIE. Once the optimum antigen/antibody ratios are established the patterns are highly reproducible for a given pool of antiserum and give both qualitative information on the numbers of membrane proteins and quantitative data since the peak areas are proportional to the antigen/antibody ratios (Axelsen *et al.*, 1973). Thus, for a given reference antiserum, individual membrane proteins can be quantified in different membrane preparations, extracts and fractions and can provide a valuable adjunct in membrane protein purification. It is of course evident that comparison of pools of antisera from different rabbits may give both quantitative and qualitative differences in the immunoprecipitate patterns, reflecting individual differences in the animal's response to immunization. Such factors have to be taken into account in the systematic analysis of membrane proteins and the detailed analysis and establishment of a full reference pattern may involve additional studies at high and low antigen to achieve detection of some minor or poorly resolved antigens. In addition the possibility that some antigens are deeply embedded in the membrane and are not expressed as immunogens should be kept in mind. However, it is of interest to note that the basic patterns for *Micrococcus lysodeikticus (M. luteus)* bacterial membranes were very similar when antibodies to untreated, isolated membranes were compared with those generated to Triton X-100-solubilized membranes (Salton, 1983). It would have been anticipated that any major antigens buried deeply in the membranes would have been unmasked by the solubilizing effects of this non-ionic detergent. Nonetheless, further studies with this and other membrane systems are clearly indicated before this possibility can be completely dismissed.

Precise technical details of buffers, detergents, type of agarose and concentration, preparation of plates and electrophoresis conditions for detergent immunoelectrophoretic analysis of membrane proteins have been extensively covered in several excellent handbooks and other publications and the reader is accordingly advised to refer to these references (Axelsen *et al.*, 1973; Bjerrum, 1977, 1983). Moreover, the properties of detergents, their advantages and disadvantages in selecting them for membrane solubilization have been critically discussed by Bjerrum (1983) and an abundance of references to original investigations will be found in this review of these important aspects of immunochemical analysis of membrane proteins. Since agarose is the only matrix so far found suitable for electroimmunoprecipitation (Laurell and McKay, 1981), the type of agarose to be used can be of technical importance (Axelsen *et al.*, 1973) and those brands giving minimal electroendosmotic effects have been found to be the most desirable in membrane protein analysis and are especially suitable for studies using the intermediate gel variation of CIE (Smyth *et al.*, 1976).

(b) *Strategies to be considered in generating reference antisera*
In developing strategies for the immunochemical analysis of membrane proteins by CIE methods several options exist for the production of suitable antisera. The form in which the membrane is presented during immunization will be important and generation of antibodies to the isolated membrane sheets or fragments has the advantage of presenting the antigens on both faces of the membrane for expression as immunogens during the immunization schedule. Thus the antiserum will contain antibody populations to inner and outer face membrane proteins. In contrast immunization with vesicles, if sealed, may present predominantly one face of the membrane depending on the right-side out or inside-out orientation of the vesicles. Those antigens exposed on the outer surface of the membrane or the intact cell can be identified specifically by carrying out adsorption of membrane antisera with protoplasts or whole cells and analysing the adsorbed antisera by CIE (Owen and Salton, 1977). Antigens expressed only on the inner membrane face will be unaffected in the CIE patterns obtained with the adsorbed antisera and solubilized reference membrane antigens (Owen and Salton, 1977). Should isolation of the membranes in sufficient quantities for immunization and subsequent CIE analysis present problems, generating antibodies to intact cells can be resorted to as an alternative. Provided that antibodies are produced primarily against surface membrane antigens and that the cell remains essentially intact during immunization, then useful information can be obtained by using detergent-solubilized whole cells as a source of surface antigens for resolution against such antisera by CIE. This method of analysis of membrane antigens is less direct than that using the isolated membranes as immunogen and any lysis of the cells during immunization could complicate the analysis by giving rise to antibodies to cytoplasmic components, thereby contaminating those formed to truly surface antigens (Smyth and Salton, 1977).

Ideally, the preparation of antisera to the cytoplasmic fraction, to the isolated membranes and to intact cells provides the best set of immunological reagents for definitive studies of membrane proteins and enables one to establish unambiguously the cellular compartments from which the antigens originate. The reference patterns of immunoprecipitates obtained with membrane-free cytoplasmic fractions are quite distinctive and differ from those obtained with detergent-solubilized membrane resolved against their homologous membrane antibodies. A typical example of this type of membrane analysis is illustrated by the results obtained with a bacterial system where the isolated membrane fraction and the corresponding cytoplasmic fraction of the Gram-positive organism, *Micrococcus lysodeikticus*, have been analysed by CIE against their respective antibodies (Figs 8.1 and 8.2). Moreover, the availability of reference antibodies for the membrane and cytoplasmic fractions of the cell permits the cross-checking of the membrane preparations for contamination with cytoplasmic components and vice versa. This can be

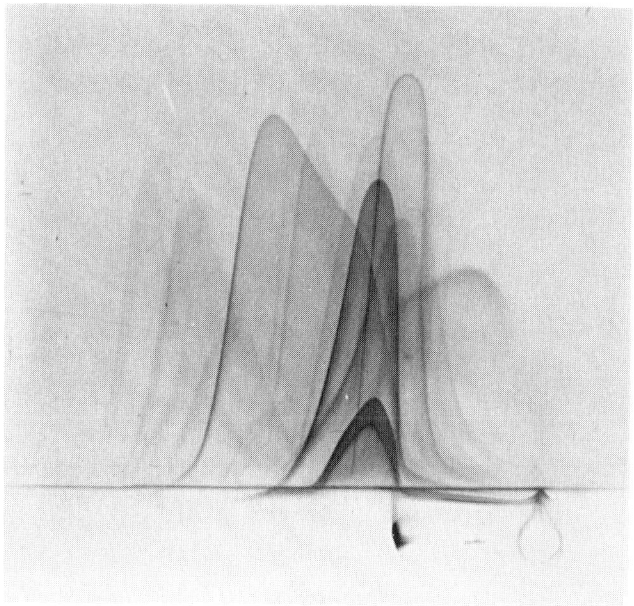

**Fig. 8.1** Two-dimensional crossed immunoelectrophoresis reference pattern of immunoprecipitates obtained with 45 $\mu$g of *M. lysodeikticus* membrane protein solubilized in Triton X-100 (4%) electrophoresed in the second dimension into agarose gel containing 200 $\mu$l of anti-membrane antibodies. In this reference system for a bacterial membrane some 30 distinct membrane antigens can be consistently resolved in the immunoprecipitate patterns. Anode was to the left and top and the slide was stained with Coomassie Blue.

readily achieved by using the intermediate gel variation of CIE by interposing, for example, anti-cytoplasm antibodies in the intermediate gel, with anti-membrane antibodies in the reference gel or by placing anti-membrane antibodies in the intermediate gel and anti-cytoplasm antibodies in the reference gel as illustrated in Fig. 8.3 from the study of Owen and Salton (1977). Inclusion of immunoprecipitates within the intermediate gel zone containing anti-membrane (as in Fig. 8.3) indicated that these antigens were probably membrane-derived and partitioned between cytoplasmic and membrane fractions during centrifugation and isolation of the membrane fractions. When the intermediate gel analysis was performed with anti-cytoplasm in the intermediate gel and anti-membrane in the reference gel and CIE analysis performed with Triton X-100 membranes, only three immunoprecipitates were found, one of which was NADH dehydrogenase and appears to partition between membrane and cytoplasmic compartments (Owen and Salton, 1977).

Thus the cytoplasm–anti-cytoplasm reference CIE system provides a valuable adjunct to the study of membrane antigens and permits the analysis

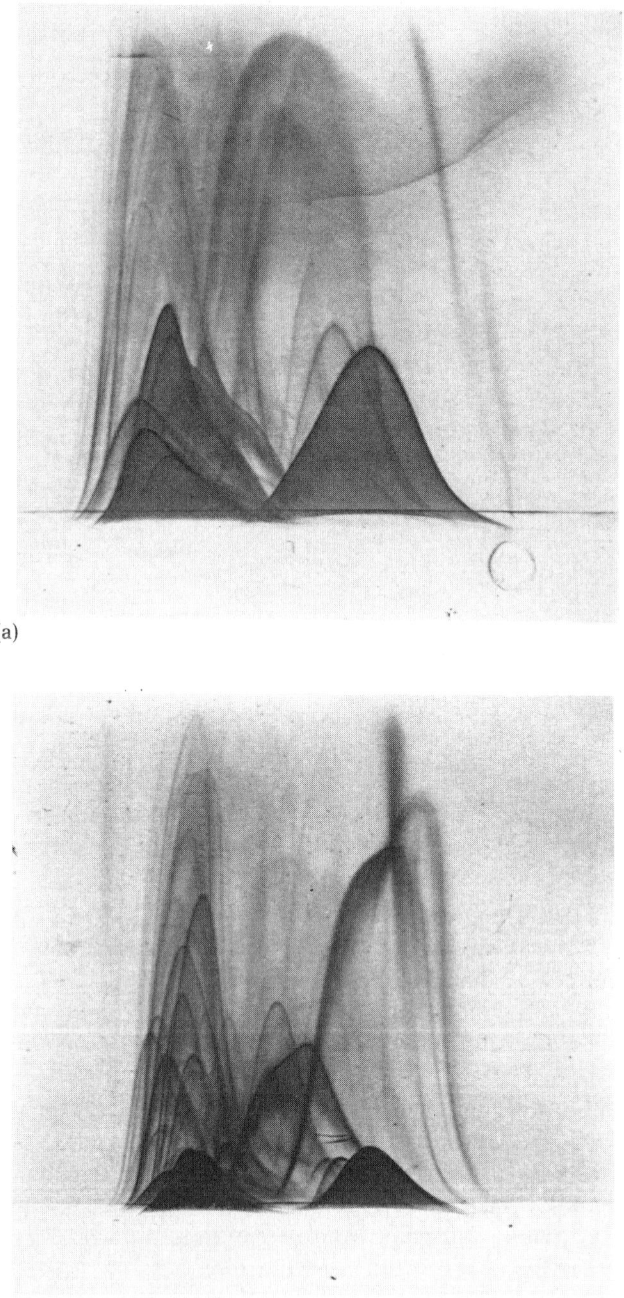

(a)

(b)

**Fig. 8.2** A complex reference pattern of immunoprecipitates is obtained with the cytoplasmic fraction of *M. lysodeikticus* examined by crossed immunoelectrophoresis against its homologous antiserum (rabbit antibodies generated to the cytoplasmic components after removal of membranes by centrifugation). The resolution of the cytoplasmic antigens (125 μg of protein) is illustrated for two levels of antiserum, 120 μl in slide a and 60 μl in slide b, added to the agarose gel. Note that the patterns of immunoprecipitates for the cytoplasmic antigens differ markedly from those seen for the Triton X-100-solubilized membrane fraction in Fig. 8.1.

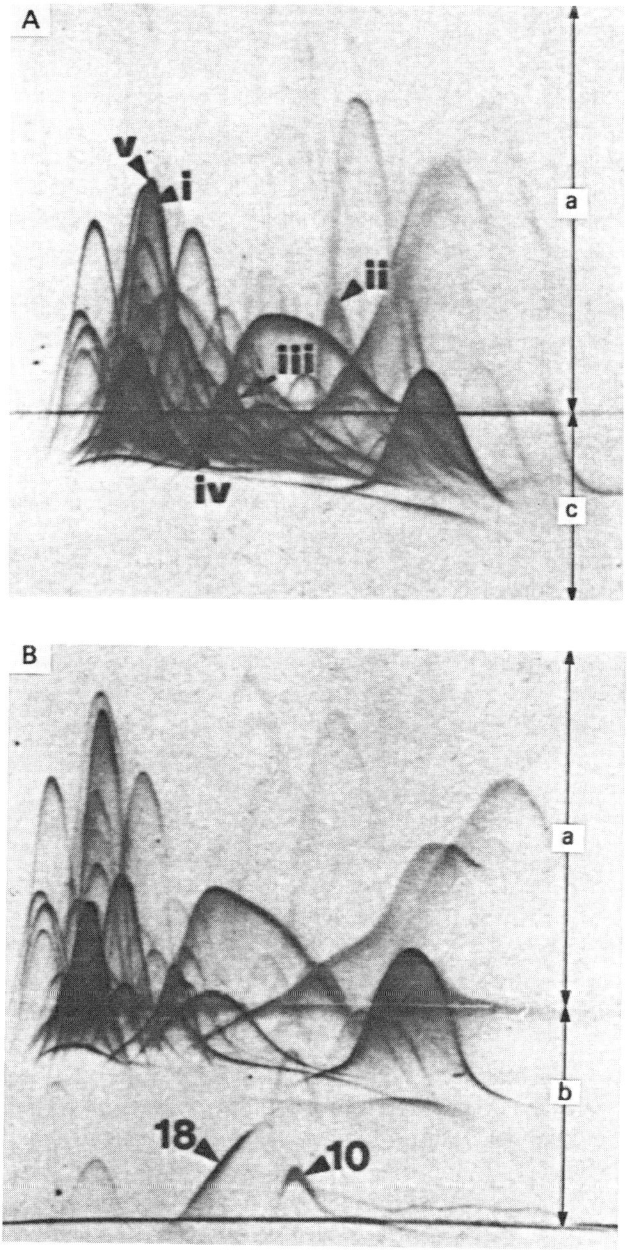


**Fig. 8.3** The use of the intermediate gel technique illustrates the detection of several membrane antigens (with anti-membrane antibodies) in the cytoplasmic fraction from *M. lysodeikticus* resolved as a complex pattern of immunoprecipitates in the reference gel containing the homologous anti-cytoplasmic antibodies (gel region a). No antiserum was added to the intermediate gel region (c) in immunoplate (A), but anti-membrane serum was added to gel region (b) in immunoplate (B). Enzymes identified in (A) were NADH dehydrogenase (i, ii), isocitrate dehydrogenase (iii), polynucleotide phosphorylase (iv) and catalase (v). Two identifiable antigens (18, lipomannan; 10, NADH dehydrogenase) were fully included as immunoprecipitates in the anti-membrane intermediate gel region (b). As in (A), the (a) gel region in (B) contained anti-cytoplasmic antibodies. From Owen and Salton (1977), reproduced with permission from the American Society of Microbiology.

of the distribution of antigens in cell fractions by the intermediate gel technique and can facilitate the monitoring of membrane isolation and determination of cross-contamination (Owen and Salton, 1977). These studies with a bacterial system therefore serve to illustrate how the generation of antibodies to the two cellular compartments, membrane and cytoplasm, can greatly assist in the immunochemical analysis of membrane proteins by defining the distributions and cellular location of antigens. The availability of antisera to vesicle, envelope and cytoplasmic fractions of *E. coli* has also greatly facilitated the detailed analysis of the vesicle system widely used in transport studies and permitted Owen and his colleagues (Owen and Kaback, 1979; Owen, 1983b) to make firm conclusions on the orientation of the vesicle membranes and the identification of vesicle antigens arising from cytoplasmic entrapment.

In preparing a reference antiserum to membranes (or other subcellular fractions for that matter) it is often quite informative to follow the course of immunization by performing CIE on individual bleedings following a series of boosters with antigen. The studies of Shulman and Karpatkin (1980) on CIE of human platelet membranes provide an excellent example of the changes occurring in the immunoprecipitate patterns after 5, 9 and 12 booster immunizations as illustrated in Fig. 8.4. The progression from a relatively simple pattern of immunoprecipitates of major and/or more strongly immunogenic antigens to a more complex reference pattern of major and weaker immunoprecipitates can be readily seen from this investigation, with the final reference patterns showing 20 distinct antigens fairly consistently (Shulman and Karpatkin, 1980). Clearly, a decision as to whether it will be advantageous to pool consecutive bleedings will depend very much on the overall objectives of a particular study. For the majority of studies, pooling the antiserum on continued immunization after the pattern has approximately reached its peak complexity will suffice and will simplify the task of purification of the immunoglobulin fractions. It is of course evident that prolonged immunization may reveal additional minor and/or weakly immunogenic antigens.

Once having generated a reference pattern, one further strategy worth mentioning is the use of the intermediate gel variation of CIE to examine bleedings from a newly immunized animal, either the same or different species of animal. By interposing the new sera in the intermediate gels and the reference immunoglobulins in the rest of the immunoplate (reference gel) the course of responses to immunization can be monitored. For such monitoring, the use of 'uncut' serum or 50% ammonium sulphate-precipitated immunoglobulins in the intermediate gel may be adequate. Moreover, the intermediate gel requires a smaller volume of antiserum than that required in the reference gel, a real advantage where the volume of blood available may be relatively small as in some animal species (e.g. guinea pigs). One further advantage of the

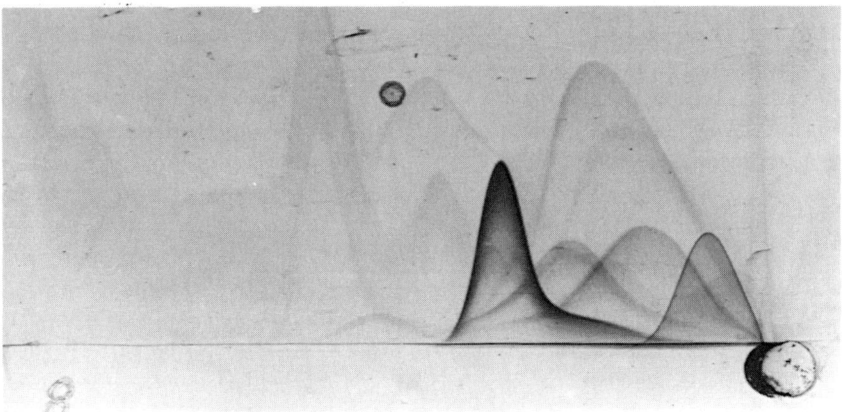

(a)

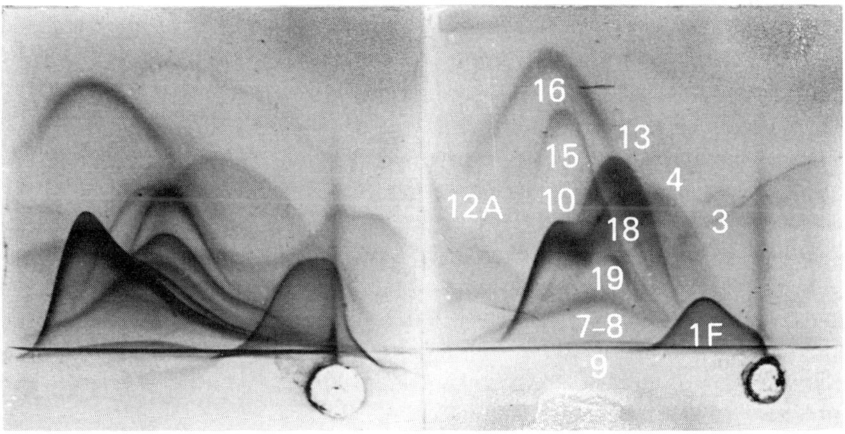


(b) (c)

**Fig. 8.4** Crossed immunoelectrophoresis of 1% Triton X-100-solubilized human platelet membrane preparations into rabbit antibody after 5(a), 9(b) and 12(c) booster immunizations with the purified platelet membranes. In these experiments 50 $\mu$g of platelet membrane antigens were subjected to electrophoresis into the partially purified rabbit antibody. Note the increasing complexity of the immunoprecipitate patterns with continued immunization. In the final reference pattern after 12 (c) booster immunizations up to 20 different antigens were fairly consistently observed. From Shulman and Karpatkin (1980), reproduced with permission of *Journal of Biological Chemistry.*

use of the intermediate gel in following the production of anti-membrane antibodies in an animal species different to the one in which the reference antiserum had been generated is that it may reveal on a single immunoplate any species differences in the immunogenicity of the membrane proteins. The availability of these intermediate gel procedures further emphasizes the great versatility of the CIE method in membrane protein analysis, and where monospecific antiserum is available to a single component, it permits its specific identification in a complex reference pattern [e.g. $F_1$-ATPase of bacterial membranes (Owen and Salton, 1975)]. Thus, the intermediate gel variation of the CIE method can be used not only in monitoring responses to immunization and comparisons of responses in different animal species it can also be used in identifying specific components in the complex array of antigens found in cell membranes and establishing their cross-reactivities. Other variations of the basic CIE method such as tandem and co-electrophoresis may also be utilized in the identification of specific membrane antigens (Owen and Salton, 1975; Owen, 1983b).

(c) *Influence of detergents on immunoelectrophoretic analysis of membrane proteins*

Two aspects of the use of detergents must be considered in the CIE analysis of membrane proteins. In the first instance detergents are required for the dissociation and solubilization of the membranes and secondly, they are required in the agarose gels during electrophoresis to keep the amphiphilic membrane antigens stabilized in solution and to prevent their aggregation except by specific precipitation with antibody molecules. Differences in the immunoprecipitate patterns can arise from the type of solubilizing detergent used and from the type of detergent (non-ionic or zwitterionic) used in the agarose gels. As pointed out by Bjerrum (1983) it is the detergent used in the gel that has the greater effect in determining the immunoprecipitate pattern, since the membrane-solubilizing detergent molecules bound to the proteins are capable of sufficiently rapid exchange to be less influential on the final precipitate patterns. Furthermore, although neutral (zwitterionic) and non-ionic detergents can be used both for solubilization and electrophoresis in the presence of antibodies without affecting the migration of antigens and antibodies, differences in the patterns can be observed due to such factors as the solubilization efficiencies, extent of dissociation of protein complexes, denaturation effects and the physical properties of detergents and the effects of temperature upon them (Bjerrum, 1983). Selecting the 'right' detergent for a particular study may thus involve investigations of the effects of a number of different detergents on the pattern of immunoprecipitates as well as the efficiency of the initial solubilizing detergent employed for membrane dissociation (Bjerrum, 1983; Collins and Salton, 1979).

Although ionic detergents are effective in solubilizing membranes their use is generally restricted to the first dimension of electrophoresis but their undesirable effects can be counteracted by use of non-ionic detergents in the gels (Bjerrum, 1983). It is also noteworthy that both cationic and anionic detergents can be used in the charge-shift electrophoresis phenomenon (Helenius and Simons, 1977) to yield valuable information on the amphiphilic character of membrane antigens. Thus the binding of mixed micelles of Triton [negatively charged with deoxycholate (DOC) or positively charged with cetyltrimethylammonium bromide (CTAB)] to the amphiphilic protein can be used to increase the resolution of the CIE patterns.

It is evident, therefore, that detergents play a pivotal role not only in dissociation of membrane complexes and solubilization of membrane proteins but they are essential for the maintenance of membrane antigens in solution during immunoelectrophoresis. They can also be manipulated to yield additional information about the amphiphilic nature of membrane proteins (Bjerrum, 1983) and the subunit reactions with antibodies following dissociation with sodium dodecyl sulphate (SDS) as shown with membrane $F_1$-ATPase by Urban and Salton (1983a,b).

### 8.2.2 ROCKET IMMUNOELECTROPHORESIS

Rocket immunoelectrophoresis, as originally devised by Laurell (1965), and its variant, fused-rocket immunoelectrophoresis (Svendsen, 1973), have found many applications in the study of membrane antigens with both monospecific and polyspecific antisera. The general considerations discussed above for CIE also apply to the rocket procedures. Rocket techniques have been particularly valuable in quantifying the extraction of membrane antigens with different detergents and chaotropic agents (Collins and Salton, 1979), and the fused-rocket variant has often been found to be the method of choice in following chromatographic separation of membrane antigens and enzymes (Svendsen, 1973; Owen and Smyth, 1977; Bjerrum and Giannaza, 1983). Both procedures have been combined with zymogram staining, radioactive-labelling and radioautography and lectin affinity electrophoresis and will be discussed further in relation to these specific methodologies.

## 8.3 Identification of membrane antigens as enzymes by zymogram staining procedures

The ability to identify antigens as enzymes by histochemical (zymogram) staining of immunoprecipitates in agarose adds yet another dimension to the technique of electroimmunochemical analysis of cell membranes. Zymogram staining for specific enzyme activities (Uriel, 1971; Owen and Smyth, 1977) not only permits identification of individual antigens as enzymes or enzyme complexes but also amplifies the detection of some immunoprecipitates not detectable by Coomassie Blue staining (Smyth *et al.*, 1978). Retention of this biological activity in addition to immunological reactivity provides valuable biochemical markers in detergent-electroimmunochemical analysis of membranes and can also be put to good effect in monitoring the fractionation of membrane enzymes by conventional column chromatographic and immuno-adsorbent procedures. Clearly, only those membrane enzymes surviving the necessary treatments with detergents for dissociation and the presence of detergent during immunoelectrophoresis will be detectable providing a suitable staining procedure is available for the specific enzyme reactions.

It has of course been well known for a long time that antibodies to enzymes can strongly inhibit enzyme reactions and indeed, in some instances they may completely inhibit the activity in solution (Cinader, 1957, 1967). Despite the inhibitory effects of antibodies on enzyme activities, it is evident that sufficient enzyme molecules remain active in the immunoprecipitates for detection by zymogram techniques. This could be due in part to the heterogeneity of antibody populations, some molecules of which are directed to antigenic sites other than the active centres of the enzymes. Perhaps of equal or greater significance is the fact that precipitation by the electroimmunochemical method in agarose often occurs in the zone of antigen excess compared to that in free solution as pointed out by Laurell and McKay (1981). This may serve to minimize the steric hindrance of the active site of the enzyme resulting from antibody molecules binding to antigen thus facilitating detection of at least some of the enzyme molecules in the precipitates.

### 8.3.1 ZYMOGRAM STAINING PROCEDURES

Enzymic staining of immunoprecipitates is based essentially on histochemical methods and both the theory and practice of zymogram staining has been extensively reviewed by Uriel (1971), Verbruggen (1975), Brogren and Bøg-Hansen (1975) and Owen and Smyth (1977). Zymogram staining has been applied to most of the variations of the electroimmunoprecipitation techniques in agarose, including rocket and crossed immunoelectrophoresis, intermediate gels with monospecific antibodies to certain membrane enzymes [e.g. $F_1$-ATPase (Owen and Salton, 1975)], and fused-rocket immunoelectrophoresis (Owen and Smyth, 1977). Following standard electrophoresis, pressing and washing procedures and where applicable, cool air-drying procedures, the immunoplates are subjected to the appropriate enzymic staining method involving incubations in buffer, substrate and enzyme detection system. The pressing and washing steps may serve to minimize interference in the enzyme reaction and/or with components of the detection systems by detergents used for membrane analysis by the immunoelectrophoresis methods. Sequential staining for different enzymes can be performed where residual components of the detection system do not interfere with the subsequent enzyme reactions, e.g. various dehydrogenases using reduction of tetrazolium dyes for detection but with different substrates (Owen and Smyth, 1977). Furthermore, in the analysis of membrane proteins by CIE or rocket immunoelectrophoresis it is possible to follow the enzyme staining with Coomassie Blue staining to reveal the rest of the immunoprecipitate patterns. Thus, a great deal of information can often be obtained by the application of these various techniques to a single immunoplate.

In general, the substrates used for zymogram staining are low-molecular-weight compounds capable of readily penetrating into the immunoprecipitates. High-molecular-weight substrates are not necessarily excluded, since there are instances where they have been used to detect hydrolases by print techniques (Uriel, 1971; Radola, 1975) or by using substrate in an agarose overlay (Wadstrom and Smyth, 1975). Where chromogenic substrates are available these can be used to generate insoluble products or poorly soluble products that may be coupled to diazonium salts giving insoluble coloured complexes (Owen and Smyth, 1977). With non-chromogenic substrates it is necessary to couple the enzyme reaction to a secondary chromogenic detection system and/or one generating an insoluble compound as a side (non-enzymic) reaction. Detection of dehydrogenases represents a typical example of how the non-chromogenic substrates (e.g. lactate, malate, succinate, NADH) can be linked to a system generating an insoluble chromogen, by using phenazine methosulphate to transfer electrons from substrate to effect the reduction of tetrazolium dyes to insoluble formazans. Moreover, by the choice of tetrazolium salts of different redox potentials it has been possible to discriminate between two membrane NADH dehydrogenases that are not only antigenically distinct but exhibit different responses depending on the level of electronegativity of the tetrazolium electron acceptors (Owen and Smyth, 1977). More complex cascade reactions can be utilized to generate reduced formazan, as for example, in the detection of enolase by the coupled reactions with pyruvate kinase, hexokinase and glucose 6-phosphate dehydrogenase (Bock and Dissing, 1975).

Formation of insoluble inorganic salts such as the reaction of inorganic phosphate liberated from ATP by adenosine triphosphatases with lead salts to yield lead phosphate can be utilized in the specific detection of ATPase (Owen and Salton, 1975, 1977). The white lead phosphate precipitated on the immunoprecipitates of $F_1$-ATPase or $F_0$–$F_1$-ATPase complex in Triton X-100-solubilized membrane can then be converted to the light-brown lead sulphide by reaction with $Na_2S$ as described by Uriel (1971). The zymogram staining of $F_1$-ATPase released from the membrane of *Micrococcus lysodeikticus* by the shock-wash procedure and analysed by CIE against membrane antiserum is illustrated in Fig. 8.5(a) and compared with the same shock-wash fraction stained with Coomassie Blue (Fig. 8.5b). Comparison of the immunoplates shows the specific detection of the $F_1$-ATPase by zymogram staining and the presence of other antigens in the crude shock-wash fraction detected by Coomassie Blue staining of immunoprecipitates (Urban and Salton, unpublished data).

It is evident from the extensive survey by Owen and Smyth (1977) of the

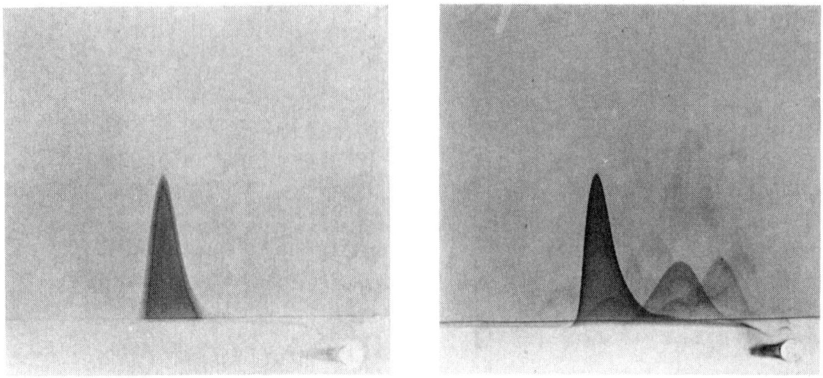

(a) (b)

**Fig. 8.5** The $F_1$-ATPase released from the membranes of *M. lysodeikticus* by a low-ionic-strength shock wash can be detected by zymogram staining for ATPase as previously described (Owen and Salton, 1975) following two-dimensional resolution of the antigens against anti-membrane antibodies. Protein (25 $\mu$g) of the shock-wash fraction was subjected to crossed immunoelectrophoresis and stained for ATPase activity (a) and with Coomassie Blue (b). Note that only a single immunoprecipitate peak ($F_1$-ATPase) was stained by the zymogram procedure and that a number of additional membrane antigens were released by the shock-wash procedure and were detectable as immunoprecipitates on staining with Coomassie Blue. The major $F_1$ immunoprecipitate in (b) corresponds very closely to the zymogram stained peak of ATPase.

enzymes analysed by quantitative immunochemical techniques that a great variety of enzyme staining procedures are available for the identification and quantification of enzymes from membranes as well as from the cytosol fractions. Such procedures range from direct staining with chromogenic substrates (e.g. alkaline and acid phosphatases, peptidases, proteases, esterases, galactosidase, hexosaminidases) to non-chromogenic substrates linked to simple or highly complex secondary detection systems generating coloured and non-coloured insoluble products outlining and precipitating on the immunoprecipitates. A limited selection of examples of the detection of membrane enzymes by zymogram staining procedures from electroimmunochemical studies is presented in Table 8.1.

In addition to the direct and indirect staining procedures for identification of enzymes in immunoprecipitates, the potential for using radioautography and radioactively labelled substrates and detection systems should not be overlooked. Such techniques would be applicable where the products of the reaction do not readily diffuse away from the precipitate (e.g. polymerization reactions or transfer of labelled substrate to a slowly diffusible acceptor) or where products can be immobilized at the site of the reaction. Moreover, the potential exists for identification of enzymes with specific inhibitors (radioactively labelled) capable of covalent or high-affinity binding to the enzymes.

Table 8.1 A selection of examples of membrane enzymes detected by CIE-zymogram procedures

| *Membrane source* | *Enzyme activities detected* | *Reference* |
|---|---|---|
| Rat liver plasma membranes | Alkaline phosphodiesterases | Berzins and Blomberg (1983) |
| | Arylamidases | |
| | NADH dehydrogenases | |
| | ATPase | |
| | Leucine aminopeptidase | |
| | Non-specific esterase | |
| Human erythrocyte membrane | Acetylcholinesterase | Bjerrum and Hagen (1983) |
| | ATPase | Bjerrum (1977); Bjerrum *et al.* (1983) |
| | Lactate dehydrogenase | |
| | Catalase | |
| | NADH oxidoreductase | |
| Microvillus membranes | Aspartate aminopeptidase | Danielsen (1983) |
| Rat brain synaptosomes | Enolase | Bock and Dissing (1975) |
| Bacterial membranes: | | |
| *Acholeplasma laidlawii* | NADH dehydrogenase | Johansson (1983) |
| *Escherichia coli* | ATPase | Smyth *et al.* (1978) |
| | Dihydrolipoyl dehydrogenase | Owen (1983b) |
| | Dihydro-orotate dehydrogenase | |
| | Glutamate dehydrogenase | |
| | D-Lactate dehydrogenase | |
| | NADH dehydrogenase | |
| | Succinate dehydrogenase | |
| | 6-Phosphogluconate dehydrogenase | |
| | Protease | |
| *Bacillus subtilis* | Succinate dehydrogenase | Rutberg *et al.* (1978) |
| *Micrococcus lysodeikticus* | ATPase | Owen and Salton (1975) |
| | Malate dehydrogenase | |
| | NADH dehydrogenase | |
| | Succinate dehydrogenase | |
| *Staphylococcus aureus* | Nitrate reductase | Burke *et al.* (1981) |

Penicillin-binding proteins (PBPs) of bacterial membranes involved in peptidoglycan synthesis and metabolism have such a potential for being identified with [$^{14}$C-]penicillin and although they can be readily identified in SDS-PAGE gels by fluorography (Spratt, 1975), they have not so far as the author is aware, been clearly detected under non-dissociating conditions in CIE studies, despite attempts in our laboratory with membranes of both Gram-positive and Gram-negative bacteria. There may be many factors mitigating against their detection by immunoelectrophoresis techniques. Little is known about the immunogenicity of PBPs and at best they are likely to be minor antigens in the membranes since the relative abundance of penicillin-binding components and numbers of molecules per cell membrane appear to be relatively low (Blumberg and Strominger, 1974). Nonetheless, further efforts to detect membrane PBPs catalysing steps in peptidoglycan synthesis should be worthwhile and may reveal new information about the organization of these unique enzymes in bacterial membranes and provide evidence as to whether they exist in the membrane as single enzyme entities or in association as multienzyme organelles.

Despite the availability of a wide range of enzymic staining procedures the functions of many membrane antigens resolved by CIE analysis remain unknown. Thus, only five of the 27 immunoprecipitates resolved in the membranes of the Gram-positive bacterium, *Micrococcus lysodeikticus*, were identified as enzymes (Owen and Salton, 1977) and of the 50 discrete antigens found in the *E. coli* ML308-225 membrane vesicles, 12 were identified as enzymes by specific histochemical stains (Owen, 1983b). These results and those of Johansson (1983) with *Acholeplasma laidlawii* for bacterial membranes are not radically different from those obtained with mammalian membranes such as rat liver microsomal membranes (Berzins and Blomberg, 1983) and erythrocyte membranes (Bjerrum *et al.*, 1983). Identification of 20–30% (or even less) of the immunoprecipitates as enzymes by zymogram staining is perhaps not entirely surprising. In the first instance, not all membrane antigens will be enzymes; some may have specific receptor or transport carrier functions. Two other reasons for the failure to identify the biochemical functions of membrane antigens are obvious, the lack of suitable enzyme detection procedures for those enzymes not readily amenable to histochemical manipulation and denaturation or inactivation of some membrane enzymes in the course of membrane isolation, solubilization and immunoelectrophoresis (e.g. inactivation by loss of necessary metal ions and cofactors). Added to these factors there is always the possibility of complete inhibition of enzyme activity by antibodies. Clearly, the identification of the enzymic functions of the complex array of antigens found in cell membranes of diverse origins is a formidable task and the resolution of the many problems that this presents the investigators of biomembrane functions will continue to tax their ingenuity and patience for years to come.

In addition to zymogram staining to detect specific membrane enzyme activities in immunoprecipitates in agarose gels, precipitates can be rendered enzymically active for detection purposes by using horseradish peroxidase-labelled immunoglobulin specific for the antibodies used in the reference system. The horseradish peroxidase can thus be visualized by incubation with an appropriate substrate and chromogen (as in enzyme linked immunosorbent assay (ELISA) tests) and immunoprecipitates can be stained with a strongly coloured deposit (Kjaervig Broe and Ingild, 1983). This procedure permits the enzymic amplification of the detection of antigens with an approximately 25- to 50-fold increase in the sensitivity of detection compared to Coomassie Blue staining of immunoprecipitates (Kjaervig Broe and Ingild, 1983). Although this amplification system was used by these authors in the detection of human $\alpha_1$-foetoprotein, it is evident that the procedure would have wide applicability to other antigens precipitated in agarose gels, including membrane antigens. Such studies could thus permit the detection of weak membrane antigens and immunoprecipitates.

## 8.4 Quantification of membrane enzymes by rocket immunoelectrophoresis

Since the introduction of rocket immunoelectrophoresis (RIE) by Laurell (1965, 1966) to study and quantify serum proteins, this technique has been widely used in the quantification of a variety of enzymes, primarily soluble enzymes in sera, urine, tissue extracts, etc. (Owen and Smyth, 1977). The RIE procedure has also been adapted for cathodal migration for quantitative studies of the basic enzyme, lysozyme (Johansson and Malmquist, 1971), thus further attesting to the great flexibility of electroimmunochemical methodology. Rocket immunoelectrophoresis therefore has considerable potential for studying both anionic and cationic membrane proteins. Indeed, Ostrovsky *et al.* (1981) detected by RIE and CIE lysozyme which had become tightly bound to the bacterial membranes during the isolation procedure. Despite its presence and detection in the Triton X-100-solubilized membranes, this cationic protein did not appear to affect the pattern of membrane immunoprecipitates, except when added in large excess. Complex formation between cationic and anionic proteins could reduce the net charge of the latter and present some complications in their quantification by RIE. Fortunately, the peak heights of the immunoprecipitates in the CIE experiments with bacterial membranes were largely unaffected and relatively insensitive to low concentrations of added lysozyme (Ostrovsky *et al.*, 1981). Nonetheless, this word of caution appears warranted as the addition of cationic compounds may interfere with the normal electrophoresis of membrane proteins, and $Ca^{2+}$-induced proteolysis of human erythrocyte membrane proteins may also result in altered immunoelectrophoretic behaviour (Bjerrum *et al.*, 1983). In spite of such complicating factors, useful information can be gained from rocket immunoelectrophoretic analysis of membrane enzymes, since as emphasized by Collins and Salton (1979) this technique can be used to quantify molecular species (including enzymes) in detergent-solubilized membranes rather than specific activities of the enzymes. Determination of molecular species of enzymes by RIE has the distinct advantage of permitting the quantification of antigenically distinct enzymes possessing the same enzyme reaction [e.g. two distinct NADH dehydrogenases identified and quantified in *M. lysodeikticus* membranes by Collins and Salton (1979)]. Measuring total enzyme activity would obviously not discriminate between the two antigenic forms of the enzyme.

Two examples will be given to illustrate how RIE can be utilized in the quantification of enzymes as antigens. In their study of the suggested glycoprotein nature of membrane $F_1$-ATPase (Andreu *et al.*, 1978), Lim and Salton (1981) were able to quantify the ATPase before and after chromatography on Con A–Sepharose 4B by RIE and determine the recovery of the ATPase in column fractions by rocket electrophoresis into either anti-(purified ATPase) antibodies or anti-membrane antibodies. The recoveries of enzyme as antigen by RIE and of ATPase activity and protein (and hence specific activities) showed excellent agreement (Lim and Salton, 1981). The elution profile of the $F_1$-ATPase from the Con A–Sepharose 4B column determined by RIE is illustrated in Fig. 8.6 and established the removal of the contaminating membrane carbohydrate component, the succinylated lipomannan from

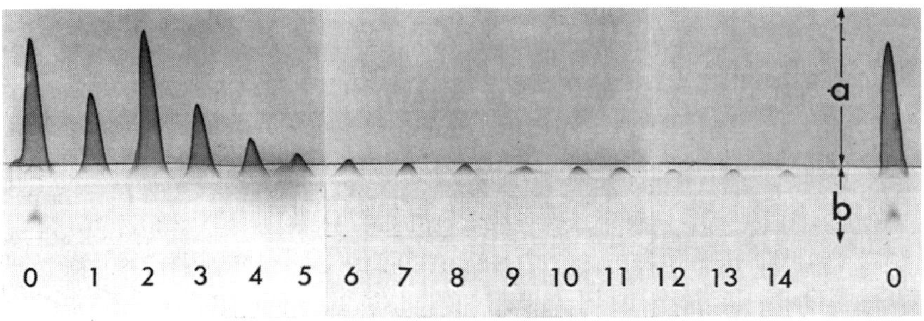


**Fig. 8.6** Rocket immunoelectrophoresis through an intermediate gel (zone b) containing concanavalin A has been used to establish the separation of contaminating membrane lipomannan from the $F_1$-ATPase of *M. lysodeikticus* by chromatography on Con A–Sepharose 4B. Well 0 contained 3.1 $\mu$g of the DEAE-Sephadex-purified $F_1$-ATPase before removal of carbohydrate by affinity chromatography. Wells 1–14 contained 8 $\mu$l of each fraction (10.5 ml) eluted from the Con A–Sepharose column. Zone a of the immunoplates contained anti-membrane immunoglobulins. Note the presence of precipitates in the intermediate gel containing concanavalin A of the $F_1$-ATPase preparation before chromatography on Con A–Sepharose and the absence of precipitates after removal of carbohydrate. Anodal migration was to the top of the plates. From Lim and Salton (1981), reproduced with permission from *Biochimica et Biophysica Acta.*

$F_1$-ATPase (Lim and Salton, 1981). RIE thus proved to be a valuable adjunct in the separation of the $F_1$-ATPase from lipomannan and facilitated the conclusion that this bacterial membrane $F_1$-ATPase was not a glycoprotein (Lim and Salton, 1981), thereby confirming a similar conclusion by Nalin *et al.* (1979) based on qualitative studies of Schiff-stained PAGE gels.

In the second example of the use of RIE in quantification of membrane enzymes, Collins and Salton (1979) determined the relative amounts of several enzymes by zymogram staining of rockets obtained by immunoelectrophoresis of detergent-solubilized bacterial membrane extracts. The recoveries of two antigenically distinct reduced nicotinamide adenine dinucleotide (NADH) dehydrogenases and succinate and malate dehydrogenases were determined in the various detergent extracts of *M. lysodeikticus* membranes by staining the rockets formed on electrophoresis into rabbit anti-membrane antiserum with specific enzyme stains. Rocket immunoelectrophoresis with monospecific antiserum was used to determine the amount of ATPase in the extracts and rocket affinoelectrophoresis into concanavalin A permitted the quantitative analysis of the membrane succinylated lipomannan. The effectiveness of various solubilizing agents could be studied quantitatively by RIE and Fig. 8.7 illustrates the detection by zymogram staining of NADH dehydrogenase I (dense precipitate) and NADH dehydrogenase II (lighter, taller precipitate) in Triton X-100 extracts and in membranes extracted with the chaotropic agent, barium thiocyanate (Collins and Salton, 1979). The rocket technique thus provided an accurate estimation of the amounts of each of these two enzymes present but such a distinction could not have been made on the basis of measurements of total enzyme activity. Moreover, it was possible by rocket analysis for the four dehydrogenases and ATPase to compare their extractability with that of total membrane protein, cytochrome and lipomannan (by rocket affinoelectrophoresis) under varying conditions of detergent concentration and for detergents of different hydrophile–lipophile balance (HLB) numbers. Of particular interest in evaluating the relationship of extractability of membrane components and HLB number was the observation of a rather sharp optimum at HLB number 13.5 (for Triton X-100) in a Triton X series for extraction of bulk protein, membrane enzymes and lipomannan as illustrated in Fig. 8.8. These studies of the specific contents of extracts obtained by treatment of membranes with a variety of detergents revealed that there was a general tendency toward uniformity of composition of the membrane extracts (Collins and Salton, 1979). In the light of the widespread use of Triton X-100 in detergent-electroimmunochemical studies, it is of interest to note that detergents with a hydrophile–lipophile balance of approximately 13–13.5 are optimal not only for extraction of bulk membrane protein, but also for the solubilization of several unrelated individual components (e.g. lipomannan membrane amphiphile, $F_0$–$F_1$-ATPase complex, dehydrogenases). Extracts prepared with chaotropic agents on the other hand differed from detergent

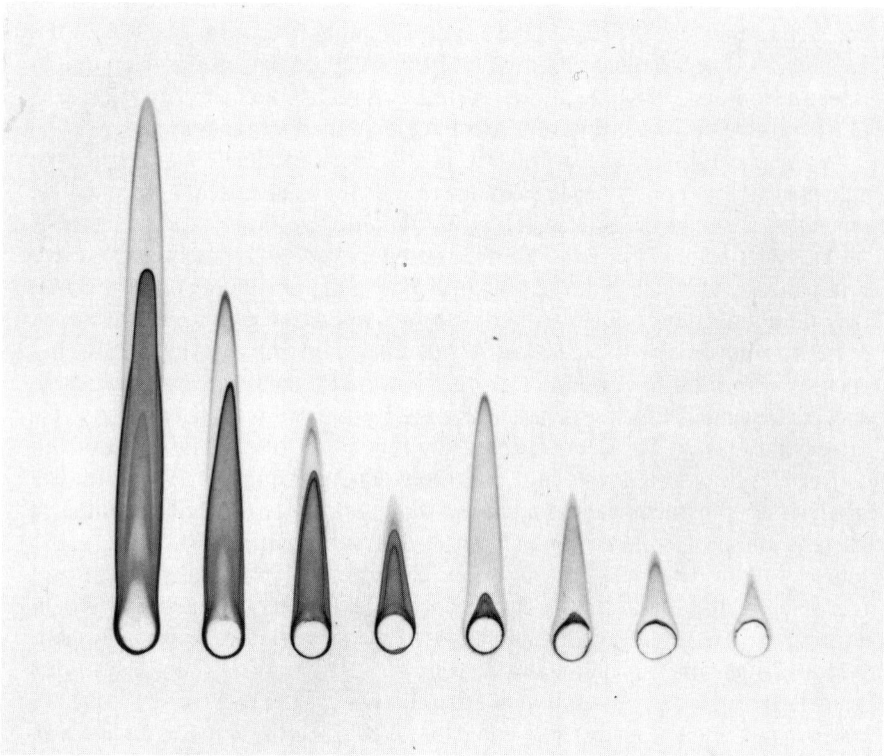

**Fig. 8.7** Rocket immunoelectrophoresis and zymogram staining for NADH dehydrogenase activity in membrane extracts from *M. lysodeikticus*. Wells contained from left to right 1% Triton X-100 extract, undiluted, diluted 1:2, 1:4 and 1:8; 0.25 M barium thiocyanate extract, undiluted, diluted 1:2, 1:4, 1:8. Note two antigenically distinct enzymes, NADH dehydrogenase I (dense precipitate) and NADH dehydrogenase II (lighter, taller precipitate) and the preferential extraction of NADH dehydrogenase II by the chaotropic agent barium thiocyanate. From Collins and Salton (1979), reproduced with permission of *Biochimica et Biophysica Acta.*

extracts both qualitatively and quantitatively when examined by RIE and other analytical procedures and although the yield of protein was low with these agents some selectivity of extraction was observed (Collins and Salton, 1979).

The availability of monospecific antiserum to purified membrane enzymes permits their RIE quantification as antigens, facilitates their identification and measurement in extracts and fractions of enzymes and can be utilized in monitoring eluate fractions during chromatography. Monospecific antisera can also be used in conjunction with lectins or other antisera to quantify more than one component on the same immunoplate (e.g. ATPase and lipomannan) by the use of intermediate gels. In addition antiserum to whole membranes can

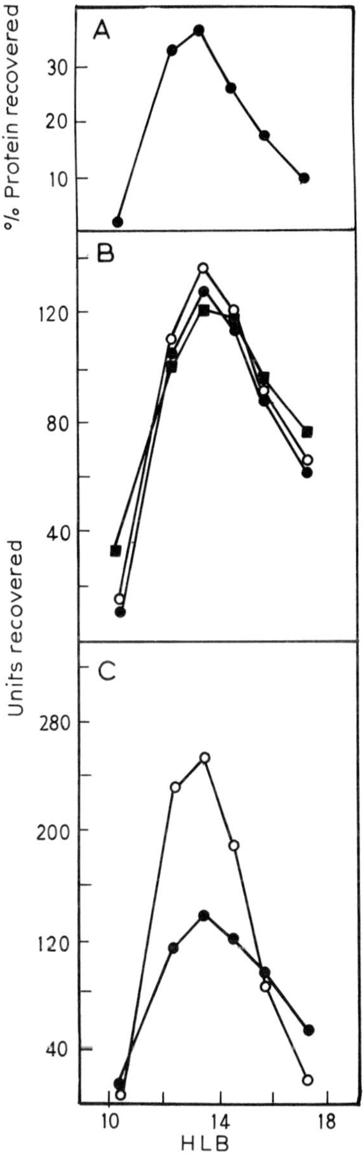


**Fig. 8.8** Influence of detergent hydrophile–lipophile balance (HLB) on the extractability of membrane components by non-ionic detergents of the Triton X series. Points represent average recovery from two extractions of *M. lysodeikticus* membranes. The components assayed were (A) protein; (B) NADH dehydrogenase I, ●; NADH dehydrogenase II, ■; malate dehydrogenase, ○; (C) ATPase, ○; lipomannan, ●. From Collins and Salton (1979), reproduced with permission of *Biochimica et Biophysica Acta*.

be utilized in the quantitative analysis of enzymes by rocket immunoelectrophoresis of membrane extracts permitting the simultaneous quantification of antigenically distinct enzymes possessing the same activities (e.g. NADH dehydrogenase). As illustrated in this section specific enzyme staining can be used as a powerful tool in optimizing the release of membrane enzymes assayed by rocket electrophoresis. Although the examples given here have been for bacterial membranes, the basic procedures are equally applicable to membranes derived from other cells and cellular organelles. Rocket immunoelectrophoresis has found wide application in the assay and study of serum proteins, antigens of body fluids and tissues and toxins of various micro-organisms. Its potential for quantitative studies of membrane antigens and enzymes is considerable and has yet to be fully realized. The simplicity of this technique, its sensitivity and specificity has much appeal and will undoubtedly find wider application to the problem of membrane analysis in the future.

## 8.5 Fused rocket electrophoresis and enzyme staining for the purification of membrane enzymes

One of the many modifications of the immunoelectrophoresis methods which have proved so valuable in the analysis of soluble and detergent-solubilized membrane proteins is the fused-rocket immunoelectrophoresis technique. This method is especially suitable for the examination of fractions from purification experiments and is particularly useful in following the elution profiles of proteins being separated by column chromatography and is equally suitable for monitoring the isolation of membrane proteins and enzymes. The principle of using the rocket method to give continuous fused lines of precipitates from fractions electrophoresed into antibodies was developed by Rose and Harboe (1970) but was modified later by Svendsen (1973) to avoid loss of resolution by allowing diffusion of the protein from the wells to occur in agarose gels free of antibodies prior to electrophoresis into the gel containing antibodies. This prevented the precipitation of proteins in the immediate vicinity of the wells in the agarose containing antibodies as first used by Rose and Harboe (1970) and allowed the clear resolution of complex mixtures of protein eluted from ion-exchange columns as fused lines and peaks of immunoprecipitates in the gels containing antibodies (Svendsen, 1973). As emphasized by Svendsen (1973) for the successful application of the fused-rocket method for purification of proteins, it is absolutely necessary that antibodies to all antigens must be present in good titre in order to detect the impurities in the course of purification. Thus antibodies have to be generated to the complex array of antigens in the starting material or to crude fractions to be followed by fused-rocket immunoelectrophoresis during separation.

This method clearly has many advantages in following the purification of membrane enzymes and can be used in combination with monospecific antiserum and/or zymogram staining to monitor the separation of enzymes in crude extracts of membranes containing a mixture of enzymes and/or membrane antigens. The potential of this method for monitoring enzyme purification has been well illustrated by the multistep purification of human serum cholinesterase by Brogren and Svendsen (1975).

The fused-rocket immunoelectrophoresis technique has a number of advantages over conventional methods in following enzyme purification, including rapidity with which it can be performed, it is less tedious than assaying a large number of individual fractions for enzyme activity, it can resolve antigenically different isozymes and it can also detect and determine the distribution of contaminating proteins. Despite these advantages it appears to have been under-utilized in studies of membrane enzyme purification. However, an example of the use of fused-rocket immunoelectrophoresis as a powerful tool in membrane enzyme purification is provided by the study of Huberman, Salton and Owen (unpublished data, quoted by Owen and Smyth, 1977). In this study fused-rocket immunoelectrophoresis was used to monitor fractions eluted from an Ultragel ACA 22 column which had been loaded with a crude fraction containing the $F_1$-ATPase released from the membranes of the Gram-positive bacterium, *Micrococcus lysodeikticus*, by the two-phase n-butanol extraction procedure (Salton and Schor, 1974). Monospecific antibodies to the purified ATPase were placed in an intermediate gel zone of the immunoplate and anti-membrane antiserum was added to the reference gel, and column fractions were examined by fused-rocket electrophoresis into the two antibody zones. Fig. 8.9 illustrates the fused-rocket profile of the initial crude fraction loaded on to the Ultragel column and the separation of ATPase from other antigens eluted from the gel filtration column. In this fused rocket elution profile the ATPase (antigen 1) was identified by its reaction with and precipitation by anti-ATPase antibodies in the intermediate gel. Other membrane antigens in the initial crude fraction and in some of the eluted fractions were detected with the anti-membrane antiserum and the extent to which they overlap in the fused-rocket profile can be seen in Fig. 8.9. Thus examination of the immunoplate indicated that fractions 95–104 would give the maximum yield of the ATPase with minimal contamination by other membrane antigens. This study also illustrates how additional information on the identity of contaminating components can be gained by the fused-rocket method. One of the antigens suspected of being released from the membranes by the n-butanol extraction method was the succinylated lipomannan, the membrane amphiphile of this organism. By including wells containing purified lipomannan (in position i on the immunoplate in Fig. 8.9) it was possible to identify antigen 2 in the fused rockets as the lipomannan and, indeed, confirm its presence in the initial crude fraction loaded on the Ultragel column and its presence as a contaminant in parts of the elution profile of the ATPase. Furthermore, zymogram staining of the fused-rocket plates for NADH

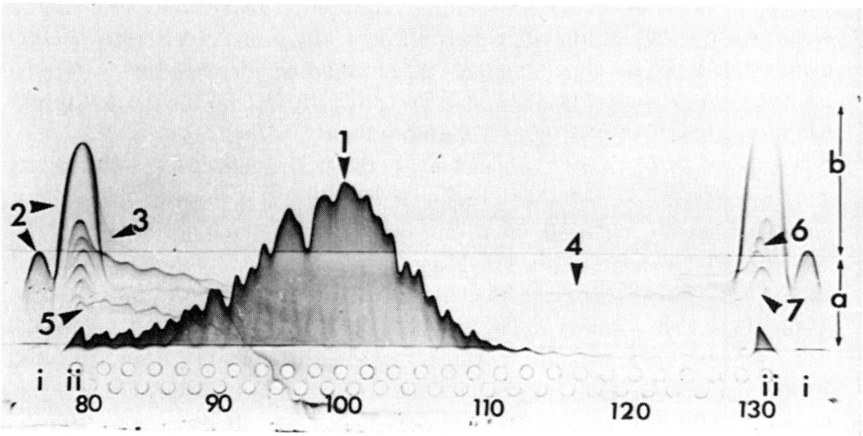


**Fig. 8.9** Fused-rocket immunoelectrophoresis of a crude $F_1$-ATPase preparation released from *M. lysodeikticus* membranes by n-butanol extraction procedure and fractionated on an Ultragel ACA 22 column. Fractions 80–130 from the column were monitored by fused-rocket immunoelectrophoresis through monospecific antiserum to purified ATPase in gel zone (a) and into anti-membrane antiserum (zone b). Wells marked i and ii contained purified lipomannan and the initial crude $F_1$-ATPase fraction respectively. The $F_1$-ATPase (antigen 1) was identified by its inclusion in the intermediate gel zone (a). Fusion of purified lipomannan precipitate identified antigen 2 in the crude fraction and antigen 3 in fractions 80–94 shows partial identity. Antigens 4 and 5 possessed NADH dehydrogenase activities detectable by zymogram staining. Antigens 6 and 7 present in the crude fraction were not detected in the ATPase fused-peak. Unpublished data of Huberman, Owen and Salton, from Owen and Smyth (1977), reproduced with permission of John Wiley and Sons, New York.

dehydrogenases could be used to determine the extent to which they were contaminants of the ATPase fractions. In parallel with the establishment of this fused-rocket profile of the gel filtration of ATPase on Ultragel ACA 22, Huberman, Salton and Owen (see Owen and Smyth, 1977) followed the elution profile by conventional protein (280 nm absorbance) and ATPase activity assay and found excellent coincidence of the activity profile and the fused-rocket immunoelectrophoretic profile.

As emphasized by Bjerrum (1983) the electroimmunochemical methods of analysis of membrane proteins have great flexibility and this is certainly the case for the fused-rocket method as illustrated by the example presented in this section and by the work of Nielsen and Bjerrum (1977) on the separation of milk fat globule membrane antigens. Many variations in the strategies used are possible. For example, in the study of the separation of the membrane ATPase, instead of using monospecific antiserum to this enzyme to establish its elution profile, it would have been equally satisfactory to use anti-membrane antiserum throughout the immunoplate and use the specific enzyme stain for the elution profile of the ATPase. The fused-rocket immunoelectrophoresis technique for separating membrane enzymes thus possesses the potential for combination with other procedures including zymogram staining, co-electrophoresis with purified components for specific fused-line identifications, radioactive labelling of components either metabolically or artificially *in vitro*, followed by radioautography of immunoplates. It should also be emphasized that these procedures are generally much more informative as to the status of purity of enzyme preparations since the reactions with antibodies and staining with Coomassie Blue amplify the detection of trace contaminants and thus provide a more sensitive index of homogeneity than would be possible with SDS–polyacrylamide gel electrophoresis alone. However, as pointed out by Bjerrum and Gianazza (1983), despite the many advantages and predictive value of RIE monitoring in membrane protein purification, exclusive use of this method may give a false impression of purity since non-antigenic impurities would not be detected. Clearly, several criteria and methods of assessment of purity are needed and reliance solely on SDS-PAGE and Coomassie Blue staining would be equally unwise since this procedure would not detect carbohydrate contaminants.

## 8.6 Lectins and affinoelectrophoresis in the analysis of membrane glycoproteins and polysaccharide complexes

Bøg-Hansen (1973) introduced the use of the lectin, concanavalin A, in a variant of the basic two-dimensional CIE procedure. By interposing the lectin in an intermediate gel between the gel containing the electrophoretically separated antigens and the reference gel containing antibodies, glycoproteins reacting with the lectin were retarded with loss or depression of corresponding immunoprecipitates in the reference antibody gel. Moreover, Bøg-Hansen and Brogren (1975) showed that glycoproteins may form precipitin rockets on electrophoresis into Con A and gave the term rocket affinoelectrophoresis to this procedure. The use of lectins has thus expanded the capabilities of electroimmunochemical methodology, yielding valuable additional information on the presence of carbohydrate moieties on protein antigens (glycoproteins) as well as providing a substitute for immunoglobulins in the quantitative analysis of membrane polysaccharide components (Bøg-Hansen and Brogren, 1975; Owen and Salton, 1976; Owen and Smyth, 1977; Bøg-Hansen *et al.*, 1981; Bjerrum and Hagen, 1983).

Although several histochemical methods including Nadi and periodic acid-Schiff staining reactions (Uriel, 1971) can be used in identifying immunoprecipitates of antigens containing carbohydrates, they are relatively non-specific and yield little information on the nature of the sugar constituents. Lectins on the other hand generally possess well defined sugar specificities (Sharon and Lis, 1972) and have established their potential as a powerful tool in identification and recognition of specific carbohydrate residues in a variety of macromolecules (Bøg-Hansen *et al.*, 1981). When binding to the lectins is accompanied by precipitation or formation of arcs of precipitates on electrophoresis, then they can be combined to great advantage with electroimmunochemical procedures. Furthermore, the fact that non-ionic detergents do not normally interfere with lectin reactivity (Lotan *et al.*, 1977) makes them ideally suited to membrane studies where the presence of such detergents is essential. However, it is worthy of note that the binding constant of concanavalin A may be decreased by detergent (Bøg-Hansen *et al.*, 1981) and that not all proteins reacting with lectins form precipitates (Harboe *et al.*, 1975). The lectins may be introduced into the electrophoresis systems either as the free lectin or in immobilized form and may be used in the agarose gel in the first dimension of electrophoresis or in an intermediate gel for electrophoresis in the second dimension where two-dimensional CIE analysis is being used. In some instances, as for example in the rocket affinoelectrophoretic quantification of a membrane lipomannan (Owen and Salton, 1976), the lectin (Con A) can be used entirely alone as a replacement for immune serum.

Identification of membrane antigens possessing carbohydrate residues can therefore be carried out conveniently by crossed affinity interaction immunoelectrophoresis as described for Triton-solubilized human platelet proteins by Hagen *et al.* (1979). In this study free Con A and wheatgerm agglutinin (WGA) were incorporated separately into the agarose gels during electrophoresis of antigens in the first direction, permitting those antigens which had not reacted with the lectins to be electrophoresed into the antibody gel in the second dimension. By comparing the immunoprecipitate patterns of the control (no lectins), Con A and WGA plates, it was possible to deduce that three of the platelet antigens had reacted with Con A and one with wheatgerm lectin. Interactions with the lectins were manifest as either affinity precipitates emerging from the wells in the first direction of electrophoresis and/or retardation in the first-dimension migration and reduction in the area of the reactive immunoprecipitated antigen (Hagen *et al.*, 1979; Bjerrum and Hagen, 1983).

Such combined lectin techniques have also been used in the biomolecular characterization of human erythrocyte membrane antigens. By using WGA immobilized on Sepharose and incorporating it into the agarose gel, interaction of glycophorin, band 3 protein, band 3 protein complex and one other antigen with WGA was evident from the electrophoretic retardation of the precipitated antigens (Bjerrum and Gianazza, 1983). Displacement of the bound proteins from the WGA-affinity matrix was observed when *N*-acetylglucosamine (as displacing agent) was incorporated into the adjoining gel (Bjerrum and Gianazza, 1983). These studies further illustrate the remarkable adaptability of the electroimmunochemical approach to the investigation of membrane antigens and determining sugar specificities of membrane glycoprotein antigens.

Essentially similar crossed affinity interaction immunoelectrophoresis techniques had also been used by Owen and Salton (1977) in the characterization of bacterial membrane antigens. At least five and possibly seven of the membrane antigens were judged to be reactive with Con A by either their total disappearance from the immunoprecipitate pattern or electrophoretic retardation when antigens were electrophoresed through an intermediate gel containing the lectin (Owen and Salton, 1975, 1977). One membrane antigen was absorbed by both ricin and soybean agglutinins but none of the antigens in the membrane reference pattern of immunoprecipitates appeared to be affected by WGA or Ulex agglutinin. Reactivity of one of the membrane antigens with ricin and soybean agglutinin suggested the possibility of the presence of D-galactosyl or *N*-acetyl-D-galactosaminyl residues. By carrying out absorption experiments with protoplasts it could be shown that five of the Con A-reactive antigens were exposed on the outer face of the bacterial plasma membrane, thereby establishing the asymmetric distribution of these carbohydrate-containing antigens (Owen and Salton, 1977). Thus, conventional antibody absorption studies combined with CIE analysis and affinoelectrophoresis can yield valuable information about the topography and asymmetry of membrane structures (Owen, 1983a). The detection of Con A-reactive antigens (other than the succinylated lipomannan) in this bacterial membrane suggested that bacteria may indeed possess glycoproteins. Doherty *et al.* (1982) have reported the resolution and *in vitro* glycosylation (mannosylation) of *M. luteus (lysodeikticus)* membrane glycoproteins. Although mannosylation of a limited number of membrane proteins was established in this study, it is not known whether these glycoproteins are strictly analogous to eukaryotic glycoproteins. The lack of reactivity of the membrane antigens with WGA may suggest the absence of *N*-acetylglucosaminyl residues and that these membrane glycoproteins are substituted with only one type of monosaccharide (mannose). Further investigations will undoubtedly clarify the nature of the Con A-reactive, glycosylated bacterial membrane antigens.

Lectins have now been used as probes for the detection and characterization of glycoproteins in a wide variety of membranes utilizing fused-rocket and crossed lectin interaction immunoelectrophoresis techniques. Con A-reactive membrane antigens have been detected in erythrocyte, human platelet, rat liver, bakers' yeast and bacterial membranes to name a few (Bjerrum, 1983). Their use in conjunction with electroimmunochemical analysis of membranes has greatly expanded our knowledge of the molecular structures and functions of membrane glycoprotein antigens and the molecular architecture of the membranes. Moreover, these analytical approaches have also been invaluable in the characterization and quantitative study of other non-protein membrane components such as the bacterial membrane lipomannan amphiphile (Owen and Salton, 1976; Lim and Salton, 1981), and immunochemical characterization of erythrocyte membrane macroglycolipids expressing I/i and ABH blood group activities (Oppenheim *et al.*, 1983). It is evident from the latter study that standard CIE and rocket electrophoretic techniques can contribute much to the resolution of non-protein, glycolipid antigens of membranes, especially where detergents such as Triton X-100 disrupt the micellar aggregates (liposomes) of the purified glycolipids. The conditions for the analysis of these membrane components by procedures that have worked well in the crossed lectin interaction immunoelectrophoresis of other membrane antigens will have to be modified to accommodate these non-protein glycolipid antigens possessing distinctly different physicochemical properties. Bacterial outer membrane lipopolysaccharides form immunoprecipitates in the presence of detergents in some cases and in other instances lipopolysaccharides from other species have failed to be precipitated in the presence of detergents (Bjerrum and Hagen, 1983). As with other membrane glycolipids, the bacterial lipopolysaccharides also need special considerations for their analysis by immunoelectrophoresis procedures. Lectins do not appear to have been used so far in the analysis of the outer membrane or vesicle antigens including lipopolysaccharides in the detailed immunoelectrophoretic studies recently reviewed by Owen (1983b). From the selected examples of the use of lectins in the immunochemical analysis of membrane antigens presented in this discussion, there is little doubt that these sensitive and specific probes will be further utilized in identifying and isolating membrane components possessing carbohydrate components.

## 8.7 Excision of immunoprecipitates for generation of monospecific antibodies and for polypeptide analysis

Monospecific antibodies to membrane proteins provide the investigator with an array of analytical approaches and are highly desirable reagents to have available for almost any type of membrane protein study. Once available they are especially useful for further purification of homologous and cross-reacting membrane proteins by immunoadsorbent chromatography and for characterization of the polypeptide constitution of membrane antigens. Several strategies for the generation of monospecific antibodies may be followed. The first and most obvious one is to purify the particular membrane protein to homogeneity and use it as immunogen for the preparation of antibodies as carried out, for example, with the bacterial $F_1$-ATPase (Whiteside and Salton, 1970). Unfortunately, not all membrane proteins are as readily amenable to purification as the water-soluble $F_1$-ATPase segment of $F_0$–$F_1$-ATPase energy-transducing complex. Thus, other strategies may have to be developed should the more conventional detergent-solubilization and chromatography procedures fail to yield a homogeneous product. Obtaining sufficient amounts of the purified membrane protein for immunization purposes can also be an additional problem. As an alternative to purification of the membrane protein and direct immunization, it is possible to generate monospecific antibodies by excision of specific immunoprecipitates from agarose gels and using these for immunization (Crowle *et al.*, 1972). This method may be less demanding in terms of the initial amounts of membrane antigen required for immunization and has the distinct advantage for membrane protein studies of being applicable to precipitated antigens in complex CIE immunoplates. Other valuable strategies that may be used alone or in combination for preparation of monospecific antisera have been discussed by Danielsen (1983) and include converting the membrane protein into a proteolytically cleaved form more readily purifiable, isolation of the membrane protein by polyacrylamide gel electrophoresis and use of crude membrane antigens to generate monoclonal antibodies by the hybridoma technique.

The two immunoelectrophoresis methods which have been used with considerable success for generating monospecific antisera are the line immunoelectrophoresis and the CIE techniques. As pointed out by Danielsen (1983) line immunoelectrophoresis is particularly suited to large-scale precipitate excision for subsequent immunization. By carrying out line immunoelectrophoresis of 600 $\mu$g of Triton X-100-solubilized microvillus (or intestine brush border) membrane proteins against the corresponding antisera, Danielsen (1983) obtained visible white lines of immunoprecipitates on the washed-pressed gels which were suitable for excision and processing for immunization. In this study of microvillus membrane aspartate aminopeptidase, a high degree of monospecificity was obtained by the immunoprecipitate excision technique. When evaluated by CIE, the antiserum from the excised precipitates showed a single immunoprecipitate of aspartate aminopeptidase compared to at least twelve precipitates in the CIE plate of the microvillus membrane protein fraction reacted against its corresponding antiserum. This study therefore shows how it is possible to generate essentially monospecific antibody against a single membrane antigen by the excision of a single line immunoprecipitate. As Danielsen (1983) has pointed out such antisera judged monospecific by CIE may contain low levels of antibodies to other membrane 'contaminants' but these may be removed by preadsorption of the antiserum with small amounts of solubilized membrane antigen (to avoid serious losses of desired antibody). Preadsorption could be performed either by test-tube precipitation or preferably by passage through an antigen immunoadsorbent column.

The same principle of immunoprecipitate excision can be adapted on a smaller scale for the separation of membrane immunogens from CIE plates. This approach was used by Rutberg *et al.* (1978) to generate specific antiserum to the *Bacillus subtilis* membrane succinate dehydrogenase (SDH). Immunoprecipitates of SDH in the CIE plates of Triton X-100-solubilized *B. subtilis* membranes were identified by zymogram staining and the top one-third of the precipitates were excised for immunization of rabbits. Antiserum so generated gave a single symmetrical immunoprecipitate by CIE and stained for SDH. This SDH-specific antiserum permitted these authors to examine SDH mutants of *B. subtilis,* some of which lacked the SDH antigen in the membranes while others possessed a soluble SDH antigen. In a subsequent study, Hederstedt *et al.* (1979) were able to use the SDH-specific antiserum to characterize the precipitated SDH by SDS-PAGE. By combined methods of immunoprecipitation, SDS-PAGE and radioautographic analysis these investigators were able to conclude that the wild-type SDH was probably attached to cytochrome *b* via a 28 000-dalton iron protein and together with a 65 000-dalton polypeptide corresponding to the flavoprotein constituted the SDH molecule. A 19 000-dalton polypeptide also detected was believed to represent the apoprotein form of the cytochrome *b* portion of the SDH complex. In some of the mutants the 65 000-dalton polypeptide was found in soluble form in the cytoplasm.

Very similar approaches have been used by Crowe and Owen (1983) in their study of SDH of *Micrococcus luteus (lysodeikticus)* membranes. Immunoprecipitate excision from CIE plates of Triton X-100-solubilized membranes was used as the source of SDH antigen for immunization. As in the work of Rutberg *et al.* (1978), well-resolved SDH precipitates free of other intersecting immunoprecipitates were excised and only a single immunoprecipitate (SDH) was obtained with the antiserum. Crowe and Owen (1983) were unable to detect any other immunoprecipitates by the intermediate gel variant of CIE, nor were any additional ones detectable by radioautography using $^{35}$S-labelled membrane fractions. Again, these results have further established the high degree of monospecificity attainable by the precipitate excision method for preparation of antisera. By the combined use of immunoprecipitation with the monospecific antiserum and SDS-PAGE analysis and radioautography of radioactively labelled membranes, Crowe and Owen (1983) found equimolar amounts of four polypeptides with molecular weights of 72 000, 30 000, 17 000 and 15 000 in the SDH complex of *M. luteus.* The 72 000 (flavoprotein) and 30 000 molecular-weight polypeptides were considered to represent SDH subunits and one (or both) of the lower-molecular-weight polypeptides attributable to the *b*-type cytochrome. Both these results for *M. luteus* and those for *B. subtilis* obtained by the application of immunochemical techniques for the resolution of the membrane enzyme, SDH, are in excellent agreement and serve to illustrate the powerful use of monospecific antiserum in the characterization of the molecular properties of membrane enzymes. Moreover, in the study of Crowe and Owen (1983) the immunoprecipitates were amenable to examination by low-temperature absorption spectroscopy thereby revealing the presence of cytochrome $b_{556}$ in the SDH-complex antigen.

The general applicability of the technique of excising immunoprecipitates from CIE plates (or line immunoprecipitates) for the production of monospecific antisera for membrane antigens is further demonstrated by its extension to structural components of viruses (Pedersen, 1983). Furthermore, excision of immunoprecipitates formed with mono- or polyspecific antisera can be widely used in the biomolecular characterization of the polypeptides of diverse membrane antigens (Bjerrum and Hagen, 1983). For the subsequent analysis of antigens excised as immunoprecipitates it is important that only the free parts of the precipitates, usually the tops and/or upper edges are cut out to avoid contamination and co-precipitation which may occur in the lower portions of the precipitates (Bjerrum and Hagen, 1983). In addition, the possibility of proteolytic modifications of antigens due to serum proteases must be kept in mind and appropriate protease inhibitors and controls run to document such changes. The polypeptides found in the SDH complex studied by Crowe and Owen (1983) were unaffected by pretreating the antiserum with a 'cocktail' of protease inhibitors thus indicating the probable absence of proteolytic modification under their conditions of immunoprecipitate excision and SDS-PAGE analysis.

Characterization of the polypeptides of excised immunoprecipitates of membrane antigens has generally been performed by means of SDS-PAGE analysis. Direct detection of polypeptides on SDS-PAGE gels by Coomassie Blue staining can be achieved by loading several excised gels as shown by Bjerrum and Hagen (1983) in their study of human erythrocyte membrane antigens. Immunoprecipitates analysed by SDS-PAGE revealed the presence of the appropriate polypeptides for the spectrin doublet, band 3 and ankyrin and were readily distinguishable from the heavy (H) and light (L) polypeptide chains of the rabbit immunoglobulins. With antigens yielding polypeptides close in molecular weight to those of H and L chains direct analysis by SDS-PAGE cannot be achieved. In such situations radioactively-labelled membranes or membrane components have to be used for unambiguous detection and/or quantification by SDS-PAGE analysis. A variety of radioactive labelling procedures can be used for analysis of the polypeptides in immunoprecipitates of the membrane antigens. Thus by radioautography of SDS-gels of [$^{125}$I]iodinated human erythrocyte membranes all of the major antigens were labelled (Bjerrum and Hagen, 1983) and corresponded to the appropriate polypeptides from the excised immunoprecipitated [$^{125}$I] labelled antigens. In addition to *in vitro* labelling by [$^{125}$I]iodination or $^{3}$H-labelling of carbohydrate residues of glycoproteins with galactose oxidase, metabolic labelling (*in vivo*) of membrane components may be achieved by the incorporation of $^{14}$C- or $^{35}$S-labelled amino acids, $^{59}$Fe, [$^{14}$C]riboflavin or other metabolites into membrane components (Crowe and Owen, 1983; Owen, 1983b). Such labelling procedures greatly facilitate the unambiguous SDS-PAGE analysis of the polypeptide constituents of excised immunoprecipitated membrane antigens and their characterization as discussed further in the following section.

## 8.8 Radiolabelling techniques and immunoelectrophoretic and SDS-PAGE analysis of membrane antigens

Both the electroimmunochemical and SDS-polyacrylamide gel electrophoresis techniques can be readily adapted to radioautography and radiolabelling procedures and as we have seen from the preceding section they have been particularly valuable in the characterization of membrane antigens excised as immunoprecipitates from CIE immunoplates. Radioautography can be performed directly on the washed, pressed and dried immunoplates prepared by the standard CIE and rocket immunoelectrophoresis procedures and the plates may be stained with Coomassie Blue prior to or subsequent to exposure to the X-ray film, depending on the nature of the labelled membrane materials under investigation. Removal of lipid-soluble, hydrophobic components in the solvents used for Coomassie Blue staining may possibly dictate a post radioautographic-exposure staining of immunoprecipitates.

It is evident from the flexibility of the immunoelectrophoretic techniques that radioactive labelling can be used to great advantage in a variety of ways. Labelling of the membranes may be: (1) of a *general* nature by supplying $^{14}C$- or $^{35}S$-labelled amino acids for incorporation into membrane proteins, i.e. *in vivo* labelling by addition of precursors for macromolecular biosynthesis; (2) it may be designed for the *specific, in vivo* labelling of defined membrane components such as, for example, sugars or amino sugars for the carbohydrate moieties of glycoproteins, [$^{14}C$]aminolaevulinic acid as precursor of haem for haem proteins, $^{59}Fe$ for haem proteins and non-haem Fe-proteins, [$^{14}C$]riboflavin for flavoproteins and [$^{14}C$]succinic acid for succinylated lipomannan (Bjerrum and Hagen, 1983; Owen, 1981, 1983b; Owen and Salton, 1976); (3) labelling may be performed *in vitro* by [$^{125}I$]iodination or [$^{3}H$]galactose oxidase labelling of membrane glycoproteins of intact cells or isolated membrane fractions (Bjerrum and Hagen, 1983); (4) labelling may be carried out by postelectrophoretic binding of a labelled ligand to antigens [e.g. binding of [$^{14}C$]serotonin to platelet antigens (Bjerrum and Hagen, 1983)]. The latter procedure clearly has great potential for the identification of cell surface and membrane receptor molecules. This limited selection of examples of the four major ways whereby radioactive labelling of membrane antigens can be applied to electroimmunochemical analysis illustrates the potential of combining the high sensitivity of radioactive labelling with the high resolving power and specificity of immunoelectrophoretic methods. When combined with SDS-PAGE analysis of excised immunoprecipitates from CIE and/or RIE resolved membrane antigens, these techniques have provided much valuable information on the molecular properties and quantification of membrane antigens as shown by selected examples presented in Table 8.2.

**Table 8.2** Characterization of molecular properties of membrane antigens by radiolabelling and electroimmunochemical and SDS-PAGE analysis

| *Type of radiolabelling* | *Membrane components characterized* | *References* |
|---|---|---|
| 1. General | | |
| E.g. metabolic labelling with $^{14}C$-, $^{35}S$-amino acids | Herpes simplex virus antigens | Norrild *et al.* (1977) |
| | $F_1$-ATPase subunits | Urban and Salton (1983b) |
| | Succinate dehydrogenase polypeptides | Crowe and Owen (1983) |
| 2. Specific | | |
| $^{59}Fe$ | Haem and non-haem Fe antigens of *E. coli* | Owen (1981, 1983b) |
| [$^{14}C$]Riboflavin | Flavoproteins; succinate dehydrogenase | Crowe and Owen (1983)<br>Owen (1983b) |
| [$^{14}C$]Aminolaevulinic | Haem; cytochromes | Owen (1983b) |
| $^{14}C$-labelled amino sugars | HSV glycoproteins | Norrild (1983) |
| 3. *In vitro* labelling | | |
| [$^{125}I$]Iodination | Erythrocyte membrane glycoproteins | Bjerrum and Hagen (1983) |
| [$^{3}H$]Galactose oxidase | Membrane glycoproteins | Bjerrum and Hagen (1983) |
| 4. Ligand binding (Postelectrophoretic labelling) | | |
| [$^{14}C$]Serotonin | Platelet membrane antigens | Bjerrum and Hagen (1983) |
| $^{125}I$-labelled lectins | Membrane glycoproteins | Plesner *et al.* (1983) |

### 8.8.1 SDS-PAGE ANALYSIS

In addition to the use of SDS-PAGE for the analysis of the polypeptide composition of membrane antigens from either excised immunoprecipitates or from detergent-solubilized antigens precipitated with monospecific antisera, this technique can also be used for the generation of antibodies to unfolded, SDS-denatured polypeptide subunits and for the identification of antigens by immunoblotting procedures. SDS-PAGE has been particularly useful in immunochemical studies of subunits of complex multimeric proteins. By excision of subunits of a multimeric protein such as the $F_1$-ATPase it has been possible to produce specific antibodies to each of the five subunits ($\alpha$, $\beta$, $\gamma$, $\delta$ and $\varepsilon$) and determine their ability to interact with the native $F_1$-ATPase molecule (Urban and Salton, 1983a,b). All subunit antibodies were capable of reaction with the native $F_1$-ATPase and indeed inhibited the ATPase activity to varying degrees (Urban and Salton, 1983b). Of particular interest was the ability of the $\delta$ and $\varepsilon$ subunit antibodies to dissociate their respective subunits from the native $F_1$-ATPase molecules (Urban and Salton, 1983b).

Immunoprecipitates of detached $\varepsilon$ and $\delta$ subunits were formed when the $F_1$-ATPase was electrophoresed sequentially through intermediate gels containing anti-($\varepsilon$ SDS-subunit) antibodies, and anti-($\delta$ SDS-subunit) antibodies, yielding a catalytically active (by zymogram staining) immunoprecipitate of the $\alpha$, $\beta$, $\gamma$ subunit complex as illustrated in Fig. 8.10 (Urban and Salton, 1983b). Smith and Sternweis (1982) also obtained antibodies to denatured $\alpha$

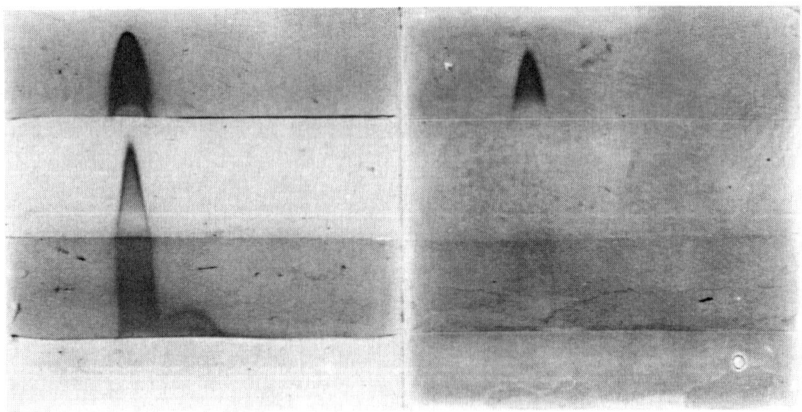

**Fig. 8.10** Electrophoresis of $F_1$-ATPase sequentially through intermediate gels containing anti-($\varepsilon$ subunit), anti-($\delta$ subunit) antibodies and then into anti-($\alpha+\beta$) antibodies results in the detachment of $\varepsilon$ and $\delta$ subunits and formation of respective $\varepsilon$ and $\delta$ subunit immunoprecipitates and the heavy immunoprecipitate of the residual $F_1$ complex with anti-($\alpha+\beta$) antiserum (left-hand slide, Coomassie Blue staining). Zymogram staining (right-hand side) shows that only the residual $F_1$ complex ($\alpha$, $\beta$, $\gamma$ subunits) immunoprecipitate retains ATPase activity. Data from Urban and Salton (1983b). Reproduced with permission of *Biochimica et Biophysica Acta.*

subunits of the *E. coli* $F_1$-ATPase but the antibodies to the denatured $\alpha$ subunit failed to react with the native $F_1$. In addition to establishing that antibodies to SDS-dissociated subunits can also react with native protein molecules, it was observed that antibodies to the native $F_1$-ATPase molecules were of two classes, those capable of reacting with denatured subunits and those reacting with the native protein (Urban and Salton, 1983a). These results illustrate that SDS-PAGE can be used for preparation of subunit antibodies which can be valuable in the immunochemical studies of such multimeric proteins.

These results with the bacterial $F_1$-ATPase are thus similar to the earlier report of two distinct antibody populations to sperm-whale myoglobin by Lando *et al.* (1982), those reacting with native myoglobin molecules and those reacting with polypeptide fragments obtained by cyanogen bromide cleavage. Until comparatively recently it had been widely believed that antibodies to determinants on denatured proteins do not react strongly with the same protein in the native structure (Crumpton, 1974). Indeed, these studies with myogoblin and $F_1$-ATPases and more recent studies with antibodies to short peptides and peptides of the influenza virus haemagglutinin (Lerner, 1982; Niman *et al.*, 1983) emphasize the ability of such antibodies to react or cross-react with intact proteins. Because of the complexity of the polyclonal immune response, the monoclonal antibodies have been invaluable in the studies with short peptides. However, a discussion of the models proposed for the protein reactivity of the anti-peptide antibodies is beyond the scope of this contribution; suffice it to say that SDS-PAGE and immunoblot techniques have played a prominent role in these and many other investigations.

Polypeptide analyses of individual immunoprecipitates or precipitated antigens have been performed by combined radioactive labelling, SDS-PAGE and radioautography. Norrild *et al.* (1977) were the first to explore the possibility of cutting out immunoprecipitates for SDS-PAGE analysis of the polypeptides by using herpes simplex virus (HSV) type 1 labelled with $^{14}C$-labelled protein hydrolysate added to infected cells and subsequent resolution of the extracted antigens by CIE and radioautography. Excised immunoprecipitates were then subjected to SDS-PAGE and radioautography for the characterization of the polypeptides of four of the HSV type 1 antigens. Human erythrocyte membrane proteins labelled by lactoperoxidase-catalysed $[^{125}I]$iodination were similarly subjected to resolution of detergent-solubilized $[^{125}I]$iodinated antigens by CIE, excision and polypeptide analysis by SDS-PAGE (Norrild *et al.*, 1977). This method showed excellent agreement with other studies on the molecular weights of the membrane proteins and enabled the detection of additional polypeptides. It is of interest to note that the efficiency of elution of labelled peptides from the immunoprecipitates in agarose gels was 65% and ultrasonic treatment of the excised gels was crucial for good recoveries. This basic methodology of SDS-PAGE analysis of labelled membrane protein antigens either excised from immunoprecipitates or precipitated with monospecific antisera as for bacterial succinate dehydrogenases (Hederstedt *et al.*, 1979; Crowe and Owen, 1983) is now a standard part of the immunochemical analytical armament and has been applied to a number of membrane and viral antigen systems (Bjerrum, 1983). Moreover, these procedures together with standard SDS-PAGE can be used to quantify the polypeptide composition of membrane antigens as, for example, in the determination of the stoichiometries of the succinate dehydrogenase complex (Crowe and Owen, 1983) and the subunits of a bacterial $F_1$-ATPase (Urban and Salton, 1983b). In addition, membrane antigens exposed on the outer face of the protoplasts of the bacterium *Micrococcus lysodeikticus* could be identified by reacting protoplasts from cells labelled metabolically with either $^{14}C$-labelled protein hydrolysate or [$^{35}S$]methionine, with anti-membrane antiserum, binding the antigen–antibody complexes from detergent-solubilized membranes to *Staphylococcus aureus* protein A and analysing the labelled polypeptides by SDS-PAGE and radioautography (Ostrovsky *et al.*, 1981). SDS-PAGE analysis can therefore be used in conjunction with other immunochemical techniques to study the asymmetry of distribution of membrane antigens on the outer and protoplasmic faces of surface membrane structures. Thus a wide variety of radioactive labelling methods combined with CIE, rocket immunoelectrophoresis, SDS-PAGE and radioautography can now be applied to the problems of molecular characterization and quantification of membrane antigens and will undoubtedly find a place in the study of cell surface and membrane receptors. Moreover, the techniques of immunoprecipitate excision of radioactively labelled antigens have been extended by Pødenphant *et al.* (1983) to the analysis of the isoelectric points of antigens by using the isoelectric focusing and two-dimensional electrophoresis method of O'Farrell (1975) combined with radioautography.

In the majority of these studies, SDS-PAGE analyses have been performed after immunoprecipitation (in gels or in solution) under non-denaturing conditions. It is also possible to perform analysis in the reverse order of carrying out SDS-PAGE first and combining it with immunoelectrophoresis for the second dimension of separation, providing the antigens retain sufficient reactivity after SDS treatment (Bjerrum, 1983). Although this sequence of analysis is not devoid of its technical problems, it has been very useful in resolution of epitopes of proteins, cross-linking studies and testing antibody specificities (Bjerrum, 1983).

### 8.8.2 IMMUNOBLOTTING

Another very useful immunochemical procedure used in conjunction with SDS-PAGE is Western blotting or immunoblotting (Burnette, 1981). By this technique it is possible to study the binding of antibodies to proteins electrophoretically transferred from SDS-PAGE gels to nitrocellulose paper.

After transfer of the polypeptides separated by SDS-PAGE, those peptides reacting with antibody molecules can be revealed on the immunoblot by peroxidase staining following reaction with peroxidase-coupled swine anti-rabbit IgG (Burnette, 1981; Bjerrum, 1983). A Western blot of human erythrocyte membrane proteins prepared in this way showed antibody binding to nearly all of the major protein bands, despite their denaturation by SDS (Bjerrum, 1983). As with other immunochemical techniques, immunoblotting has considerable potential for the identification of membrane proteins, the study of antibody specificities, cross-reactions of polypeptide antigens and it has been applied to such diverse membrane systems as human erythrocytes, envelopes of Gram-negative bacteria and measles virus structural protein antigens (Bjerrum, 1983; Owen, 1983b; Pedersen, 1983). Immunoblotting procedures also share the flexibility of electroimmunochemical method and can be combined with radioactive labelling of antigens or antibodies. Although peroxidase-coupled antibody to IgG has been used as a staining detection system for immunofixation to transferred (blotted) antigens from SDS-PAGE, other means of detecting antibodies reacting with the polypeptides are possible. Thus, Norrild (1983) used [$^{125}$I]iodinated protein A binding following reaction of immobilized (blotted) HSV type 1 proteins with hyperimmune serum, thereby permitting radioautographic detection of the HSV polypeptides. There is little doubt that many more variations in the immunoblotting detection systems will be conceived in the future and will expand the utility of this system for membrane antigen and antibody analysis.

## 8.9 Immunoadsorbent and affinity adsorbent chromatography of membrane proteins

Purification of membrane proteins by conventional chromatographic procedures has presented many problems due to their amphiphilic nature and in some instances a strongly hydrophobic character necessitating the presence of detergents or chaotropic agents during chromatography and subsequent handling of the proteins to prevent aggregation. Thus in contrast to peripheral membrane proteins that have generally been more amenable to 'standard' biochemical fractionation procedures, the purification of integral membrane proteins has frequently required proteolytic cleavage of hydrophobic domains and complex schemes for isolation with consequent losses of yields and activities. The introduction of immunoadsorbents and other non-antibody, affinity adsorbents for chromatographic purification has provided more efficient and simpler affinity separation techniques for the isolation of membrane proteins. The general principles involved in the preparation and use of antigens, antibodies, lectins and other ligands in affinity chromatography have been extensively reviewed (Cuatrecasas and Tell, 1973;

Ruoslahti, 1976; Kristiansen, 1976; Danielsen, 1983) and for many of the technical details the reader should refer to these and other sources.

### 8.9.1 IMMUNOADSORBENTS

Immunoadsorbents for column purification are generally prepared by immobilizing the antigens or antibodies to form a matrix-bound immunospecific adsorbent. This is usually achieved by either direct polymerization to yield a water-insoluble matrix, or by physical adsorption or entrapment in inert polymer supports or by covalent bonding of antibody or antigen molecules to a carrier polymer (Kristiansen, 1976). The latter has become the method of choice by covalently immobilizing the IgG (or protein antigen) on a support such as Sepharose 4B by activation with cyanogen bromide (March *et al.*, 1974). The antibodies coupled to the immunoadsorbent support are usually monospecific, purified IgG antibodies which have been checked for their specificity and the presence of any antibodies to contaminants in the immunizing antigen. Some of the electroimmunochemical techniques discussed earlier in this contribution have proved to be of considerable utility in preparation and monitoring of monospecific antisera for immunoadsorbent chromatography. As noted by Ruoslahti (1976) low avidity antibodies are preferable in antigen purification since they permit the dissociation of antibody–antigen complexes under milder conditions. For such purposes either early bleedings of antisera or high immunizing dose of antigen can be used generally to provide lower avidity antisera. It is, however, well to keep in mind that use of high-antigen doses may induce antibodies against impurities. The CIE method of monitoring antibodies is particularly valuable in this respect as it amplifies the detection of contaminants by immunoprecipitation and Coomassie Blue staining (compare with SDS-PAGE and detection only of contaminant polypeptide bands).

Once the immunoadsorbent support is available, many factors have to be taken into consideration for its effective use in membrane protein purification. Theoretically, and perhaps in practice under optimal conditions, immunoadsorbent chromatography should provide a one step purification of a membrane protein following solubilization of the isolated membranes. However, the complexity of the starting fractions and the potentially wide choice of elution systems may make this goal difficult to achieve. Thus, the investigator will have to decide whether to use a crude membrane fraction, a partially purified preparation or a fraction enriched in the particular antigen, choose a suitable non-denaturing detergent for solubilization, decide whether proteolytic cleavage of hydrophobic portions of integral proteins is needed and finally select a suitable desorbing buffer (Danielsen, 1983). A wide variety of conditions have been used for the dissociation of the antigen–antibody complexes and recovery of biologically active antigen molecules. In the past low pH, high concentrations of chaotropic ions such as thiocyanate and iodide, or 4–8 M urea have been used (Ruoslahti, 1976), in addition to high pH (10.6) and more recently hypotonic elution (Danielsen, 1983). Some of these dissociating treatments (low pH, chaotropic agents, urea) are too drastic for certain antigens and it is therefore of considerable interest that elution with hypotonic buffer (2 mM Tris-HCl, pH 8.0, containing 0.1% Triton X-100) permitted the desorption of several brush-border membrane enzymes from immunoadsorbent columns (Danielsen, 1983).

Although no clear explanation in molecular terms is available for the dissociation of the antigen–antibody complexes by hypotonic buffer in the presence of 0.1% Triton X-100, elution of immunoadsorbent columns or excised immunoprecipitates with 2 mM Tris-HCl, pH 8.0, and 1 mM potassium phosphate, pH 7.4, have given enzyme recoveries ranging from 32 to 70% and purification factors of 5.6 to 182 (Danielsen, 1983). Such results are indeed very promising and it will be of considerable interest to see if they are more widely applicable to the problems of membrane protein isolation by immunoadsorbent chromatography. However, despite the success of these mild elution conditions, Danielsen (1983) cautiously adds that 'no general recommendation as to which type of buffer to choose can be given here; one has to proceed according to the method of trial and error.'

Immunoadsorbent chromatography has not only been used for the purification of specific proteins but has also been used for removal of contaminants in preparations. Such a procedure is clearly more applicable where a single, predominant contaminant has been identified but where contamination is of a heterogeneous character the generation of a battery of monospecific antisera would be time consuming and necessitate extra chromatographic steps in the procedure. As an alternative, polyvalent antisera to crude membrane preparations could be utilized and columns and conditions adjusted to remove contaminants while giving negligible losses of the desired protein (Danielsen, 1983).

In addition to the high specificity of the immunoadsorbent antibody used in chromatographic purification of membrane proteins, it has the distinct advantage of continued use simply by regenerating the antibodies to the antigen-adsorbing state by washing the columns with buffer. Removal of tightly bound antigen from the column by short exposure to 2 M guanidinium chloride (pH 7.2) may be possible but could lead in some instances to irreversible changes in the capacity and specificity of the immunoadsorbent (Danielsen, 1983). Once the appropriate antibodies are generated, the immunoadsorbents are reusable and can be kept in stable condition with 15 mM $NaN_3$ at 4° C for years (Danielsen, 1983) and can thus greatly facilitate extended investigations of specific membrane protein isolation.

### 8.9.2 STAPHYLOCOCCAL PROTEIN A ANTIBODY-ADSORBENT TECHNIQUES

The discovery that certain strains of *Staphylococcus aureus* (e.g. Cowan strain 1) possess a surface antigen (protein A) capable of binding to the Fc fragment of immunoglobulins (Sjöquist *et al.*, 1967) paved the way for the development of a whole new battery of immunoadsorbent methods based on the IgG-binding capacity of protein A. As in immunoadsorbent chromatography, purified staphylococcal protein A can be linked covalently to a solid phase for adsorption of IgG antibodies or antigen–antibody complexes and it has been used in radioimmunoassays of proteins and membrane antigens (Clausen, 1981). The studies of Kessler (1976) have established the value of protein A adsorbents in binding to certain IgG subclasses of cell surface immunoglobulins of mouse and human lymphocytes, as well as binding and complexing with $\beta_2$-microglobulin and human HL-A-alloantigens. Kessler's (1975, 1976) investigations have thus provided rapid and specific isolation of cell membrane antigens (e.g. surface immunoglobulins, HL-A and mouse H-2 alloantigens, murine leukaemic virus glycoprotein) by radiolabelling (iodination), detergent solubilization of membrane components, immune complex formation by addition of specific antisera and insolubilization with staphylococcal protein A adsorbent.

Protein A can therefore be used as an adsorbent either linked to solid matrices or *in situ* on the surface of treated whole bacterial cells and both can be used to bind either antibody or antibody–antigen complexes. Whole staphylococcal cells (azide-treated) have been especially useful in identifying surface membrane antigens by combined labelling, reaction with specific antisera, non-ionic detergent solubilization of membranes and by sequestering the immune complexes on staphylococcal adsorbents and recovering the bound complexes after centrifugation and detergent removal from the adsorbent surface. Once recovered the immune complexes can be further characterized and labelled antigens identified by combined SDS-PAGE and radioautography as performed in other immunochemical procedures. Sequestering antigen–antibody complexes on protein A adsorbent has found wide application for the identification of surface membrane antigens of human, mouse, bacterial and parasitic cells.

The antibody adsorbent properties of staphylococcal cells have also provided the basis of diagnostic tests by co-agglutination reactions of micro-organisms. Thus staphylococcal cells with bound antibodies to surface antigens of another bacterial cell will co-agglutinate on mixing with the bacterium possessing the antigen(s) on its surface. Co-agglutination reactions could therefore be used for rapid identification of unknown isolates or they could be used to detect antibody responses to known infectious bacterial agents and have potentials for recognizing cross-reactive cell surface antigens and cross-reactive antibodies.

Protein A and its adsorbent capacity for certain IgG classes of antibodies has provided us with yet another valuable general (Fc fragment) affinity adsorbent and has found a very useful role in the studies of membrane protein antigens from cells of diverse origin.

### 8.9.3 AFFINITY ADSORBENTS

Affinity adsorbents of a wide variety of molecular structures have been used in developing affinity chromatography procedures for the fractionation of membrane proteins. Many of these procedures have been used in conjunction with immunochemical studies of membranes so that brief mention of selected examples of affinity supports used in membrane protein purification seems appropriate. Indeed, rocket, fused-rocket and crossed immunoelectrophoresis methods have been widely used in following the separation of membrane antigens by affinity chromatography and the intermediate gel technique with affinity adsorbents has been used for screening potentially useful affinity systems for subsequent chromatography (Bjerrum and Gianazza, 1983). Thus affinity adsorbents other than antibodies and affinity chromatography have been combined with most of the standard electroimmunochemical analytical procedures in the investigation of membrane proteins and cell surface and organelle receptors. This powerful combination of affinity adsorbents and immunochemical reactions has greatly expanded the resolution, specific identification and isolation of membrane enzymes, antigens, hormones and drug and toxin receptors.

Bjerrum and Gianazza (1983) have placed affinity adsorbents used in combination with CIE into two categories: (1) *group-specific adsorbents* such as lectins, hydrophobic compounds, metal chelators, thiol groups, dyes and nucleotides; and (2) *specific adsorbents*, antibodies, hormones/toxins, receptors/ligands and substrates/inhibitors. Staphylococcal protein A with its affinity for the Fc fragment of immunoglobulins would fall within the latter category.

The two principal affinity adsorbents have been antibodies and lectins, and of all the lectins, concanavalin A has been the most widely used, either in free solution in agarose for rocket and two-dimensional affinoelectrophoresis, in intermediate gels for CIE and fused-rocket immunoelectrophoresis or linked to Sepharose 4B as a solid affinity support for column chromatography and in intermediate agarose gels. Wheatgerm agglutinin with its specificity for *N*-acetylglucosaminyl residues has also found valuable application in affinity chromatography, isolation of erythrocyte membrane glycophorin and in combination with immunoelectrophoretic studies. Other lectins with different sugar specificities have been used similarly and in line immunoelectrophoresis experiments to determine the affinity profiles of human erythrocyte antigens for lectin and other chromatography matrices (Bjerrum and Gianazza, 1983).

In addition to their use in the identification and isolation of membrane components, lectins have been especially valuable in the resolution of membrane proteins co-separating with membrane glycolipids. Con A–Sepharose 4B columns could be used to separate the $F_1$-ATPase protein from contaminating lipomannan by combining affinity chromatography to retain the lipomannan and establishing the elution profile of the $F_1$-ATPase freed from the contaminant lipomannan as determined by rocket electrophoresis through an intermediate gel of Con A and then into antibodies precipitating the $F_1$-ATPase (Lim and Salton, 1981). This study with affinity adsorbents not only demonstrated their utility in separating two distinct types of membrane macromolecules but also provided definitive evidence that the membrane $F_1$-ATPase was not a glycoprotein as suggested by other investigators (Andreu *et al.*, 1978). The combined use of lectins as affinity adsorbents and immunochemical studies illustrates how the lectins can be used in predicting useful fractionation procedures and resolving problems as to the precise protein or glycoprotein nature of specific membrane antigens.

The hydrophobicity of many membrane proteins has made their separation feasible on hydrophobic matrices. Unlike other affinity chromatography matrices which are little affected by the presence of non-ionic detergents, the hydrophobic interaction chromatography systems are markedly affected. Octyl- and phenyl-Sepharose have proved useful affinity adsorbents, and Bjerrum and Gianazza (1983) have demonstrated increased binding of human erythrocyte membrane antigens to hydrophobic-agarose matrices of increasing alkyl chain length. Hydrophobic matrices have, for example, been used in various stages of membrane enzyme purification and n-octylamine-Sepharose 4B has been used successfully in the early steps of purification of microsomal membrane cytochrome *P*-450 solubilized in cholate and eluted from the column with Lubrol PX (Kaminsky *et al.*, 1981).

One very useful adaptation of affinity adsorbents has been the development of 'table-top' prediction methods with a variety of group-specific adsorbents including hydrophobic and metal-chelating affinity matrices, both of which have been tested in concert with immunoelectrophoresis to determine the abilities of membrane antigens to bind to such adsorbents (Bjerrum and Gianazza, 1983). These studies have yielded valuable information as to the types of affinity adsorbents that may be used for the chromatographic separation of specific membrane proteins.

The basic principles of affinity adsorption and chromatography have tremendous potentials for application to a wide variety of membrane protein and membrane receptor studies and they have already been extensively used as, for example, in the affinity of bungarotoxin for isolation of acetylcholine receptors (Lindstrom *et al.*, 1981), affinity chromatography of insulin receptors on immobilized insulin (Jacobs and Cuatrecasas, 1981), the use of the inhibitor oxaloacetate as an affinity ligand for purification of membrane succinate dehydrogenase (Linder *et al.*, 1975) and ampicillin–Sepharose for the purification of a $\beta$-lactam-sensitive D-alanine carboxypeptidase (Linder and Salton, 1975) and D-alanine carboxypeptidase-transpeptidase from *E. coli* membranes (Pollock *et al.*, 1974). Thus the affinity of these enzyme inhibitors has provided ligands suitable for coupling to matrices for affinity chromatography of their corresponding enzymes. These studies serve to illustrate the variety of affinity ligands that can be coupled to provide affinity matrices for chromatographic separation of membrane proteins, and the tremendous potential of combining such affinity techniques with immunochemical resolution of column fractions is abundantly illustrated by the examples where group-specific and specific affinity adsorbents have been used together with CIE, rocket and fused-rocket immunoelectrophoresis. With membrane investigations turning more and more to identification and isolation of specific hormone, drug and toxin receptors, it can be anticipated that combined affinity adsorbent and immunochemical methods will play a pivotal role in the successful execution of this important area of biomembrane research, a conclusion amply illustrated in Table 8.3 by the selection of membrane proteins isolated by these techniques of immunoadsorbents and affinity adsorbent chromatography.

## 8.10 Conclusions

Immunochemistry has become an integral part of the investigation of the structure and function of biological membranes and the application of the wide variety of immunochemical analytical techniques has greatly expanded our knowledge of membrane proteins as antigens, enzymes and receptors for hormones, drugs, toxins and other biologically active components. Immunoelectrophoretic and immunoadsorbent techniques in combination with radioactive labelling, radioautography, polypeptide analysis by SDS-PAGE and lectin affinity studies have all provided powerful approaches for the analysis of membrane proteins. Electroimmunochemical precipitation in agarose gels has led to the unravelling of the complexity of membranes of diverse origins and provided sensitive, high-resolution techniques for membrane analysis as well as immunoprecipitate excision methodologies for the generation of monospecific antibodies to membrane antigens. These procedures in combination with zymogram staining and lectin affinities have facilitated the identification of many membrane proteins and contributed to our understanding of the molecular architecture and asymmetry of membrane structures.

It can be anticipated that these highly adaptable immunochemical methodologies will contribute significantly to the future analysis of cell membranes and to the newer and exciting areas of recognition, isolation and understanding the functions of cell surface and organelle membrane receptors.

**Table 8.3** Selected examples of membrane protein isolation by immunoadsorbent and affinity adsorbent purification techniques

| | *Enzymes/antigens* | *Reference* |
|---|---|---|
| Immunoadsorbent chromatography | Microvillus membrane aspartate aminopeptidase, maltase, lactase | Danielsen (1983) |
| Protein A adsorbent | $\beta_2$-Microglobulin | Kessler (1975, 1976) |
| | HL-A-alloantigens | |
| | Murine leukaemia virus glycoprotein | |
| | *M. lysodeikticus* protoplast membrane surface antigens | Ostrovsky *et al.* (1981) |
| Affinity Chromatography | | |
| Wheatgerm agglutinin (lectin) | Erythrocyte glycophorin | Bjerrum and Gianazza (1983) |
| Bungarotoxin–Separose 4B | Acetylcholine receptors | Lindstrom *et al.* (1981) |
| Insulin–Sepharose 4B | Insulin receptors | Jacobs and Cuatrecasas (1981) |
| n-Octylamino-Sepharose 4B (hydrophobic affinity) | Rat liver cytochrome *P*-450 | Kaminsky *et al.* (1981) |
| Ampicillin–Sepharose 4B (inhibitor affinity) | *E. coli* carboxypeptidase-transpeptidase | Pollock *et al.* (1974) |
| | *M. lysodeikticus* D-alanine carboxypeptidase | Linder and Salton (1975) |
| Oxaloacetate–Sepharose 4B (inhibitor affinity) | *M. lysodeikticus* succinate dehydrogenase | Linder *et al.* (1975) |

## Acknowledgements

I wish to thank Josephine Markiewicz for all her help throughout the preparation of the manuscript and her expert typing. I also wish to thank Carl Urban for generously providing a number of the CIE plates used in this article, Kwang S. Kim for help with photography and Simon Karpatkin for providing the illustration in Fig. 8.4.

## 8.11 References

Andreu, J. M., Warth, R. and Muñoz, E. (1978) *FEBS Lett.*, **86**, 1–5.

Axelsen, N. H., Kroll, J. and Weeke, B. (1973) in *A Manual of Quantitative Immunoelectrophoresis* (eds N. H. Axelsen, J. Kroll and B. Weeke), Universitetforlaget, Oslo, pp. 1–169.

Berzins, K. and Blomberg, F. (1983) in *Electroimmunochemical Analysis of Membrane Proteins* (ed O. J. Bjerrum), Elsevier, Amsterdam, pp. 223–41.

Bjerrum, O. (1977) *Biochim. Biophys. Acta*, **472**, 135–95.

Bjerrum, O. J. (ed) (1983) in *Electroimmunochemical Analysis of Membrane Proteins*, Elsevier, Amsterdam, pp. 3–44.

Bjerrum, O. J., Bjerrum, P. J., Larsen, K. P., Norrild, B. and Bhakdi, S. (1983) in *Electroimmunochemical Analysis of Membrane Proteins* (ed O. J. Bjerrum), Elsevier, Amsterdam, pp. 173–200.

Bjerrum, O. J. and Gianazza, E. (1983) in *Electroimmunochemical Analysis of Membrane Proteins* (ed O. J. Bjerrum), Elsevier, Amsterdam, pp. 127–55.

Bjerrum, O. J. and Hagen, I. (1983) in *Electroimmunochemical Analysis of Membrane Proteins* (ed O. J. BJerrum), Elsevier, Amsterdam, pp. 77–115.

Blumberg, P. M. and Strominger, J. L. (1974) *Bacteriol. Rev.*, **38**, 291–335.

Bock, E. and Dissing, J. (1975) *Scand. J. Immunol.*, **4**, Suppl. 2, 31–6.

Bøg-Hansen, T. C. (1973) *Anal. Biochem.*, **56**, 480–8.

Bøg-Hansen, T. C. and Brogren, C.-H. (1975) *Scand. J. Immunol.*, **4**, Suppl. 2, 135–9.

Bøg-Hansen, T. C., Jensen, P., Hinnerfeldt, F. and Takeo, K. (1981) in *Lectins. Biology, Biochemistry, Clinical Biochemistry* (ed T. C. Bøg-Hansen), de Gruyter, Berlin, Vol. 1, pp. 241–58.

Brogren, C.-H. and Bøg-Hansen, T. C. (1975) *Scand. J. Immunol.*, **4**, Suppl. 2, 37–51.

Brogren, C.-H. and Svendsen, P. J. (1975) *Protides Biol. Fluids*, **22**, 685–9.

Burke, K. A., Brown, A. E. and Lascelles, J. (1981) *J. Bacteriol.*, **148**, 724–7.

Burnette, W. N. (1981) *Anal. Biochem.*, **112**, 195–203.

Cinader, B. (1957) *Annu. Rev. Microbiol.*, **11**, 371–90.

Cinader, B. (ed) (1967) *Antibodies to Biologically Active Molecules*, Pergamon Press, Oxford, pp. 85–137.

Clausen, J. (1981) *Immunochemical Techniques for the Identification and Estimation of Macromolecules*, 2nd edn, Elsevier/North Holland Biomedical Press, New York, pp. 144–5.

Collins, M. L. P. and Salton, M. R. J. (1979) *Biochim. Biophys. Acta*, **553**, 40–53.

Crowe, B. A. and Owen, P. (1983) *J. Bacteriol.*, **153**, 1493–501.

Crowle, A. J., Revis, G. J. and Jarrett, K. (1972) *Immunol. Commun.*, **1**, 325–36.

Crumpton, M. J. (1974) in *The Antigens* (ed M. Sela), Academic Press, New York, pp. 87–159.

Cuatrecasas, P. and Tell, G. P. E. (1973) *Proc. Natl. Acad. Sci. USA*, **70**, 485–9.

Danielsen, E. M. (1983) in *Electroimmunochemical Analysis of Membrane Proteins* (ed O. J. Bjerrum), Elsevier, Amsterdam, pp. 157–69.

Doherty, H., Condon, C. and Owen, P. (1982) *FEMS Microbiol. Lett.*, **15**, 331–6.

Downie, J. A., Gibson, F. and Cox, G. B. (1979) *Annu. Rev. Biochem.*, **48**, 103–31.

Hagen, I., Bjerrum, O. J. and Solum, N. O. (1979) *Eur. J. Biochem.*, **99**, 9–22.

Harboe, M., Saltvedt, E., Closs, O. and Olsnes, S. (1975) *Scand. J. Immunol.*, **4**, Suppl. 2, 125–34.

Hederstedt, T., Holmgren, E. and Ruthberg, L. (1979) *J. Bacteriol.*, **138**, 370–6.

Helenius, A. and Simons, K. (1977) *Proc. Natl. Acad. Sci. U.S.A.*, **74**, 529–32.

Jacobs, S. and Cuatrecasas, P. (1981) *Methods Enzymol.*, **74**, 471–8.

Johansson, B. G. and Malmquist, J. (1971) *Scand. J. Clin. Lab. Invest.*, **27**, 255–61.

Johansson, K.-E. (1983) in *Electroimmunochemical Analysis of Membrane Proteins* (ed O. J. Bjerrum), Elsevier, Amsterdam, pp. 321–46.

Kaminsky, L. S., Fasco, M. J. and Guengerich, F. P. (1981) *Methods Enzymol.*, **74**, 262–72.

Kessler, S. W. (1975) *J. Immunol.*, **115**, 1617–24.

Kessler, S. W. (1976) *J. Immunol.*, **117**, 1482–90.

Kjaervig Broe, M. and Ingild, A. (1983) *Scand. J. Immunol.*, **17**, Suppl **10**, 255–8.

Kristiansen, T. (1976) in *Immunoadsorbents in Protein Purification* (ed E. Ruoslahti), University Park Press, Baltimore, MD, pp. 19–27.

Lando, G., Berzofsky, J. A. and Reichlin, M. (1982) *J. Immunol.*, **129**, 200–11.

Laurell, C.-B. (1965) *Anal. Biochem.*, **10**, 358–61.

Laurell, C.-B. (1966) *Anal. Biochem.*, **15**, 45–52.

Laurell, C.-B. and McKay, E. J. (1981) *Methods Enzymol.*, **73**, 338–69.

Lerner, R. A. (1982) *Nature (London)*, **299**, 592–6.

Lim, S. and Salton, M. R. J. (1981) *Biochim. Biophys. Acta*, **638**, 275–81.

Linder, R. and Salton, M. R. J. (1975) *Eur. J. Biochem.*, **55**, 291–7.

Linder, R., Salton, M. R. J. and Pollock, J. J. (1975) *Prep. Biochem.*, **5**, 349–57.

Lindstrom, J., Einarson, B. and Tzartos, S. (1981) *Methods Enzymol.*, **74**, 432–60.

Lotan, R., Beattie, G., Hubbell, W. and Nicolson, G. L. (1977) *Biochemistry*, **16**, 1787–94.

March, S. C., Parikh, I. and Cuatrecasas, P. (1974) *Anal. Biochem.*, **60**, 149–52.

Marchesi, V. T. and Steers, E., Jr. (1968) *Science*, **159**, 203–4.

Nalin, C. M., Cross, R. L., Lucas, J. L. and Kohlbrenner, W. E. (1979) *FEBS Lett.*, **104**, 209–14.

Nielsen, C. S. and Bjerrum, O. J. (1977) *Biochim. Biophys. Acta*, **466**, 496–509.

Niman, H. L., Houghton, R. A., Walker, L. E., Reisfeld, R. A., Wilson, I. A., Hogle, J. M. and Lerner, R. A. (1983) *Proc. Natl. Acad. Sci. USA*, **80**, 4949–53.

Norrild, B. (1983) in *Electroimmunochemical Analysis of Membrane Proteins* (ed O. J. Bjerrum), Elsevier, Amsterdam, pp. 395–407.

Norrild, B., Bjerrum, O. J. and Vestergaard, B. F. (1977) *Anal. Biochem.*, **81**, 432–41.

O'Farrell, P. H. (1975) *J. Biol. Chem.*, **250**, 4007–21.

Oppenheim, J. D. and Nachbar, M. S. (1977) in *Immunochemistry of Enzymes and Their Antibodies* (ed M. R. J. Salton), John Wiley and Sons, New York, pp. 89–124.

Oppenheim, J. D., Nachbar, M. S. and Blank, M. (1983) *Electrophoresis*, **4**, 53–62.
Ostrovsky, D. N., Urban, C., Reshtnyak, V. I. and Salton, M. R. J. (1981) *Biochemistry (Russian)*, **46**, 1321–30.
Owen, P. (1981) in *Organization of Prokaryotic Cell Membranes* (ed B. K. Ghosh), CRC Press, Boca Ratan, FL, Vol. 1, pp. 73–164.
Owen, P. (1983a) in *Electroimmunochemical Analysis of Membrane Proteins* (ed O. J. Bjerrum), Elsevier, Amsterdam, pp. 55–76.
Owen, P. (1983b) in *Electroimmunochemical Analysis of Membrane Proteins* (ed O. J. Bjerrum), Elsevier, Amsterdam, pp. 347–73.
Owen, P. and Kaback, H. R. (1979) *Biochemistry*, **18**, 1413–22.
Owen, P. and Salton, M. R. J. (1975) *Proc. Natl. Acad. Sci. USA*, **72**, 3711–15.
Owen, P. and Salton, M. R. J. (1976) *Ann. Biochem.*, **73**, 20–6.
Owen, P. and Salton, M. R. J. (1977) *J. Bacteriol.*, **132**, 974–85.
Owen, P. and Smyth, C. J. (1977) in *Immunochemistry of Enzymes and Their Antibodies* (ed M. R. J. Salton), John Wiley and Sons, New York, pp. 147–202.
Pedersen, I. R. (1983) in *Electroimmunochemical Analysis of Membrane Proteins* (ed O. J. Bjerrum), Elsevier, Amsterdam, pp. 409–18.
Pedersen, P. L. (1975) *Bioenergetics*, **6**, 243–75.
Plesner, T., Wilken, M. and Avnstrøm, S. (1983) in *Electroimmunochemical Analysis of Membrane Proteins* (ed O. J. Bjerrum), Elsevier, Amsterdam, pp. 45–54.
Pødenphant, J., Bock, E., Thyman, M. and Gozes, I. (1983) *Scand. J. Immunol.*, **17**, Suppl. **10**, 283–7.
Pollock, J. J., Nguyen-Disteche, M., Ghuysen, J.-M., Coyette, J., Linder, R., Salton, M. R. J., Kim, K.-S., Perkins, H. R. and Reynolds, P. (1974) *Eur. J. Biochem.*, **41**, 439–46.
Poulik, M. D. and Bron, C. (1969) in *Red Cell Membrane, Structure and Function* (eds G. A. Jamieson and T. J. Greenwalt), J. B. Lippincott Co., Philadelphia, PA, pp. 131–53.
Radola, B. (1975) in *Isoelectric Focusing* (eds J. B. Arbuthnott and J. A. Beeley), Butterworths, London, 182 pp.
Reynolds, J. A. and Trayer, H. (1971) *J. Biol. Chem.*, **246**, 7337–42.
Rose, C. and Harboe, N. M. G. (1970) *Protides Biol. Fluids*, **17**, 397–400.
Ruoslahti, E. (1976) in *Immunoadsorbents in Protein Purification* (ed E. Ruoslahti), University Park Press, Baltimore, MD, pp. 3–7.
Rutberg, B., Hederstedt, L., Holmgren, E. and Rutberg, L. (1978) *J. Bacteriol.*, **136**, 304–11.
Salton, M. R. J. (1974) *Adv. Microb. Physiol.*, **11**, 213–83.
Salton, M. R. J. (1983) *J. Gen. Microbiol.*, **129**, 2685–702.
Salton, M. R. J. and Schor, M. T. (1974) *Biochim. Biophys. Acta*, **345**, 74–82.
Sharon, N. and Lis, H. (1972) *Science*, **177**, 949–59.
Shulman, S. and Karpatkin, S. (1980) *J. Biol. Chem.*, **225**, 4320–7.
Singer, S. J. and Nicolson, G. L. (1972) *Science*, **175**, 720–31.
Sjöquist, J., Forsgren, A., Gustafson, G. T. and Stålenheim, G. (1967) *Cold Spring Harbor Symp. Quant. Biol.*, **32**, 577–81.
Smith, J. B. and Sternweis, P. C. (1982) *Arch. Biochem. Biophys.*, **217**, 376–87.
Smyth, C. J., Friedman-Kien, A. E. and Salton, M. R. J. (1976) *Infect. Immunol.*, **13**, 1273–88.
Smyth, C. J. and Salton, M. R. J. (1977) in *The Gonococcus* (ed R. B. Roberts), John Wiley and Sons, New York, pp. 303–31.

Smyth, C. J., Siegel, J., Salton, M. R. J. and Owen, P. (1978) *J. Bacteriol.*, **133**, 306–19.
Spratt, B. G. (1975) *Proc. Natl. Acad. Sci. USA*, **72**, 2999–3003.
Svendsen, P. J. (1973) in *A Manual of Quantative Immunoelectrophoresis* (eds N. H. Axelsen, J. Kroll and B. Weeke), Universitetforlaget, Oslo, pp. 69–70.
Urban, C. and Salton, M. R. J. (1983a) *FEMS Microbiol. Lett.*, **19**, 97–101.
Urban, C. and Salton, M. R. J. (1983b) *Biochim. Biophys. Acta*, **724**, 230–40.
Uriel, J. (1971) *Methods Immunol. Immunochem.*, **3**, pp. 294–321.
Verbruggen, R. (1975) *Clin. Chem.*, **21**, 5–43.
Wadstrom, T. and Smyth, C. J. (1975) in *Isoelectric Focusing* (eds. J. P. Arbuthnott and J. A. Beeley), Butterworths, London, pp. 152–77.
Whiteside, T. L. and Salton, M. R. J. (1970) *Biochemistry*, **9**, 3034–40.

# 9

# Sequence analysis of membrane proteins

JOHN E. WALKER and IAN M. FEARNLEY



## 9.1 Scope of the review

The primary structures of proteins can be determined by two distinct methods: directly by sequence analysis of the protein or indirectly by DNA sequence analysis of the corresponding gene or c-DNA. The direct protein sequence analysis methods are strongly influenced by the nature of the protein. They have been largely developed to accommodate water-soluble globular proteins, and methods for isolation and sequence analysis of hydrophobic proteins are relatively deficient and underdeveloped. The indirect sequence analysis by DNA sequence analysis is not influenced by the nature of the protein.

This advantage of the indirect method notwithstanding, the necessity for the direct analysis of membrane proteins (as for other classes of protein) remains for the following reasons:

(1) Information about post-translational processing of proteins can only be obtained by direct sequence analysis. These processing events include phosphorylation, methylation, glycosylation, acylation and removal of parts of the polypeptide chain by proteolysis.
(2) A prominent method for isolation of genes from libraries in bacteria is based upon the use of synthetic oligonucleotide hybridization probes. These are designed on the basis of protein sequence obtained by direct means.
(3) The interpretation of DNA sequence is aided by an independent knowledge of the sequence of the protein encoded therein. Such knowledge for instance can help to identify start and stop codons and intervening

Techniques for the Analysis of Membrane Proteins. Edited by C. I. Ragan & R. J. Cherry.
Published in 1986 by Chapman and Hall, 11 New Fetter Lane, London EC4P 4EE

sequences; it was essential in establishing the existence of overlapping genes (Sanger *et al.*, 1977; Shaw *et al.*, 1978). In analysis of complete genomes (e.g. bacteriophages $\phi$X174 and lambda, human and bovine mitochondrial DNA) protein sequences were used to identify the genes (Sanger *et al.*, 1977; Shaw *et al.*, 1978; Walker *et al.*, 1980, 1982a).

(4) Indirect and direct protein sequences are obtained independently and so they serve as a useful mutual checking procedure.

Therefore, the first part of this review will describe the extant methods for direct sequencing of membrane proteins including the isolation of the protein, production and fractionation of derived fragments, and their sequence analysis: it will discuss the deficiencies and difficulties of these methods. The second part will outline current methods for isolation of genes for membrane proteins and for the determination of their DNA sequences. It will particularly emphasize the close relationship between the indirect and direct methods and the benefits that accrue from the use of both methods. The review will conclude with a discussion of procedures for prediction of secondary structures of membrane proteins from their amino acid sequences.

## 9.2 Direct sequence analysis of membrane proteins

### 9.2.1 WHAT IS A MEMBRANE PROTEIN?

Membrane proteins fall into two broad classes, extrinsic and intrinsic (see Fig. 9.1). The extrinsic proteins are globular structures that lie outside the lipid bilayer and do not penetrate the lipid bilayer at all (e.g. $\beta_2$-microglobulin, mitochondrial $F_1$-ATPase; Type I in Fig. 9.1). A second class of extrinsic membrane proteins (see Table 9.1) are anchored totally or partly via covalent lipid (Type II in Fig. 9.1). The intrinsic membrane proteins are themselves of several kinds. Some have extensive globular domains that lie outside the membrane and are anchored by a hydrophobic membrane sector (Types III, IV

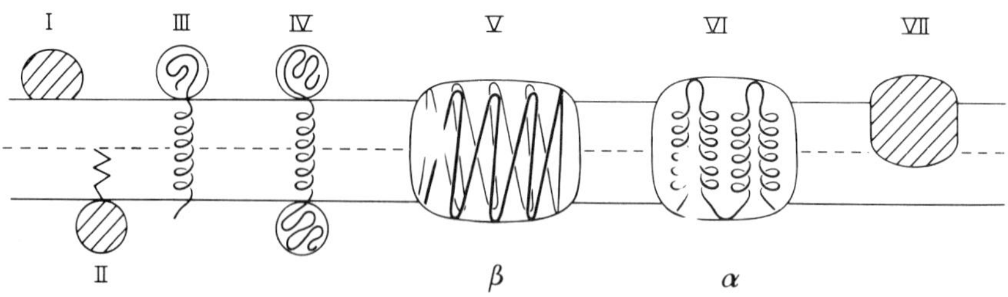


**Fig. 9.1** Types of membrane proteins. I. Superficially associated extrinsic membrane protein. II. Extrinsic protein anchored via covalently attached lipid. III and IV. Intrinsic membrane proteins with α-helical membrane span and extensive extramembrane domains. V. Intrinsic membrane protein folded in β-sheets and VI in α-helices with minor extramembrane domains. VII. Partially penetrating membrane protein.

**Table 9.1** Membrane-associated proteins with covalent lipid

| *Protein* | *Lipid* | *Footnotes* |
|---|---|---|
| *E. coli* outer membrane lipoprotein | Palmitic acid, *cis*-vaccenic acid, 9,10-methylenehexadecanoic acid, palmitoleic acid | 1 |
| c-AMP-dependent protein kinase | α-*N*-Myristoyl | 2 |
| Calcineurin B | α-*N*-Myristoyl | 3 |
| Murine retrovirus $P15^{gag}$ | α-*N*-Myristoyl | 4 |
| Rous sarcoma virus p60src | α-*N*-Myristoyl | 5–8 |
| Harvey sarcoma virus p21 | Palmitoyl | 6 |
| Lipophilin | Palmitic acid, stearic acid, oleic acid | 9 |
| HLA-B, HLA-DR heavy chains | Palmitic acid | 10 |
| Transferrin receptor | Palmitic acid | 11 |
| Cytochrome $b_5$ reductase | α-*N*-Myristoyl | 12 |
| *Plasmodium falciparum* variant surface antigen | *sn*-1,2-Dimyristoylglycerol | 13 |
| *Trypanosome brucei* variant surface antigen | *sn*-1,2-Dimyristoylglycerol | 14, 15 |

*Footnotes.* 1. Attached via amide to *N*-terminal cysteine and ester linked to glycerol, itself thioether linked to cysteine-1 (Hantke and Braun, 1973). 2. Carr *et al.* (1982). 3. Membrane association has not been demonstrated (Aitken *et al.*, 1982). 4. Henderson *et al.* (1983); Schultz and Oroszlan (1984). 5. Buss and Sefton (1985a,b); Schultz *et al.* (1985). 6. Sefton *et al.* (1982). 7. Garber *et al.* (1983). 8. Pellman *et al.* (1985). 9. Esterified at threonine-198; Stoffel *et al.* (1983). 10. Via thioester bond; Kaufman *et al.* (1984). 11. Omary and Trowbridge (1981). 12. Ozols *et al.* (1984). 13. Haldar *et al.* (1985). 14. Attachment to *C*-terminal oligosaccharide via phosphodiester bond between *sn*-3-glycerolhydroxyl and a sugar hydroxyl; phospholipase *c* releases protein. Ferguson *et al.* (1985). 15. Thy-1, acetylcholinesterase, 5′-nucleotidase and alkaline phosphatase are attached in a similar way. Kolata (1985).

in Fig. 9.1), others are predominantly hydrophobic and are found almost totally within the lipid bilayer (Types V and VI in Fig. 9.1). The extrinsic proteins and the extrinsic protein domains of intrinsic proteins present few difficulties for direct protein sequencing; the major problems are associated with the hydrophobic intrinsic membrane sectors. They are of necessity largely composed of hydrophobic amino acids and are insoluble in aqueous solutions. This discussion is directed towards the primary structural determination of this class of membrane proteins (see Table 9.2).

### 9.2.2 ISOLATION OF MEMBRANE PROTEINS

(a) *Native proteins*

The methods for the isolation of individual membrane proteins and membrane

**Table 9.2** Membrane protein sequences determined by direct protein sequence analysis

| *Protein* | *Source* | *Footnotes* |
|---|---|---|
| Glycophorin | Red cell membrane | 1 |
| Cytochrome $b_5$ | Liver microsomes | 2,3,4 |
| ATP synthase proteolipid | Mitochondria, chloroplasts, bacteria | 5 |
| ATP synthase-associated protein | Yeast mitochondria | 6 |
| Lipophilin | Human and bovine brain | 7–11 |
| Bacteriorhodopsin | *Halobacterium halobium* | 12–14 |
| Rhodopsin | Bovine and ovine retina | 15,16 |
| Light-harvesting polypeptides | *Rhodospirillaceae* | 17–20 |
| Bacteriochlorophyll *c* binding | *Chloroflexus aurantiacus* | 21 |
| Cytochrome oxidase subunits | Bovine mitochondria | 22–25 |
| Cytochrome $c_1$ | Bovine mitochondria | 26 |
| ADP/ATP translocase | Bovine mitochondria | 27 |
| Uncoupling protein | Hamster mitochondria | 28 |
| H-2K | Mouse lymphocytes | 29 |
| Anion transport protein | Human erythrocytes | 30 |
| *Omp*F | *E. coli* | 31 |
| Transducin $\gamma$-subunit | Cow | 32 |
| Surface glycoprotein | *Trypanosoma brucei* | 33 |
| Asialoglycoprotein receptor | Rat liver | 34 |

*Footnotes.* 1. Tomita and Marchesi (1975). 2. Takagaki *et al.* (1980). 3. Kondo *et al.* (1979). 4. Ozols and Heinemann (1982). 5. Sebald and Hoppe (1981). 6. Velours *et al.* (1984). 7. Laursen *et al.* (1984). 8. Lees *et al.* (1983). 9. Stoffel *et al.* (1983). 10. Jollès *et al.* (1979). 11. Stoffel *et al.* (1985). 12. Gerber *et al.* (1979). 13. Khorana *et al.* (1979). 14. Ovchinnikov *et al.* (1979). 15. Ovchinnikov (1982). 16. Brett and Findlay (1983). 17. Brunisholz *et al.* (1985). 18. Theiler *et al.* (1984). 19. Brunisholz *et al.* (1984a). 20. Brunisholz *et al.* (1981). 21. Wechsler *et al.* (1985). 22. Tanaka *et al.* (1979). 23. Steffens and Buse (1979). 24. Biewald and Buse (1982). 25. Erdweg and Buse (1985). 26. Wakabayashi *et al.* (1980, 1982). 27. Aquila *et al.* (1982). 28. Aquila *et al.* (1985). 29. Coligan *et al.* (1978). 30. Hydrophobic sector only; Brock *et al.* (1983). 31. Chen *et al.* (1982). 32. McConnell *et al.* (1984). 33. Allen *et al.* (1982). 34. Drickamer *et al.* (1984).

complexes are extremely diverse and a summary of such procedures is beyond the scope of this review (see Nelson and Robinson, 1983). Many of these procedures depend upon the differential extraction of the protein from the membrane under particular conditions of detergent, salt concentration and pH. The choice of detergent is particularly important as is an understanding of the properties of the detergent. For such information see Chapter 3 and reviews by Helenius and Simons (1975), Tanford and Reynolds (1976), Hjelmel and Chrambach (1984) and Zulauf and Rosenbusch (1983).

(b) *Denatured membrane proteins*

The methods for isolation of membrane proteins in their denatured state depend upon two distinctly different approaches to their solubilization. These are firstly, solubilization in chaotropes (urea, guanidinium hydrochloride) and strong detergents [sodium dodecyl sulphate (SDS)] and secondly, solubilization in organic solvents. The solubilization procedure will often limit and thereby dictate the subsequent purification steps.

(i) *Chaotropes and strong detergents.* Urea is a relatively mild chaotrope and incapable of solubilizing effectively many membrane proteins (e.g. bacteriorhodopsin). However, if solubilization can be achieved, the normal gamut of chromatographic methods that are applicable to globular proteins (gel filtration, ion-exchange chromatography, reverse phase chromatography) can be employed. Guanidinium hydrochloride is a more effective chaotrope but its use precludes many purification methods, and the choice is restricted to size fractionation. However, this may be effective for the resolution of simple mixtures. The use of sodium dodecyl sulphate also excludes some methods (ion exchange) but it has been used to effect separation of subunits of mitochondrial complexes such as cytochrome oxidase (Steffens and Buse, 1976) and ATP synthase (Steffens *et al.*, 1982). In combination with polyacrylamide gel electrophoresis and isoelectric focusing it provides an extremely powerful method with high resolution for the production of small quantities of highly purified proteins that can be successfully sequenced in modern instruments. The rest of this section will assess this approach.

The method (see Fig. 9.2) consists of the application of the sample dissolved

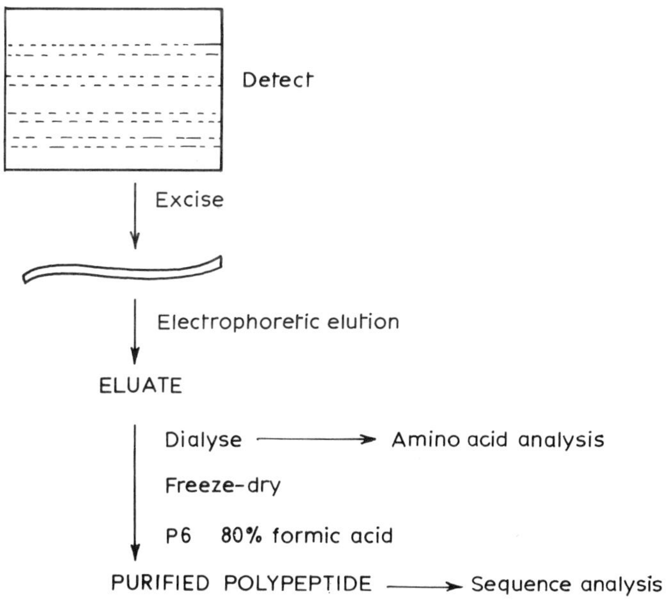


**Fig. 9.2** Preparative polyacrylamide gel electrophoresis.

in the presence of SDS to a polyacrylamide slab separating gel surmounted by a stacking gel. The constituent proteins are separated under standard conditions, detected, excised and eluted by electrophoresis into a dialysis bag or similar device. Following dialysis and removal of low-molecular-weight contaminants, the recovered protein can then be sequenced (Bhown *et al.*, 1980; Walker *et al.*, 1982b; Hunkapiller *et al.*, 1983a).

The main considerations in this type of procedure are to minimize chemical modification of the protein during the procedures, to avoid cleavage of acid-sensitive peptide bonds during staining procedures to detect the protein, to maximize recovery and to remove contaminants that would interfere with the subsequent sequencing reactions.

Chemical modifications can be avoided in part by using highly purified reagents for electrophoresis. The introduction of reducing agents (0.1 mM sodium thioglycolate) into the upper chamber (Hunkapiller *et al.*, 1983a) helps to scavenge amino-reactive contaminants, oxidants and free radicals that remain in the gel after polymerization. Excess persulphate can be removed by pre-electrophoresis but only at the cost of the loss of stacking gel and its attendant advantages.

Many methods have been tried for visualization of proteins. These include tagging the protein with fluorescent reagents (fluorescamine) or with radioactivity, by intrinsic labelling or by $^{125}$I-iodination. These methods may not be generally applicable (intrinsic labelling) or may alter the mobility of the partially reacted protein relative to the unreacted protein (fluorescamine; Talbot and Yphantis, 1971). Other mild staining procedures employ sodium acetate (Higgins and Dahmus, 1979) or potassium chloride (Nelles and Bamburg, 1976) but are relatively insensitive (Walker *et al.*, 1980). The staining of side bands with Coomassie Blue dye can also be employed, but swelling or shrinkage of the stained side band relative to the unstained gel and irregularities in unstained bands can make precise localization of unstained bands difficult. The method of choice is probably brief staining with Coomassie Blue dye (5 min) followed by destaining for 10–15 min. This allows rapid and precise localization of protein bands which can then be excised with a razor.

The protein dye complex can then be recovered from the gel by passive elution in buffer or, better, by electroelution. Maceration of gel prior to these procedures leads to release of acrylamide oligomers from the gel. Therefore, it should be avoided and the integrity of the gel slices maintained.

Following elution, contaminants should be removed prior to analysis of protein samples. Dialysis against 10% ethanol helps to remove low-molecular-weight components and SDS. However, for small amounts of protein it may lead to further loss on the dialysis membrane surface.

An alternative procedure is to remove the contaminants from the freeze-dried electroeluent by gel filtration in 80% formic acid on a small column of Biogel P6 (Walker *et al.*, 1980). Alternatively, an ethanol precipitation step

(90% ethanol at $-20°$ C; M. Hunkapiller, personal communication) or a chloroform step (Wessel and Flügge, 1984) have been recommended to purify proteins prior to sequence analysis on a gas phase sequencer.

Another procedure used in conjunction with solid phase sequencing (see below) selectively immobilizes the protein by covalent attachment to glass derivatives leaving the low-molecular-weight contaminants in solution. They can then be removed by a simple washing step (Walker *et al.*, 1982b).

Recent developments in this methodology have concentrated on the direct elution of the protein from the gel on to a solid matrix (such as the glass-fibre disc employed as a support in the gas phase sequencer; Aebersold *et al.*, 1986). Such methods avoid a number of steps, are therefore simpler and lead to an overall improvement in recovery and, in addition, allow even smaller amounts of protein to be manipulated. Such methods clearly hold much promise and will be developed further.

(ii) *Gel permeation chromatography in organic solvents.* In 1951 Folch and Lees reported that a group of proteins could be extracted from brain tissue into mixtures of chloroform and methanol. In this property these proteins resemble lipids and hence the proteins were called proteolipids. Later, Cattell *et al.* (1970, 1971) demonstrated that proteolipids could be extracted from bovine mitochondria in a similar manner. Fractionation of this brain extract by gel permeation chromatography on hydroxypropylated Sephadex G-25 (Sephadex LH-20) was first reported by Mokrasch (1967). The proteins themselves were not resolved from each other and were found in the excluded volume of the column. However, further fractionation of the proteins was achieved by partition chromatography (see below) on the same support. Sandermann and Strominger (1972) showed that hydroxypropylated Sephadex G-50 (Sephadex LH-60) had a more effective separation range for proteins in their purification of a lipophilic enzyme $C_{55}$-isoprenoid alcohol phosphokinase. Later Blondin (1979a) showed that the chloroform/methanol-soluble extract of bovine mitochondria could be separated into three fractions by size fractionation on Sephadex LH-60. These fractions themselves are still very complex (I. M. Fearnley and J. E. Walker, unpublished work).

This simple gel filtration technique has proved to be very useful in the purification of a wide range of proteins extracted from membranes with chloroform and methanol. These include a proteolipid associated with kidney $(Na^+ + K^+)$-ATPase (Reeves *et al.*, 1980) and components of chromatophores of the *Rhodospirillaceae* (Tonn *et al.*, 1977; Theiler *et al.*, 1984; Brunisholz *et al.*, 1981, 1984a,b 1985). The range of the technique has also been extended by Theiler *et al.* (1983) who separated the L and M subunits of photosynthetic reaction centres of *Rhodospirillum rubrum* by gel filtration in organic solvents on hydroxypropylated Sephadex G-100 prepared according to the procedure of Ellingboe *et al.* (1969). Other solvents have also been used for protein purification in combination with lipophilic Sephadex. For example phospholamban and another proteolipid were isolated from sarcoplasmic reticulum proteolipids by gel filtration on Sephadex LH-60 in 88% formic acid/95% ethanol (1:3, v/v) (Collins *et al.*, 1981).

(iii) *Ion exchange chromatography in organic solvents.* Fractionation of chloroform/methanol-soluble proteolipids has also been achieved by ion-exchange chromatography in the same solvent to which buffer (ammonium acetate) has been added. The procedure is derived from a method devised for fractionation of lipids (Rouser *et al.*, 1969). It is likely that the separation achieved depends not only on ionic interactions but also on partition chromatography. It is important that the ion-exchange matrices are pretreated as recommended by Rouser *et al.* (1969). (See also Fillingame, 1976, and Theiler *et al.*, 1984). The method has been used as a purification step in the isolation of the carbodiimide-reactive proteolipid (Fillingame, 1976; Graf and Sebald, 1978; Sebald *et al.*, 1979) and a second proteolipid (Velours *et al.*, 1982, 1984) from ATP synthases, and in the purification of light-harvesting proteins from *Rhodospirillaceae* (Theiler *et al.*, 1983, 1984).

(iv) *Normal phase and reverse phase chromatography.* Lipophilic Sephadex swells only in polar organic solvents, so polar components of a solvent mixture become more strongly associated with it thereby creating a marked difference in composition between stationary and mobile phases. Therefore, separations can be achieved because the more polar components of a mixture of solutes are more strongly associated with the stationary phase. LH-20 and LH-60 Sephadex are suitable materials for partition (normal phase) chromatography. They have been used in this way to fractionate proteolipids from brain (Soto *et al.*, 1969) and from beef mitochondria (Cattell *et al.*, 1970, 1971). Earlier, brain proteolipids were also fractionated by partition chromatography on silicic acid (Matsumoto *et al.*, 1964). Rather surprisingly, this kind of chromatography has been little used for membrane protein fractionation under the highly controlled conditions of high-performance liquid chromatography. Rubinstein (1979) has emphasized the usefulness of the approach for chromatography of hydrophobic proteins. He demonstrated the use of Lichrosorb Diol as a suitable support in the fractionation of hydrophobic protein components of foetal calf serum and also demonstrated a wider applicability of the procedure in the isolation of human leucocyte interferon (Rubinstein *et al.*, 1978).

Reverse phase chromatography, which employs hydrophobic matrices to preferentially bind the less polar components of solvent mixtures, has been used to great effect for purification of globular proteins and subfragments. However, the strong and often irreversible binding of hydrophobic membrane proteins to this kind of support has severely restricted its usefulness as a technique for membrane protein fractionation. Nonetheless, a small number of chloroform/methanol-soluble proteolipids have been purified on $C_{18}$ reverse phase supports. Examples are small proteolipids associated with yeast (Velours *et al.*, 1982, 1984) and bovine ATP synthases (Blondin, 1979a,b; I. M. Fearnley and J. E. Walker, unpublished work) and proteolipids from cardiac membrane (Capony *et al.*, 1983). Tarr and Crabb (1983) found that cytochrome *P*-450 and rhodopsin could be recovered from a CN-bonded phase with ternary solvents containing acetonitrile, propan-1-ol and 0.1% trifluoroacetic acid.

### 9.2.3 FRAGMENTATION OF MEMBRANE PROTEINS

(a) *Proteolytic enzymes*

Proteolytic enzymes can be employed in two distinctly different ways for degradation of membrane proteins. (1) By treating native membrane proteins (either in the membrane itself or in a dispersed form in mild detergent) limited proteolysis will occur such that the extrinsic membrane sectors are degraded and removed, leaving intact the membrane sectors which are protected by the lipid bilayer. (2) Extensive proteolysis of the entire protein (including the intrinsic membrane sector) can be achieved in the presence of strong detergents (e.g. sodium dodecyl sulphate) or chaotropes (urea, guanidinium hydrochloride).

(i) *Limited proteolysis.* Treatment of glycophorin with trypsin leads to the formation of an insoluble, hydrophobic precipitate. This proved to be the membrane-spanning segment of 32 amino acids in the *C*-terminal half of the molecule. It was shown to anchor the protein in the erythrocyte membrane (Segrest *et al.*, 1972; Tomita and Marchesi, 1975). Since then a growing number of membrane proteins anchored by a single membrane-spanning segment have been identified by similar means. In a number of cases these segments have been isolated and their sequences determined by direct sequence analysis. In numerous examples where the sequences of the proteins have been determined by DNA sequencing (see Table 9.3) the positions and sequences of membrane anchors have been proposed by looking for hydrophobic spans within the predicted protein sequences (see Section 9.4.1). Examples are influenza virus neuraminidase (Fields *et al.*, 1981) the *b* subunit of ATP synthase (Gay and Walker, 1981) and surface glycoproteins of Epstein–Barr virus (Baer *et al.*, 1984; Hudson *et al.*, 1985).

Proteolytic cleavage of proteins in the membrane proved to be a helpful procedure in the elucidation of the structure of rhodopsin. A number of studies had shown that a range of proteases cleaved the native protein into two large membrane-bound fragments. (For example see Pober and Stryer, 1975; Sale *et al.*, 1977; Towner *et al.*, 1977; Hargrave *et al.*, 1980; Mullen and Akhtar, 1981). Brett and Findlay (1983) and Ovchinnikov *et al.* (1983) in their sequence analysis of ovine and bovine rhodopsins put this finding to good effect. They cleaved the native rhodopsin with proteases to produce two membrane-bound fragments of $M_r$ 27 K and 12 K respectively. These were separated by chromatography, further fragmented and sequenced. A similar approach was also valuable in the structural analysis of bacteriorhodopsin (Gerber *et al.*, 1979; Ovchinnikov *et al.*, 1979).

Limited proteolysis of membrane proteins is not only valuable in the generation of large fragments for sequence analysis, but also gives valuable information about the topography of the protein molecule: degraded segments are accessible and lie outside the membrane, protected (undegraded) segments lie within the lipid bilayer. In the case of bacteriorhodopsin, proteolysis experiments showed that the *C*-terminal 23 amino acids, seven residues at the *N*-terminal end of the chain and a short segment between residues 70 and 77 were accessible to proteolysis and therefore exposed to the aqueous milieu (Gerber *et al.*, 1979; Ovchinnikov *et al.*, 1979; Walker *et al.*, 1979).

(ii) *Proteolysis in denaturing conditions.* Many of the specific endoproteinases commonly used for digestion of globular proteases are active in the presence of SDS (0.1%) or urea (2–8 M); so they can be employed for degradation of membrane proteins in the presence of denaturants. A demonstration of the stability of many endoproteases in the presence of SDS is proteolytic digestion of proteins in polyacrylamide gels (Cleveland *et al.*, 1977). The list of proteinases that can be used in this way include trypsin, chymotrypsin, staphylococcal V8 proteinase, subtilisin, thermolysin, endoproteinase lysC and clostripain. Of the exoproteinases carboxypeptidase Y is most stable, being active in the presence of 0.1% SDS. So providing susceptible bonds are present, and the proteins can be solubilized, the extensive proteolysis of membrane proteins is not a serious problem; the difficulty is the purification of the peptides that are generated (see Section 9.2.4). Other procedures have also been recommended (Allen, 1981). For example, succinylation, maleylation or citraconylation of amino groups in membrane proteins produces polyanionic derivatives that often are soluble at alkaline pH in the absence of denaturants and so can be proteolysed easily. Another procedure that has found occasional use for digestion of membrane proteins is to first solubilize in concentrated (90–100%) formic acid and then to digest with pepsin (Allen, 1980a; Hayes *et al.*, 1986).

(b) *Chemical cleavage*

Membrane proteins are usually soluble in formic and trifluoroacetic acids (70–100%) and so cleavage on the *C*-terminal side of methionyl bonds with cyanogen bromide has been widely used for their fragmentation. (For examples see Ovchinnikov *et al.*, 1979; Gerber *et al.*, 1979; Brett and Findlay, 1983.) Cleavage on the *C*-terminal side of tryptophanyl bonds has also proved useful. A number of different reagents have been employed for this purpose. For example, Green and Toms (1985) cleaved $Ca^{2+}$-ATPase according to the procedure of Savige and Fontana (1977). The protein was first carboxymethylated, delipidated in the presence of SDS and then the freeze-dried protein–SDS complex dispersed by sonication in the cleavage reagent (containing acetic acid, 12 M HCl, dimethyl sulphoxide and phenol). Brunisholz *et al.* (1985) used iodosobenzoic acid to cleave light-harvesting proteins from *Rps. viridis* and Lees *et al.* (1983) employed 2-(2-nitrophenylsulphenyl)-3-methyl-3′-bromoindolenine (NBS-skatole) to produce fragments of a brain proteolipid. Additional cleavages after tyrosine as well as tryptophan residues can be made with the more powerful reagent, *N*-bromosuccinimide (Khorana *et al.*, 1979).

Cleavage of asparagine–glycine bonds with hydroxylamine can be performed in the presence of 6 M-guanidinium hydrochloride and partial hydrolysis with acid (predominant cleavage of aspartyl–X bands) could also in principle be used for cleavage of membrane proteins.

### 9.2.4 PURIFICATION OF FRAGMENTS DERIVED FROM MEMBRANE PROTEINS

The purification of fragments from digests of membrane proteins has been a major obstacle in their sequence analysis. The difficulties are to solubilize and to disaggregate the fragments under conditions that do not chemically modify the peptides and which permit high-resolution chromatography. The range of chromatographic methods available is in essence that described in Section 9.2.2 for the purification of membrane proteins. However, the problems of disaggregation and solubilization associated with the complex mixtures of peptides of rather similar chemical characteristics that result from fragmentations have often proved to be more severe than those encountered in protein purification. So greater effort has been invested in this problem, and a wider range of chromatographic parameters have been investigated. These methods are described in the following sections.

#### (a) *Gel permeation chromatography*

Size fractionations have been performed successfully on mixtures of hydrophobic peptides with combinations of Sephadex and 6 M-guanidinium hydrochloride (Ovchinnikov *et al.*, 1979) and polyamide supports (Biogels) with various dilutions with water or acetic acid (Erdweg and Buse, 1985; Brunisholz *et al.*, 1985; Lees *et al.*, 1983) or formic acid (Steffens and Buse, 1976; Jolles *et al.*, 1979). Lipophilic Sephadex (LH-20 or LH-60) has been employed with mixtures of formic acid and ethanol (Gerber *et al.*, 1979; Takagaki *et al.*, 1980; Green and Toms, 1985) or with a mixture of formic acid, acetic acid, chloroform and ethanol (Brett and Findlay, 1983). Allen (1977) used solvents containing phenol, acetic acid and water in attempts to disaggregate and resolve on Sephadex mixtures of hydrophobic peptides from $Ca^{2+}$-ATPase, and phenol/water/formamide mixtures for subsequent chromatographic steps. Unfortunately the peptides recovered at the end of these purification steps had blocked *N*-terminals, having been modified during chromatography.

(b) *High-performance liquid chromatography*

Formic acid, in water and in combination with various other solvents (usually alcohols), has been used as a solvent for purification of hydrophobic peptides by high-performance chromatography. It was introduced by Gerber *et al.* (1979) and Takagaki *et al.* (1980) for isolation of fragments of bacteriorhodopsin and cytochrome $b_5$ by reverse phase chromatography on $C_{18}$ supports. Subsequently, Velours *et al.* (1984) used formic acid and propanol with a $C_{18}$ matrix in their sequence analysis of an ATPase-associated proteolipid from yeast. Stoffel *et al.* (1982a,b, 1985) in their work on lipophilins have preferred silica supports with similar solvents. Ternary solvents containing formic acid, trimethylamine and propanol with cyanopropyl supports have been advocated by Tarr and Crabb (1983) for isolation of large hydrophobic peptides.

The major drawback of all these methods is the inimical nature of formic acid to protein, instruments, columns and experimenters. Tarr and Crabb (1983) and Walker *et al.* (1985) have noted that exposure of proteins and peptides to this acid causes chemical modification and thus leads to additional chemical complexity. For example, serine and threonine residues become esterified, a reaction that may be reversed by treatment with aminoethanol (Tarr and Crabb, 1983).

A somewhat different approach was used by Aquila *et al.* (1982). They maleylated the ADP/ATP carrier of bovine mitochondria and citraconylated a fragment of it, and then performed partition chromatography of blocked fragments on silica. Peptides not purified by this method were rechromatographed on sulphopropyl-Sephadex in pyridine acetate gradients containing propan-1-ol.

It has been possible to isolate some hydrophobic peptides under conditions employed for more hydrophilic fragments. For example the four cyanogen bromide fragments of the A6L proteolipid from bovine mitochondria can be purified by reverse phase chromatography on a $C_{18}$ support in trifluoracetic acid and acetonitrile (I. M. Fearnley and J. E. Walker, unpublished work).

(c) *Gas chromatography of polyaminoalcohols*

Volatility of peptides can be increased by methylation, trifluoroacetylation and reduction with $LiAl[^2H]_4$ or $B_2[^2H]_6$ to the polyaminoalcohols (Kelley *et al.*, 1975; Nau and Biemann, 1976; Frank *et al.*, 1978). The resultant derivatives can then be purified by gas chromatography which can be linked directly to a mass spectrometer (see Section 9.2.5.c). This method has been used in the sequence analysis of bacteriorhodopsin (Gerber *et al.*, 1979) and lipophilin (Stoffel *et al.*, 1982a).

### 9.2.5 SEQUENCE ANALYSIS

Most protein sequences determined directly on the protein (see Table 9.2) have depended largely upon the Edman degradation whereby amino acids are removed sequentially from the *N*-terminal of the protein or polypeptide (for reviews see Allen, 1981; Edman and Henschen, 1975). This is achieved by a cycle of reactions containing two steps (see Fig. 9.3):

(1) *coupling* of the *N*-terminal residue with phenylisothiocyanate (PITC);
(2) *cleavage* of the *N*-terminal residue by a cyclization reaction in acid;

In a third reaction, *conversion*, the cleavage product, the anilinothiazolinone (ATZ), is rearranged to a more stable form, the phenylthiohydantoin (PTH). This derivative is then identified, usually by high-performance liquid chromatography. A manual version of the procedure has been widely used in various modified forms (see Allen, 1981) particularly for degradation of small peptides. However, these procedures have been largely superseded by automated procedures performed in specially designed instruments, protein (or peptide) sequencers. They are of three types: the liquid phase (or spinning cup) sequencer, the solid phase sequencer and the gas phase sequencer.

(a) *The spinning cup sequencer*

This instrument was designed by Edman and Begg (1967). The heart of the instrument is a glass beaker (or cup) rotated about its vertical axis in a controlled environment. The protein is introduced in solution and forms a film on the wall of the cup. It is dried *in vacuo*, redissolved in a non-volatile buffer, Quadrol, and reacted in this film with PITC. Excess reagents, buffer and products are then removed by extraction of the film with a solvent (a mixture of benzene and ethyl acetate). The protein film is again dried. Cleavage is then effected with an anhydrous acid and then the ATZ removed by extraction with butyl chloride. This extract is collected, dried, converted to the PTH in a second reaction chamber (the convertor) and then collected for subsequent identification. Meanwhile, the film in the cup, now containing the protein with one amino acid removed from its *N*-terminal continues into the next cycle of degradation.

From the point of view of sequence analysis of membrane proteins the major disadvantage of this instrument is that the hydrophobic membrane proteins have a propensity for dissolving in the organic solvents (ethyl acetate, benzene and butyl chloride) and so these proteins tend to be extracted from the cup and lost. This problem can be alleviated by making the protein more polar, for

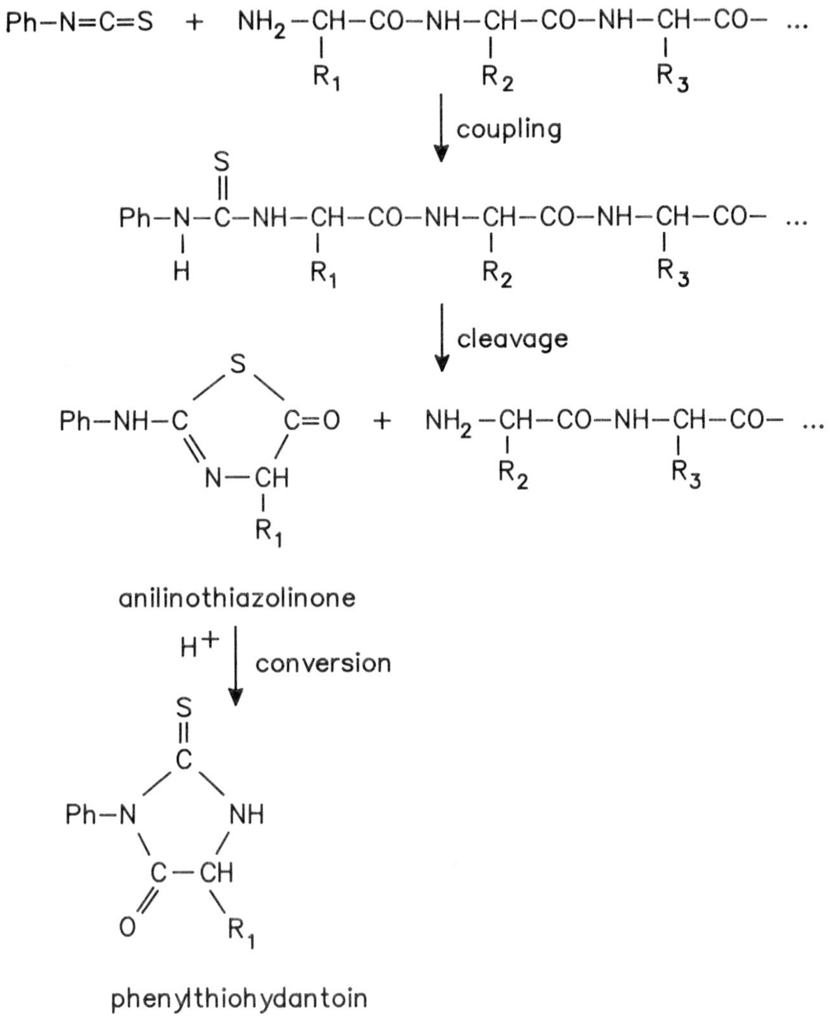


Fig. 9.3 The Edman degradation.

example by reaction in the first cycle of the Edman degradation with polar sulphonated isothiocyanates that will thereby derivatize ε-amino groups with polar residues. The α-$NH_2$ group also reacts but is removed by the cleavage reaction (Braunitzer *et al.*, 1970, 1971). Other hydrophilic derivatives can also be made by carbodiimide coupling of carboxyl groups with hydrophilic amines (Foster *et al.*, 1973) or by reaction of the C-terminal homoserine lactone of CNBr peptides with polar derivatives (Braunitzer and Pfletschinger, 1978). However, the charged amino acids tend to be rare in membrane proteins and so coupling of polar groups to carboxyls and ε-amino groups have been little used amongst this group of proteins. However, the reaction of the *C*-terminal lactone of CNBr peptides with solid supports (rather than polar derivatives) has been widely used for immobilization of cyanogen bromide peptides for solid phase sequence analysis (see Section 9.2.5.b).

An important device for reducing the wash out of proteins and peptides from the film of the spinning cup sequencer is the addition of the polycationic compound, Polybrene (Tarr *et al.*, 1978; Klapper *et al.*, 1978). Its use has extended greatly the range of peptides and proteins that can be degraded in the spinning cup sequencer, and it is equally important in the gas phase instrument (Section 9.2.5.c).

These potential difficulties notwithstanding the liquid phase sequencer has been used successfully for sequencing a number of membrane proteins (e.g. Brunisholz *et al.*, 1984a,b, 1985). These experiments probably owe their success to the fact that although these proteins contain long hydrophobic stretches, they do also have polar amino acids in their sequences and particularly arginine residues near to their *C*-terminals. This, and the presence of Polybrene, apparently are sufficient to retain the protein in the reaction cup. However, hydrophobic proteins lacking such polar features tend not to be retained in the reaction cup and other means have to be employed to determine their sequences.

(b) *The solid phase sequencer*

An alternative solution to the problem of sample loss in extraction solvents incurred in the spinning cup sequencer, is the covalent attachment of the protein or peptide to an insoluble support. This is the basis of the solid phase sequencer (Laursen, 1971). The support with the attached peptide is placed in a small column (the counterpart of the cup in the spinning cup sequencer). Reagents can then be introduced sequentially as required by the Edman degradation. The major advantage is that by-products can be removed by solvent washes, without concomitant danger of loss of the sample.

A range of supports has been proposed (see Laursen *et al.*, 1975). However, those based on porous glass are most valuable and have found widest use. They have excellent physical properties and can be easily derivatized to give covalently bound amino groups (see Fig. 9.4). These in turn can be activated to isothiocyanato derivatives which will react with amino groups of a protein or can be acylated with the *C*-terminal homoserine lactone (Horn and Laursen, 1973) or spirolactone (Wachter and Werhahn, 1979) generated by chemical cleavage on the *C*-terminal sides of methionyl and tryptophanyl groups, respectively. Attachment via carboxyl groups activated with water-soluble carbodiimides has proved to be more problematical, but has found some use.

This approach to sequence analysis of membrane proteins has been particularly successful and has been employed extensively in the sequence

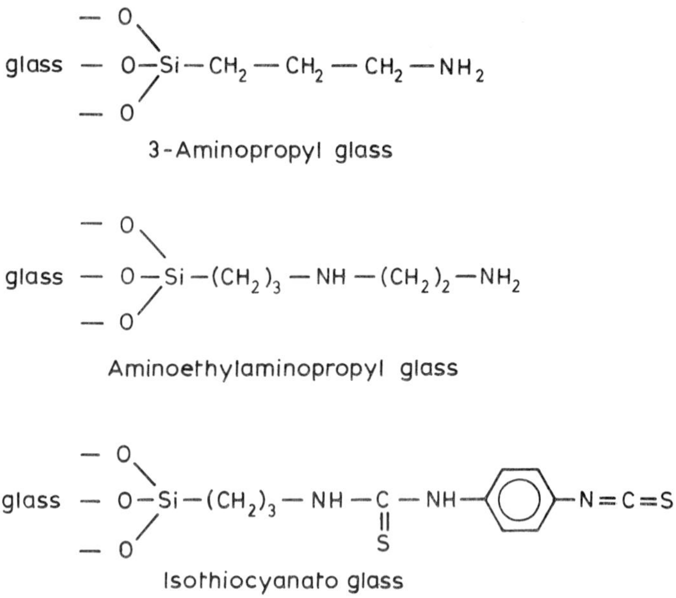


**Fig. 9.4** Glass derivatives for protein immobilization. See Wachter *et al.* (1975).

analysis of the most hydrophobic proteins such as bacteriorhodopsin (Gerber *et al.*, 1979), the membrane segment of cytochrome $b_5$ (Takagaki *et al.*, 1980) and the proteolipids associated with ATP synthases from various sources (Hoppe and Sebald, 1980; Sebald and Hoppe, 1981; Velours *et al.*, 1984; I. M. Fearnley and J. E. Walker, unpublished results).

(c) *The gas phase sequencer*

The gas phase sequencer (Hunkapiller and Hood, 1980; Hewick *et al.*, 1981, Hunkapiller *et al.*, 1983b) differs from the liquid phase and solid phase sequencers in two important aspects. Firstly, two reagents, the buffer (trimethylamine) for the coupling reaction and the acid for cleavage (trifluoroacetic acid) are delivered as vapours in an argon stream. Thereby large excesses of reagents are avoided and side products and contaminants are minimized. As in the other designs of sequencer the PITC is delivered as a liquid, and washes with appropriate solvents (ethyl acetate, heptane, butyl chloride) are interposed between coupling and cleavage and after cleavage to remove reaction by-products and ATZ respectively. Secondly, the reaction chamber in this instrument is a cartridge in which a glass fibre disc is sandwiched. The disc is impregnated with Polybrene and the samples applied to this matrix. This instrument has been particularly successful for micro-sequence analysis in the range of 10 pmol–1 nmol. Hitherto it has been little used for analysis of hydrophobic proteins and direct comparisons of sequencing efficiency of the same hydrophobic proteins in gas and solid phase instruments have not been made. Successful attempts have been made to sequence peptides covalently attached to glass beads in this instrument, as a way of avoiding sample loss (Strickler *et al.*, 1984).

(d) *C-terminal sequences*

In the case of membrane proteins with exposed *C*-terminal domains, *C*-terminal sequence can be deduced by treatment with carboxypeptidases A and B (Ambler, 1972), C (Tschesche and Kupfer, 1972) or Y (Martin *et al.*, 1977; see Allen, 1981). An example of this is the determination of the sequence of the *C*-terminal region of bacteriorhodopsin (Gerber *et al.*, 1979; Ovchinnikov *et al.*, 1979). Denatured membrane proteins can be investigated in the presence of SDS with carboxypeptidase Y.

(e) *Mass spectrometry*

Electron impact mass spectrometry has been used to deduce sequences of peptides and in rare cases of complete proteins (see Morris and Dell, 1975). The major problem has been to convert the peptides to volatile derivatives; this has been achieved by acetylation and permethylation (Morris *et al.*, 1971), reduction to polyaminoalcohols (Nau, 1976; Frank and Desiderio, 1978) and trimethylsilylation. In combination with gas chromatographic separation of polyaminoalcohols the method has been successfully used in the sequence analyses of the membrane proteins bacteriorhodopsin and lipophilin (see Section 9.2.4.c). The more recently developed technique of fast atom bombardment mass spectrometry (Barber *et al.*, 1981) was employed to identify myristic acid as the hydrophobic membrane anchor of a number of proteins that become associated with membrane (Aitken *et al.*, 1982; Carr *et al.*, 1982). Gas chromatography/electron impact mass spectrometry was employed to identify the same blocking groups in retrovirus proteins (Henderson *et al.*, 1983). Amongst membrane proteins, as with other proteins, these techniques have much promise and utility, particularly when used in combination with other sequencing methods (Gibson and Biemann, 1984).

## 9.3 Indirect methods of protein sequencing

An ever increasing number of protein sequences are being determined by DNA sequence analysis of genes and complementary DNA. The major difficulty in this approach is the isolation of the desired clones containing the DNA sequence encoding the protein of interest. (For detailed reviews on these techniques see Maniatis *et al.*, 1982.) The methods employed to do this depend strongly upon the origin of the protein (gene) as summarized in Fig. 9.5. DNA from organelles (mitochondria, chloroplasts) and some viruses is often not very large and so the entire genome can be cloned and sequenced. In the case

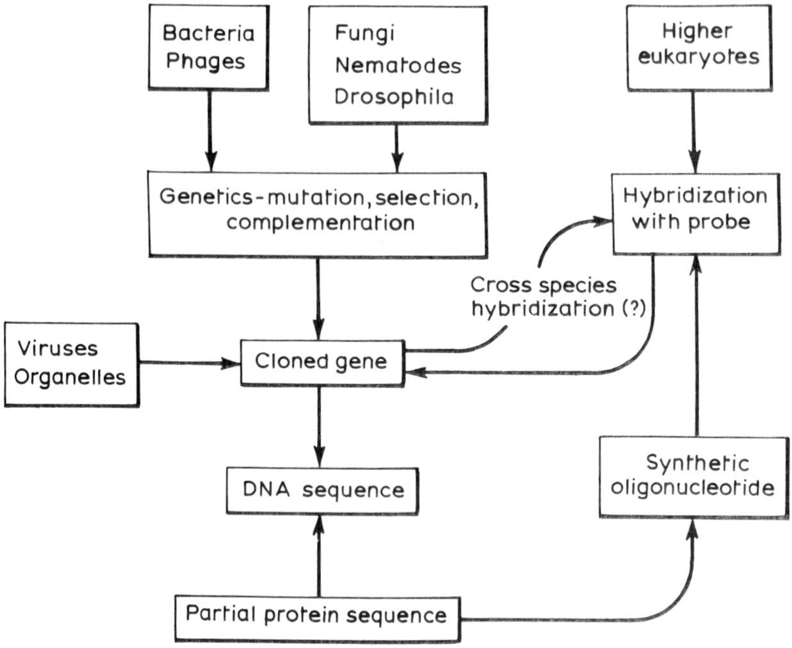


**Fig. 9.5** Strategies for gene cloning.

of genes from prokaryotic organisms and bacteriophages, viruses and simpler eukaryotes (e.g. fungi, nematodes, *Drosophila*), the powerful tools of genetics can be employed to isolate genes. However, these tools cannot be applied to the isolation of nuclear genes from higher eukaryotes and so different techniques are used. These can also be equally well applied to prokaryotic DNA. Three general methods are available.

(1) Expression methods in which the expression of a gene *in vivo* or *in vitro* is detected by biological assay or by immunochemical techniques. (For a discussion of these methods see Maniatis *et al.*, 1982.)
(2) Mixed oligonucleotide probes, 11–17 nucleotides in length, are synthesized to correspond to (fragmentary) protein sequence information. The degeneracy of the genetic code is accounted for by incorporation of a mixture of two to four bases (if necessary) at particular positions. These oligonucleotides are labelled with [$^{32}$P]phosphate and used as hybridization probes under stringent hybridization conditions (Wood *et al.*, 1985) to identify, in libraries, individual recombinants containing the sequences of interest (reviewed by Smith, 1983).
(3) An alternative procedure employs long (50–90 bases) unique oligonucleotide hybridization probes (Anderson and Kingston, 1983). In this case the

**Table 9.3** Membrane protein sequences determined by DNA sequencing

| *Function* | *Protein* | *Source* | *Comments* | *Footnotes* |
|---|---|---|---|---|
| 1. Energy transduction | | | | |
| (a) ATP synthase components | Subunit *a* | E. coli | Component of proton channel | 1 |
| | ATPase-6 | *Synechococcus* 6301 | Equivalent to bacterial *a* | 2 |
| | | *D. melanogaster* (mt) | | 3,4 |
| | | Mammals (mt) | | 5–7 |
| | | *A. nidulans* (mt) | | 8 |
| | | *S. cerevisiae* (mt) | | 9 |
| | Subunit X | Pea (ch) | Equivalent to bacterial *a* | 10 |
| | Subunit *b* | *E. coli* | Component of $H^+$-channel, binds $F_1$ | 1 |
| | | *Synechococcus* 6301 | Also has a duplicated version, b′. | 2 |
| | Subunit I | Wheat (ch) | Equivalent to bacterial *b* | 11 |
| | Subunit *c* | *E. coli* | Component of proton channel | 1 |
| | | *Synechococcus* 6301 | | 2 |
| | ATPase-9 | Cow (n) | Equivalent to bacterial *c* | 12 |
| | | *A. nidulans* (mt) | | 13 |
| | | *N. crassa* (n) | | 14 |
| | | *S. cerevisiae* (mt) | | 15 |
| | Subunit III | Spinach (ch) | Equivalent to bacterial *c* | 16 |
| | | Wheat (ch) | | 17 |
| | aap I | *S. cerevisiae* (mt) | Associated with ATP synthase | 18 |
| | | *A. nidulans* (mt) | | 19 |
| | A6L | *D. melanogaster* (mt) | Associated with ATP synthase | 4 |
| | | Mammals (mt) | | 5–7 |
| (b) Cytochrome $bc_1$ | Cytochrome *b* | *A. nidulans* (mt) | | 19 |
| | | Mammals (mt) | | 5–7 |
| | | *S. cerevisiae* (mt) | | 20 |
| | | *Z. mays* (mt) | | 21 |
| | Cytochrome $c_1$ | *S. cerevisiae* (n) | | 22 |

**Table 9.3** (continued)

| *Function* | *Protein* | *Source* | *Comments* | *Footnotes* |
|---|---|---|---|---|
| (c) Cytochrome oxidase | Subunit I | *A. nidulans* (mt) | | 23 |
| | | *D. melanogaster* (mt) | | 4 |
| | | Mammals (mt) | | 5–7 |
| | | *S. cerevisiae* (mt) | | 24 |
| | Subunit II | *D. melanogaster* (mt) | | 4 |
| | | Mammals (mt) | | 5–7 |
| | | *Z. mays* (mt) | | 25 |
| | | *S. cerevisiae* (mt) | | 26,27 |
| | Subunit III | *D. melanogaster* (mt) | | 4 |
| | | Mammals (mt) | | 5–7 |
| | | *S. cerevisiae* (mt) | | 28 |
| (d) Oxidoreductases | | | | |
| NADH dehydrogenase | | Mammals (Mt) | 6 subunits of complex | 5–7,29 |
| | | *E. coli* | One chain only | 30 |
| Succinate dehydrogenase | 4 subunits | *E. coli* | sdhA, *sdh*B, *sdh*C, *sdh*B | 31,32 |
| Fumarate reductase | 4 subunits | *E. coli* | *frd*A, *frd*B, *frd*C, *frd*D | 33,34 |
| (e) Photosynthesis | Reaction centre | | | 35 |
| | subunits L,H,M | *Rps. sphaeroides* | | |
| | light-harvesting | *Rps. capsulata* | | 35a |
| | subunits $\alpha,\beta$ | | | |
| | Chlorophyll a | Pea (n) | Chlorophyll binding | 36,37 |
| | Apoprotein | Spinach (ch) | | 38 |
| | P680 chloro- | | | |
| | phyll | Spinach (ch) | | 39 |
| | a apoprotein | | | |

| Function | Protein | Source | Comments | Footnotes |
|---|---|---|---|---|
| | Ps II 32 kD protein | Tobacco (ch) | Psb or herbicide-binding protein of | 40 |
| | | Spinach (ch) | PsII plastoquinone reductase | 40 |
| | PsII 39 kD protein | Spinach (ch) | homologous to PsbA | 38 |
| (e) Photosynthesis | Cytochrome *f* | pea (ch) | Component of $b_6f$ complex | 41 |
| | | Spinach (ch) | | 42 |
| | | Wheat (ch) | | 43 |
| | 15.2 kD protein | Pea (ch) | Component of $b_6f$ complex | 44 |
| | | Spinach (ch) | | 45 |
| | Apocytochrome $b_6$ | Spinach (ch) | Component of $b_6f$ complex | 45 |
| | Bacteriorhodopsin | *H. halobium* | Light-driven proton pump | 46 |
| 2. Immune response | | | | |
| | IgM | | | 47 |
| MHC antigen complex | class I | Human, mouse | | 48 |
| | class II | Human, mouse | | 49 |
| Cell surface antigens | MRC OX-2 glycoprotein | Mouse | | 50 |
| | Thy-1 | Rat, mouse | | 51 |
| 3. Receptors | | | | |
| (a) Acetylcholine | $\alpha,\beta,\gamma,\delta$ | Calf muscle | | 52–56 |
| | $\alpha,\gamma$ | Human | | 52,57 |
| | $\alpha,\beta,\gamma$ | *Torpedo californica* | | 58–61 |
| | $\alpha$ | *Torpedo marmorata* | | 62,63 |
| (b) Asialoglycoprotein | Asialoglycoprotein receptor | | Glycoprotein transport | 64,65 |
| (c) Epidermal growth factor | EGF-receptor | Human | | 66 |
| (d) Insulin | Insulin receptor | Human | | 67 |
| (e) Interleukin 2 | Interleukin 2 receptor | Human | | 68 |

**Table 9.3** (continued)

| Function | Protein | Source | Comments | Footnotes |
|---|---|---|---|---|
| (f) LDL | Low-density lipoprotein receptor | Human | Cholesterol transport | 69 |
| (g) T-cell | α,β | Human | | 70–72 |
| (h) Transferrin | Transferrin receptor | Human | | 73 |
| | | Murine | | 74 |
| (i) IgR | Immunoglobulin receptor | Rabbit | Transcellular transporter of IgA and IgM | 75 |
| 4. Transport | | | | |
| (a) Cations | $Ca^{2+}$-ATPase | Rabbit sarcoplasmic reticulum | $Ca^{2+}$ and $Mg^{2+}$ vectorial ion transport | 76 |
| | $(Na^{+}+K^{+})$-ATPase | Sheep kidney | | 77 |
| | α subunit | *Torpedo californica* | $Na^{+}$ and $K^{+}$ vectorial exchange | 78 |
| | Kdp-ATPase | *E. coli* | ATP-driven $K^{+}$ transport | 79 |
| | $Na^{+}$ channel | *Electrophorus* | $Na^{+}$ permeability of electrically excitable membranes | 80 |
| (b) Anions | Band III protein | Mouse | $Cl^{-}/HCO_3^{-}$ exchange in erythrocyte membrane | 81 |
| | *Pst*A, *Pst*B | *E. coli* inner membrane | Phosphate-specific transport | 82 |
| (c) Amino acids | *His*P, *His*Q, *His*M | *E. coli* inner membrane | High-affinity histidine transport | 83 |
| (d) Sugars | Lactose permease | *E. coli* inner membrane | *lac*Y | 84 |
| | Glucose transporter | Human erythrocyte | | 85 |
| | Maltose transporter | *E. coli* inner membrane | *mal*F | 86 |

| *Function* | *Protein* | *Source* | *Comments* | *Footnotes* |
|---|---|---|---|---|
| 4. Transport | | | | |
| (e) Nucleotides | ADP/ATP translocase | *N. crassa* mitochondria (n) | Translocation of ADP and ATP in and out of mitochondria | 87 |
| | | *Z. mays* mitochondria (n) | | 88 |
| (f) Pores | *OmpC* | *E. coli* outer membrane | | 89 |
| | *OmpF* | *E. coli* outer membrane | Porin | 90 |
| | *LamB* | *E. coli* outer membrane | Permeation of maltose and maltodextrins, used as lambda $\lambda$ receptor | 91 |
| | *phoE* | *E. coli* outer membrane | | 92 |
| | Mitochondrial porin | *S. cerevisiae* outer mt-membrane (n) | | 93 |
| 5. Viral | Haemagglutinin | Human influenza | | 94–96 |
| | | Fowl plague influenza | | 97 |
| | Neuraminidase | Human influenza | | 98,99 |
| | Glycoprotein | Vesicular stomatitis | | 100 |
| | Glycoproteins | Epstein–Barr | | 101,102 |
| | Variable surface glycoprotein | *Trypanosoma brucei* | | 103 |
| | E1 glycoprotein | Corona virus | | 104 |
| | E1, E2 glycoproteins | Semliki Forest virus | | 105 |
| 6. Miscellaneous | | | | |
| (a) Cytochrome *P*-450 | P450MC | Rat | | 106 |
| | $P_1450$ | Mouse, human | | 107,108 |
| | $P_3450$ | Mouse | | 107 |
| | P450 (SSC) | Cow | | 109 |
| (b) Surface antigens | | *Plasmodium falciparum* | | 110 |
| (c) Visual proteins | Transducin $\alpha$-subunit | Bovine | | 111 |
| | Rhodopsin | Human | | 112 |

**Table 9.3** (continued)

| *Function* | *Protein* | *Source* | *Comments* | *Footnotes* |
|---|---|---|---|---|
| (d) Myelin sheath protein | Po glycoprotein | Rat Schwann cell | | 113 |
| (e) Flagellar rotation | *mot*A | *E. coli* | | 114 |
| 7. Unknown | Glycoproteins | Epstein–Barr virus | | 101 |
| | Interferon-induced P16 | Human | | 115 |
| | *unc*I | *E. coli* | | 1 |

*Footnotes.* 1. Gay and Walker (1981); 2. A. L. Cozens and J. E. Walker, unpublished work; 3. mt, mitochondrial gene; n, nuclear gene; ch, chloroplast gene; 4. de Bruijn (1983); 5. Anderson *et al.* (1981); 6. Anderson *et al.* (1982); 7. Bibb *et al.* (1981); 8. Grisi *et al.* (1982); 9. Macino and Tzagoloff (1980); 10. Cozens *et al.* (1986); 11. Bird *et al.* (1985); 12. Gay and Walker (1985); 13. Brown *et al.* (1984); 14. Viebrok *et al.* (1982); 15. Macino and Tzagoloff (1979); 16. Alt *et al.* (1983); 17. Howe *et al.* (1982); 18. Macreadie *et al.* (1983); 19. Waring *et al.* (1981); 20. Nobrega and Tzagoloff (1980); 21. Dawson *et al.* (1984); 22. Sadler *et al.* (1984); 23. Waring *et al.* (1984); 24. Bonitz *et al.* (1980); 25. Fox and Leaver (1981); 26. Fox (1979); 27. Coruzzi and Tzagoloff (1979); 28. Thalenfield and Tzagoloff (1980); 29. Chomyn *et al.* (1985); 30. Young *et al.* (1981); 31. Wood *et al.* (1984); 32. Darlison and Guest (1984); 33. Grundström and Jaurin (1982); 34. Cole *et al.* (1982); 35. Sutton *et al.* (1982); 35a. Youvan *et al.* (1984); 36. Corruzzi *et al.* (1983); 37. Cashmore (1984); 38. Alt *et al.* (1984); 39. Morris and Herrmann (1984); 40. Zurawski *et al.* (1982); 41. Willey *et al.* (1984a). 42. Alt and Herrmann (1984); 43. Willey *et al.* (1984b); 44. Phillips and Gray (1984); 45. Heinemeyer *et al.* (1984); 46. Dunn *et al.* (1981); 47. Rogers *et al.* (1980); 48. Ploegh *et al.* (1981); 49. Kaufman *et al.* (1984); 50. Clark *et al.* (1985); 51. Seki *et al.* (1985); 52. Noda *et al.* (1983c); 53. Tanabe *et al.* (1984); 54. Takei *et al.* (1984); 55. Takei *et al.* (1985); 56. Kubo *et al.* (1985); 57. Shibahara *et al.* (1985); 58. Noda *et al.* (1982); 59. Noda *et al.* (1983a); 60. Noda *et al.* (1983b); 61. Claudio *et al.* (1983); 62. Devillers-Thiery *et al.* (1983); 63. Sumikawa *et al.* (1982); 64. Spiess and Lodish (1985); 65. Spiess *et al.* (1985); 66. Ullrich *et al.* (1984); 67. Ullrich *et al.* (1985); 68. Cosman *et al.* (1984); 69. Sudhof *et al.* (1985); 70. Sim *et al.* (1984); 71. Yanagi *et al.* (1984); 72. Hedrick *et al.* (1984); 73. Schneider *et al.* (1984); 74. Stearne *et al.* (1985); 75. Mostov *et al.* (1984); 76. MacLennan *et al.* (1985); 77. Shull *et al.* (1985); 78. Kawakami *et al.* (1985); 79. Hesse *et al.* (1984); 80. Noda *et al.* (1984). 81. Kopito and Lodish (1985); 82. Surin *et al.* (1984); 83. Higgins *et al.* (1982); 84. Büchel *et al.* (1980); 85. Mueckler *et al.* (1985); 86. Froshauer and Beckwith (1984); 87. Arrends and Seband (1984); 88. Baker and Leaver (1985); 89. Mizuno *et al.* (1983); 90. Inokuchi *et al.* (1982); 91. Clement and Hofnung (1981); 92. Overbeeke *et al.* (1983); 93. Mihara and Sato (1985); 94. Winter *et al.* (1981); 95. Verhoeyen *et al.* (1980); 96. Gething *et al.* (1980); 97. Porter *et al.* (1979); 98. Blok and Air (1982); 99. Fields *et al.* (1981); 100. Rose *et al.* (1980); 101. Baer *et al.* (1984); 102. Hudson *et al.* (1985); 103. Boothroyd *et al.* (1982); 104. Armstrong *et al.* (1984); 105. Garoff *et al.* (1980); 106. Yabusaki *et al.* (1984); 107. Kimura *et al.* (1984); 108. Jaiswal *et al.* (1985); 109. Morohashi *et al.* (1984); 110. Holder *et al.* (1985); 111. Tanabe *et al.* (1985); 112. Nathans and Hogness (1984); 113. Lemke and Axel (1985); 114. Dean *et al.* (1984); 115. T. Evans and D. S. Secher, unpublished.

degeneracy of the genetic code is accounted for by 'guessing' the most likely unique codon for a particular amino acid. The guesses are influenced by a knowledge of codon usage (codon strategy) in the organism from which the DNA originates, of the relative stability of G-T versus A-C mismatches and of the infrequency of the dinucleotide dCpG in eukaryotic genes (Caruthers, 1985).

(4) A recent development is the use of synthetic probes containing deoxyinosine at ambiguous codon positions. It is proposed that this analogue is an inert base that neither destabilizes nor contributes at mismatched sites towards formation of a DNA duplex (Takahashi *et al.*, 1985).

These methods, and particularly methods 2–4, require an input of protein sequence independently determined by direct means.

Having obtained a clone of interest, the determination of the recombinant DNA sequence by the dideoxy method coupled with cloning in bacteriophage M13 (Sanger *et al.*, 1977; Biggin *et al.*, 1983; Bankier and Barrell, 1983) or by the chemical method (Maxam and Gilbert, 1977) is extremely rapid.

Thus, a modern, efficient approach to sequence analysis of proteins (irrespective of their nature) combines partial direct protein sequence information with the indirectly determined information deduced from DNA sequences. This approach has revolutionized the sequence analysis of membrane proteins (as other proteins) and the majority of the sequences of membrane proteins now known have been determined by this approach (see Table 9.3).

## 9.4 Secondary structures of membrane proteins

### 9.4.1 SECONDARY STRUCTURE PREDICTION

Membrane proteins contact two distinct environments, the lipid phase in which they are embedded, and the aqueous phase into which they protrude. It is to be expected that the parts of the protein in contact with the lipid will be made of hydrophobic amino acids and that those protruding into the aqueous environment will be more polar. However, some polar amino acids may be functional (for example in an ion channel) or structural, and so may also be within the lipid bilayer. In the latter case they will probably pair with opposite charges to form salt bridges. It is also to be expected that the parts of membrane proteins buried in the hydrophobic interior of the membrane will be largely folded in $\alpha$-helices or $\beta$-sheets. This is because these arrangements allow the hydrophilic CO and NH groups of the polypeptide backbone to form hydrogen bonds. In the absence of such bonds to each other they would need to form hydrogen bonds with water. By the same argument, less regular parts of the chain containing bends required to link α-helices or β-sheets should be either at the hydrophilic membrane surface or in the aqueous environment, where the groups can form hydrogen bonds (Unwin and Henderson, 1984). Two kinds of membrane proteins have been observed so far; bacteriorhodopsin and photosynthetic reaction centres are made largely of transmembrane α-helices, and bacterial porin is made of β-sheets.

These simple considerations and the physical thickness of the lipid bilayer (ca. 40 Å) impose constraints on membrane protein structure that can be exploited in the prediction of their secondary structures from primary sequences. This is particularly true for transmembrane α-helices. They will be made of stretches of predominantly hydrophobic acids. With a vertical rise in an α-helix of 1.5 Å/amino acid residue these stretches will be 25–27 amino acids in length. It can also be anticipated that these α-helical segments will be joined by more hydrophilic structures containing turns (β-turns).

Various methods of calculating hydrophobicity along a sequence have been proposed (Segrest and Feldmann, 1974; Kyte and Doolittle, 1982; Engelman *et al.*, 1982). The most widely used today employs a computer program SOAP to calculate hydrophobicity (Kyte and Doolittle, 1982; for a FORTRAN listing see Rose *et al.*, 1985). For this purpose a hydropathy scale, an amalgam of experimental observations has been composed that takes into consideration the hydrophilic and hydrophobic properties of each amino acid. The program continuously determines the average hydropathy within a segment of predetermined length as it advances through the sequence, and the consecutive scores are plotted from *N*- to *C*-terminus. As an example, the program finds seven hydrophobic segments in the sequence of bacteriorhodopsin (Fig. 9.6). These are assumed to correspond to α-helical membrane spans.

Numerous other examples of the use of this program to predict the positions of transmembrane segments in amino acid sequences are to be found in the literature. Some predictions can be developed further by correlation with other biochemical and structural features (see the ADP/ATP translocase, Saraste and Walker, 1982; cytochromes *b* and $b_{563}$/f, Saraste, 1984; Widger *et al.*, 1984; Mansfield and Anderson, 1985; $Ca^{2+}$-ATPase, MacLennan *et al.*, 1985; bacteriorhodopsin, Engleman *et al.*, 1980).

This approach to looking for membrane-spanning segments does not work in the case of the *E. coli* outer membrane protein, porin, as the protein does not contain sizeable hydrophobic domains and consists predominantly of β-structures. About two-thirds of the polypeptide backbone is arranged in antiparallel β-pleated sheet in an orientation approximately normal to the membrane plane, the average strand length being 10–12 residues (Kleffel *et al.*, 1985). However, the extents of these segments can be predicted by looking in the primary structure for sequences that promote turns and thereby reverse the direction of the polypeptide chain (Paul and Rosenbusch, 1985).

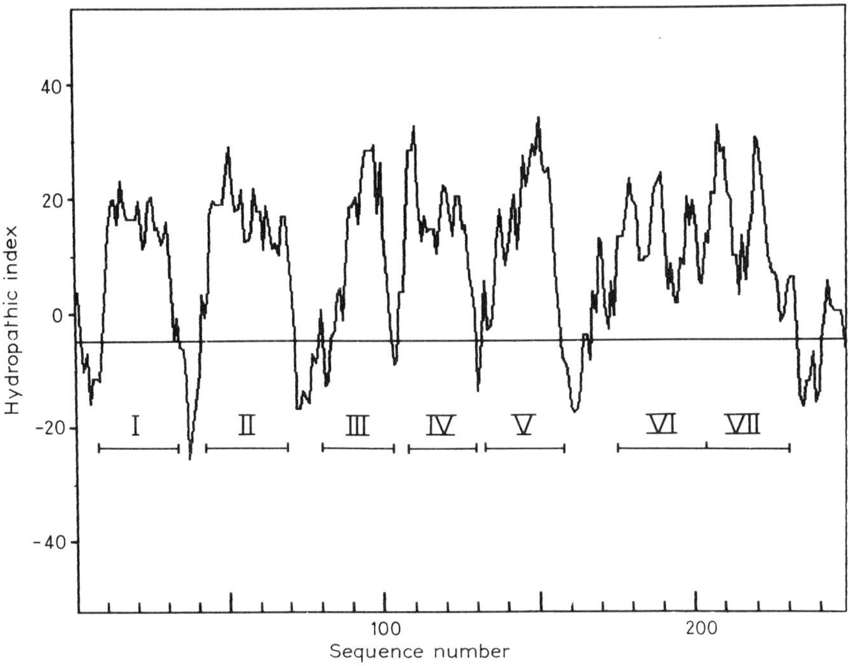


**Fig. 9.6** Hydrophobic profile of bacteriorhodopsin. The calculation was made with a span of seven amino acids using the program SOAP (Kyte and Doolittle, 1982). The bars indicate the seven hydrophobic stretches which are assumed to correspond to membrane-spanning segments.

Application of this approach to bacteriorhodopsin produces a series of segments joined by $\beta$-turns. Mostly, the segments correspond quite well to hydrophobic segments predicted by the Kyte and Doolittle method (Paul and Rosenbusch, 1985). So this procedure can be viewed as a useful complement to the hydrophobicity calculations. The prediction of $\beta$-turns depends upon a knowledge of the amino acids observed in $\beta$-turns in known structures, an approach used also for $\alpha$-helices and $\beta$-sheets by Chou and Fasman (1978). It should be noted that the Chou and Fasman method depends upon an analysis of structures of *globular* proteins and so makes falacious predictions of secondary structure when applied to membrane proteins.

The problem of predicting from the proposed helices of bacteriorhodopsin the most likely packing of the helices in the membrane has been discussed by Engleman *et al.* (1980, 1982). An important consideration in their analysis was that buried charges should be satisfied by proximal charges with opposite sign.

In order to gain insight into the possible formation of the ion channel in the acetylcholine receptor, amphipathic features of the predicted membrane-spanning segments were detected by Fourier analysis of the hydrophobicities of the proteins (Finer-Moore and Stroud, 1984). Amphipathic features are also evident in helical wheel plots (Schiffer and Edmundson, 1967).

### 9.4.2 CIS α-HELICES

So far in this discussion it has been assumed that protein domains that penetrate the lipid bilayer will span it. The reason for this assumption is that membrane proteins, in which the polypeptide chain penetrates only part way into the membrane and then exits on the same side of the membrane, necessarily require turns or bends to occur in the centre of membrane. This would expose hydrogen-bonding groups in the turn to lipid, an energetically unfavourable arrangement (Henderson, 1981).

Models of lipophilin deduced from primary sequences (Laursen *et al.*, 1984; Stoffel *et al.*, 1984) have been proposed which contain such *cis*-membrane domains. It is argued that this arrangement is partly justified by the presence of turns in the hydrophobic interiors of some globular proteins (Rose *et al.*, 1983). However, these particular turns have bound buried water to satisfy the hydrogen bonds in the buried turns. It has also been proposed that the membrane anchor of cytochrome $b_5$ has a *cis* arrangement (Takagaki *et al.*, 1983). However, the available data can also be reconciled with a different structure in which the membrane anchor forms an amphipathic helix which lies in the surface of the membrane, its hydrophobic face towards the lipid. Melittin has been shown to bind to membranes in this way (Eisenberg, 1984).

The major value of this predictive approach to secondary structural analysis of membrane proteins is that it proposes experimentally testable models. It also provides the basis for the development of more refined models such as those for bacteriorhodopsin (Engleman *et al.*, 1980, 1982) or cytochrome *b* (Saraste, 1984) which incorporate biochemical as well as primary sequence information. The most generally relevant biochemical information is of two kinds: firstly that which defines the exposed regions of the membrane protein, and secondly, that which defines the regions of the polypeptide that are within the lipid bilayer. The use of proteolytic enzymes for defining exposed regions has already been described (Section 9.2.3.a.i); the following section describes chemical labelling studies of topography of membrane proteins.

### 9.4.3 CHEMICAL STUDIES OF TOPOGRAPHY

Surface labelling was introduced to study the red cell membrane. In such experiments, the intention is to chemically label only the exposed parts of membrane proteins on the side of the membrane to which the reagent is applied. It is based upon the use of chemical reagents that can label proteins but whose properties prevent them from diffusing through the lipid bilayer. This procedure was originally used by Maddy (1964) who employed a fluorescent agent, stilbene-4-acetamido-4′-thiocyanate disulphonate (SITS) to label red blood cells. Then Berg (1969) developed $^{35}$S-labelled diazobenzene sulphonate, and Bretscher (1971) used [$^{35}$S]formylmethionylsulphone methyl phosphate to investigate the erythrocyte membrane; pyridoxal phosphate was also used for the same purpose (Rifkin *et al.*, 1972). Lactoperoxidase-generated $I^+$ has also been employed, but doubt has been cast on its usefulness as extensive labelling of lipids has been reported (Bretscher, 1973).

A more recent and important development has been the introduction of photoactivatable hydrophobic or amphipathic reagents for the general labelling of the hydrophobic core of membranes (for reviews see Brunner, 1981; Bayley, 1982, 1983). The hydrophobic reagents dissolve and react with the entire lipid phase whereas the amphipathic probes react with the membrane such that only the non-polar portion of the molecule penetrates the lipid bilayer. Thus, labelling is restricted to a certain depth in the membrane. The aim is to identify polypeptide segments of integral membrane proteins that are buried within the lipid bilayer.

An example of a reagent that is finding extensive use is 3-trifluoromethyl-3-(*m*-[$^{125}$I]iodophenyl)diazirine, [$^{125}$I]TID (Brunner *et al.*, 1979; Brunner and Semenza, 1981; Frielle *et al.*, 1982; Spiess *et al.*, 1982; Jorgensen and Brunner, 1983; Stieger *et al.*, 1984; Hoppe *et al.*, 1984; Kahan and Moscarello, 1985). The basic photoreactive unit, 3-trifluoromethyl-3-phenyl-diazirine, has also been incorporated into a fatty acyl chain of a phospholipid which can be used to label from a single leaflet of the lipid bilayer.

The main relevance of this technique to this review is that once the protein has been labelled, the reacted side chains of amino acids must be identified. Once again this requires isolation, fragmentation and direct sequence analysis of the protein. The labelling information may then be used to refine secondary structure predictions.

Further information on topographical studies of membrane proteins is contained in Chapters 6 and 7 of this volume.

## 9.5 Conclusions and perspectives

An important objective of this review is to demonstrate that the difficulties of direct sequence analysis of membrane proteins can be largely circumvented by the use of a combination of direct protein sequencing and analysis of appropriate DNA fragments. Nonetheless, as emphasized both at the outset and in the preceding section some kinds of information can only be obtained by direct study of the protein. However, the enormous recent growth of primary structural information of membrane proteins is not being paralleled by an increase in secondary and tertiary structural data. Methods for predicting secondary structures of membrane proteins can provide useful information, but lack the precision and conviction of true experimental methods. Significant advances have been made recently in membrane protein structure, particularly in the development of methods for crystallization of membrane proteins. This has led to the elucidation of a high-resolution model for bacterial photosynthetic reaction centres (Deisenhofer *et al.*, 1984). It can be confidently expected that other structures will follow in the next few years and that they will greatly extend our understanding of membrane biochemistry.

## Acknowledgement

We thank R. Henderson for discussions and Miss M. Whittaker for her help in preparation of the manuscript.

## 9.6 References

Aebersold, R. H., Teplow, D. B., Hood, L. E. and Kent, S. B. H. (1986) *J. Biol. Chem.*, **261**, 4229–38.

Aitken, A., Cohen, P., Santikarn, S., Williams, D. H., Calder, A. G., Smith, A. and Klee, C. B. (1982) *FEBS Lett.*, **150**, 314–18.

Allen, G. (1977) *FEBS 11th Symp.*, **45**, 159–68.

Allen, G. (1980) *Biochem. J.*, **187**, 545–63.

Allen, G. (1981) *Sequencing of Proteins and Peptides* (eds T. S. Work and R. H. Burdon), North-Holland, Amsterdam, New York and Oxford, pp. 266–7.

Allen, G., Gurnett, L. P. and Cross, G. A. M. (1982) *J. Mol. Biol.*, **157**, 527–46.

Alt, J. and Herrmann, R. G. (1984) *Curr. Genet.*, **8**, 551–7.

Alt, J., Morris, J., Werlhoff, P. and Herrmann, R. G. (1984) *Curr. Genet.*, **8**, 597–606.

Alt, J., Winter, P., Sebald, W., Moser, J. G., Schedel, R., Westhoff, P. and Herrmann, R. G. (1983) *Curr. Genet.*, **7**, 129–33.

Ambler, R. (1972) *Methods Enzymol.*, **25**, 143–54.

Anderson, S., Bankier, A. T., Barrell, B. G., de Bruijn, M. H. L., Coulson, A. R., Drouin, J., Eperon, I. C., Nierlich, D. P., Roe, B. A., Sanger, F., Schreier, P. H., Smith, A. J. H., Staden, R. and Young, I. G. (1981) *Nature (London)*, **290**, 457–65.

Anderson, S., de Bruijn, M. H. L., Coulson, A. R., Eperon, I. C., Sanger, F. and Young, I. G. (1982) *J. Mol. Biol.*, **156**, 683–717.

Anderson, S. and Kingston, I. B. (1983) *Proc. Natl. Acad. Sci. USA*, **80**, 6838–42.

Aquila, H., Link, T. A. and Klingenberg, M. (1985) *EMBO J.*, **4**, 2369–76.

Aquila, H., Misra, D., Eulitz, M. and Klingenberg, M. (1982) *Hoppe Seyler's Z. Physiol. Chem.*, **363**, 345–9.

Armstrong, J., Nieman, H., Smeekens, S., Rottier, P. and Warren, G. (1984) *Nature (London)*, **308**, 751–2.

Arrends, H. and Sebald, W. (1984) *EMBO J.*, **3**, 377–82.

Baer, R., Bankier, A. T., Biggin, M. D., Deininger, P. L., Farrell, P. J., Gibson, T. J., Hatfull, G., Hudson, G. S., Satchwell, S. C., Seguin, C., Tuffnell, P. S. and Barrell, B. G. (1984) *Nature (London)*, **310**, 207–11.

Baker, A. and Leaver, C. J. (1985) *Nucleic Acids Res.*, **13**, 5857–67.

Bankier, A. and Barrell, B. G. (1983) In *Techniques in Nucleic Acid Biochemistry* (ed R. A. Flavell), Elsevier Scientific Publishers, Amsterdam, pp. B508/1-B508/31.

Barber, M., Bordoli, R. S., Sedgwick, R. D. and Tyler, A. N. (1981) *Nature (London)*, **293**, 270–5.
Bayley, H. (1982) *Membr. Transp.*, **1**, 185–94.
Bayley, H. (1983) *Photogenerated Reagents in Biochemistry and Molecular Biology*, Elsevier Biomedical Press, Amsterdam, pp. 1–187.
Berg, H. C. (1969) *Biochim. Biophys. Acta*, **183**, 65–78.
Bhown, A. S., Mole, J. E., Hunter, F. and Bennett, C. J. (1980) *Anal. Biochem.*, **103**, 184–90.
Bibb, M. J., van Etten, R. A., Wright, C. T., Walberg, M. W. and Clayton, D. A. (1981) *Cell*, **26**, 167–80.
Biewald, R. and Buse, G. (1982) *Hoppe-Seylers Z. Physiol. Chem.*, **363**, 1141–53.
Biggin, M. D., Gibson, T. J. and Hong, G. F. (1983) *Proc. Natl. Acad. Sci. USA*, **80**, 3963–5.
Bird, C. R., Koller, B., Auffret, A. D., Huttly, A. K., Howe, C. J., Dyer, T. A. and Gray, J. C. (1985) *EMBO J.*, **4**, 1381–8.
Blok, J. and Air, G. M. (1982) *Biochemistry*, **21**, 4001–7.
Blondin, G. A. (1979a) *Biochem. Biophys. Res. Commun.*, **87**, 1087–94.
Blondin, G. A. (1979b) *Biochem. Biophys. Res. Commun.*, **90**, 355–61.
Bonitz, S. G., Corruzzi, G., Thalenfield, B. E., Tzagoloff, A. and Macino, G. (1980) *J. Biol. Chem.*, **255**, 11927–41.
Boothroyd, J. C., Paynter, A., Coleman, S. L. and Cross, G. A. M. (1982) *J. Mol. Biol.*, **157**, 547–56.
Braunitzer, G. and Pfletschinger, J. (1978) *Hoppe Seyler's Z. Physiol. Chem.*, **359**, 1015–18.
Braunitzer, G., Schrank, B. and Ruhfus, A. (1970) *Hoppe Seyler's Z. Physiol. Chem.*, **351**, 1589–90.
Braunitzer, G., Schrank, B., Ruhfus, A., Petersen, S. and Petersen, U. (1971) *Hoppe Seyler's Z. Physiol. Chem.*, **352**, 1730–2.
Bretscher, M. S. (1971) *J. Mol. Biol.*, **58**, 775–81.
Bretscher, M. S. (1973) *Science*, **181**, 622–9.
Brett, M. and Findlay, J. B. C. (1983) *Biochem. J.*, **211**, 661–70.
Brock, C. J., Tanner, M. J. A. and Kempf, C. (1983) *Biochem. J.*, **213**, 577–86.
Brown, T. A., Ray, J. A., Waring, R. B., Scazzocchio, C. and Davies, R. W. (1984) *Curr Genet.*, **8**, 489–92.
Brunner, J. (1981) *Trends Biochem. Sci.*, **6**, 44–6.
Brunner, J., Hauser, H., Braun, G., Wilson, K. J., Wacker, H., O'Neill, B. and Semenza, G. (1979) *J. Biol. Chem.*, **254**, 1821–8.
Brunner, J. and Semenza, G. (1981) *Biochemistry*, **20**, 7174–82.
Brunisholz, R. A., Cuendet, P. A., Theiler, R. and Zuber, H. (1981) *FEBS Lett.*, **129**, 150–4.
Brunisholz, R. A., Jay, F., Suter, F. and Zuber, H. (1985) *Biol. Chem. Hoppe-Seyler*, **366**, 87–98.
Brunisholz, R. A., Suter, F. and Zuber, H. (1984a) *Hoppe-Seyler's Z. Physiol. Chem.*, **365**, 675–88.
Brunisholz, R. A., Wiemken, V., Suter, F., Bachofen, R. and Zuber, H. (1984b) *Hoppe-Seyler's Z. Physiol. Chem.*, **365**, 689–701.
Büchel, D. E., Gronenborn, B. and Müller-Hill, B. (1980) *Nature (London)*, **283**, 541–5.

Buss, J. E. and Sefton, B. M. (1985a) *J. Virol.*, **53**, 7–12.
Buss, J. E. and Sefton, B. M. (1985b) *Biochemistry*, **20**, 7174–82.
Capony, J. P., Rinaldi, M. L., Guilleux, F. and Demaille, J. G. (1983) *Biochim. Biophys. Acta*, **728**, 83–91.
Carr, S. A., Biemann, K., Shoji, S., Parmelee, D. C. and Titani, K. (1982) *Proc. Natl. Acad. Sci. USA*, **79**, 6128–31.
Caruthers, M. H. (1985) *Science*, **230**, 281–5.
Cashmore, A. R. (1984) *Proc. Natl. Acad. Sci.*, **81**, 2960–4.
Cattell, K. J., Knight, I. G., Lindop, C. R. and Beechey, R. B. (1970) *Biochem. J.*, **117**, 1011–13.
Cattell, K. J., Lindop, C. R., Knight, I. G. and Beechey, R. B. (1971) *Biochem. J.*, **125**, 169–77.
Chen, R., Krämer, C., Schmidmayr, W., Chen-Schmeisser, U. and Henning, U. (1982) *Biochem. J.*, **203**, 33–43.
Chomyn, A., Mariottini, P., Cleeter, M. W. J., Ragan, C. I., Matsuno-Yagi, A., Hatefi, Y., Doolittle, R. F. and Attardi, G. (1985) *Nature (London)*, **314**, 529–97.
Chou, P. Y. and Fasman, G. D. (1978) *Annu. Rev. Biochem.*, **47**, 251–76.
Clark, M. J., Gagnon, J., Williams, A. F. and Barclay, A. N. (1985) *EMBO J.*, **4**, 113–18.
Claudio, T., Ballivet, M., Patrick, J. and Heinemann, S. (1983) *Proc. Natl. Acad. Sci. USA*, **80**, 1111–15.
Clement, J. M. and Hofnung, M. (1981) *Cell*, **27**, 507–18.
Cleveland, D. W., Fischer, S. G., Kirschener, M. W. and Laemmli, U. K. (1977) *J. Biol. Chem.*, **252**, 1102–6.
Cole, S. T., Grundström, T., Jaurin, B., Robinson, J. J. and Weiner, J. H. (1982) *Eur. J. Biochem.*, **126**, 211–16.
Coligan, J. E., Kindt, T. J., Ewenstein, B. M., Uehara, H., Nisizawa, T. and Nathenson, S. G. (1978) *Proc. Natl. Acad. Sci. USA*, **75**, 3390–4.
Collins, J. H., Kramias, E. G., Reeves, A. S., Bilezikjian, L. M. and Schwartz, A. (1981) *Biochem. Biophys. Res. Commun.*, **99**, 796–803.
Corruzzi, G., Broglie, R., Cashmore, A. and Chua, N-H. (1983) *J. Biol. Chem.*, **258**, 1399–402.
Corruzzi, G. and Tzagoloff, A. (1979) *J. Biol. Chem.*, **254**, 9324–30.
Cosman, D., Gerretti, D. P., Larsen, A., Park, L., March, C., Dower, S., Gillis, S. and Urdel, D. (1984) *Nature (London)*, **312**, 768–71.
Cozens, A. L., Walker, J. E., Phillips, A. L., Huttly, A. K. and Gray, J. C. (1986) *EMBO J.*, **5**, 217–22.
Darlison, M. G. and Guest, J. R. (1984) *Biochem. J.*, **223**, 507–17.
Dawson, A. J., Jones, V. P. and Leaver, C. J. (1984) *EMBO J.*, **3**, 2107–13.
de Bruijn, M. H. L. (1983) *Nature (London)*, **304**, 234–41.
Dean, G. E., MacNab, R. M., Stader, J., Matsumara, P. and Burks, C. (1984) *J. Bacteriol.*, **159**, 991–9.
Deisenhofer, J., Epp, O., Miki, K., Huber, R. and Michel, H. (1984) *J. Mol. Biol.*, **180**, 385–98.
Devillers-Thiery, A., Giraudet, J., Bentaboulet, M. and Changeux, J-P. (1983) *Proc. Natl. Acad. Sci. USA*, **80**, 2067–71.
Drickamer, K., Mamon, J. F., Binns, G. and Leung, J. O. (1984) *J. Biol. Chem.*, **259**, 770–8.

Dunn, R., McCoy, J., Simsek, M., Majumdar, A., Chang, S. H., Rajbhandary, U. L. and Khorana, H. G. (1981) *Proc. Natl. Acad. Sci. USA*, **78**, 6744–8.
Edman, P. and Begg, G. (1967) *Eur. J. Biochem.*, **I**, 80–91.
Edman, P. and Henschen, A. (1975) in *Protein Sequence Determination* (ed S. B. Needleman), Springer-Verlag, Berlin, Heidelberg and New York, pp. 232–79.
Eisenberg, D. (1984) *Annu. Rev. Biochem.*, **53**, 595–623.
Ellingboe, J., Nyström, E. and Sjövall, J. (1969) *Methods Enzymol.*, **14**, 317–29.
Engelman, D. M., Goldman, A. and Steitz, T. A. (1982) *Methods Enzymol.*, **88**, 81–8.
Engelman, D. M., Henderson, R., McLachlan, A. D. and Wallace, B. A. (1980) *Proc. Natl. Acad. Sci. USA*, **77**, 2023–7.
Erdweg, M. and Buse, G. (1985) *Biol. Chem. Hoppe Seyler*, **366**, 257–63.
Ferguson, M. A. J., Haldar, K. and Cross, G. A. M. (1985) *J. Biol. Chem.*, **260**, 4963–8.
Fields, S., Winter, G. and Brownlee, G. G. (1981) *Nature (London)*, **290**, 213–17.
Fillingame, R. H. (1976) *J. Biol. Chem.*, **251**, 6630–7.
Finer-Moore, J. and Stroud, R. M. (1984) *Proc. Natl. Acad. Sci. USA*, **81**, 155–9.
Folch, J. and Lees, M. (1951) *J. Biol. Chem.*, **191**, 807–17.
Foster, J. A., Bruenger, G. L., Hu, C. L., Albertson, K. and Franzblau, C. (1973) *Biochem. Biophys. Res. Commun.*, **53**, 70–4.
Fox, T. D. (1979) *Proc. Natl. Acad. Sci. USA*, **76**, 6534–8.
Fox, T. D. and Leaver, C. J. (1981) *Cell*, **26**, 315–23.
Frank, H., Chaves das Neves, H. J. and Bayer, E. (1978) *J. Chromatogr.*, **152**, 357–62.
Frank, H. and Desiderio, D. M. (1978) *Anal. Biochem.*, **90**, 413–19.
Frielle, T., Brunner, J. and Curthoys, N. P. (1982) *J. Biol. Chem.*, **257**, 14979–82.
Froshauer, S. and Beckwith, J. (1984) *J. Biol. Chem.*, **259**, 10896–903.
Garber, E. A., Krueger, J. G., Hanfusa, H. and Goldberg, A. R. (1983) *Nature (London)*, **302**, 161–3.
Garoff, H., Frischauf, A-M., Simons, K., Lehrach, H. and Delius, H. (1980) *Nature (London)*, **288**, 236–41.
Gay, N. J. and Walker, J. E. (1981) *Nucleic Acids Res.*, **9**, 3919–26.
Gay, N. J. and Walker, J. E. (1985) *EMBO J.*, **4**, 3519–24.
Gerber, G. E., Anderegg, R. J., Herlihy, W. C., Gray, C. P., Biemann, K. and Khorana, H. G. (1979) *Proc. Natl. Acad. Sci. USA*, **76**, 227–31.
Gething, M. J., Bye, J., Skehel, J. and Waterfield, M. (1980) *Nature (London)*, **287**, 301–6.
Gibson, B. W. and Biemann, K. (1984) *Proc. Natl. Acad. Sci. USA*, **81**, 1956–60.
Graf, T. and Sebald, W. (1978) *FEBS Lett.*, **94**, 218–22.
Green, N. M. and Toms, E. J. (1985) *Biochem. J.*, **231**, 425–9.
Grisi, E., Brown, T. A., Waring, R. B., Scazzocchio, C. and Davies, R. W. (1982) *Nucleic Acids Res.*, **10**, 3531–9.
Grundström, T. and Jaurin, B. (1982) *Proc. Natl. Acad. Sci. USA*, **79**, 1111–15.
Haldar, K., Ferguson, M. A. J. and Cross, G. A. M. (1985) *J. Biol. Chem.*, **260**, 4969–74.
Hantke, K. and Braun, V. (1979) *Eur. J. Biochem.*, **34**, 284–96.
Hargrave, D. A., Fong, S. L., McDowell, J. H., Mas, M. T., Curtis, D. R., Wang, J. K., Siemiatkowski-Juszczak, E. and Smith, D. P. (1980) *Neurochem. Int.*, **1**, 231–44.
Hayes, P. K., Walsby, A. E. and Walker, J. E. (1986) *Biochem. J.*, **236**, 31–36.
Hedrick, S. M., Nielsen, E. A., Kavaler, J., Cohen, D. I. and Davis, M. M. (1984) *Nature (London)*, **308**, 153–8.
Heinemeyer, W., Alt, J. and Hermann, R. G. (1984) *Curr. Genet.*, **8**, 543–9.
Helenius, A. and Simons, K. (1975) *Biochim. Biophys. Acta*, **415**, 29–79.

Henderson, L. E., Krutzsch, H. C. and Oroszlan, S. (1983) *Proc. Natl. Acad. Sci. USA*, **80**, 339–43.

Henderson, R. (1981) in *Membranes and Intracellular Communication* (eds R. Balian *et al.*), North-Holland, Amsterdam, pp. 232–49.

Hesse, J. E., Wieczorek, L., Altendorf, K., Reicin, A. S., Dorus, E. and Epstein, W. (1984) *Proc. Natl. Acad. Sci. USA*, **81**, 4746–50.

Hewick, R. M., Hunkapiller, M. W., Hood, L. E. and Dreyer, W. J. (1981) *J. Biol. Chem.*, **256**, 7990–7.

Higgins, C. F., Haag, P. D., Nikaido, K., Ardeshiri, F., Garcia, G. and Gerro-Luzzi-Ames, G. (1982) *Nature (London)*, **298**, 723–7.

Higgins, R. C. and Dahmus, M. E. (1979) *Anal. Biochem.*, **93**, 257–60.

Hjelmel, L. M. and Chrambach, A. (1984) *Methods Enzymol.*, **104**, 305–47.

Holder, A. A., Lockyer, M. J., Odink, K. G., Sandhu, J. S., Riveros-Moreno, V., Nicholls, S. C., Hillman, Y., Davey, L. S., Tizard, M. L. V., Schwartz, R. T. and Freeman, R. R. (1985) *Nature*, **317**, 270–3.

Hoppe, J., Brunner, J. and Jorgensen, B. B. (1984) *Biochemistry*, **23**, 5610–16.

Hoppe, J. and Sebald, W. (1980) *Eur. J. Biochem.*, **107**, 57–65.

Horn, M. and Laursen, R. A. (1973) *FEBS Lett.*, **36**, 285–9.

Howe, C. J., Auffret, A. D., Doherty, A., Bowman, C. M., Dyer, T. A. and Gray, J. C. (1982) *Proc. Natl. Acad. Sci. USA*, **79**, 6903–7.

Hudson, G. S., Farrell, P. J. and Barrell, B. G. (1985) *J. Virol.*, **53**, 528–35.

Hunkapiller, M. W. and Hood, L. E. (1980) *Science*, **207**, 523–5.

Hunkapiller, M. W., Lujan, E., Ostrander, F. and Hood, L. E. (1983a) *Methods Enzymol.*, **91**, 227–36.

Hunkapiller, M. W., Hewick, R. M., Dreyer, W. J. and Hood, L. E. (1983b) *Methods Enzymol.*, **91**, 399–13.

Inokuchi, K., Mutoh, N., Matsuyama, S. and Mizushima, S. (1982) *Nucleic Acids Res.*, **10**, 6957–68.

Jaiswal, A. K., Gonzalez, F. J. and Nebert, D. W. (1985) *Nucleic Acids Res.*, **13**, 4503–20.

Jollès, J., Schoentgen, F., Jollès, P., Vacher, M., Nicot, C. and Alfsen, A. (1979) *Biochem. Biophys. Res. Commun.*, **87**, 619–26.

Jorgensen, P. and Brunner, J. (1983) *Biochim. Biophys. Acta*, **735**, 291–6.

Kahan, I. and Moscarello, M. A. (1985) *Biochemistry*, **24**, 538–44.

Kaufman, J. F., Auffrey, C., Korman, A. J., Shackelford, D. A. and Strominger, J. (1984) *Cell*, **36**, 1–13.

Kawakami, K., Noguchi, S., Noda, M., Takahashi, H., Ohta, T., Kawamura, M., Nojima, H., Hirose, T., Inayama, S., Hayashida, H., Miyata, T. and Numa, S. (1985) *Nature (London)*, **316**, 733–6.

Kelley, J. A., Nau, N., Forster, H. J. and Biemann, K. (1975) *Biomed. Mass Spectrom.*, **2**, 313–25.

Khorana, H. G., Gerber, G. E., Herlihy, W. C., Gray, C. P., Anderegg, R. J., Nihei, K. and Biemann, K. (1979) *Proc. Natl. Acad. Sci. USA*, **76**, 5046–50.

Kimura, S., Gonzalez, F. J. and Nebert, D. W. (1984) *J. Biol. Chem.*, **259**, 10705–13.

Klapper, D. G., Wilde, C. E. and Capra, J. D. (1978) *Anal. Biochem.*, **85**, 126–31.

Kleffel, B., Garavito, R. M., Baumeister, W. and Rosenbusch, J. P. (1985) *EMBO J.*, **4**, 1589–92.

Kolata, G. (1985) *Science*, **229**, 850.

Kondo, K., Tajima, S., Sato, R. and Narita, K. (1979) *J. Biochem. (Tokyo)*, **86**, 1119–28.

Kopito, R. R. and Lodish, H. F. (1985) *Nature (London)*, **316**, 234–8.

Kubo, T., Noda, M., Takai, T., Tanabe, T., Kayano, T., Shimizu, S., Tanaka, K., Takahishi, H., Hirose, T., Inayama, S., Kikuno, R., Miyata, T. and Numa, S. (1985) *Eur. J. Biochem.*, **149**, 5–13.

Kyte, J. and Doolittle, R. F. (1982) *J. Mol. Biol.*, **157**, 105–32.

Laursen, R. A. (1971) *Eur. J. Biochem.*, **20**, 89–102.

Laursen, R. A., Bonner, A. G. and Horn, M. J. (1975) in *Instrumentation in Amino Acid Sequence Determination* (ed R. N. Perham), Academic Press, London, New York and San Francisco, pp. 73–110.

Laursen, R. A., Samiullah, M. and Lees, M. B. (1984) *Proc. Natl. Acad. Sci. USA*, **81**, 2912–16.

Lees, M. B., Chao, B. H., Leu-Fen, H. L., Samiullah, M. and Laursen, R. A. (1983) *Arch. Biochem. Biophys.*, **226**, 643–56.

Lemke, G. and Axel, R. (1985) *Cell*, **40**, 501–8.

Macino, G. and Tzagoloff, A. (1979) *Proc. Natl. Acad. Sci. USA*, **76**, 131–5.

Macino, G. and Tzagoloff, A. (1980) *Cell*, **20**, 507–17.

MacLennan, D. H., Brandl, C. J., Korczak, B. and Green, N. M. (1985) *Nature (London)*, **316**, 696–700.

Macreadie, I. G., Novitski, C. E., Maxwell, R. J., John, U., Ooi, B. G., McMullen, G. L., Lukins, H. B., Linnane, A. W. and Nagley, P. (1983) *Nucleic Acids Res.*, **11**, 4435–51.

Maddy, A. H. (1964) *Biochim. Biophys. Acta*, **88**, 390–9.

Maniatis, T., Fritsch, E. F. and Sambrook, J. (1982) *Molecular Cloning: A Laboratory Manual*. Cold Spring Harbor.

Mansfield, R. W. and Anderson, J. M. (1985) *Biochim. Biophys. Acta*, **809**, 435–44.

Martin, B., Svendsen, I. and Ottesen, M. (1977) *Carlsberg Res. Commun.*, **42**, 99–102.

Matsumoto, M., Matsumoto, R. and Folch-Pi, J. (1964) *J. Neurochem.*, **11**, 829–38.

Maxam, A. M. and Gilbert, W. (1977) *Proc. Natl. Acad. Sci. USA*, **74**, 560–64.

McConnell, D. G., Kohnker, R. E. and Smith, A. J. (1984) *Fed. Proc. Fed. Am. Soc. Exp. Biol.*, **43**, 1585.

Mihara, K. and Sato, R. (1985) *EMBO J.*, **4**, 769–74.

Mizuno, T., Chou, M-Y. and Inouye, M. (1983) *J. Biol. Chem.*, **258**, 6932–40.

Mokrasch, L. C. (1967) *Life Sci.*, **6**, 1905–9.

Morohashi, K., Fujii-Kuriyama, Y., Okada, Y., Sogana, K., Hirose, T., Inayama, S. and Omura, T. (1984) *Proc. Natl. Acad. Sci. USA*, **81**, 4647–51.

Morris, H. R. and Dell, A. (1975) in *Amino Acid Sequence Determination* (ed R. N. Perham), Academic Press, London, New York and San Francisco, pp. 147–91.

Morris, H. R., Williams, D. H. and Ambler, R. P. (1971) *Biochem. J.*, **125**, 189–201.

Morris, J. and Herrmann, R. G. (1984) *Nucleic Acids Res.*, **12**, 2837–49.

Mostov, K. E., Friedlander, M. and Blobel, G. (1984) *Nature (London)*, **308**, 37–43.

Mueckler, M., Caruso, C., Baldwin, S. A., Panico, M., Blench, I., Morris, H. R., Allard, W. J., Lienhard, G. E. and Lodish, H. F. (1985) *Science*, **229**, 941–5.

Mullen, E. and Akhtar, M. (1981) *FEBS Lett.*, **132**, 261–4.

Nathans, J. and Hogness, D. S. (1984) *Proc. Natl. Acad. Sci. USA*, **81**, 4851–5.

Nau, H. (1976) *Angew. Chem. (Int. Edn)*, **15**, 75–86.

Nau, H. and Biemann, K. (1976) *Anal. Biochem.*, **73**, 139–86.

Nelles, L. P. and Bamburg, J. R. (1976) *Anal. Biochem.*, **73**, 522–31.

Nelson, D. R. and Robinson, N. C. (1983) *Methods Enzymol.*, **97**, 571–618.
Nobrega, F. G. and Tzagoloff, A. (1980) *J. Biol. Chem.*, **255**, 9828–37.
Noda, M., Furutani, Y., Takahashi, H., Toyosato, M., Tanabe, T., Shimizu, S., Kikyotani, S., Kayano, T., Hirose, T., Inayama, S. and Numa, S. (1983c) *Nature (London)*, **305**, 818–23.
Noda, M., Shimizu, S., Tanabe, T., Takei, T., Kayano, T., Ikeda, T., Takahashi, H., Nakayama, H., Kanaoka, Y., Minamino, N., Kangawa, K., Matsuo, H., Raftery, M. A., Hirose, T., Inayama, S., Hayashida, H., Miyata, T. and Numa, S. (1984) *Nature (London)*, **312**, 121–7.
Noda, M., Takahashi, H., Tanabe, T., Toyosato, M., Furutani, Y., Hirose, T., Asai, M., Inayama, S., Miyata, T. and Numa, S. (1982) *Nature (London)*, **299**, 793–7.
Noda, M., Takahashi, H., Tanabe, T., Toyosato, M., Kikyotani, S., Furutani, Y., Hirose, T., Takashima, H., Inayama, S., Miyata, T. and Numa, S. (1983b) *Nature (London)*, **302**, 528–31.
Noda, M., Takahashi, H., Tanabe, T., Toyosato, M., Kikyotani, S., Hirose, T., Asai, M., Takashima, H., Inayama, S., Miyata, T. and Numa, S. (1983a) *Nature (London)*, **301**, 251–5.
Omary, M. B. and Trowbridge, I. S. (1981) *J. Biol. Chem.*, **256**, 4715–18.
Ovchinnikov, Yu. A., Abdulaev, N. G., Feigina, M. Yu., Artamonov, I. D., Bogachuk, A. S., Zolotarev, A. S., Eganyan, E. R. and Kostetsky, P. V. (1983) *Bioorg. Khim.*, **9**, 1331–40.
Ovchinnikov, Yu. A., Abdulaev, N. G., Feigina, M. Yu, Kiselev, A. V. and Lobanov, N. A. (1979) *FEBS Lett.*, **100**, 219–24.
Overbeeke, N., Bergmans, H., van Mansfield, F. and Lugtenberg, B. (1983) *J. Mol. Biol.*, **163**, 513–32.
Ozols, J., Carr, S. A. and Strittmatter, P. (1984) *J. Biol. Chem.*, **259**, 13349–54.
Ozols, J. and Heinemann, F. S. (1982) *Biochim. Biophys. Acta*, **704**, 163–73.
Paul, C. and Rosenbusch, J. P. (1985) *EMBO J.*, **4**, 1593–7.
Pellman, D., Garber, E. A., Cross, F. R. and Hanafusa, H. (1985) *Nature (London)*, **314**, 374–7.
Phillips, A. L. and Gray, J. C. (1984) *Mol. Gen. Genet.*, **194**, 477–84.
Ploegh, H. L., Orr, H. T. and Strominger, J. (1981) *Cell*, **24**, 287–99.
Pober, J. S. and Stryer, L. (1975) *J. Mol. Biol.*, **95**, 477–81.
Porter, A. G., Barber, C., Carey, N. H., Hallewell, R. A., Threlfall, G. and Emtage, J. S. (1979) *Nature (London)*, **282**, 471–7.
Reeves, A. S., Collins, J. H. and Schwartz, A. (1980) *Biochem. Biophys. Res. Commun.*, **95**, 1591–8.
Rifkin, D. B., Compans, R. W. and Reich, E. (1972) *J. Biol. Chem.*, **247**, 6432–7.
Rogers, J., Early, P., Carter, C., Calame, K., Bond, M., Hood, L. and Wall, R. (1980) *Cell*, **20**, 303–12.
Rose, G. D., Gierasch, L. M. and Smith, J. A. (1985) *Adv. Protein Chem.*, **37**, 1–109.
Rose, G. D., Young, W. B. and Gierasch, L. M. (1983) *Nature (London)* **304**, 654–7.
Rose, J. K., Welch, W. J., Sefton, B. M., Esch, F. S. and Ling, N. C. (1980) *Proc. Natl. Acad. Sci. USA*, **77**, 3884–8.
Rouser, G., Kritchevsky, G., Yamamoto, A., Simon, G., Galli, L. and Bauman, A. J. (1969) *Methods Enzymol.*, **14**, 272–317.
Rubinstein, M. (1979) *Anal. Biochem.*, **98**, 1–7.

Rubinstein, M., Stein, S. and Udenfriend, S. (1978) *Proc. Natl. Acad. Sci. USA*, **75**, 669–71.

Sadler, I., Sadu, K., Schatz, G., Kaudewitz, F. and Haid, A. (1984) *EMBO J.*, **3**, 2137–43.

Sale, G. J., Towner, P. and Akhtar, M. (1977) *Biochemistry*, **16**, 5641–8.

Sanderman, H. and Strominger, J. L. (1972) *J. Biol. Chem.*, **247**, 5123–31.

Sanger, F., Nicklen, S. and Coulson, A. R. (1977) *Proc. Natl. Acad. Sci. USA*, **74**, 5463–7.

Sanger, F., Air, G. M., Barrell, B. G., Brown, N. L., Coulson, A. R., Fiddes, J. C., Hutchinson III, C. A., Slocombe, P. M. and Smith, M. (1979) *Nature (London)*, **265**, 687–95.

Saraste, M. (1984) *FEBS Lett.*, **166**, 367–72.

Saraste, M. and Walker, J. E. (1982) *FEBS Lett.*, **144**, 250–4.

Savige, W. E. and Fontana, A. (1977) *Methods Enzymol.*, **47**, 459–69.

Schiffer, M. and Edmundson, A. B. (1967) *Biophys. J.*, **7**, 121–35.

Schneider, C., Owen, M. J., Banville, D. and Williams, J. G. (1984) *Nature (London)*, **311**, 675–8.

Schultz, A. M., Henderson, L. E., Oroszlan, S., Garber, E. A. and Hanfusa, S. (1985) *Science*, **227**, 427–9.

Schultz, A. M. and Oroszlan, S. (1984) *Virology*, **133**, 431–7.

Sebald, W. and Hoppe, J. (1981) *Curr. Top. Bioenerget.*, **12**, 1–64.

Sebald, W., Graf, T. and Lukins, H. B. (1979) *Eur. J. Biochem.*, **93**, 587–99.

Sefton, B. M., Trowbridge, I. S., Cooper, J. A. and Scolnick, E. M. (1982) *Cell*, **31**, 465–74.

Segrest, J. P. and Feldmann, R. J. (1974) *J. Mol. Biol.*, **87**, 853–8.

Segrest, J. P., Jackson, R. L., Marchesi, V. T., Guyer, R. B. and Terry, W. (1972) *Biochem. Biophys. Res. Commun.*, **49**, 964–9.

Seki, T., Chang, H. C., Moriuchi, T., Denome, T., Ploegh, H. and Silver, J. (1985) *Science*, **227**, 649–51.

Shaw, D. C., Walker, J. E., Northrop, F. D., Barrell, B. G., Godson, G. N. and Fiddes, J. C. (1978) *Nature (London)*, **272**, 510–15.

Shibahara, S., Kubo, T., Perski, H. J. Takahashi, H., Noda, M. and Numa, S. (1985) *Eur. J. Biochem.*, **146**, 15–22.

Shull, G. E., Schwartz, A. and Lingrel, J. B. (1985) *Nature (London)*, **316**, 691–5.

Sim, G. K., Vagüe, J., Nelson, J., Marrack, P., Palmer, E., Augustin, A. and Kappler, J. (1984) *Nature (London)*, **312**, 771–5.

Smith, M. (1983) in *Methods of DNA and RNA Sequencing* (ed S. M. Weismann), Praeger, New York, pp. 23–68.

Soto, E. F., Pasquiri, J. M., Placido, R. and La Torre, J. L. (1969) *J. Chromatogr.*, **41**, 400–9.

Spiess, M., Brunner, J. and Semenza, G. (1982) *J. Biol. Chem.*, **257**, 2370–7.

Spiess, M. and Lodish, H. F. (1985) *Proc. Natl. Acad. Sci. USA*, **82**, 6465–9.

Spiess, M., Schwartz, A. L. and Lodish, H. F. (1985) *J. Biol. Chem.*, **260**, 1979–82.

Stearne, P. A., Pietersz, G. A. and Goding, J. W. (1985) *J. Immunol.*, **134**, 3474–9.

Steffens, G. and Buse, G. (1976) *Hoppe-Seyler's Z. Physiol. Chem.*, **357**, 1125–37.

Steffens, G. J. and Buse, G. (1979) *Hoppe-Seyler's Z. Physiol. Chem.*, **360**, 613–19.

Steffens, K., Kiltz, H. H., Schneider, E., Schmid, R. and Altendorf, K. (1982) *FEBS Lett.*, **142**, 151–4.

Stieger, S., Brodbeck, U., Reber, B. and Brunner, J. (1984) *FEBS Lett.*, **168**, 231–4.

Stoffel, W., Giersiefen, H., Hillen, H., Schroeder, W. and Tunggaal, B. (1985) *Biol. Chem. Hoppe-Seyler*, **366**, 627–35.

Stoffel, W., Hillen, H. and Giersiefen (1984) *Proc. Natl. Acad. Sci. USA*, **81**, 5012–16.

Stoffel, W., Hillen, H., Schröder, W. and Deutzmann, R. (1982a) *Hoppe-Seyler's Z. Physiol. Chem.*, **363**, 855–64.

Stoffel, W., Hillen, H., Schröder, W. and Deutzmann, R. (1983) *Hoppe Seyler's Z. Physiol. Chem.*, **364**, 1455–66.

Stoffel, W., Schröder, W., Hillen, H. and Deutzmann (1982b) *Hoppe Seyler's Z. Physiol. Chem.*, **363**, 1117–31.

Strickler, J. E., Hunkapiller, M. W. and Wilson, K. J. (1984) *Anal. Biochem.*, **140**, 553–66.

Sudhoff, T. C., Goldstein, J. L., Brown, M. S. and Russell, D. W. (1985) *Science*, **288**, 815–22.

Sumikawa, K., Houghton, M., Smith, J. C., Richards, B. M. and Barnard, E. A. (1982) *Nucleic Acids Res.*, **10**, 5809–12.

Surin, B. P., Jans, D. A., Fimmel, A. L., Shaw, D. C., Cox, G. B. and Rosenburg, H. (1984) *J. Bacteriol.*, **157**, 772–8.

Sutton, M. R., Rosen, D., Feher, G. and Steiner, L. A. (1982) *Biochemistry*, **21**, 3842–9.

Takagaki, Y., Gerber, G. E., Nihei, K. and Khorana, H. G. (1980) *J. Biol. Chem.*, **255**, 1536–41.

Takagaki, Y., Radhakrishnan, R., Gupta, C. M. and Khorana, H. G. (1983) *J. Biol. Chem.*, **258**, 9128–35.

Takahashi, Y., Kato, K., Hayashizaki, Y., Wakabayashi, T., Ohtsuka, E., Matsuki, S., Ikehara, M. and Matsubara, K. (1985) *Proc. Natl. Acad. Sci. USA*, **82**, 1931–5.

Takei, T., Noda, M., Furutani, Y., Takahashi, H., Notaka, M., Shimizu, S., Kayano, T., Tanabe, T., Tanaka, K., Hirose, T., Inayama, S. and Numa, S. (1984) *Eur. J. Biochem.*, **143**, 109–15.

Takei, T., Noda, M., Mishina, M., Shimizu, S., Furutani, Y., Kayano, T., Ikeda, T., Kubo, T., Takahashi, H., Takahashi, T., Kuno, M. and Numa, S. (1985) *Nature*, **315**, 761–4.

Talbot, D. N. and Yphantis, D. A. (1971) *Anal. Biochem.*, **44**, 246–53.

Tanabe, T., Noda, M., Furutani, Y., Takei, T., Takahashi, H., Tanaka, K., Hirose, T., Inayama, S. and Numa, S. (1984) *Eur. J. Biochem.*, **144**, 11–17.

Tanabe, T., Nukada, T., Nishikawa, Y., Sugimoto, K., Suzuki, H., Takahashi, H., Noda, M., Hagg, T., Ichiyama, A., Kangawa, K., Minamino, N., Matsuo, H. and Numa, S. (1985) *Nature (London)*, **315**, 242–5.

Tanaka, M., Hanio, M., Yasunobo, K. T., Yu, C. A., Yu, L., Wei, Y-H. and King, T. E. (1979) *J. Biol. Chem.*, **254**, 3879–85.

Tanford, C. and Reynolds, J. A. (1976) *Biochim. Biophys. Acta*, **457**, 132–70.

Tarr, G., Beecher, J. F., Bell, M. and McKean, D. J. (1978) *Anal. Biochem.*, **84**, 622–7.

Tarr, G. E. and Crabb, J. W. (1983) *Anal. Biochem.*, **131**, 99–107.

Thalenfield, B. E. and Tzagoloff, A. (1980) *J. Biol. Chem.*, **255**, 6173–80.

Theiler, R., Suter, F. and Zuber, H. (1983) *Hoppe-Seyler's Z. Physiol. Chem.*, **364**, 1765–76.

Theiler, T., Suter, F., Wiemken, V. and Zuber, H. (1984) *Hoppe-Seyler's Z. Physiol. Chem.*, **365**, 703–19.

Tomita, M. and Marchesi, V. T. (1975) *Proc. Natl. Acad. Sci. USA*, **72**, 2964–8.
Tonn, S. J., Gogel, G. E. and Loach, P. A. (1977) *Biochemistry*, **16**, 877–85.
Towner, P., Sale, G. J. and Akhtar, M. (1977) *FEBS Lett.*, **76**, 51–5.
Tschesche, H. and Kupfer, S. (1972) *Eur. J. Biochem.*, **26**, 33–6.
Ullrich, A., Bell, J. R., Chen, E. Y., Herrera, T., Petruzzelli, L. M., Dull, T. J., Gray, A., Coussens, L., Liao, Y.-C., Tsubokawa, M., Mason, A., Seeburg, P. H., Grunfeld, C., Rosen, O. M. and Ramachandran, J. (1985) *Nature (London)*, **313**, 756–61.
Ullrich, A., Coussens, L., Hayflick, J. S., Dull, T. J., Gray, A., Tam, A. W., Lee, J., Yarden, Y., Liberman, T. A., Schlessinger, J., Downward, J., Mayes, E. L. V., Whittle, N., Waterfield, M. D. and Seeburg, P. H. (1984) *Nature (London)*, **309**, 418–25.
Unwin, N. and Henderson, R. (1984) *Sci. Am.*, **250**, 56–66.
Velours, J., Esparza, M. and Guerin, B. (1982) *Biochem. Biophys. Res. Commun.*, **109**, 1192–9.
Velours, J., Esparza, M., Hoppe, J., Sebald, W. and Guerin, B. (1984) *EMBO J.*, **3**, 207–12.
Verhoeyen, M., Fang, R., Jou, W. H., Deuos, R., Huylebroeck, D., Saman, E. and Fiers, W. (1980) *Nature (London)*, **286**, 771–6.
Viebrok, A., Perz, A. and Sebald, W. (1982) *EMBO J.*, **1**, 565–71.
Wachter, E., Hofner, H. and Macleidt, W. (1975) in *Solid Phase Methods in Protein Sequence Analysis* (ed R. A. Laursen), Pierce Chemical Co., Rockford, Illinois, pp. 31–46.
Wachter, E. and Werhahn, R. (1979) *Anal. Biochem.*, **97**, 56–64.
Wakabayashi, S., Matsubara, H., Kim, C. H., Kawa, K. and King, T. S. (1980) *Biochem. Biophys. Res. Commun.*, **97**, 1548–54.
Wakabayashi, S., Matsubara, H., Kim, C. H. and King, T. E. (1982) *J. Biol. Chem.*, **257**, 9335–44.
Walker, J. E., Auffret, A. D., Carne, A., Gurnett, A., Hanisch, P., Hill, D. and Saraste, M. (1982b) *Eur. J. Biochem.*, **123**, 253–60.
Walker, J. E., Auffret, A. D., Carne, A. F., Naughton, M. A. and Runswick, M. J. (1980) in *Methods in Peptide and Protein Sequence Analysis* (ed Ch. Birr), Elsevier/North-Holland Biomedical Press, Amsterdam, pp. 257–65.
Walker, J. E., Carne, A. F. and Schmitt, H. W. (1979) *Nature (London)* **278**, 653–4.
Walker, J. E., Eberle, A., Gay, N. J., Hanisch, P., Saraste, M. and Runswick, M. J. (1982a) in *Methods in Peptide and Protein Sequence Analysis* (ed M. Elzinga), Humana, Clifton, New Jersey, pp. 337–51.
Walker, J. E., Fearnley, I. M., Gay, N. J., Gibson, B. W., Northrop, F. D., Powell, S. J., Runswick, M. J., Saraste, M. and Tybulewicz, V. L. J. (1985) *J. Mol. Biol.*, **184**, 677–701.
Waring, R. B., Brown, T. A., Ray, J. A., Scazzocchio, C. and Davies, R. W. (1984) *EMBO J.*, **3**, 2121–8.
Waring, R. B., Davies, R. W., Lee, S., Grisi, E., Berks, M. and Scazzocchio, C. (1981) *Cell*, **27**, 4–11.
Wechsler, T., Suter, F., Fuller, C. and Zuber, H. (1985) *FEBS Lett.*, **181**, 173–8.
Wessel, D. and Flügge, U. I. (1984) *Anal. Biochem.*, **138**, 141–3.
Widger, W. R., Cramer, W. A., Herrman, R. G. and Trebst, A. (1984) *Proc. Natl. Acad. Sci. USA*, **81**, 674–8.
Willey, D. L., Auffret, A. D. and Gray, J. C. (1984a) *Cell*, **36**, 555–62.

Willey, D. W., Howe, C. H., Auffret, A. D., Bowman, C. M., Dyer, T. A. and Gray, J. C. (1984b) *Mol. Gen. Genet.*, **194**, 416–22.

Winter, G., Fields, S. and Brownlee, G. G. (1981) *Nature (London)*, **292**, 72–5.

Wood, D., Darlison, M. G., Wilde, R. J. and Guest, J. R. (1984) *Biochem. J.*, **222**, 519–34.

Wood, W. I., Gitschier, J., Lasky, L. A. and Lawn, R. M. (1985) *Proc. Natl. Acad. Sci. USA*, **82**, 1585–8.

Yabusaki, Y., Shimizu, M., Murakami, H., Nakamura, K., Oeda, K. and Ohkawa, H. (1984) *Nucleic Acids Res.*, **12**, 2929–38.

Yanagi, Y., Yoshikai, Y., Leggett, K., Clark, S. P., Aleksander, I. and Mak, T. W. (1984) *Nature (London)*, **308**, 145–9.

Young, I. G., Rogers, B. L., Campbell, H. D., Jaworowski, A. and Shaw, D. C. (1981) *Eur. J. Biochem.*, **116**, 165–70.

Youvan, D. C., Bylina, E. J., Alberti, M., Begusch, H. and Hearst, J. E. (1984) *Cell*, **37**, 949–57.

Zulauf, M. and Rosenbusch, J. P. (1983) *J. Phys. Chem.*, **87**, 856–62.

Zurawski, G., Bohnert, H. J., Whitfeld, P. R. and Bottomley, W. (1982) *Proc. Natl. Acad. Sci. USA*, **79**, 7699–703.

# 10

# Freeze–fracture and freeze–etch electron microscopy of membrane proteins

JENA K. KHODADAD, JEROME M. LOEW and RONALD S. WEINSTEIN



## 10.1 Introduction

The development of the freeze–fracture electron microscopy preparative technique (Moor *et al.*, 1961; Moor and Mühlethaler, 1963) and its successful application to biological systems (Branton and Moor, 1964; Bullivant and Ames, 1966) has revolutionized the study of biomembrane structure (Branton, 1966; Steck *et al.*, 1970; McNutt and Weinstein, 1970; Branton *et al.*, 1975). Fortuitously, biological membranes fracture along cleavage planes positioned within the membranes (Fig. 10.1). Fracture faces produced by fracturing can be replicated *in vacuo* at a level of resolution near to molecular (Slayter, 1976; Weinstein *et al.*, 1979). Such replicas faithfully represent at least some aspects of the supramolecular structure within biomembranes (Caspar *et al.*, 1977; Makowski *et al.*, 1977; Heuser and Salpeter, 1979; Weinstein *et al.*, 1980; Hirokawa and Heuser, 1982a). Using freeze–fracturing, investigators have visualized a broad spectrum of membrane morphologies. These data have served to refine and, in many instances, change earlier concepts of membrane structure (Singer and Nicolson, 1972; Khodadad and Weinstein, 1983), particularly with respect to the organization of membrane proteins (Steck, 1974, 1978).

The intent of this chapter is to survey freeze-fracture electron microscopy

Techniques for the Analysis of Membrane Proteins. Edited by C. I. Ragan & R. J. Cherry. Published in 1986 by Chapman and Hall, 11 New Fetter Lane, London EC4P 4EE

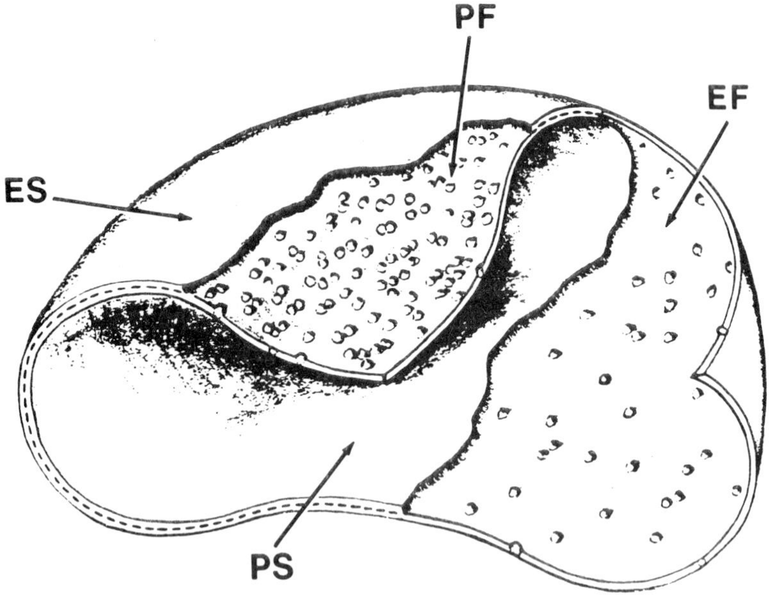


**Fig. 10.1** Diagram of a red cell membrane used to demonstrate the two natural surfaces (ES and PS) and the two fracture faces (PF and EF) represented in electron micrographs of freeze–fractured and freeze–etched membranes. Adapted from Weinstein (1974) and labelled according to the nomenclature of Branton *et al.* (1975). EF, extracellular (outer half-membrane) fracture face; ES, extracellular surface; PF, protoplasmic (inner half-membrane) fracture face; PS, protoplasmic (inner) surface.

studies of membrane proteins. Our major emphasis will be on a family of morphological entities, intramembrane particles (IMPs), which generally represent integral membrane proteins. Several studies employing freeze–fracture and deep-etch electron microscopy have examined the extracellular poles of integral membrane proteins and certain peripheral membrane proteins (Cartaud *et al.*, 1978; Weinstein *et al.*, 1978a,b; Heuser and Salpeter, 1979; Costello and Frey, 1982; Grohovaz *et al.*, 1982; Hirokawa and Heuser, 1982a,b; Roof and Heuser, 1982; Roof *et al.*, 1982).

Despite current interest in these topics, space does not permit detailed consideration of this literature.

## 10.2 Methodological considerations

Freeze–fracturing or freeze–etching is a multistep procedure (Fig. 10.2). First the specimen is rapidly frozen by immersion in liquid nitrogen or Freon-22 cooled with liquid nitrogen. In a recent modification, the specimen is cooled by slamming on to a copper block cooled with liquid helium (Heuser and

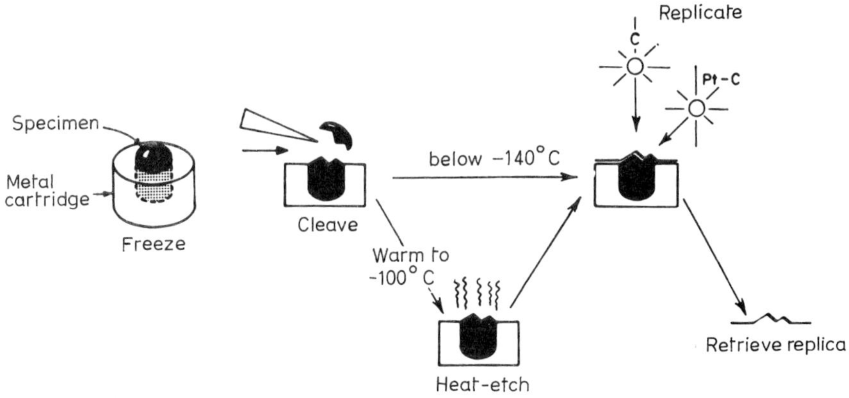


**Fig. 10.2** Diagram demonstrating steps in freeze–fracturing and freeze–etching. The specimen is held in a specimen holder for rapid freezing, then is cleaved. Freeze–etching, if desired, is achieved by warming the specimens to −100° C before replication. Once the replica is made, it is cleaned and retrieved for examination in a transmission electron microscope.

Kirschner, 1980). Rapid freezing is necessary to minimize ice crystal formation and distortion of the specimen. This necessitates small specimens (McNutt and Weinstein, 1973). The frozen specimen is fractured by one of two general methods to reveal the fracture face. Moor and co-workers (1961) developed the apparatus in which the specimen is fractured *in vacuo* with a cold knife in a microtome. The frozen cleaved surface is protected prior to replication by a combination of high vacuum and the cold knife, which remains near the specimen as a cryotrap (Rash *et al.*, 1979). The prototype of the other fracturing procedure was developed by Bullivant and Ames (1966). In this method, the specimen is fractured in liquid nitrogen, placed in a liquid nitrogen cooled brass block for transfer to a high vacuum chamber, and placed under vacuum while remaining frozen. The brass block functions as a cold trap for specimen protection. Once the specimen is fractured, etching may follow. If the specimen is warmed to −100° C, ice will sublime at the vacuums obtainable in a good commercial high vacuum apparatus. The sublimation of the ice further exposes membrane and any other non-volatile structures or contaminants (Koehler, 1972; McNutt and Weinstein, 1973). To permit visualization of the fractured, etched surface, a replica is made by evaporation of a heavy metal, usually platinum with a carbon backing for mechanical stability, on to the specimen. The specimen is then brought to room temperature, removed from the vacuum, and the tissue removed from the replica by flotation and digestion with diluted sodium hypochlorite. After cleaning with distilled water, the replica is picked up on an EM grid and examined in a transmission electron microscope.

Criteria for identifying IMP in freeze–fracture replicas have not been standardized. Unfortunately, precise criteria for IMP identification are not included in most scientific publications. Differing criteria compounded by the use of non-standardized quantitation sampling protocols and differences in technique may account, in part, for the large disparities in IMP numerical densities reported for specific cell membranes, such as that of the human erythrocyte (Pricam *et al.*, 1977; Weinstein *et al.*, 1980; Glaubensklee *et al.*, 1982).

Weinstein *et al.* (1979) have used computer-generated synthetic electron optical images of prototypic IMPs (Fig. 10.3) under an assortment of defined shadowing conditions to: (1) examine the effects of variables (i.e. replica thickness, IMP shape, etc.) on the electron optical appearances of IMP; (2) account for anomalous or unexplained images that are commonly encountered at real fracture faces when attempting to categorize every IMP-like image; and (3) assess the contributions of artifacts that are introduced during specimen preparation, freeze–fracturing, and specimen photography in the electron microscope (Misra and Das Gupta, 1965; Abermann *et al.*, 1972; Henderson and Griffiths, 1972; Koehler, 1972; Slayter, 1976). These studies have emphasized the large dimensional distortions that are inevitably

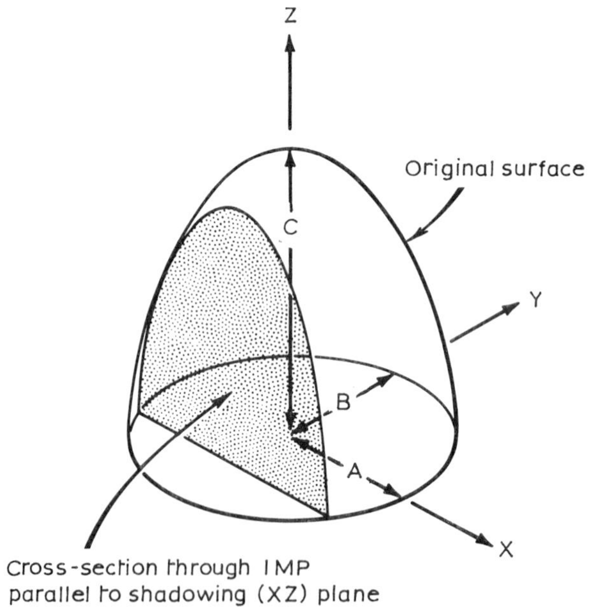


$$\frac{X^2}{A^2} - \frac{Y^2}{B^2} - \frac{Z^2}{C^2} = 1, \ Z > 0$$

**Fig. 10.3** A prototypic intramembrane particle (IMP) is illustrated schematically as an oblate hemispheroid in isometric projection. The cross-section parallel to the shadowing plane (illustrated in Fig. 10.4) is stippled. (From Weinstein *et al.*, 1979, used with permission.)

introduced by unidirectional heavy metal shadowing, a necessary step in the freeze–fracture technique (Fig. 10.4). Other distortions are inherent in rotary replication (Margaritis *et al.*, 1977; Kuszak *et al.*, 1982). Correction factors can be used to overcome some of these distortions if replica thickness is known, as is the case when a quartz film thickness analyser is used in the preparation of freeze-fracture replicas.

Using computer simulation techniques, the effects of incremental increases in replica thickness on the electron optical image of a hemispheroidal particle were examined in detail since this configuration appears to approximate the geometry of many IMPs. Four major features of the IMP image were identified with unidirectional shadowing: (1) there is an increase in metal cap length with increased replica thickness; (2) the relative shadow remains unchanged in area and shape but the absolute shadow decreases progressively in length;

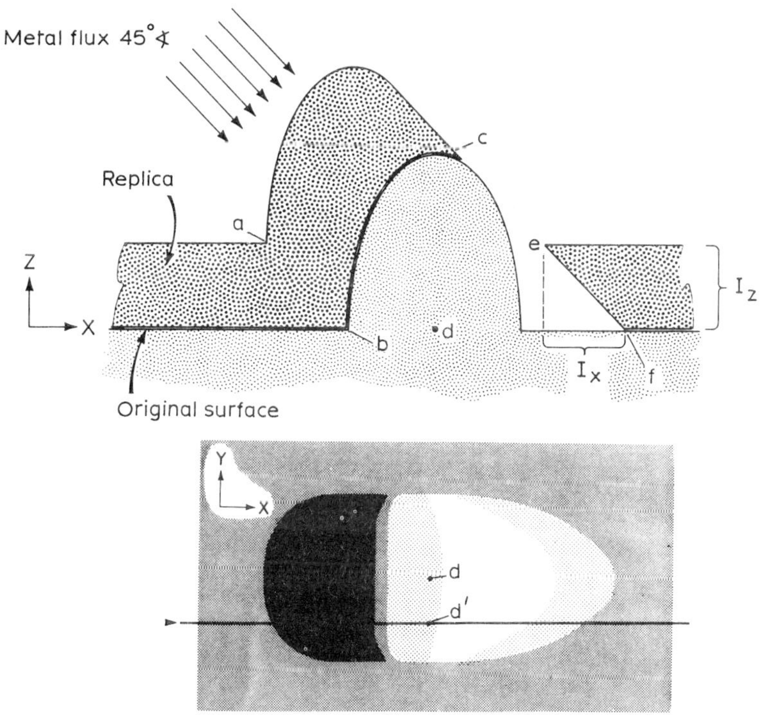


**Fig. 10.4** Top: Schematic representation of an eccentric axial cross-section of a unidirectionally shadowed hemispheroidal IMP in the *x–y* plane. Bottom: Simulated image of the replica of the above IMP, viewed in the *x–y* plane. The drawing is based on a computer-generated 'synthetic electron optical image' (at four grey levels) of the IMP. 'a' to 'b' is the edge of the particle. 'e' to 'f' is the edge of the shadow. c, point of tangency of the replicating flux on the particle. d, true geometric centre of the IMP. d′ is the projection of 'd' in the plane of section. Simulation conditions: $I_x = I_z = 0.84$ B (see Fig. 10.3 for definition of B). Shadowing angle is 45°. From Weinstein *et al.*, 1979, used with permission.

(3) where replica thickness significantly exceeds particle height, the metal cap and the shadow region on the opposite side of the IMP appear to dissociate; and (4) at all replica thicknesses, particle width is unaltered (Weinstein *et al.*, 1979). Computer simulations showing that IMP width remains constant require the invoking of two assumptions. The first assumption is that the atoms in the flux of replicating materials stick where they hit the specimen surface. This is unlikely since, at the molecular level of resolution, platinum probably decorates rather than coats (Ruben and Telford, 1975; Slayter, 1976). The second assumption is that the lateral margins of the metal cap adsorb only replicating atoms from the metal flux that result in a direct hit and that atoms grazing the sides of the cap will bounce off or pass by. This assumption is an obvious simplification of what occurs at the growth front (i.e. site of metal cap accumulation) of a metal replica of an IMP. At present, there is a general consensus that some cap widening is an inevitable consequence of metal shadowing (Lessin *et al.*, 1969; Lickfeld *et al.*, 1972; Weinstein *et al.*, 1979). Fracture face contamination *in vacuo* prior to and during specimen replication would also contribute significantly to cap widening.

One approach to the problem of compensating for the dimensional distortions inherent in freeze–fracture replication has been to introduce empirically derived correction factors in order to calculate actual dimensions (Lessin *et al.*, 1969). Another approach has been to devise freeze–fracture methodology that would minimize these distortions. Early studies on the relationship of metal replication to cap widening were carried out under conditions which compounded the problem. Hall (1960), and Misra and Das Gupta (1965) used platinum wire wrapped around a long tungsten filament as an extended unshielded source for their metal flux. An elongated source tends to widen the cap by virtue of a dispersed incident flux of metal. Heating may have served as an additional important contributing factor to cap widening in these experiments. Unshielded metal sources would result in the heating of the supporting surface by radiation. A hot surface increases the time required to dissipate the kinetic energy of the impacting atoms. It is possible that the replicating atoms would re-evaporate under these conditions, with the replica surface then acting as a secondary source of replicating atoms (Preuss, 1965).

Modern freeze–fracture methodology helps to minimize some of the dimensional distortion of replicated IMP. First, in freeze–fracturing, the replicated surface is cold, typically below $-100°$ C, thereby reducing secondary source emission. Second, early electron microscope studies on metal-shadowed globular proteins used specimens resting on a supporting film. These molecules were very different from the IMP in freeze fracturing which are not isolated particles resting unattached on a support film, but rather are projections or mounds at the surface of an irregular fracture face. Indeed, the IMPs are not true particles as defined in the thin-film literature; they only superficially resemble true particles by virtue of some of their metal shadowing characteristics. Since the IMPs of biological membranes are a physical component of the underlying frozen tissue, the heat of replication is dissipated more efficiently than by isolated particles resting on a support film. Third, modern electron guns used as a source for the metal flux also help reduce dimensional artifacts. Currently, a small aperture is used to reduce the size of the metal source as well as radiant heating. A magnet traps ions within the electron gun, preventing the new surface from becoming charged.

Unidirectional shadowing has advantages for demonstrating pores or channels extending through membrane protein complexes (McNutt and Weinstein, 1970, Hirokawa and Heuser, 1982a; Kuszak *et al.*, 1982) but has serious drawbacks for examining the fine structure of membrane proteins, particularly when it is desirable to visualize subunits in protein arrays such as the connexons of gap junctions (Goodenough, 1975; Hirokawa and Heuser, 1982a; Kuszak *et al.*, 1982), or non-specialized plasma membranes with a very high protein content (Khodadad and Weinstein, 1980, 1983). Rotary shadowing can be useful in some of these settings but is not universally desirable because IMP heights are more difficult to measure in rotary shadowed replicas. Another drawback is that the area of membrane suitable for quantitative analysis of the IMPs is severely limited in rotary shadowed replicas of curved membranes. The reason for this is that rotary shadowing is uniform over 360 degrees only in areas where the membrane is co-planar with the specimen stage (Weinstein *et al.*, 1979).

Recently, newly developed labelling techniques have provided a reasonable approach to the direct identification of IMP which may resolve many unanswered questions. Pioneered independently by Pinto da Silva (Pinto da Silva *et al.*, 1981a,b,c), Rash (Rash *et al.*, 1980, 1981), and their associates these techniques combine freeze–fracture with other electron microscopy preparative techniques. The method of Pinto da Silva is a logical extension of the Bullivant–Ames simple freeze–fracture method (Bullivant and Ames, 1966). With the 'fracture-label' method of Pinto da Silva, frozen specimens are fractured under a protective layer of liquid nitrogen and thawed. The exposed fracture faces are cytochemically labelled and then the specimen is processed either for thin sectioning or for critical-point drying. The method does not involve replication of the freeze–fractured specimen. The Rash method utilizes a combination of freeze–fracture replication and thin sectioning. Freeze–fracture replicas are embedded and processed for thin sections without digesting away the underlying tissue. With this method, electron-dense labels may be applied either before or after fracturing. To date, these methods have been used to visualize anionic and lectin binding sites. Rash has presented some data on the correlation of acetylcholine receptor proteins with a distinctive class of IMPs (Rash *et al.*, 1981, 1982). Surface labelling experiments present many technical problems in interpretation which must be resolved in the future (Severs and Robenek, 1983).

Finally, it should be noted that certain steps in the freeze–fracture technique may pose additional problems affecting the artifactual creation and redistribution of IMPs, altering particle partition coefficients between the two fracture faces and inducing aggregation of proteins. In some instances particles can arise from contamination (Bullivant and Ames, 1966). In the nuclear membrane and mitochondrial membrane changes are induced in the distribution of intramembrane particles by temperature shifts during initial tissue preparation (Maul, 1979). Correction of this problem is by fixation with glutaraldehyde. Glutaraldehyde fixation, however, as well as cryoprotection with glycerol can affect the particle-partition coefficient as demonstrated in red cells (Pricam *et al.*, 1977) plant protoplasts and the alga Oocystis (see Willison and Brown, 1979). Cryoprotection with glycerol can also affect the particle-partition coefficients for P-face and E-face (Pricam *et al.*, 1977). In certain instances, IMPs may arise from artifactual aggregation of proteins. This has been shown for glycophorin reconstituted into liposomes (Grant and McConnell, 1974; Segrest *et al.*, 1974) and for rhodopsin (see Roof and Heuser, 1982).

## 10.3 The nature of IMP

### 10.3.1 GENERAL CONSIDERATIONS

A central dogma in the freeze–fracture literature has been that the IMPs visualized on the two fracture faces of membranes, exoplasmic (E) and protoplasmic (P), represent proteins that are intercalated into the membrane's lipid bilayer (Singer and Nicolson, 1972). Further, it is held that linear intramembrane strands such as those which characterize tight junctions (Fig. 10.5) represent linear aggregates of IMPs (Staehelin, 1973, 1974; Van Deurs and Luft, 1979). This central dogma was challenged when it was shown that IMPs also can be demonstrated in freeze–fractured membranes totally devoid of protein. Both pits and particles were demonstrated in liposomes (Verkleij *et al.*, 1979a,b; Miller, 1980; Verkleij *et al.*, 1980). Several interpretations for such 'lipidic particles' have been offered. One interpretation is that 'lipidic particles' represent inverted micellar configurations of lipid molecules within a single bilayer (Verkleij *et al.*, 1979b; Sen *et al.*, 1981; Van Venetie and Verkleij, 1981). Borovjagin *et al.* (1982) identified the lipidic particle as the intermediate stage when the bilayer lamellar phase of phospholipids undergoes transition to the hexagonal phase. An alternative interpretation is that lipidic particles represent sites of intermembrane attachment (Miller, 1980; Hui and Stewart, 1981) or contacts (Rand *et al.*, 1981).

The relevance to cellular systems of lipidic particles in artificial membranes is not known. If the $H_{II}$ phase results in the appearance of lipidic particles and is

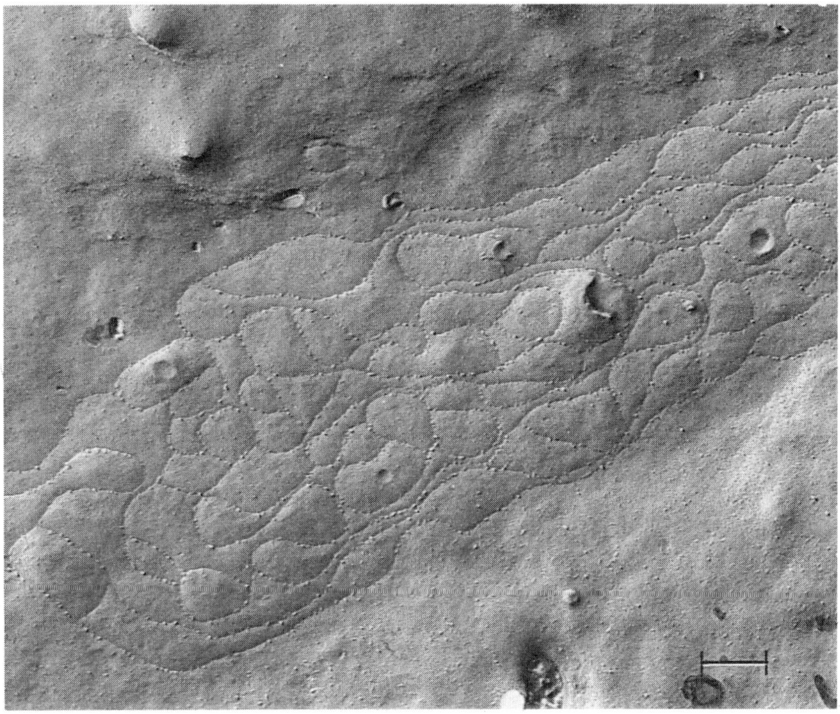

**Fig. 10.5** Replica of the exoplasmic fracture face of a tight junction. An extensive network of interconnected furrows is demonstrated. Numerous discrete intramembrane particles are present in these furrows. (Courtesy of Dr Stanley Bullivant.) Bar = 0.2 μm.

an important pathway for lipid movement and ion transfer in biological membranes, they may be more common in freeze–fracture replicas than is currently appreciated (Verkleij, 1984). If however, they only occur naturally at intermembrane attachment or fusion sites, they would not account for the IMPs in many biological membranes (Miller, 1980). Bearer and co-workers (1982) did not find these particles to be obligate structures in fusion of phospholipid vesicles.

Recently, it has been shown that reticulate arrays of particles resembling the linear arrays of IMPs in unfixed tight junctions can be produced in lipid vesicles containing *N*-acetylphosphatidylethanolamine in the presence of 10 mM $Ca^{2+}$, supporting the hypothesis that lipid micelles might be a constituent of the intramembrane strands of tight junctions and that tight junctions can be regarded as an outcome of membrane fusion at lipid-rich sites (Pinto da Silva and Kachar, 1982). However, the lipid hypothesis is not supported by observations that the intramembrane strands are stabilized by glutaraldehyde, which cross-links membrane proteins (Weinstein *et al.*, 1970) and that their prevalence in some cell systems is apparently related to cell surface protease activity (Pauli and Weinstein, 1982). Experimentally, it has been shown that proteins can trigger conversion of lipids from the bilayer to the $H_{II}$ phase (De Kruijff and Cullis, 1980). If the $H_{II}$ phase gives the appearance of lipidic IMP and proteins are involved in the formation of $H_{II}$ phase regions in cell membranes, the distinction between lipidic and protein IMP in biological membranes may be arbitrary. What is clear is that it cannot be assumed that IMPs are protein on the basis of ultrastructural observations alone.

IMPs have also been attributed to lipopolysaccharides in the outer membrane of *E. coli* (Ververgaert and Verkleij, 1978). On the basis of complementarity of pits and particles on two fracture faces, Verkleij and Ververgaert (1981) attribute the *E. coli* outer membrane particles to lipopolysaccharides and conversely, on the basis of lack of complementarity, the particles of the red cell membrane to proteins.

Another possibility is that IMPs can arise artifactually. Robertson and Vergara (1980) attribute the globular intramembrane particles in mammalian urinary bladder membrane to plastic deformation artifact although this is controversial (Pauli *et al.*, 1983). Particles may be artifactually produced as a result of ice crystal condensation on fracture faces (Gross *et al.*, 1978).

The amount of evidence in favour of the view that IMPs are composed of proteins is substantial (for review see Zingsheim and Plattner, 1976). It is important to note that the proteins which intercalate into the hydrophobic core of the membrane are associated with 'boundary lipids' which together with the protein core give rise to IMPs (see Zingsheim and Plattner, 1976 for references). The labelled replica techniques of Rash *et al.* (1982) permit immunologic and receptor-specific identification of the association of IMP with specific membrane proteins. These do not prove that the identified protein is the sole component of the IMP nor that it is solely responsible for the freeze–fracture morphology of the particle. Perhaps the most direct evidence for the presence of proteins in IMPs comes from reconstitution of defined proteins into pure lipid membranes and the consequent presence of IMPs in freeze–fracture replicas (Hong and Hubbell, 1972; Grant and McConnell, 1974; Vail *et al.*, 1974; Segrest *et al.*, 1974; Yu and Branton, 1976; Skriver *et al.*, 1980a,b). Furthermore in certain systems it has been shown that dilution, selective extraction or enrichment of certain proteins results in respective decrease or increase in the numbers of IMPs (Schneider *et al.*, 1980; Skriver *et al.*, 1980a; Kawato *et al.*, 1982; Khodadad and Weinstein, 1983). It should be noted, however, that it is difficult to compare results of studies on liposomes obtained by different investigators, even when they are studying the same protein. Incorporation and distribution of an integral protein in the lipid bilayer can be affected by the lipid composition, the method used to produce the liposome, the degree of protein purification and the amount of native membrane lipid co-purified with the protein. In several systems it has been found that incorporation into the membrane occurs with non-denatured protein (Darszon *et al.*, 1980; Lindstrom *et al.*, 1980). In many instances the asymmetry in the distribution of IMPs between the two fracture faces characteristic of the native membrane is lost in reconstituted systems. There is some correlation between the particle-partition coefficient and insertion and function of the protein (Satir and Satir, 1979).

Substantial data are available on certain membrane systems providing strong evidence in support of a correspondence between specific proteins and IMPs. In almost all of these systems, the transmembrane disposition of the protein has been shown and in many such cases the respective polypeptides have been isolated, the amino acid sequence determined, and the various fragments generated by cleavage studies isolated. These studies, together with electron microscopy and X-ray diffraction studies have provided much information on the subunit structure of the protein and its disposition into the bilayer. In a few of these cases actual models for the disposition of the proteins into the membrane have been proposed. In several of these cases the question of how many protein protomers give rise to an IMP remains unresolved. The membrane systems for which data establishing a relationship between a specific protein and a respective IMP are most compelling are the band 3 protein and the red cell membrane, bacteriorhodopsin and the purple patch membrane of *Halobacterium halobium*, cytochrome *c* oxidase and the inner mitochondrial membrane, connexon and the gap junction membranes, rhodopsin and the photoreceptor outer segment and disk membranes, the Light Harvesting Complex and the chloroplast thylakoid membrane, $Ca^{2+}$, $Mg^{2+}$ activated ATPase and the sarcoplasmic reticulum membrane, ($Na^{+}$-$K^{+}$)-ATPase and a variety of membranes, and acetylcholine receptors and the postsynaptic membrane. Data on each of these membrane systems will be summarized.

## 10.3.2 INTEGRAL MEMBRANE PROTEINS

### (a) *Integral proteins of the red cell membrane*

In the human red cell membrane, two glycoproteins designated band 3 and glycophorin span the membrane (Bretscher 1971a,b; Pinto da Silva and Nicolson, 1974; Shin and Carraway, 1974; Singer, 1974), with band 3 having a greater aggregate mass than glycophorin (see Steck, 1974). Both have been implicated in the formation of erythrocyte IMPs.

(i) *Glycophorin.* The complete amino acid sequence of glycophorin has been determined (Segrest *et al.*, 1973; Tomita and Marchesi, 1975). The membrane penetrating segment of glycophorin has been reconstituted into phospholipid bilayers resulting in the formation of intramembranous particles of approximately 80 Å diameter (Segrest *et al.*, 1974) suggesting that in the red cell membrane glycophorin can give rise to IMP. In the Segrest study, only the transmembrane portion of the molecule was used and, furthermore, the concentration of glycophorin was several times higher than in the native membrane. Hence the IMP could have resulted from the aggregation of the intramembrane portion of glycophorin molecules. Incorporating the complete glycophorin molecules into liposomes, Grant and McConnell (1974) observed that glycophorin occurs as round bumps of roughly 40 Å diameter which they suggest represent clusters of the glycoprotein. Additional evidence for a relationship between IMPs and glycophorin has been based on observations of similarities in the patterns of aggregation of IMP and surface markers for glycophorin under experimental conditions (Tillack *et al.*, 1972; Nicolson, 1973). Bächi (1980) cautions in drawing a direct relationship between the topography of a surface moiety of an integral protein and IMP based on their coaggregation since, when IMPs are aggregated, all of the major red cell membrane proteins may be coaggregated with them (Shotton *et al.*, 1978). Bächi (1980) in experiments with intact red cells using limulin, a lectin specific for sialic acids, and anti-glycophorin, a surface label for glycophorin, does not observe a correspondence between surface marker distribution and the distribution of IMP. These results further cast doubt on the notion that glycophorin is a major constituent of IMP. Additional problems arise from drawing quantitative relationships between IMP and glycophorin. Several investigators have noted that the size of the glycophorin molecule (approximately 50 000 $M_r$) and the number of protomers (500 000 per cell) are not sufficient to account for the number and size of IMP (see Bretscher, 1973; Steck, 1974). Studies on En(a−) human erythrocytes, which lack glycophorin, show no change in IMP number or morphology (Bächi *et al.*, 1977). These data seem to disqualify glycophorin as a major constituent of the intramembrane particle. It is possible that glycophorin together with another protein species, i.e. band 3, may contribute to the electron microscope image of IMP (Nigg *et al.*, 1980). The results of Bächi (1980), however, using specific surface labels for glycophorin on intact red cells cast doubt on this notion as well. At the present time it is not possible to conclusively prove that glycophorin is absent from red cell IMP, but the evidence that another protein, band 3, is the major constituent is strong.

(ii) *Band 3.* Band 3 is the anion transport protein of the human red blood cell (Cabantchik and Rothstein, 1974; for reviews see Steck, 1974 and Cabantchik *et al.*, 1978). It is the predominant protein of this membrane constituting approximately one quarter of the total membrane protein mass (see Steck, 1974). It has an approximate molecular weight of 90 000, spans the membrane asymmetrically, and is exposed at both membrane surfaces. There are $1 \times 10^6$ copies per ghost (see Steck, 1978). The carboxyl group and the carbohydrates are confined to the extracellular surface (see Steck, 1978), whereas the amino terminus is confined to the protoplasmic surface

(Drickamer, 1976; Fukuda *et al.*, 1978; see also Steck, 1978). On the basis of mapping of fragments generated by proteolysis of band 3 *in situ* with chymotrypsin and trypsin and cleavage with NTCB (2-nitro-5-thiocyanobenzoate), a scheme for the disposition of this polypeptide in the membrane has been presented (Steck *et al.*, 1978).

Early studies (Branton, 1971) suggested that IMP in the red cell membrane may be related to proteins since a reduction in IMP number was shown in cell membranes treated with protease. Freeze–fracture electron microscopy reveals 80 Å IMPs on fracture faces of red cells (Weinstein, 1974). It had been postulated that band 3 is the predominant component of the red cell IMP (Guidotti, 1972; Steck, 1974; Cherry *et al.*, 1976; Yu and Branton, 1976). When anion transport activity has been studied in a reconstituted system (Wolosin *et al.*, 1977), anion transport rates have correlated with protein incorporation. In a correlated biochemical and ultrastructural study, Weinstein and co-workers utilized the findings obtained from proteolytic dissection of band 3 polypeptide and the binding of glyceraldehyde 3-phosphate dehydrogenase (G3PD) by its cytoplasmic domain for elucidating the ultrastructure of the band 3 polypeptide *in situ* (Weinstein *et al.*, 1978a,b, 1980). Proteolytic cleavage of band 3 *in situ* results in the generation of three fragments; a 38 000-dalton outer segment fragment which remains associated with the membrane bilayer, a 17 000-dalton membrane spanning fragment, and a cytoplasmic water soluble 40 000-dalton fragment which is released from the membrane (Steck *et al.*, 1976, 1978). The cytoplasmic domain of band 3 has the binding site for the enzyme G3PD (Yu and Steck, 1975). Weinstein and co-workers (1978a,b, 1980) identified the IMPs of the protoplasmic fracture face (PF) as the morphological representations of intramembrane segments of band 3 polypeptide. This is supported by the findings of Edwards and co-workers (1979) who by polyacrylamide gel electrophoresis (SDS-PAGE) of half membranes showed that band 3 is associated with the protoplasmic fracture face of the membrane. We have shown the cytoplasmic domain of band 3 as a granulofibrillar meshwork at the protoplasmic surface (PS) in ghosts and inside-out vesicles which can be decorated with G3PD, and the extracellular aspect of band 3 as extracellular surface (ES) particles revealed after the unmasking of this surface by removal of sialic acids. Thus there is evidence that in the red cell membrane we are able to visualize the three domains of an integral membrane transport protein by freeze–fracture and freeze–etch electron microscopy (Weinstein *et al.*, 1978a,b, 1980).

Studies on the llama red blood cell membrane (Fig. 10.6) support the idea that the subgroup of red cell IMPs that partition with the protoplasmic fracture face represent band 3 protein. SDS-PAGE of the llama red cell membrane shows the same major proteins as the human (Fig. 10.6a). The band 3 polypeptide, however, is over two and a half times more concentrated in the

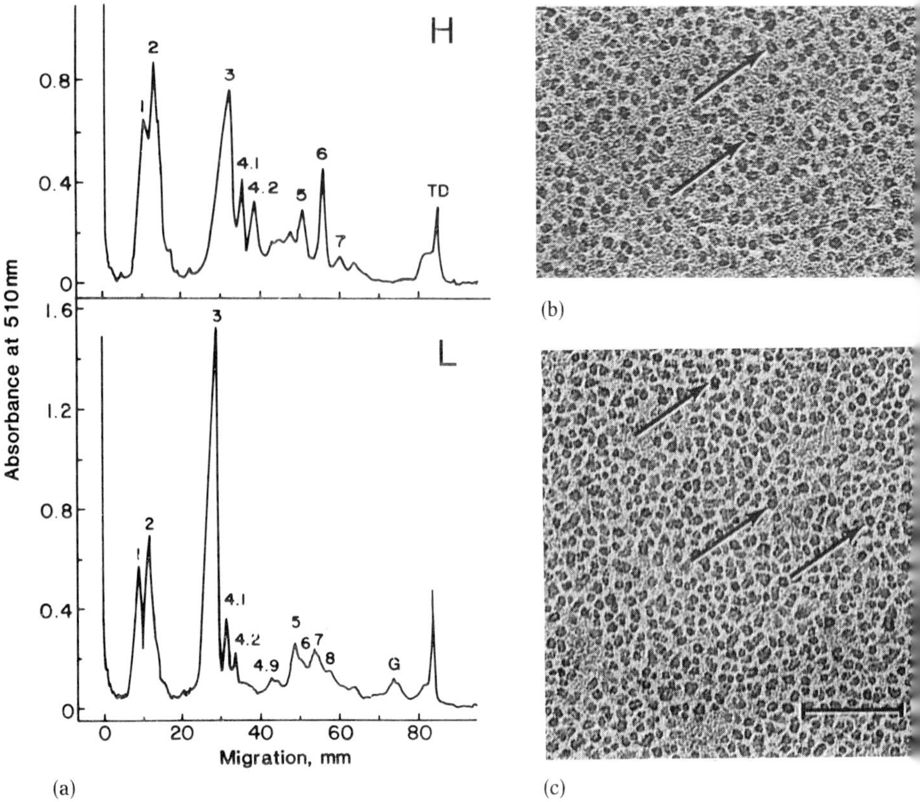


**Fig. 10.6** Comparative studies on the red cell membranes of the human and llama. (a) Densitometric scans of Coomassie Blue-stained gels of membranes of the human (H) and llama (L) erythrocytes. The band 3 polypeptide is relatively more abundant in the llama. Protein (40 $\mu$g) solubilized in SDS was loaded on gels. TD marks the position of the tracking dye; G is globin. The protoplasmic fracture faces of the human (b) and llama (c). Replicas were obtained from freeze–fractured, glutaraldehyde-fixed and glycerinated red blood cells rotary shadowed at a 25° angle. The P-face IMP (arrows) are more abundant in the llama membrane than in the human. (Adapted from Khodadad and Weinstein, 1983.) Bar = 0.1 $\mu$m.

llama than in the human. The llama band 3 polypeptide is an integral membrane protein (Khodadad and Weinstein, 1983) with cytoplasmic (Khodadad and Weinstein, 1985) and extracellular domains which can be excised with proteolysis (Khodadad and Weinstein, unpublished results). Hence the llama band 3 polypeptide is also a transmembrane protein. Concomitant with the high concentration of band 3 in the llama there is an increase of more than two and a half times in the numerical density of IMPs of the protoplasmic fracture face (Fig. 10.6b and c) as determined by quantitative electron microscopy of rotary shadowed replicas. The numbers of intramembrane particles on the exoplasmic fracture face of the llama are similar to that of the human. These findings provide a strong argument that the intramembrane particles of the protoplasmic fracture face arise from the band 3 polypeptide (Khodadad and Weinstein, 1983).

The question of how many band 3 protomers constitute an intramembrane particle has been considered. Studies based on cross linking of band 3 *in situ* in the membrane by copper *o*-phenanthroline suggest that band 3 exists in the membrane as a stable non-covalent dimer (Steck, 1972; Wang and Richards, 1974; and see Steck, 1978). Also its predominant form in Triton X-100 solutions is a dimer (Clarke, 1975; Yu and Steck, 1975), whereas at higher protein concentrations in acetic acid solutions, larger oligomers have been generated reversibly (Dörst and Schubert, 1979). Work by several investigators suggests that band 3 polypeptide exists in the membrane as a tetramer (Weinstein *et al.*, 1980; Khodadad and Weinstein, 1983; see also Dörst and Schubert, 1979 for references). There is a strong correlation between P-face IMP numerical densities in freeze–fracture replicas and the number of band 3 monomers if they are organized predominantly as tetramers.

(b) *Bacteriorhodopsin of the purple patch membrane*

Bacteriorhodopsin is found in the purple patch membrane of *Halobacterium halobium* (Stoeckenius and Rowen, 1967). Bacteriorhodopsin is the only protein in the purple patch membrane and consists of a polypeptide with a molecular weight of 26 000, constituting 75% of the total membrane mass (Osterhelt and Stoeckenius, 1971). Freeze–fracture electron microscopy of purple patch membrane demonstrates particles arranged in regular arrays (Blaurock and Stoeckenius, 1971). Incorporation of detergent solubilized bacteriorhodopsin into phosphatidylcholine vesicles results in the formation of IMPs (Mueller and Cherry, 1981) providing evidence that these IMPs and bacteriorhodopsin are related.

The amino acid sequence of bacteriorhodopsin has been determined (Gerber *et al.*, 1979; Khorana *et al.*, 1979; Ovchinnikov *et al.*, 1979; Walker *et al.*, 1979) and a model for the disposition of the polypeptide chain in the membrane bilayer has been presented (Engelman *et al.*, 1980). There are seven α-helical segments, generally hydrophobic residues separated by short hydrophilic segments (Ovchinnikov *et al.*, 1979). These seven α-helices conform to the seven rods which are demonstrated by Henderson and Unwin (1975) to span the membrane. These investigators performed image analysis of electron micrographs of crystalline arrays of unstained isolated purple patch membrane and elucidated the three-dimensional structure of the bacteriorhodopsin molecule to a resolution of 7 Å. Their work demonstrates that the individual protein molecule consists of seven rods, corresponding to seven α-helical segments, 35–40 Å long, extending all the way through the membrane and packed 10–12 Å apart. Three of the rods are perpendicular to the plane of the membrane and the other four are slightly tilted (Henderson and Unwin,

1975; also see Unwin and Henderson, 1984). The model of these membrane crystals shows intimate contact between three protein molecules grouped around a three fold axis and describes the arrangement of the α-helices in this group of three (Henderson and Unwin, 1975). There had been uncertainties as to how many bacteriorhodopsin molecules constitute an IMP. It is reasonable that bacteriorhodopsin in the membrane is arranged in clusters of three (Blaurock, 1975). A trimer is demonstrated by the work of Henderson and Unwin on isolated purple patch membrane (Henderson, 1975; Henderson and Unwin, 1975). In addition, a study using a high resolution freeze replica method with rotary shadowing of unfixed fresh membranes shows a trimeric structure for bacteriorhodopsin (Usukura *et al.*, 1980).

(c) *Connexons of the gap junction*

The ultrastructure of the apposing membranes in a synapse was first described by thin section electron microscopy (Robertson, 1963). Kreutziger (1968) was the first to demonstrate by freeze–fracture electron microscopy hexagonal arrays of particles on one fracture face and corresponding pits on the other fracture face of gap junction membranes. On the basis of the work of a number of investigators (Chalcroft and Bullivant, 1970; McNutt and Weinstein, 1970, 1973; Robertson, 1981), the IMPs were assigned to the protoplasmic fracture face (PF) in vertebrates. The subunit of the hexagonal lattice of the gap junction has been termed 'connexon' (Goodenough, 1976). Lanthanum impregnation demonstrates connexons in hexagonal arrays with annuli in a plane perpendicular to the surface of the membrane and protruding into the extracellular space (Revel and Karnovsky, 1967). In replicas of quick-frozen and deep-etched gap junctions central pores are observed in surface protrusions as well as in PF particles (Hirokawa and Heuser, 1982a). Central pores are also seen in unidirectionally shadowed freeze–fracture replicas of gap junctions in the lens (Fig. 10.7).

Gap junctions have been isolated by several groups of investigators (Bennedeti and Emmelot, 1968; Evans and Gurd, 1972; Goodenough and Stoeckenius, 1972). The major gap junction membrane protein component is a 26 000-dalton polypeptide (Henderson *et al.*, 1979; see Finbow *et al.*, 1980). Based on freeze fracture electron microscopy, (McNutt and Weinstein, 1970) and X-ray diffraction of gap junction hexagonal arrays (Makowski *et al.*, 1977) models for the three dimensional distribution of proteins, lipid and the water channel in the gap junction have been presented. It is suggested in these models that connexons consist of protein oligomers which traverse the bilayer. Zampighi and Unwin (1979) studied the isolated gap junctions by electron microscopy and negative staining and identified two different conformational forms of connexons with the same hexagonal lattice constants. Unwin and Zampighi (1980) studied the structure of gap junctions to a resolution of 18 Å with electron microscopy and image analysis. They presented the three dimensional structure of connexons, as annular oligomeric pairs with a

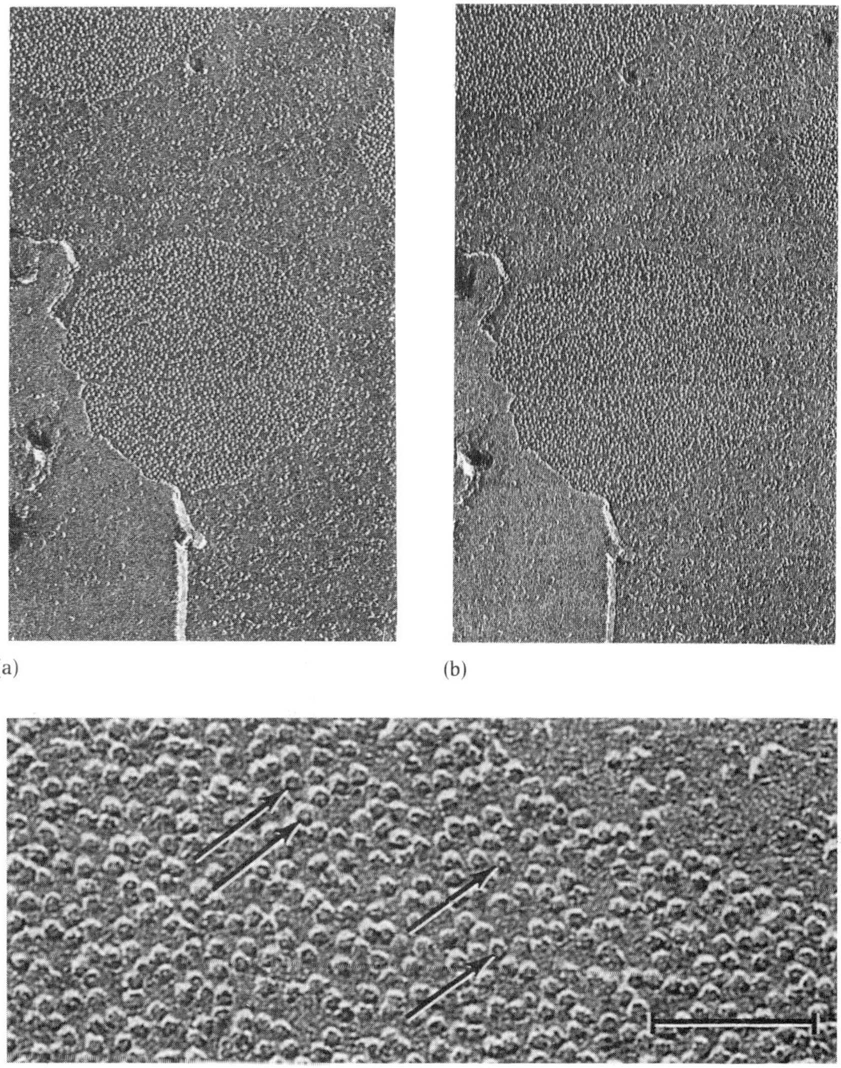


**Fig. 10.7** Top: Stereo-pair electron micrographs of P face of gap junctions from the rat lens. The fracture face to the left of the junction is EF, that to the right and above it is PF. The narrow zone between the bottom of the field and the gap junction is the intercellular space. This pair shows the raised nature of the plaque, the narrowing of the intercellular space at the site of the plaque, and the uniform appearance of the connexons in the junction. Magnification before reproduction 52 000, goniometer tilt $\pm 12.5°$. Bottom: High magnification of a unidirectionally shadowed replica of a lens gap junction showing the central hydrophilic pore of the connexon (arrow). (Courtesy of Dr Jerome Kuszak.) Bar $= 0.1\ \mu m$.

central channel embedded in a lipid bilayer. Each connexon consists of six protein subunits extending out into the extracellular space. Similar conclusions could be deduced from the results of electron microscope studies and low angle X-ray diffraction (Makowski *et al.*, 1977). Recently, Unwin and Henderson (1984) updated their three dimensional model of the connexon. On the basis of electron microscope image analysis and other considerations, connexons appear as cylindrical oligomeric proteins embedded in the membrane and extending into the extracellular space where each connexon is linked to the apposing connexon. Individual connexons consist of six subunits arranged around a central channel and extending approximately 15–20 Å into the extracellular space and approximately 10 Å into the cytoplasm of the cells.

(d) *Cytochrome* c *oxidase of the inner mitochondrial membrane*

Cytochrome *c* oxidase is the enzyme complex found in inner mitochondrial membranes of eukaryotes which catalyses the final step in oxidative phosphorylation. A number of studies have established the transmembrane disposition of cytochrome oxidase (Eytan *et al.*, 1975; Hackenbrock and Hammon, 1975; Ruben *et al.*, 1976; Ludwig *et al.*, 1979). The monomer of cytochrome oxidase is composed of a minimum of seven polypeptide chains (Downer *et al.*, 1976; also see Capaldi *et al.*, 1977), to a maximum of 12 (see Azzi, 1980). The polypeptides of cytochrome oxidase have been isolated, purified and characterized (see Azzi, 1980, for review). Using osmotically swollen mitochondria, Höchli and Hackenbrock (1979) showed a relationship between the clustering of cytochrome oxidase induced by cross linking through an immunoglobulin lattice and a redistribution of a category of IMPs in freeze–fracture replicas. This work thus provided evidence for a relationship between cytochrome oxidase and IMPs. Work with liposomes further supported this view. In liposomes reconstituted from purified phospholipids and cytochrome oxidase, with high oxidase activity and respiratory control, Ruben and co-workers (1976) have demonstrated IMPs. These particles were similar to those observed in swollen mitochondrial membranes. Furthermore, in these liposomes intramembrane particles were observed traversing the edge of the fracture plane (Ruben *et al.*, 1976) consistent with the transmembrane disposition of the protein.

It is possible to selectively enrich cytochrome oxidase in membranes by use of detergents. Two different two dimensional crystalline arrays of the enzyme can be obtained by treatment of the mitochondrial membrane with Triton X-100 or deoxycholate. The bidimensional crystals obtained from deoxycholate (Seki *et al.*, 1970) contain the monomeric form of the enzyme with no continuous lipid bilayer whereas the enzyme in bidimensional vesicle crystals obtained with Triton X-100 (Vanderkooi *et al.*, 1972) is organized as dimers traversing a lipid bilayer (see Azzi, 1980). Cytochrome oxidase from beef heart has been studied with electron microscopy and image reconstruction on preparations obtained with deoxycholate (Fuller *et al.*, 1979) and Triton X-100 (Henderson *et al.*, 1977; Frey *et al.*, 1978, 1982; Deatherage *et al.*, 1980). Fuller and co-workers (1979) have described the structure of the monomer in these preparations as a lopsided Y of 110 Å length, with a length of 55 Å for the arms of the Y, and a centre to centre separation of 40 Å. The arm domains are embedded into the cytoplasmic side. Deatherage and co-workers (1980) have proposed that the dimers in vesicle crystals are formed from such monomers described by Fuller *et al.* (1979). Frey and co-workers (1978) have shown that the orientation of cytochrome oxidase in vesicle crystals is opposite to that of intact mitochondria; the side of the vesicle bearing the arms of the Y is the matrix or M side. Costello and Frey (1982) have studied membranous cytochrome oxidase prepared from beef heart mitochondria by extraction with Triton X-100. These preparations contained both crystalline and noncrystalline vesicles. Freeze–fracture and freeze–etch studies of these preparations show that crystalline lattices are derived from portions of the enzyme protruding from the membrane bilayer. Comparing the 100 Å IMPs to the three dimensional arrangement of cytochrome oxidase described by Frey *et al.* (1982) they suggest that these intramembranous particles consist of two dimers paired along the *a*-axis. A dimeric complex of cytochrome oxidase is shown by Unwin and Henderson (1984).

Attempts have been made to measure the dimensions of active cytochrome oxidase. Ruben and Telford (1980) have measured the dimensions of active cytochrome oxidase in quick frozen hydrated specimens in reconstituted liposomes using a gold ball shadow width standard. They report dimensions of $58 \pm 3$ (S.D.) Å for the average diameter and $82 \pm 3$ (S.D.) Å for the maximum length. These values are in agreement with the values reported earlier on regular two dimensional arrays of membranous Triton X-100-extracted cytochrome oxidase (see Ruben and Telford, 1980).

(e) *Acetylcholine receptors of the postsynaptic membrane*

In the neuromuscular junctions of vertebrates, clusters of large IMPs have been observed in the postsynaptic region (Peper *et al.*, 1974; Rash and Ellisman, 1974; Ellisman *et al.*, 1976). Data obtained from fluorescence light microscopy (Anderson and Cohen, 1974) and electron microscopy (Fertuck and Salpeter, 1974, 1976; Lentz *et al.*, 1977) demonstrate that these regions are sites of α-bungarotoxin binding and hence sites of localization of acetylcholine receptors. Similarly, in cultured muscle cells of chick (Yee *et al.*, 1978; Cohen and Pumplin, 1979) and *Xenopus* embryos (Peng and Nakajima, 1978) clusters of IMPs have been observed. The work with developing chick myotube has shown a correspondence of these IMPs to the regions of α-bungarotoxin binding (Cohen and Pumplin, 1979) or acetylcholine receptors as determined by microiontophoretic mapping (Yee *et al.*, 1978). In the

*Xenopus* muscle cells in tissue culture it is proposed that the clusters of IMPs (Peng and Nakajima, 1978) correspond to the distribution of acetylcholine receptors demonstrated by $\alpha$-bungarotoxin binding (Anderson and Cohen, 1977). These findings demonstrate that in the adult vertebrate postsynaptic region and in the cultured muscle cell systems the IMPs correspond to sites with a high concentration of acetylcholine receptors.

Freeze–fracture and deep-etch electron microscopy have been used in several laboratories to demonstrate surface protrusions of approximately 80–90 Å diameter on the postsynaptic membrane of vertebrate neuromuscular junctions (Grohovaz *et al.*, 1982; Hirokawa and Heuser, 1982b). These are attributed to the outer aspect of acetylcholine receptors. Applying quick-freeze deep-etch rotary shadowing to their preparations, Hirokawa and Heuser (1982b) demonstrate that these protrusions are organized on the apices of the postsynaptic fold and suggest that they may be heads of acetylcholine receptors as this is the specific region where the receptors have been demonstrated (Fertuck and Salpeter, 1976). Grohovaz and co-workers (1982) and Hirokawa and Heuser (1982b) suggest that the close correspondence between their values of 9000–10 000/$\mu m^2$ for the numerical densities of these surface protrusions and the estimates of acetylcholine receptors of 10 000/$\mu m^2$ (see Heuser and Salpeter, 1979, footnote 1) provides further support for the view that these represent the extracellular surface (ES) protrusions of acetylcholine receptor. It is important to note, however, that the numbers of IMPs on both fracture faces are considerably fewer than the ES protrusions (Rash *et al.*, 1978).

In the electric organ of *Torpedo*, cholinergic receptor proteins account for up to 30% of the total membrane protein (see Dupont *et al.*, 1974; for references). Freeze–fracture electron microscopy demonstrates IMPs in the postsynaptic membranes (Orci *et al.*, 1974; Rosenbluth, 1975). Circumstantial evidence that IMPs in this system arise from acetylcholine receptors comes from work with reconstituted systems. Functionally intact acetylcholine receptors from the electric organ of *Torpedo californica* have been reconstituted into lipid vesicles and planar lipid bilayers (see Anholt *et al.*, 1982 for references). These produce IMPs that correspond in size to IMPs of the electric organ membrane *in situ*. The pharmacological function of acetylcholine receptor in the reconstituted vesicles is similar to that of the receptor in the native membrane (Anholt *et al.*, 1982).

Many data are available on the organization and structure of the acetylcholine receptors of the postsynaptic membranes of the electrocytes of the electric fish from freeze–fracture, deep-etch and negative stain electron microscopy (Cartaud *et al.*, 1973, 1978; Nickel and Potter, 1973; Heuser and Salpeter, 1979), X-ray diffraction and electron microscopy (Ross *et al.*, 1977), and neutron scattering together with negative staining (Wise *et al.*, 1981). A lattice organization of the receptor enriched membranes has been shown by freeze–etching and negative staining (Cartaud *et al.*, 1978) and by X-ray diffraction (Dupont *et al.*, 1974). By immunospecific staining and negative stain electron microscopy, Klymkowsky and Stroud (1979) have studied the three-dimensional structure of the acetylcholine receptor molecule in membrane fragments. Their work, together with X-ray diffraction studies of Ross and co-workers (1977), identifies the acetylcholine receptor as a molecule 110 Å in length, with a maximum in plane diameter of 85 Å, extending above the postsynaptic cleft by approximately 55 Å with minimal extension on the cytoplasmic side of the membrane. A molecular weight of 250 000–310 000 is estimated for the receptor, consistent with values for the molecular weight of detergent solubilized monomer (see Klymkowsky and Stroud, 1979). In receptor enriched membranes of *Torpedo*, the receptor is inferred to be a disc of 70–90 Å across with a central hole (see Wise *et al.*, 1981).

The monomeric form of acetylcholine receptor complex of *Torpedo californica* is composed of five glycopeptides in the subunit stoichiometry of $\alpha_2\beta\gamma\delta$ (Reynolds and Karlin, 1978; and see Anholt *et al.*, 1982 for references). The predominant native form of acetylcholine receptor from the electric organ of *Torpedo californica* is a dimer of $M_r = 500\ 000$ cross-linked by a disulphide link between the $\delta$ chains (for references see Wise *et al.*, 1981). Small-angle neutron scattering of the Triton X-100 solubilized and purified receptor dimers together with electron microscope negative staining of these dimers have been employed to further elucidate the structure of the acetylcholine receptor dimers (Wise *et al.*, 1981). Based on this work, a model has been constructed for the acetylcholine receptor dimer of *Torpedo californica*, showing two disc shaped monomers each approximately 90 Å across with a central pit (Wise *et al.*, 1981).

The extracellular surface (ES) of acetylcholine receptors of *Torpedo* postsynaptic membrane has been visualized with deep etching and negative staining; arrays of repeating particles with a subunit structure and central pit are observed at this surface (Cartaud *et al.*, 1978). Heuser and Salpeter (1979) have investigated, by quick-freeze, deep-etch and rotary shadowing the relationship between the IMP and 85 Å surface projections on these membranes; they observed that the protrusions are twice as numerous as the intramembrane particle category believed to represent acetylcholine receptors. The number of surface protrusions, however, matched estimates for acetylcholine receptors obtained by physiological and biochemical methods.

(f) *Rhodopsin and the outer segment membrane of the rod photoreceptor cell*
Visual pigment is an intrinsic component of vertebrate rod photoreceptor outer segment plasma membrane and disc membrane (Jan and Revel, 1974;

Basinger *et al.*, 1976), where it appears to give rise to IMPs on the P-fracture face (Raubach *et al.*, 1974; Corless *et al.*, 1976; Krebs and Kühn, 1977; Andrews and Cohen, 1979).

Rhodopsin is the principal protein of the vertebrate rod outer segment membranes, and comprises approximately 85% of the total membrane protein mass. The molecular weight of the protein is between 35 000 and 40 000 (for references see Hargrave, 1977). Rhodopsin is a glycoprotein (Heller, 1968) containing two asparagine-linked oligosaccharide chains (Fukuda *et al.*, 1979). Proteolytic digestion and chemical labelling studies have shown that rhodopsin spans the bilayer (Hubbell and Fung, 1979). It should be noted also that there have been conflicting reports on the transmembrane disposition of rhodopsin (for discussion and references see Corless *et al.*, 1976). Neutron diffraction studies show that the bulk of the rhodopsin mass is within the hydrophobic core of the disc membrane (Saibil *et al.*, 1976).

The extracellular surface of the rhodopsin molecule has been demonstrated by rapid freezing, freeze–fracture and deep-etching (Roof and Heuser, 1982). Particles 60 Å in diameter with a numerical density of 30 000/$\mu m^2$ are present at the membrane surface in *Bufo marinus*. Roof and Heuser (1982) suggest that these protrusions represent the outer aspect of rhodopsin, which projects into the intradisc space. It is known that the two carbohydrate chains are covalently linked to rhodopsin (Hargrave, 1977) and penetrate into the internal (intradisc) surface as demonstrated by enzymatic labelling (Adams *et al.*, 1979) and electron microscope histochemical localization (Röhlich, 1976).

As with other membrane systems the strongest evidence in support of the view that the IMPs arise from rhodopsin is derived from work with reconstituted systems. Artificial membranes reconstituted from purified rhodopsin and phospholipid bilayers show intramembrane particles whereas liposomes lacking rhodopsin do not (Hong and Hubbell, 1972). In this study IMPs are seen in fracture faces but not on the membrane surface. Similarly, intramembrane particles are shown in both fracture faces in liposomes reconstituted with rhodopsin. In rod outer segment vesicles, the particles are seen only in one fracture face (Chen and Hubell, 1973). Experiments of Chabre *et al.* (1972) with reconstituted systems further support this finding. In addition, it has been demonstrated that in the vertebrate vitamin deprivation reduces rhodopsin content as well as numbers of intramembrane particles on the disc membrane (Jan and Revel, 1974). In the rhabdomeric microvillar membrane of the photoreceptor of the invertebrate numerous intramembranous particles are demonstrated (for references see Schinz *et al.*, 1982). Recent studies of Schinz and co-workers (1982) with mutants of *Drosophila melanogaster* with mutations decreasing both rhodopsin content and numbers of intramembrane particles also support the view that IMPs are related to rhodopsin although a quantitative relationship between rhodopsin and IMP has not been established.

Rhodopsin and IMP cannot be correlated on a one to one basis (Corless *et al.*, 1976; Krebs and Kühn, 1977). In the disc membranes, estimates of numbers of rhodopsin molecules per unit of membrane are four to five times higher than the number of IMPs (see Corless *et al.*, 1976). Similarly, Roof and Heuser (1982) using the rapid-freezing technique have noted that the numbers of IMPs cannot be related to ES protrusions or numbers of rhodopsin molecules. It is proposed that IMPs are post-fracture entities which arise artifactually as a result of aggregation of rhodopsin molecules (Roof and Heuser, 1982; see also Clark and Branton, 1968 and Corless *et al.*, 1976). X-ray diffraction studies of Chabre (1975; see also Corless *et al.*, 1976) demonstrate that in the frog native disc membranes, rhodopsin aggregates corresponding to freeze–fracture particles are not present.

The amount of data relating IMP to the intrinsic transmembrane protein, rhodopsin, are impressive. The association number of rhodopsin molecules in the native membrane, however, is not known.

(g) *($Na^+$-$K^+$)-ATPase*

($Na^+$-$K^+$)-ATPase is present in membranes in numerous mammalian systems (see Jorgensen, 1980). It is possible to prepare ($Na^+$-$K^+$)-ATPase in pure and undenatured form from mammalian kidney (Jorgensen, 1975). Purified and membrane-bound ($Na^+$-$K^+$)-ATPase can be obtained by selective extraction of extraneous proteins of cell membranes of mammalian kidney outer medulla (Jorgensen, 1974; see Jorgensen, 1980) and salt gland of ducks (Hopkins *et al.*, 1976). Such membrane preparations from the kidney have densely packed particles and show the retained sidedness of the pump (see Jorgensen, 1980). Studies on such preparations have resulted in detailed data on pure renal ($Na^+$-$K^+$)-ATPase.

In purified membranes ($Na^+$-$K^+$)-ATPase is believed to be represented as extracellular surface particles after negative staining and as intramembrane particles after freeze–fracturing in kidney (Van Winkle *et al.*, 1976; Deguchi *et al.*, 1977; Maunsbach *et al.*, 1979; Giradet *et al.*, 1981; see Jorgensen, 1980) and in avian salt gland (Gassner and Komnick, 1983). Gassner and Komnick (1983) demonstrated that solubilization of ($Na^+$–$K^+$)-ATPase from purified membranes results in disappearance of IMPs. This provides strong evidence that IMPs are related to the ($Na^+$-$K^+$)-ATPase pump. Reconstitution studies based on the incorporation of ($Na^+$-$K^+$)-ATPase from porcine renal medulla (Skriver *et al.*, 1980a,b) and from duck salt gland (Gassner and Komnick, 1983) into liposomes further strengthens the view that intramembrane particles correspond to the ($Na^+$-$K^+$) pump complex. Incorporation of purified ($Na^+$-$K^+$)-ATPase into liposomes results in the appearance of IMPs.

Furthermore, both the capacity for cation transport in these liposomes and the number of particles present are proportional to the amount of ($Na^+$-$K^+$)-ATPase incorporated (Skriver *et al.*, 1980a). However, unlike the native membrane, particles are seen in both fracture faces. Recently it has been shown that reconstitution of purified ($Na^+$-$K^+$)-ATPase from the avian salt gland into liposomes results in 80–100 Å IMPs (Gassner and Komnick, 1983).

In the purified fractions of kidney medulla, ($Na^+$-$K^+$)-ATPase contains two subunits, $\alpha$ and $\beta$ in a 1:1 molar ratio. The $\alpha$-subunit has an approximate molecular weight of 104 000 and constitutes 66–77% of the protein. The $\beta$-subunit is a sialoglycoprotein of 40 000 molecular weight and a 17 000-dalton carbohydrate moiety (see Jorgensen, 1980). In purified fractions from the avian salt gland, the two polypeptides of the ($Na^+$-$K^+$)-ATPase consist of a 94 000–98 000-dalton larger catalytic subunit and a 54 000–60 000-dalton smaller glycoprotein subunit (Hopkins *et al.*, 1976; Stewart *et al.*, 1976; Lingham *et al.*, 1980). The transmembrane nature of the larger ($\alpha$) subunit polypeptide has been shown in purified kidney membranes (Ruoho and Kyte, 1974; Kyte, 1975; see Jorgensen, 1980). In reconstituted systems, evidence has been provided that the $\beta$ subunit is exposed at the extracellular face of the membrane (Hokin, 1979). Recently, Girardet and co-workers (1981) have provided immunochemical evidence for the transmembrane orientation of both subunits in kidney of *Bufo marinus*.

Much data is available to support the view that the functional unit of the pure renal enzyme consists of $2\alpha$ and $2\beta$ subunits ($\alpha_2\beta_2$) with a molecular weight of approximately 280 000 for the protein moieties and 34 000 for the carbohydrate moieties (for discussion and references see Jorgensen, 1980). Electron micrographs of these preparations show surface particles of 30–50 Å with a numerical density of 12 500/$\mu m^2$ seen after negative staining. Deguchi and co-workers (1977) suggest that each surface particle corresponds to a monomer containing one $\alpha$-subunit. They note that in these preparations, IMPs observed after freeze–fracturing of membranes are much larger than the surface projections. They are approximately 90–100 Å in diameter and are far fewer, and assumed to be oligomers of two or more of the protein units (Deguchi *et al.*, 1977). Rotary shadowed replicas obtained from such purified renal ($Na^+$-$K^+$)-ATPase suggest a dimeric substructure for the intramembrane particle (Maunsbach *et al.*, 1979). Based on analysis of dimensions and numbers of IMPs and sizes of the protein subunits, it has been suggested that the particles may represent the 280 000-dalton dimeric ($\alpha_2\beta_2$) functional unit of the pump (Maunsbach *et al.*, 1979). The three-dimensional structure and the association number of the ($Na^+$-$K^+$)-ATPase in the native membrane, however, remains to be clarified.

### (h) *($Ca^{2+}$-$Mg^{2+}$)-ATPase of sarcoplasmic reticulum*

The active transport of calcium through the membranes of sarcoplasmic reticulum is accomplished by a ($Ca^{2+}$-$Mg^{2+}$)-activated ATPase. This transport enzyme consists of a protein which can be solubilized into a monomeric form with the retention of full enzymatic activity (Dean and Tanford, 1978; Andersson *et al.*, 1982). X-ray diffraction studies indicate an asymmetric disposition for this protein in the membrane lipid bilayer (Dupont *et al.*, 1973; Herbette *et al.*, 1977 also see Moller *et al.*, 1982) and that a substantial portion of the protein mass protrudes on to the exterior of the membrane vesicle (Herbette *et al.*, 1977) as the active site of the enzyme (Thorley-Lawson and Green, 1973). These findings confirm the early suggestion of Hasselbach and Elfvin (1967) and the later findings of Saito *et al.* (1978) based on ultrastructural studies of the distribution of the ($Ca^{2+}$-$Mg^{2+}$)-ATPase in the membrane.

Work on molecular weight and amino acid sequence of the polypeptide has been performed in several laboratories (see Moller *et al.*, 1982 for references). This polypeptide has a relatively high proportion of polar residues, consistent with the thinking that the large part of the protein is located on the outside of the lipid core of the membrane. On the basis of the relationship between the amino acid sequence and tryptic fragments of the polypeptide, a model for the disposition of the protein has been presented (see Moller *et al.*, 1982). Structural studies on detergent solubilized ATPase indicate that it is an elongated molecule approximately 110 Å in length, with the major part of the protein located at one end (see Moller *et al.*, 1982).

Freeze–fracture electron microscopy of fragmented sarcoplasmic reticulum membrane preparations has demonstrated numerous intramembrane particles of 75–90 Å diameter on the concave (PF) face and scant intramembrane particles on the convex (EF) face (Deamer and Baskin, 1969; Tillack *et al.*, 1974; Jilka *et al.*, 1975; Scales and Inesi, 1976a,b). Scales and Inesi (1976a) propose that the protein of the outer leaflet (PF) of the sarcoplasmic reticulum membrane is mostly ($Ca^{2+}$-$Mg^{2+}$)-ATPase.

A close correlation has been shown in sarcoplasmic reticulum membrane during development between IMP numerical density and $Ca^{2+}$ transport activity (Tillack *et al.*, 1974). Reconstitution of purified ($Ca^{2+}$-$Mg^{2+}$)-ATPase into phospholipid bilayers is accompanied by the appearance of 75 Å intramembrane particles. Liposomes lacking the enzyme show no IMP. This constitutes circumstantial evidence that intramembrane particles arise from ($Ca^{2+}$-$Mg^{2+}$)-ATPase (Jilka *et al.*, 1975). In reconstituted systems, however, the anisotropy in the distribution of the IMP observed in the native membrane is not retained (Wang *et al.*, 1979); there are essentially equal numbers of IMP on both fracture faces of reconstituted liposomes (Jilka *et al.*, 1975). In systems consisting of purified ($Ca^{2+}$-$Mg^{2+}$)-ATPase reconstituted into vesicles with defined lipid composition, numbers of intramembrane particles correlate with protein content (Wang *et al.*, 1979) further providing evidence for a correlation of intramembrane particles and the calcium pump.

Negative staining has been used for the visualization of the surface of the sarcoplasmic reticulum membrane. With this technique outer surface projections or particles have been reported by many investigators (Saito *et al.*, 1978; Stewart and McLennan, 1974; see Moller *et al.*, 1982 for references). These projecting particles can be removed by prolonged digestion with trypsin (Inesi and Scales, 1974) and are believed to represent the outer pole of the pump protein. These surface particles of 40 Å are similarly observed in reconstituted systems (Thorley-Lawson and Green, 1973; Jilka *et al.*, 1975; Saito *et al.*, 1978; Wang *et al.*, 1979) and demonstrated at both inner and outer surfaces of such membranes (Saito *et al.*, 1978). Their presence and that of IMP are dependent on the incorporation of ($Ca^{2+}$-$Mg^{2+}$)-ATPase and thus they are related to the pump enzyme, and believed to represent the outer pole of the pump (Thorley-Lawson and Green, 1973; Jilka *et al.*, 1975). In both native and reconstituted membrane preparations of sarcoplasmic reticulum the surface particles are far more numerous than the intramembrane particles (Jilka *et al.*, 1975). The estimated number of pump molecules in native sarcoplasmic reticulum and in reconstituted systems (see Wang *et al.*, 1979) is greater by a factor of two than the sum of intramembrane particles at both fracture faces suggesting that the IMP may consist of two or more molecules of pump protein (Wang *et al.*, 1979).

IMPs occur isolated in membranes of liposomes reconstituted from ($Ca^{2+}$-$Mg^{2+}$)-ATPase and excess lipids. These single particles are capable of optimal $Ca^{2+}$ transport (Andersen *et al.*, 1983) suggesting that a single IMP is functional. There is a large amount of data available on the association number of the pump protein (see Moller *et al.*, 1982). These are based on studies on the functional properties and the aggregation state of detergent solubilized ($Ca^{2+}$-$Mg^{2+}$)-ATPase; peptide cross linking *in situ* in the membrane of sarcoplasmic reticulum; correlation of numbers of ($Ca^{2+}$-$Mg^{2+}$)-ATPase protomers and numbers of intramembrane particles; and spectroscopy studies. Some of these data have provided conflicting results. Self association of the protein is indicated by a majority of these studies and the ($Ca^{2+}$-$Mg^{2+}$)-ATPase may be associated in the membrane as at least a dimer (Wang *et al.*, 1979), a trimer (Malan *et al.*, 1975), trimer or tetramer (le Maire, 1976; also see Moller *et al.*, 1982), or tetramer (Murphy, 1976). Fluorescence energy-transfer experiments provide evidence for the existence of ATPase oligomers in reconstituted systems (Vanderkooi *et al.*, 1977; for further discussion see Moller *et al.*, 1982). The association number of ($Ca^{2+}$-$Mg^{2+}$)-ATPase still remains unresolved.

(i) *The light harvesting complex of the thylakoid membrane*

Higher plants and green algae contain a large amount of chlorophyll *a*/*b* protein complex on their photosynthetic membranes responsible for harvesting solar energy (for reviews see Thornber, 1975; Bennett, 1979b; Thornber and Barber, 1979; Thornber *et al.*, 1979). Park and Pheifhofer (1968) proposed that 'quantasomes', the particles associated with the outer surface of thylakoid membranes and first described by Park and Biggins (1964), may represent the photosynthetic units. The photochemical activity of the pigment of isolated particles was demonstrated supporting this notion (see Wessels, 1977).

Freeze–fracture electron microscopy has been used to demonstrate two types of IMP on the thylakoid membranes differing in their average size, distribution and partitioning into the two fracture faces (see Anderson, 1975; Staehelin, 1975). The large particles ($\approx$160–170 Å) are found on the internal or luminal fracture face (EF) in stacked regions while the much smaller ($\approx$80 Å) particles are found on the external fracture face (PF) in both stacked and non-stacked regions (Goodenough and Staehelin, 1971; Staehelin, 1975; Staehelin *et al.*, 1976). Characterization of chloroplast lamella membrane fractions (Sane *et al.*, 1970; Arntzen *et al.*, 1969) suggests a relationship between photosystem I activity and the small particles of the external leaflet, and photosystem II activity and the large particles of the internal leaflet. It is now generally believed that the IMPs of the external leaflet correspond to photosystem I and the IMPs of the internal leaflet correspond to photosystem II reaction centres and their associated light harvesting antenna pigments respectively (Armond *et al.*, 1977). Deep-etching reveals particles at the luminal surface (ES) of stacked regions of spinach chloroplast membranes. The large particles of the E surface have been correlated with the large intramembrane EF particles (Staehelin *et al.*, 1976). A close correlation has been shown between the light harvesting complex and the large intramembrane particles. Working with inner, middle, and outer leaves of lettuce which vary in their content of light harvesting chlorophyll-protein complex as well as with chlorophyll *b*-deficient mutants of barley, Henriques and Park (1976) show a correlation between the light harvesting chlorophyll protein complex and the large intramembrane particles in stacked regions. Furthermore, it has been shown that during membrane differentiation of pea plastids by light greening, the light-harvesting chlorophyll proteins are inserted into chloroplast lamella. Concomitant with this there is an increase in particle size of intramembrane particles of internal fracture leaflets (EF) in stacked regions (Armond *et al.*, 1977).

The purified light-harvesting complex of spinach chloroplast membranes has been quantitatively incorporated into liposomes containing the natural mixture of the four major chloroplast diacyl lipids and studied with a freeze–thaw technique (Ryrie *et al.*, 1980); in this study incorporation of light harvesting complex resulted in the appearance of intramembrane particles 80–100 Å in diameter. Highly purified chlorophyll *a*/*b* light-harvesting complex obtained from Triton X-100-solubilized pea and barley chloroplasts were reconstituted into liposomes by a different procedure (McDonnel and

Staehelin, 1980). This resulted in the appearance of intramembranous particles of approximately 80 Å diameter. The findings of both of these reconstitution studies provide evidence that the light-harvesting complex can give rise to intramembrane particles.

The light-harvesting complex spans the membrane and has different antigenic sites on the two surfaces (Andersson *et al.*, 1982). Proteolysis of membranes with trypsin results in cleavage from one end of the molecule (Süss *et al.*, 1976). The observed effects of the removal of this protruding peptide suggest that it is involved in the ability of light-harvesting complex to form higher order structures (see Bennett, 1979b).

It is possible to isolate the light harvesting complex from chloroplasts and precipitate them into stacked membrane crystals (see Kühlbrandt *et al.*, 1983). The structure of such crystalline arrays from a variety of sources have been studied with electron microscopy as well as X-ray diffraction (see Kühlbrandt *et al.*, 1983). Recently, these investigators have studied the structure of such membrane crystals obtained from the light harvesting complex of pea chloroplast by electron microscopy and image analysis. They have identified in the crystalline fraction two polypeptides of 25 000 and 27 000 daltons identical to the polypeptides isolated from pea thylakoids (Bennett, 1979a; Bennett *et al.*, 1981). Based on their studies, Kühlbrandt and co-workers (1983) suggest that the complex consists of a heterodimer of two polypeptides, and that three of these subunits form the two structural units. Two structural units of threefold symmetry and opposite handedness form a crystallographic unit cell. These investigators suggest that each triangular unit corresponds to one light harvesting complex of the thylakoid membrane. It has been suggested that the native light harvesting complex is an oligomer (Takahashi and Gross, 1978, Dunkley and Anderson, 1979; McDonnel and Staehelin, 1980, Ryrie *et al.*, 1980). Caution must be exercised in the extrapolation of findings from these crystalline arrays to the native membrane. It should be noted, however, that in the stacked regions of the chloroplast membrane, the intramembrane particles of the internal leaflet are similar in shape and size (Goodenough and Staehelin, 1971) to those of the crystalline membranes.

## 10.4 Peripheral membrane proteins

### 10.4.1 GENERAL CONSIDERATIONS

Peripheral membrane proteins are loosely associated with the membrane bilayer and do not intercalate into its hydrophobic core (Singer and Nicolson, 1972). The tendency for fracture planes to split phospholipid bilayers prohibits visualization of peripheral proteins by freeze–fracturing alone. An additional step, freeze–etching, must be employed. Generally, the membranes must be rapidly frozen after soluble cytosol components have been extracted. Because of these restrictions, most studies of peripheral proteins have been in the red cell or other specialized systems in which membranes with adherent proteins can be readily isolated free of cytoplasm. Exploration of the potential of the ultra-low temperature rapid-freeze technique (Heuser and Kirshner, 1980) to study the periphery of cell membranes has only begun.

### 10.4.2 RED CELL MEMBRANES

Freeze–etching techniques have been of some value in studies of peripheral proteins of the mammalian erythrocyte. These cells are readily obtained, the membranes can easily be freed of cytoplasmic components, and monolayers of uniformly oriented membranes can be produced.

Shotton *et al.* (1978) using conventional freeze–fracture methods, examined the protoplasmic surface (PS) of erythrocyte ghosts and inside-out vesicles in a study of distribution of peripheral proteins and intramembranous particles. They found particulate proteins, specifically labelled and aggregated by a ferritin–anti-spectrin conjugate, on the PS. These proteins appeared to be 500 Å in length and composed of three roughly uniform globular domains. Actin filaments were not identified, and as only small areas of PS were seen the inter-relationship of the elements observed could not be ascertained.

Weinstein and co-workers (1978a, 1980) deep-etched red cell 'ghosts' and demonstrated a coarse granulofibrillar array on the P-surface. When they extracted the peripheral proteins including spectrin and actin, they uncovered an underlying fine granulofibrillar array which they interpreted as representing cytoplasmic domains of the integral protein, band 3, in a partially extended configuration. This interpretation was supported by experiments in which the fine granulofibrillar array was decorated with rabbit muscle G3PD which has a strong affinity for the cytoplasmic pole of band 3. G3PD binding was abolished by trypsinization which excises the cytoplasmic pole of band 3 and, correspondingly, the fine granulofibrillar array.

Other studies have produced comparable results. Using the rapid-freeze, deep-etch technique, Shelton and co-workers (1979) identified a network on the PS of erythrocyte 'ghosts'. This network was also removed by conditions known to extract spectrin, and restored by incubation with a spectrin–actin mixture under conditions known to allow re-association of the proteins with spectrin-depleted vesicles. The network was composed of 60 Å and 30 Å diameter elements approximately 1000 Å long. The thicker elements often joined end-to-end with one to four other thick elements, but were occasionally free. Rarely, pairs of 30 Å elements joined to form 60 Å elements. Actin filaments were not observed, although they have been identified by this technique in other cell types (Heuser and Kirshner, 1980). Espevik and

Elgsaeter (1981) used a liquid propane rapid-freeze technique to study haemolysing red cells and 'ghosts'. They found evidence of fibre-like structures, some over 1000 Å long, in fracture faces. Under conditions favouring spectrin and action release from the membranes, continuous fibre-like structures were seen extending from the E-face and P-surface, but the authors could not conclusively prove that these were not artifact.

Nermut and co-workers have developed techniques for producing red cell monolayers and lysing or freeze–fracturing cells on those monolayers (Lang *et al.*, 1981; Nermut, 1981). Subsequent freeze–etching reveals large areas of PS. These techniques have demonstrated particles and interconnecting filaments, up to 800 Å long and roughly 80 Å thick but of variable dimensions, although the latter could have also been composed of linear aggregates of fine particles. The filaments interconnect and could be longer but bending of filaments at points where they join could not be distinguished from filament terminations. Nermut has recently stressed that the networks seen are discontinuous (Nermut, 1983). We have employed one of Nermut's methods to examine ageing changes in red cells, and found a continuous network of interconnecting filaments, which change in appearance in red cells aged *in vivo* (Loew *et al.*, in press).

Peripheral proteins have only been identified by freeze-etch techniques in a few other systems. A light-activated GTP binding protein has been identified as a peripheral particle associated with rod disc membranes in the retina (Roof *et al.*, 1982). This protein can be specifically extracted and allowed to re-associate with these membranes. Protein quantitation by gel electrophoresis and particle counting of freeze–etch micrographs yield parallel results.

In a few systems it has been possible to visualize the attachments of cytoskeletons of nucleated cells to the plasma membrane. In hair cells of the chick inner ear, several classes of connections between actin filaments of the stereocilium and the cuticular plate have been visualized by the quick-freeze deep-etch technique (Hirokawa and Tilney, 1982). These are poorly seen in thin-section electron microscopy and may be labile during aldehyde fixation and subsequent processing. In the mouse intestinal brush border, fibrils have been demonstrated connecting actin filaments to the plasma membrane, and identified as fodrin using antibody conjugated to ferritin (Hirokawa *et al.*, 1983).

Future use of freeze–etch techniques for studying peripheral proteins and attachments of cytoskeletons to cell membranes may require improvement in technique in order to permit proper orientation of cells and sufficiently rapid freezing to avoid ice crystal artifacts. However, the study of cell structure without exposure to fixatives, organic solvents, heavy metal stains, or embedding resin permits detection, localization and even immunological identification of proteins poorly visualized by and even removed in preparation for thin-section transmission electron microscopy.

## 10.5 Conclusion

Freeze–fracture and freeze–etch electron microscopy have provided important new information on the structure of biological membranes. The freeze–fracture method has uncovered a novel intramembrane structure, the intramembrane particle or IMP. The exact nature of IMPs has been the subject of intense investigation. While lipids may be a constituent of IMPs in some settings, extensive research supports the view that some, if not most of the intramembrane particles seen in cell membranes represent integral membrane proteins.

Several factors inherent in modern freeze–fracture methodology tend to minimize problems of dimensional distortion present in early freeze–fracture studies. Combining the advantages of unidirectional and rotary shadowing in identification, quantitation and subunit structure analysis of particles provides a more comprehensive approach to the ultrastructure of membrane proteins. In spite of these considerations, however, the extent to which the observed morphology of IMP reflects the dimensions and ultrastructure of integral membrane proteins remains unresolved. Artifacts could be due to plastic deformation as the membrane is fractured, thermal distortion during shadowing, and dimensional distortion due to non-ideal deposition of shadowing metal or complex irregularities in particle shape not resolved in the replica.

Despite these limitations, freeze–fracture techniques have proven useful in the study of integral membrane proteins. In a number of systems, the putative principle protein component of the major class of IMP has been identified. In many systems a direct correlation between intramembrane particles and a defined protein has been demonstrated. Some of the most direct evidence is based on the reconstitution of defined proteins into liposomes. In many systems, the transmembrane nature of the respective membrane proteins has been demonstrated. Data obtained from biochemical studies, ultrastructural studies, quantitative electron microscopy, and X-ray diffraction studies have elucidated the organization of these integral membrane proteins in the bilayer. In these systems it can be assumed that the IMPs may be attributed to specific species of integral membrane proteins. The association number of the proteins forming an IMP, however, remains unresolved in many cases.

Freeze–etch techniques may permit visualization of some peripheral membrane proteins although the images obtained to date are generally less than optimal. While not subject to fracture deformation artifacts, such replicas are subject to ice crystal artifacts, sublimation artifacts and the effects of residual salts or adherent non-membrane protein. Recently developed ultralow temperature freezing and deep-etching techniques yield dramatic pictures. The limiting factor in the use of this technology has been the paucity of available techniques for cell preparation.

Within the past few years techniques have been developed to permit immunological or receptor-specific recognition of individual protein molecules within IMPs. In the future such advances should permit more detailed study of the arrangement of both integral and peripheral membrane proteins and their associations *in situ*.

## Acknowledgements

The authors thank Ms Elaine Sundin for expert technical assistance and aid with the preparation of the manuscript, Ms Sandra Velasco for the typing of the manuscript, and Ms Denise Wiler for photographic work.

## 10.6 References

Abermann, R., Salpeter, M. M. and Bachmann, L. (1972) in *Principles and Techniques of Electron Microscopy* (ed M. A. Hayat), Von Nostrand Reinhold, New York, Vol. 2, pp. 197–217.

Adams, A. J., Somers, R. L. and Shichi, H. (1979) *Photochem. Photobiol.*, **29**, 687–92.

Andersen, J. P., Skriver, E., Mahrous, T. S. and Moller, J. V. (1983) *Biochim. Biophys. Acta*, **728**, 1–10.

Anderson, J. M. (1975) *Biochim. Biophys. Acta*, **416**, 191–235.

Anderson, M. J. and Cohen, M. W. (1974) *J. Physiol., London*, **237**, 385–400.

Anderson, M. J. and Cohen, M. W. (1977) *J. Physiol., London*, **268**, 757–73.

Andersson, B., Anderson, J. M. and Ryrie, I. J. (1982) *Eur. J. Biochem.*, **123**, 465–72.

Andrews, L. D. and Cohen, A. I. (1979) *J. Cell Biol.*, **81**, 215–28.

Anholt, R., Fredkin, D. R., Derrinck, T., Ellisman, M., Montal, M. and Lindstrom, J. (1982) *J. Biol. Chem.*, **257**, 7122–34.

Armond, P. A., Staehelin, L. A. and Arntzen, C. J. (1977) *J. Cell Biol.*, **73**, 400–18.

Arntzen, C. J., Dilley, R. A. and Crane, F. L. (1969) *J. Cell Biol.*, **43**, 16–31.

Azzi, A. (1980) *Biochim. Biophys. Acta*, **594**, 231–52.

Bächi, Th. (1980) *Acta Histochem.* Suppl., **XXII**, S, 291–7.

Bächi, Th., Whiting, K., Tanner, M. J. A., Metaxas, M. N. and Anstee, D. J. (1977) *Biochim. Biophys. Acta*, **464**, 635–9.

Basinger, S., Bok, D. and Hall, M. (1976) *J. Cell Biol.*, **69**, 29–42.

Bearer, E. L., Düzgünes, N., Friend, D. S. and Papahadjopoulos, D. (1982) *Biochim. Biophys. Acta*, **693**, 93–8.

Benedetti, E. L. and Emmelot, P. (1968) *J. Cell Biol.*, **38**, 15–24.

Bennett, J. (1979a) *Eur. J. Biochem.*, **99**, 133–7.

Bennett, J. (1979b) *Trends Biochem. Sci.*, **4**, 268–71.

Bennett, J., Markwell, J. P., Skrdla, M. P. and Thornber, J. P. (1981) *FEBS Lett.*, **131**, 325–30.

Blaurock, A. E. (1975) *J. Mol. Biol.*, **93**, 139–58.

Blaurock, A. E. and Stoeckenius, W. (1971) *Nature (London) New Biol.*, **233**, 152–5.

Borovjagin, V. L., Vergara, J. A. and McIntosh, T. J. (1982) *J. Membr. Biol.*, **69**, 199–212.

Branton, D. (1966) *Proc. Natl. Acad. Sci. USA*, **55**, 1048–56.

Branton, D. (1971) *Philos. Trans. R. Soc. London Ser. B. Biol. Sci.*, **261**, 133–8.

Branton, D., Bullivant, S., Gilula, N. B., Karnovsky, M. Y., Moor, H., Mühlethaler, K., Northcote, D. H., Packer, L., Satir, B., Satir, P., Speth, V., Staehelin, L. A., Steere, R. L. and Weinstein, R. S. (1975) *Science*, **190**, 54–6.

Branton, D. and Moor, H. (1964) *J. Ultrastruct. Res.*, **11**, 401–11.

Bretscher, M. S. (1971a) *J. Mol. Biol.*, **59**, 351–7.

Bretscher, M. S. (1971b) *Nature (London) New Biol.*, **231**, 229–32.

Bretscher, M. S. (1973) *Science*, **181**, 622–9.

Bullivant, S. and Ames, A. (1966) *J. Cell Biol.*, **29**, 435–47.

Cabantchik, Z. I., Knauf, P. A. and Rothstein, A. (1978) *Biochim. Biophys. Acta*, **515**, 239–302.

Cabantchik, Z. I. and Rothstein, A. (1974) *J. Membr. Biol.*, **15**, 227–48.

Capaldi, R. A., Bell, R. L. and Branchek, T. (1977) *Biochem. Biophys. Res. Commun.*, **74**, 425–433.

Cartaud, J., Benedetti, E. L., Cohen, J. B., Meunier, J-C. and Changeux, J-P. (1973), *FEBS Lett.*, **33**, 109–13.

Cartaud, J., Benedetti, E. L., Sobel, A. and Changeux, J.-P. (1978) *J. Cell Sci.*, **29**, 313–37.

Caspar, D. L. D., Goodenough, D. A., Makowski, L. and Phillips, W. C. (1977) *J. Cell Biol.*, **74**, 605–28.

Chabre, M. (1975) *Biochim. Biophys. Acta*, **382**, 322–35.

Chabre, M., Cavaggioni, A., Osborne, H. B. and Gulik-Krzywicki, T. (1972) *FEBS Lett.*, **26**, 197–202.

Chalcroft, J. P. and Bullivant, S. (1970) *J. Cell Biol.*, **47**, 49–60.

Chen, Y. S. and Hubbell, W. L. (1973) *Exp. Eye Res.*, **17**, 517–32.

Cherry, R. J., Buerkli, A., Busslinger, M., Schneider, G. and Parish, G. R. (1976) *Nature (London)*, **263**, 389–93.

Clark, A. W. and Branton, D. (1968) *Z. Zellforsch*, **91**, 586–603.

Clarke, S. (1975) *J. Biol. Chem.*, **250**, 5459–69.

Cohen, S. A. and Pumplin, D. W. (1979) *J. Cell Biol.*, **82**, 494–516.

Corless, J. M., Cobbs, III, W. H., Costello, M. J. and Robertson, J. D. (1976) *Exp. Eye Res.*, **23**, 295–324.

Costello, M. J. and Frey, T. G. (1982) *J. Mol. Biol.*, **162**, 131–56.

Darszon, A., Vandenberg, C. A., Schonfeld, M., Ellisman, M. H., Spitzer, N. C. and Montal, M. (1980) *Proc. Natl. Acad. Sci. USA*, **77**, 239–43.

Deamer, D. W. and Baskin, R. J. (1969) *J. Cell Biol.*, **42**, 296–307.

Dean, W. L. and Tanford, C. (1978) *Biochemistry*, **17**, 1683–90.

Deatherage, J. F., Henderson, R. and Capaldi, R. A. (1980) in *Electron Microscopy of Molecular Dimensions* (eds W. Baumelster and W. Vogell), Springer-Verlag, New York, pp. 91–100.

Deguchi, N., Jorgensen, P. L. and Maunsbach, A. B. (1977) *J. Cell Biol.*, **75**, 619–34.

DeKruijff, B. and Cullis, P. R. (1980) *Biochim. Biophys. Acta*, **602**, 477–90.

Dörst, H-J. and Schubert, D. (1979) *Hoppe-Seyler's Z. Physiol. Chem.*, **360**, S. 1605–18.

Downer, N. W., Robinson, N. C. and Capaldi, R. A. (1976) *Biochemistry*, **15**, 2930–6.

Drickamer, L. K. (1976) *J. Biol. Chem.*, **251**, 5115–23.

Dunkley, P. R. and Anderson, J. M. (1979) *Biochim. Biophys. Acta*, **545**, 174–87.

Dupont, Y., Cohen, J. B. and Changeux, J-P. (1974) *FEBS Lett.*, **40**, 130–3.

Dupont, Y., Harrison, S. C. and Hasselbach, W. (1973) *Nature (London)*, **244**, 555–8.
Edwards, H. H., Mueller, T. J. and Morrison, M. (1979) *Science*, **203**, 1343–6.
Ellisman, M. K., Rash, J. E., Staehelin, L. A. and Porter, K. R. (1976) *J. Cell Biol.*, **68**, 752–74.
Engelman, D. M., Henderson, R., McLachlan, A. D. and Wallace, B. A. (1980) *Proc. Natl. Acad. Sci. USA*, **77**, 2023–7.
Espevik, T. and Elgsaeter, A. (1981) *J. Microsc.*, **122**, 159–63.
Evans, W. H. and Gurd, J. W. (1972) *Biochem. J.*, **128**, 691–700.
Eytan, G. D., Carroll, R. C., Schatz, G. and Racker, E. (1975) *J. Biol. Chem.*, **250**, 8598–603.
Fertuck, H. C. and Salpeter, M. M. (1974) *Proc. Natl. Acad. Sci. USA*, **71**, 1376–8.
Fertuck, H. C. and Salpeter, M. M. (1976) *J. Cell Biol.*, **69**, 144–58.
Finbow, M., Yancey, S. B., Johnson, R. and Revel, J-P. (1980) *Proc. Natl. Acad. Sci. USA*, **77**, 970–4.
Frey, T. G., Chan, S. H. P. and Schatz, G. (1978) *J. Biol. Chem.*, **253**, 4389–95.
Frey, T. G., Costello, M. J., Karlsson, B., Haselgrove, J. C. and Leigh, J. S. (1982) *J. Mol. Biol.*, **162**, 113–30.
Fukuda, M., Eshdat, Y., Tarone, G. and Marchesi, V. T. (1978) *J. Biol. Chem.*, **253**, 2419–28.
Fukuda, M. N., Papermaster, D. S. and Hargrave, P. A. (1979) *J. Biol. Chem.*, **254**, 8201–7.
Fuller, S. D., Capaldi, R. A. and Henderson, R. (1979) *J. Mol. Biol.*, **134**, 305–27.
Gassner, D. and Komnick, H. (1983) *Eur. J. Cell. Biol.*, **29**, 226–35.
Gerber, G. E., Anderegg, R. J., Herlihy, W. C., Gray, C. P., Biemann, K. and Khorana, H. G. (1979) *Proc. Natl. Acad. Sci. USA*, **76**, 227–31.
Girardet, M., Geering, K., Frantes, J. M., Geser, D., Rossier, B. C., Kraehenbuhl, J-P. and Bron, C. (1981) *Biochemistry*, **20**, 6684–91.
Glaubensklee, C. S., Evan, A. P. and Galey, W. R. (1982) *Vox Sang.*, **42**, 262–71.
Goodenough, D. A. (1975) in *Methods in Membrane Biology*, (ed E. D. Korn), Plenum Press, New York, Vol. 3, pp. 51–80.
Goodenough, D. A. (1976) *J. Cell Biol.*, **68**, 220–31.
Goodenough, D. A. and Stoeckenius, W. (1972) *J. Cell Biol.*, **54**, 646–56.
Goodenough, U. W. and Staehelin, L. A. (1971) *J. Cell Biol.*, **48**, 594–619.
Grant, C. W. M. and McConnell, H. M. (1974) *Proc. Natl Acad. Sci. USA*, **71**, 4653–7.
Grohovaz, F., Limbrick, A. R. and Miledi, R. (1982) *Proc. R. Soc. London Ser., B*, **215**, 147–54.
Gross, H., Kuebler, O., Bas, E. and Moor, H. (1978) *J. Cell Biol.*, **79**, 646–56.
Guidotti, G. (1972) *Annu. Rev. Biochem.*, **41**, 731–52.
Hackenbrock, C. G. and Hammon, K. M. (1975) *J. Biol. Chem.*, **250**, 9185–97.
Hall, C. E. (1960) *J. Biophys. Biochem. Cytol.*, **7**, 613–18.
Hargrave, P. A. (1977) *Biochim. Biophys. Acta*, **492**, 83–94.
Hasselbach, W. and Elfvin, L.-G. (1967) *J. Ultrastruct. Res.*, **17**, 598–622.
Heller, J. (1968) *Biochemistry*, **7**, 2906–13.
Henderson, D., Eibl, H. and Weber, K. (1979) *J. Mol. Biol.*, **132**, 193–218.
Henderson, R. (1975) *J. Mol. Biol.*, **93**, 123–38.
Henderson, R., Capaldi, R. A. and Leigh, S. J. (1977) *J. Mol. Biol.*, **112**, 631–48.
Henderson, R. and Unwin, P. N. T. (1975) *Nature (London)*, **257**, 28–32.

Henderson, W. J. and Griffiths, K. (1972) in *Principles and Techniques of Electron Microscopy* (ed. M. A. Hayat), Von Nostrand Reinhold, New York, Vol. 2, pp. 151–93.
Henriques, F. and Park, R. B. (1976) *Proc. Natl. Acad. Sci. USA,* **73,** 4560–4.
Herbette, L., Marquardt, J., Scarpa, A. and Blasie, J. K. (1977) *Biophys. J.,* **20,** 245–72.
Heuser, J. E. and Kirschner, M. W. (1980) *J. Cell Biol.,* **86,** 212–34.
Heuser, J. E. and Salpeter, S. R. (1979) *J. Cell Biol.,* **82,** 150–73.
Hirokawa, N., Cheney, R. E. and Willard, M. (1983) *Cell,* **32,** 953–65.
Hirokawa, N. and Heuser, J. (1982a) *Cell,* **30,** 395–406.
Hirokawa, N. and Heuser, J. E. (1982b) *J. Neurocytol.,* **11,** 487–510.
Hirokawa, N. and Tilney, L. G. (1982) *J. Cell Biol.,* **95,** 249–61.
Höchli, M. and Hackenbrock, C. R. (1979) *Proc. Natl. Acad. Sci. USA,* **76,** 1236–40.
Hokin, L. E. (1979) in *Function and Molecular Aspects of Biomembrane Transport,* (eds E. Quagliariello, F. Palmieri, S. Papa and M. Klinkenberg), Elsevier, North Holland, pp. 495–503.
Hong, K. and Hubbell, W. L. (1972) *Proc. Natl. Acad. Sci. USA,* **69,** 2617–21.
Hopkins, B. E., Wagner, H., Jr. and Smith, T. W. (1976) *J. Biol. Chem.,* **251,** 4365–71.
Hubbell, W. L. and Fung, K-K. (1979) in *Membrane Transduction Mechanisms* (eds R. A. Cone and J. E. Dowling), Raven Press, New York, pp. 17–25.
Hui, S. W. and Stewart, T. P. (1981) *Nature (London),* **290,** 427.
Inesi, G. and Scales, D. (1974) *Biochemistry,* **13,** 3298–306.
Jan, L. Y. and Revel, J-P. (1974) *J. Cell Biol.,* **62,** 257–73.
Jilka, R. L., Martonosi, A. N. and Tillack, T. W. (1975) *J. Biol. Chem.,* **250,** 7511–24.
Jorgensen, P. L. (1974) *Biochim. Biophys. Acta,* **356,** 36–67.
Jorgensen, P. L. (1975) *Q. Rev. Biophys.,* **7,** 239–79.
Jorgensen, P. L. (1980) *Physiol. Rev.,* **60,** 864–917.
Kawato, S., Lehner, C., Mueller, M. and Cherry, R. J. (1982) *J. Biol. Chem.,* **257,** 6470–6.
Khodadad, J. K. and Weinstein, R. S. (1980) *Proc. 38th Annu. Mtg. Electron Microsc. Soc. Amer.,* 788–9.
Khodadad, J. K. and Weinstein, R. S. (1983) *J. Membr. Biol.,* **72,** 161–71.
Khodadad, J. K. and Weinstein, R. S. (1985), *Biochem. Biophys. Res. Commun.,* **130,** 493 9.
Khorana, H. G., Gerber, G. E., Herlihy, W. C., Gray, C. P., Anderegg, R. J., Nihei, K. and Biemann, K. (1979) *Proc. Natl. Acad. Sci. USA,* **76,** 5046–50.
Klymkowsky, M. W. and Stroud, R. M. (1979) *J. Mol. Biol.,* **128,** 319–34.
Koehler, J. K. (1972) in *Principles and Techniques of Electron Microscopy* (ed M. A. Hayat) Von Nostrand Reinhold, New York, Vol. 2, pp. 53–98.
Krebs, W. and Kühn, H. (1977) *Exp. Eye Res.,* **25,** 511–26.
Kreutzinger, G. O. (1968) *Proc. 26th Annu. Mtg. Electron Microsc. Soc. Amer.,* 234–5.
Kühlbrandt, W., Thaler, Th. and Wehrli, E. (1983) *J. Cell Biol.,* **96,** 1414–24.
Kuszak, J. R., Rae, J. L., Pauli, B. U. and Weinstein, R. S. (1982) *J. Ultrastruct. Res.,* **81,** 249–56.
Kyte, J. (1975) *J. Biol. Chem.,* **250,** 7443–9.
Lang, R. D. A., Nermut, M. V. and Williams, L. D. (1981) *J. Cell Sci.,* **49,** 383–99.
Le Maire, M., Moller, J. V. and Tanford, C. (1976) *Biochemistry,* **15,** 2336–42.
Lentz, T. L., Mazurkiewicz, J. E. and Rosenthal, J. (1977) *Brain Res.,* **132,** 423–42.

Lessin, L. S., Jensen, W. N. and Ponder, E. (1969) *J. Exp. Med.*, **130**, 443–66.
Lickfeld, K. G., Achterrath, M. and Hentrich, F. (1972) *J. Ultrastruct. Res.*, **38**, 279–87.
Lindstrom, J., Anholt, R., Einarson, B., Engel, A., Osame, M. and Montal, M. (1980) *J. Biol. Chem.*, **255**, 8340–50.
Lingham, R. B., Stewart, D. J. and Sen, A. K. (1980) *Biochim. Biophys. Acta*, **601**, 229–34.
Loew, J. M., Marikovsky, Y., Tazelaar, H. D. and Weinstein, R. S. in *Contractile Proteins in Muscle and Non-muscle Cell Systems: Biochemistry, Physiology and Pathology*, (eds. E. E. Alia, N. Arena and M. A. Russo), Praeger Scientific Publishing, New York (in press).
Ludwig, B., Downer, N. W. and Capaldi, R. A. (1979) *Biochemistry*, **18**, 1401–7.
Makowski, L., Caspar, D. L. D., Phillips, W. C. and Goodenough, D. A. (1977) *J. Cell Biol.*, **74**, 629–45.
Malan, N. T., Sabbadini, R., Scales, D. and Inesi, G. (1975) *FEBS Lett.*, **60**, 122–5.
Margaritis, L. H., Elgsaeter, A. and Branton, D. (1977) *J. Cell Biol.*, **72**, 47–56.
Maul, G. G. (1979) in *Freeze Fracture: Methods, Artifacts, and Interpretations* (eds J. E. Rash and C. S. Hudson), Raven Press, New York, pp. 37–42.
Maunsbach, A. B., Skriver, E. and Jorgensen, P. L. (1979) in *Na, K-ATPase: Structure and Kinetics* (eds J. C. Skou and J. G. Norby), Academic Press, London, pp. 3–13.
McDonnel, A. and Staehelin, L. A. (1980) *J. Cell Biol.*, **84**, 40–56.
McNutt, N. S. and Weinstein, R. S. (1970) *J. Cell Biol.*, **47**, 666–88.
McNutt, N. and Weinstein, R. S. (1973) *Prog. Biophys. Mol. Biol.*, **26**, 45–101.
Miller, R. G. (1980) *Nature (London)*, **287**, 166–7.
Misra, D. N. and Das Gupta, N. N. (1965) *J. R. Microsc. Soc.*, **84**, 373–84.
Moller, J. V., Andersen, J. P. and le Maire, M. (1982) *Mol. Cell. Biochem.*, **42**, 83–107.
Moor, H. and Mühlethaler, K. (1963) *J. Cell Biol.*, **17**, 609–28.
Moor, H., Mühlethaler, K., Waldner, H. and Frey-Wissling, A. (1961) *J. Biophys. Biochem. Cytol.*, **10**, 1–13.
Mueller, U. and Cherry, R. J. (1981) *Acta Histochem.*, Suppl., **XXIII**, S 205–9.
Murphy, A. J. (1976) *Biochem. Biophys. Res. Commun.*, **70**, 160–6.
Nermut, M. V. (1981) *Eur. J. Cell Biol.*, **25**, 265–71.
Nermut, M. V. (1983) *Trends Biochem. Sci.*, **8**, 303–6.
Nickel, E. and Potter, L. T. (1973) *Brain Res.*, **57**, 508–17.
Nicolson, G. L. (1973) *J. Cell Biol.*, **57**, 373–87.
Nigg, E. A., Bron, C., Girardet, M. and Cherry, R. J. (1980) *Biochemistry*, **19**, 1887–93.
Orci, L., Perrelet, A. and Dunant, Y. (1974) *Proc. Natl. Acad. Sci. USA*, **71**, 307–10.
Osterhelt, D. and Stoeckenius, W. (1971) *Nature (London), New Biol.*, **233**, 149–52.
Ovchinnikov, Yu. A., Abdulaev, N. G., Feigina, M. Yu., Kiselev, A. V. and Lobanov, N. A. (1979) *FEBS Lett.*, **100**, 219–24.
Park, R. B. and Biggins, J. (1964) *Science*, **144**, 1009–11.
Park, R. B. and Pfeifhofer, A. O. (1968) *Proc. Natl. Acad. Sci. USA*, **60**, 337–43.
Pauli, B. U., Alroy, J. and Weinstein, R. S. (1983) in *The Pathology of Bladder Cancer* (eds S. Cohen and G. T. Bryan), CRC Press, Boca Raton, Florida, Vol. 2, pp. 41–140.
Pauli, B. U. and Weinstein, R. S. (1982) *Cancer Res.*, **42**, 2289–97.
Peng, H. B. and Nakajima, Y. (1978) *Proc. Natl. Acad. Sci. USA*, **75**, 500–4.
Peper, K. Dreyer, F., Sandri, C., Akert, K. and Morr, H. (1974) *Cell Tissue Res.*, **149**, 437–55.
Pinto da Silva, P. and Kachar, B. (1982) *Cell*, **28**, 441–50.

Pinto da Silva, P. and Nicolson, G. L. (1974) *Biochim. Biophys. Acta*, **363**, 311–19.
Pinto da Silva, P., Kachar, B., Torrisi, M. R., Brown, C. and Parkinson, C. (1981a) *Science*, **213**, 230–3.
Pinto da Silva, P., Parkinson, C. and Dwyer, N. (1981b) *J. Histochem. Cytochem.*, **29**, 917–28.
Pinto da Silva, P., Parkinson, C. and Dwyer, N. (1981c) *Proc. Natl. Acad. Sci. USA*, **78**, 343–7.
Preuss, L. E. (1965) *Lab. Invest.*, **14**, 919–32.
Pricam, C., Fisher, K. A. and Friend, D. S. (1977) *Anat. Rec.*, **189**, 595–608.
Rand, R. P., Reese, T. S. and Miller, R. G. (1981) *Nature (London)*, **293**, 237–8.
Rash, J. E. and Ellisman, M. H. (1974) *J. Cell Biol.*, **63**, 567–86.
Rash, J. E., Graham, W. F. and Hudson, C. S. (1979) in *Freeze Fracture: Methods, Artifacts, and Interpretations* (eds J. E. Rash and C. S. Hudson), Raven Press, New York, pp. 111–22.
Rash, J. E., Hudson, C. S., Albuquerque, E. X., Eldefrawi, M. E., Mayer, R. F., Graham, W. F., Johnson, T. J. A. and Giddings, F. D. (1981) *Ann. N.Y. Acad. Sci.*, **377**, 38–60.
Rash, J. E., Hudson, C. S. and Ellisman, M. H. (1978) in *Cell Membrane Receptors for Drugs and Hormones: A Multidisciplinary Approach* (eds R. W. Straub and L. Bolis), Raven Press, New York, pp. 47–68.
Rash, J. E., Johnson, T. J. A., Hudson, C. S., Copio, D. S., Graham, W. F., Eldefrawi, M. E. and Giddings, F. D. (1980) *Proc. 38th Annu. Mtg. Electron Microsc. Soc. Amer.*, 692–5.
Rash, J. E., Johnson, T. J. A., Hudson, C. S., Giddings, F. D., Graham, W. F. and Eldefrawi, M. E. (1982) *J. Microsc.*, **128**, 121–38.
Raubach, R. A., Nemes, P. P. and Dratz, E. A. (1974) *Exp. Eye Res.*, **18**, 1–12.
Revel, J-P. and Karnovsky, M. J. (1967) *J. Cell Biol.*, **33**, C7–C12.
Reynolds, J. A. and Karlin, A. (1978) *Biochemistry*, **17**, 2035–8.
Robertson, J. D. (1963) *J. Cell Biol.*, **19**, 201–21.
Robertson, J. D. (1981) *J. Cell Biol.*, **91**, 189s–204s.
Robertson, J. D. and Vergara, J. (1980) *J. Cell Biol.*, **86**, 514–28.
Röhlich, P. (1976) *Nature (London)*, **263**, 789–91.
Roof, D. J. and Heuser, J. E. (1982) *J. Cell Biol.*, **95**, 487–500.
Roof, D. J., Korenbrot, J. I. and Heuser, J. E. (1982) *J. Cell Biol.*, **95**, 501–9.
Rosenbluth, J. (1975) *J. Neurocytol.*, **4**, 697–712.
Ross, M. J., Klymkowsky, M. W., Agard, D. A. and Stroud, R. M. (1977) *J. Mol. Biol.*, **116**, 635–59.
Ruben, G. C. and Telford, J. N. (1975) *Proc. 33rd Annu. Mtg. Electron Microsc. Soc. Amer.*, 282–3.
Ruben, G. C. and Telford, J. N. (1980) *J. Microsc.*, **118**, 191–216.
Ruben, G. C., Telford, J. N. and Carroll, R. C. (1976) *J. Cell Biol.*, **68**, 724–39.
Ruoho, A. and Kyte, J. (1974) *Proc. Natl. Acad. Sci. USA*, **71**, 2352–6.
Ryrie, I. J., Anderson, J. M. and Goodchild, D. J. (1980) *Eur. J. Biochem.*, **107**, 345–54.
Saibil, H., Chabre, M. and Worcester, D. (1976) *Nature (London)*, **262**, 266–70.
Saito, A., Wang, C-T. and Fleischer, S. (1978) *J. Cell Biol.*, **79**, 601–16.
Sane, P. V., Goodchild, D. J. and Park, R. B. (1970) *Biochim. Biophys. Acta*, **216**, 162–78.
Satir, B. and Satir, P. (1979) in *Freeze Fracture: Methods, Artifacts, and Interpretations* (eds J. E. Rash and C. S. Hudson), Raven Press, New York, pp. 43–9.
Scales, D. and Inesi, G. (1976a) *Biophys. J.*, **16**, 735–51.

Scales, D. J. and Inesi, G. (1976b) *Arch. Biochem. Biophys.*, **176**, 392–4.
Schneider, H., Lemasters, J. J., Höchli, M. and Hackenbrock, C. R. (1980) *J. Biol. Chem.*, **255**, 3748–56.
Schinz, R. H., Lo, M-V.C., Larrivee, D. C. and Pak, W. L. (1982) *J. Cell Biol.*, **93**, 961–9.
Segrest, J. P., Gulik-Krzywicki, T. and Sardet, C. (1974) *Proc. Natl. Acad. Sci. USA*, **71**, 3294–8.
Segrest, J. P., Kahane, I., Jackson, R. L. and Marchesi, V. T. (1973) *Arch. Biochem. Biophys.*, **155**, 167–83.
Seki, S., Hayashi, H. and Oda, T. (1970) *Arch. Biochem. Biophys.*, **138**, 110–21.
Sen, A., Williams, W. P., Brain, A. P. R., Dickens, M. J. and Quinn, P. J. (1981) *Nature (London)*, **293**, 488–90.
Severs, N. J. and Robenek, H. (1983) *Biochim. Biophys. Acta*, **737**, 373–408.
Shelton, D., Wyatt, J. and Heuser, J. (1979) *J. Cell Biol.*, **83**, 274a.
Shin, B. C. and Carraway, K. L. (1974) *Biochim. Biophys. Acta*, **345**, 141–53.
Shotton, D., Thompson, K., Wolfsy, L. and Branton, D. (1978) *J. Cell Biol.*, **76**, 512–31.
Singer, S. J. (1974) *Annu. Rev. Biochem.*, **43**, 805–33.
Singer, S. J. and Nicolson, G. L. (1972) *Science*, **175**, 720–31.
Skriver, E., Maunsbach, A. B. and Jorgensen, P. L. (1980a) *J. Cell Biol.*, **86**, 746–54.
Skriver, E., Maunsbach, A. B., Anner, B. M. and Jorgensen, P. L. (1980b) *Cell Biol. Int. Rep.*, **4**, 585–91.
Slayter, H. S. (1976) *Ultramicroscopy*, **1**, 341–57.
Staehelin, L. A. (1973) *J. Cell Sci.*, **13**, 763–86.
Staehelin, L. A. (1974) *Int. Rev. Cytol.*, **39**, 191–283.
Staehelin, L. A. (1975) *Biochim. Biophys. Acta*, **408**, 1–11.
Staehelin, L. A., Armond, P. A. and Miller, K. R. (1976) *Brookhaven Symp. Biol.*, **28**, 278–315.
Steck, T. L. (1972) *J. Mol. Biol.*, **66**, 295–305.
Steck, T. L. (1974) *J. Cell Biol.*, **62**, 1–19.
Steck, T. L. (1978) *J. Supramol. Struct.*, **8**, 311–24.
Steck, T. L., Koziarz, J. J., Singh, M. K., Reddy, G. and Köhler, H. (1978) *Biochemistry*, **17**, 1216–22.
Steck, T. L., Ramos, B. and Strapazon, F. (1976) *Biochemistry*, **15**, 1154–61.
Steck, T. L., Weinstein, R. S., Straus, J. H. and Wallach, D. F. H. (1970) *Science*, **168**, 255–7.
Stewart, D. J., Semple, E. W., Swart, G. T. and Sen, A. K. (1976) *Biochim. Biophys. Acta*, **419**, 150–63.
Stewart, P. S. and MacLennan, D. H. (1974) *J. Biol. Chem.*, **249**, 985–93.
Stoeckenius, W. and Rowen, R. (1967) *J. Cell Biol.*, **34**, 365– 393.
Süss, K-H., Schmidt, O. and Machold, O. (1976) *Biochim. Biophys. Acta*, **448**, 103–13.
Takahashi, M. and Gross, E. L. (1978) *Biochemistry*, **17**, 806–10.
Thorley-Lawson, D. A. and Green, N. M. (1973) *Eur. J. Biochem.*, **40**, 403–13.
Thornber, J. P. (1975) *Annu. Rev. Plant Physiol.*, **26**, 127–58.
Thornber, J. P. and Barber, J. (1979) in *Photosynthesis in Relation to Model Systems* (ed J. Barber), Elsevier/North Holland Biomedical Press, Amsterdam, pp. 37–70.
Thornber, J. P., Markwell, J. P. and Reinman, S. (1979) *Photochem. Photobiol.*, **29**, 1205–16.
Tillack, T. W., Boland, R. and Martonosi, A. (1974) *J. Biol. Chem.*, **249**, 624–33.

Tillack, T. W., Scott, R. E. and Marchesi, V. T. (1972) *J. Exp. Med.*, **135**, 1209–27.

Tomita, M. and Marchesi, V. T. (1975) *Proc. Natl. Acad. Sci. USA*, **72**, 2964–8.

Unwin, N. and Henderson, R. (1984) *Sci. Amer.*, **250**, 78–94.

Unwin, P. N. T. and Zampighi, G. (1980) *Nature (London)*, **283**, 545–9.

Usukura, J., Yamada, E., Tokunaga, F. and Yoshizawa, T. (1980) *J. Ultrastruct. Res.*, **70**, 204–19.

Vail, W. J., Papahadjopoulos, D. and Moscarello, M. A. (1974) *Biochim. Biophys. Acta*, **345**, 463–7.

Van Duers, B. and Luft, J. H. (1979) *J. Ultrastruct. Res.*, **68**, 160–72.

Van Venetie, R. and Verkleij, A. J. (1981) *Biochim. Biophys. Acta*, **645**, 262–9.

Van Winkle, W. B., Lane, L. K. and Schwartz, A. (1976) *Exp. Cell Res.*, **100**, 291–6.

Vanderkooi, G., Senior, A. E., Capaldi, R. A. and Hayashi, H. (1972) *Biochim. Biophys. Acta*, **274**, 38–48.

Vanderkooi, J. M., Ierokomas, A., Nakamura, H. and Martonosi, A. (1977) *Biochemistry*, **16**, 1262–7.

Verkleij, A. J. (1984) *Biochim. Biophys. Acta*, **779**, 43–63.

Verkleij, A. J., Mombers, C., Gerritsen, W. J., Leunissen-Bijvelt, L. and Cullis, P. R. (1979a) *Biochim. Biophys. Acta*, **555**, 358–61.

Verkleij, A. J., Mombers, C., Leunissen-Bijvelt, J. and Ververgaert, P. H. J. Th. (1979b) *Nature (London)*, **279**, 162–3.

Verkleij, A. J., Van Echteld, C. J. A., Gerritsen, W. J., Cullis, P. R. and DeKruijff, B. (1980) *Biochim. Biophys. Acta*, **600**, 620–4.

Verkleij, A. J. and Ververgaert, P. H. J. Th. (1981) *Acta Histochem. Suppl.*, **XXIII**, S137–43.

Ververgaert, P. H. J. Th. and Verkleij, A. J. (1978) *Experientia*, **34**, 454–6.

Walker, J. E., Carne, A. F. and Schmitt, H. W. (1979) *Nature (London)*, **278**, 653–4.

Wang, C-T., Saito, A. and Fleischer, S. (1979) *J. Biol. Chem.*, **254**, 9209–19.

Wang, K. and Richards, F. M. (1974) *J. Biol. Chem.*, **249**, 8005–18.

Weinstein, R. S. (1974) in *The Red Blood Cell* (ed. D. MacN. Surgenor), Academic Press, New York, pp. 213–68.

Weinstein, R. S., Benefiel, D. J. and Pauli, B. U. (1979) in *Freeze Fracture: Methods, Artifacts, and Interpretations* (eds J. E. Rash and C. S. Hudson), Raven Press, New York, pp. 175–83.

Weinstein, R. S., Khodadad, J. K. and Steck, T. L. (1978a) *Prog. Clin. Biol. Res.*, **21**, 413–27.

Weinstein, R. S., Khodadad, J. K. and Steck, T. L. (1978b) *J. Supramol. Struct.*, **8**, 325–35.

Weinstein, R. S., Khodadad, J. K. and Steck, T. L. (1980) in *Membrane Transport in Erythrocytes* (eds U. V. Lassen, H. H. Ussing and J. O. Wieth), Munksgaard, Copenhagen, pp. 35–46.

Weinstein, R. S., McNutt, N. S., Nielsen, S. L. and Pinn, V. W. (1970) *Proc. 28th Annu. Mtg. Electron Microsc. Soc. Amer.*, 108–9.

Wessels, J. S. C. (1977) in *Photosynthesis* (eds A. Trebst and M. Auron), Springer-Verlag, Berlin, pp. 563–73.

Willison, J. H. and Brown, R. M. (1979) in *Freeze Fracture: Methods, Artifacts, and Interpretations* (eds J. E. Rash and C. S. Hudson), Raven Press, New York, pp. 51–7.

Wise, D. S., Schoenborn, B. P. and Karlin, A. (1981) *J. Biol. Chem.*, **256**, 4124–6.

Wolosin, J. M., Ginsburg, H. and Cabantchik, Z. I. (1977) *J. Biol. Chem.*, **252**, 2419–27.
Yee, A. G., Fishback, G. D. and Karnovsky, M. J. (1978) *Proc. Natl. Acad. Sci. USA*, **75**, 3004–8.
Yu, J. and Branton, D. (1976) *Proc. Natl. Acad. Sci. USA*, **73**, 3891–5.
Yu, J. and Steck, T. L. (1975) *J. Biol. Chem.*, **250**, 9176–84.
Zampighi, G. and Unwin, P. N. T. (1979) *J. Mol. Biol.*, **135**, 451–64.
Zingsheim, H. P. and Plattner, H. (1976) *Methods Membr. Biol.*, **7**, 1–146.

# 11

# Three-dimensional structure of membrane proteins

SVEN HOVMÖLLER



## 11.1 Introduction

### 11.1.1 WHAT IS MEANT BY A THREE-DIMENSIONAL STRUCTURE DETERMINATION?

The term structure is used with different meanings in the biochemical literature. Sometimes it is used to describe the detailed atomic arrangement, as determined by X-ray diffraction; sometimes it is used for less detailed information. An example of the latter is topological information such as subunit A is close to subunit B, as may be deduced from, for example, cross-linking experiments.

In this chapter the concept structure will be used strictly in the meaning of three-dimensional arrangement, including co-ordinates in the three directions of space $x$, $y$ and $z$. A structure determination, then, is a procedure to find the arrangement of atoms within a polypeptide, or in the case of multisubunit complexes, also the relation between the different subunits in terms of co-ordinates in space. This criterion is far stricter than a statement such as A is close to B, in which case B may be situated anywhere on a sphere of a certain radius surrounding A.

### 11.1.2 DIFFERENT WAYS TO VISUALIZE ATOMS AND MOLECULES

In order to determine the three-dimensional structure of a membrane protein, a method is required to see its fine details at a very high magnification. The first thought may be to use a powerful light microscope, but unfortunately this is

Techniques for the Analysis of Membrane Proteins. Edited by C. I. Ragan & R. J. Cherry.
Published in 1986 by Chapman and Hall, 11 New Fetter Lane, London EC4P 4EE

not sufficient since the size of the smallest details that can be seen with a light microscope, i.e. its resolution, is limited to about the wavelength of visible light. This wavelength is between 4000 Å and 8000 Å, and light microscopes may be used to see bacteria, but not even the largest biological macromolecules can be seen in a light microscope. A ribosome has a diameter of about 200 Å and membrane proteins are even smaller. Thus a radiation with a much shorter wavelength than visible light must be used for viewing such molecules. There are basically three choices: X-rays, electrons and neutrons. They all have their advantages and disadvantages.

(a) *X-ray diffraction*

X-ray diffraction has been, and will for the foreseeable future be, the main method for structure determination. X-ray diffraction was the first method to be used for structure determination, and the first structure determination was the elucidation of the atomic arrangement in NaCl crystals in 1912 by Bragg and Bragg. The principle of this method is unchanged today, but with the development of electronic equipment and computers it is possible to solve the structures of molecules of far greater complexity than was possible at the beginning of this century.

X-rays interact weakly with matter. Visible light is stopped by thin black paper, but X-rays can pass right through several centimetres of organic matter. This is well known through the widely used medical practice of 'taking an X-ray'. The principle of X-ray crystallography is, however, quite different from that of a medical X-ray. In the medical approach only the transmitted X-rays are being studied. The skeleton inside a human body can be seen since X-rays interact more strongly with elements with higher atomic numbers, and thus penetrate the organic matter, which is mainly carbon, hydrogen, oxygen and nitrogen, more than they penetrate the heavier atoms calcium and phosphorus present in bone tissue. In X-ray crystallography, or X-ray diffraction, all X-rays are used *except* those that pass straight through the crystal. Rather than studying the attenuation of the direct beam the X-ray crystallographer is interested in the intensities of the diffracted beams. Since X-rays interact rather weakly with matter, and especially so with organic matter, rather big crystals are needed for X-ray diffraction experiments. The most used kind of X-ray is Cu $k_\alpha$, with a wavelength of 1.54 Å. In order to diffract about 50% of all Cu $K_\alpha$ X-ray photons which are directed at a protein crystal the crystal should be about 1 mm thick. This means that it should be about 100 000 unit cells thick, since unit cells in protein crystals have dimensions of about 100 Å. In order to get an acceptable signal of diffracted X-rays the other two dimensions of the crystal should also be about 1 mm. The whole crystal will then contain more than $10^{15}$ (one million billion) molecules. This can present difficulties in cases in which the supply of protein is limited (1 $mm^3$ with a typical protein concentration of 50% contains 0.5 mg of protein) and because the growth of such large, well-ordered crystals is a process which is poorly understood and may prove to be very difficult. On the other hand, if such crystals can be grown, there is a great advantage in the large number of protein molecules per crystal. The structure determined by X-ray diffraction is the average over all the molecules in the crystal, and the very large number of molecules gives a very high signal-to-noise ratio. If the order of the crystal is sufficiently good, a three-dimensional electron density map of a protein molecule can be determined in which the resolution is sufficiently high to allow the positions of individual amino acid side chains to be determined.

Indissolubly related to the property of X-rays to penetrate deeply into matter is their inability to be focused by any type of lens. Thus we cannot see an enlarged image of the atoms in a crystal, we can only recognize the patterns of the diffracted X-ray photons, that is, the positions and intensities of the diffracted beams. The following parallel may be useful to appreciate the difficulty of reconstructing an image from diffracted beams without having a lens with which to focus the beams. Imagine that you are in a completely dark room with a statue in the middle. You want to know what the statue looks like, i.e. its structure, but you cannot see it or touch it with your hands. The only thing you may do is to spray water on to the statue with a garden hose. You also have a friend in the room, and as you direct the water jet from different positions in the room on to the statue from all possible directions, your friend moves around in the room and feels how the water sprinkles as it hits the statue. Two different statues will give different results, but it is clear that it is a difficult task to derive the structure of the statue from the deflected water jets only. This is much the same situation as that of X-ray crystallography.

(b) *Electron microscopy*

From the foregoing discussion it is clear that it would be a great advantage to use a type of radiation other than X-rays with a wavelength about or below 1 Å, but one which could be deflected by lenses. Electrons accelerated to very high velocities fulfil these criteria. Every moving particle has a distinct wave associated with it, of a wavelength related to the mass and speed of the particle. An electron accelerated in an electric field of between 10 000 and 100 000 V reaches approximately half the speed of light, and the associated wave has then a wavelength of much less than 1 Å. Since the electron is charged it can be deflected by electromagnetic lenses, and thus focused. This gives the basic principle for the electron microscope: electrons are accelerated by an electric field, they interact with the specimen, and are focused by electromagnetic lenses. There are two main difficulties in making electron microscopes capable of giving information about the structure of molecules. Electrons interact more than 10 000 times more strongly with matter than do X-rays, and thus would be totally absorbed by 1 mm of air. The electron microscope must be equipped with a powerful vacuum system if any of the electrons are to reach the detector. The other difficulty is to produce a set of lenses with such power and precision that two electrons that have been deflected by atoms only 10 Å apart reach the detector at distinctly different and well-defined positions. In the fifty years of electron microscopy there has been a steady improvement in these respects and modern electron microscopes are now capable of producing images of single atoms at a magnification of more than 500 000 times.

The limitations of the electron microscope are coupled to its advantages. The strong interaction between electrons and matter makes it possible to focus the electrons into an image. However, a disadvantage of such a strong interaction is that it is only possible to look at extremely thin objects in an ordinary transmission electron microscope. An organic object suitable for analysis in the electron microscope is therefore typically at the most a few hundred ångströms thick and less than 1 $\mu m^2$ in area, or more than a billion times smaller than typical crystals used in X-ray diffraction experiments. Electron microscopy may be the only possible method for structure investigations of molecules of which only small quantities are available. The small size of the specimen in the electron microscope carries with it a fundamental disadvantage; that of radiation damage. Both X-rays and electrons damage the molecules severely. Both types of radiation interact with the specimen in two ways: elastic and inelastic scattering. There is about a 50% probability for an interaction with an atom to be elastic and 50% chance for it to be inelastic. Only the elastically scattered radiation, the radiation which comes from a 'bounce' off the electrons of the atoms without energy loss, is useful for structural investigations. Inelastically scattered radiation deposits energy into the molecule, often enough to eject an electron from an atom, leaving it with a positive charge. This will result in a rearrangement of the molecule, and the next time this molecule is hit by an electron or a photon it is no longer the same molecule. Such radiation damage is a serious limitation in the structure determination of organic molecules at atomic resolution and is more serious in the case of electron microscopy than in X-ray diffraction because of the smaller size of the specimen. The comparatively large size of the crystals used in X-ray crystallography ensures that even after a day of exposure in an intense X-ray beam, most protein molecules in the crystal have not undergone a single interaction with an X-ray photon, and thus most molecules are still intact. In the electron microscope the situation is radically different. The dose a biological specimen receives during focusing and exposure is equivalent to what it would receive if it were 10 cm away from an exploding nuclear bomb!

Although the optics of electron microscopes are now sufficiently advanced to allow individual atoms to be seen (Hovmöller *et al.*, 1984), radiation damage makes it impossible to determine the structure of a single protein molecule at atomic resolution. There are two ways around this problem. One is to stain the molecule with an inorganic salt, which is much less sensitive to radiation damage, and, since the atoms are heavier, gives more contrast. By this method, however, only the outline or contours of the protein molecule can be visualized, the inside being burnt to ashes by the electron dose required to focus the image. The other method of overcoming the problem of radiation damage is to use the crystallographic approach. If the molecule under study is crystalline, we may average over many identical molecules, each of which will then receive only a small fraction of the total dose. Both these methods will be discussed in detail in this chapter. An excellent review of three-dimensional structure determination by electron microscopy of two-dimensional crystals has been published by Amos *et al.* (1982).

(c) *Neutron scattering*

If neutrons are accelerated they also are associated with a characteristic wave, and they may interact with matter. This interaction is even weaker than that of X-rays, so a suitable size for crystals used in neutron scattering investigations is about 1 $cm^3$, or 1000 times larger than the size of crystals used for X-ray diffraction. This may be impossible to achieve for a protein, and the results from neutron diffraction are similar to those obtained from X-ray diffraction. The main advantage with neutrons is that hydrogens can be localized just as easily as other elements, whereas in X-ray diffraction hydrogen atoms are almost impossible to see in the electron density map, especially for large molecules such as proteins. Another use of neutron diffraction is that of contrast matching. Different isotopes of hydrogen scatter neutrons with different powers, whereas X-rays do not distinguish between hydrogen and deuterium. It is possible, by exchanging $H_2O$ for $D_2O$, to match out the scattering of one component in a sample, leaving the other component(s) unaffected. This has been used mainly for the study of protein–nucleic acid complexes such ribosomes, viruses and chromatin but also for the membrane protein cytochrome reductase (Perkins and Weiss, 1983). A further disadvantage with neutrons is the price and size of the equipment required. The only available neutron sources of sufficient intensity are nuclear reactors, and a collimator of several tens of metres is required to obtain a parallel beam. On the other hand, there are several neutron sources around the world and it is usually possible for the scientist with a project to get access to the equipment without cost.

(d) *Low-angle solution scattering*

Both X-rays and neutrons are used for scattering experiments on non-crystalline material in solution. In principle this gives the same information as diffraction from a crystal, since each individual molecule scatters similarly whether it is in a crystal or in solution, but the random orientation of the protein molecules imposes a severe limitation on the resolution which can be obtained. A solution spectrum is equivalent to a three-dimensional diffraction pattern, but smeared out by rotating 360 degrees around the three orthogonal directions in space. The result is that most diffraction intensities are too diffuse to detect, and we get a tremendous overlap of reflections. This limits neutron and X-ray scattering experiments to the determination of the coarse features of molecules, since at low resolution the overlap is less severe. These coarse features are the overall size and shape of the protein molecule. One can distinguish between solid and hollow spheres, ellipsoids, cylinders, and so on, and it is possible to determine the length of the three principal axes of an ellipsoidal molecule.

### 11.1.3 RESOLUTION AND MAGNIFICATION

The concepts of resolution and magnification are central to the structural chemist. There is a fundamental difference between resolution and magnification, yet many would not know the difference. One reason for this confusion may be that the optical systems of a microscope or a pair of binoculars are such that the resolution increases in parallel with the magnification. However, this is not always the case. If you take a photograph of the spectators in the opposite stand of a sports stadium, and then make a very high magnification print of a fine detail, such as a person's head, you will appreciate the difference between magnification and resolution. Although the head will be natural size in the enlargement, the fine details will not be visible. You will not see the hairs on the head, but only the silver grains of the negative. The same principle applies for electron microscope images. It is not difficult to construct a microscope which enlarges a million times, but the technical difficulties involved in constructing an electron microscope of very high resolution (a few ångströms) are tremendous.

There are several different definitions of resolution. It may be defined as the smallest distance between two points for which they can still be seen. Resolution may also be defined by features of the diffraction pattern. For a crystalline object the resolution is defined as the *d*-value corresponding to the outermost significant reflection in the diffraction pattern.

The resolution range of interest for the study of proteins is between 1 and 50 Å. Since most protein molecules are up to 100 Å long a resolution of better than 50 Å is required to provide more information about the protein than that it is a blob of less than 100 Å diameter. Since the distance between atoms is about 1.5 Å for covalently bound atoms, and 3.5 Å for atoms in van der Waals contact, the chemist is not usually interested in resolutions better than 1 Å. In order to see the overall shape of a protein, and whether it is a monomer, a dimer or a more complicated oligomer, a resolution of about 25 Å is required. At a resolution of about 7 Å $\alpha$-helices can be seen, at 3–4 Å the side chains of amino acids appear, and at 1–2 Å resolution individual atoms can be located.

X-ray diffraction will give 2 Å resolution for proteins, even for such large molecules as viruses, if the crystals are of sufficiently high quality. Electron microscopy of tissue embedded in plastic and sectioned will give a resolution of 50–30 Å. Electron microscope images of individual molecules will not give a resolution better than 100–50 Å if unstained, and about 40 Å if stained with a heavy metal stain. Thin negatively stained protein crystals can give a resolution as high as 12 Å if precautions are taken not to irradiate the specimen before the actual picture is taken, but a procedure involving focusing on the object of interest gives only 25 Å resolution. The method of staining limits the resolution in two ways. Firstly, the main contrast in the micrograph will be between stain and stain-excluding matter (such as proteins) and the structure within a protein molecule will give no observable contrast. Secondly, unlike the protein molecules, the stain is disordered, and the staining will be different for different molecules in the crystal, giving a deterioration in resolution. If the crystals are consistently highly ordered it is possible to achieve at least 7 Å resolution using unstained, glucose-embedded crystals. This allows us to see internal features in proteins using electron microscopy.

It may be emphasized that the difference in information content between 25 Å and 5 Å resolution is not a factor of 5 but a factor of 125 ($5 \times 5 \times 5$) since the resolution increases in all three directions of space.

## 11.2 Electron microscopy

### 11.2.1 SINGLE MOLECULES OR CRYSTALS?

When membrane proteins are studied by electron microscopy it is a great advantage to have them in crystalline form. An individual membrane protein molecule will go undetected if stain is not used, but the use of negative stain decreases the attainable resolution to such an extent that it can be difficult to distinguish one membrane protein from another. It is very important to have a pure preparation, in order to overcome the difficulties of identification. In addition to this, the molecules are lying in random orientations relative to the grid, and the uncertainty concerning the orientation of the molecule leads to difficulties in interpreting the observed structure. These problems of identification and orientation can be overcome if the membrane protein molecules are embedded in a lipid bilayer, and ordered in two dimensions on a repeating lattice. In the optimum case only one layer of molecules will be present. The term membrane crystal is used to describe such an ordered array of molecules, one or a few layers thick. A membrane crystal of one protein can be distinguished from another by the shape, size and symmetry of the unit cell alone.

An additional advantage of studying membrane proteins in the form of crystals is the possibility of calculating the average image of all the protein molecules in the crystal, thus increasing the signal-to-noise ratio or decreasing the electron dose per molecule, or both. The crystals need not be very large, 10 by 10 molecules in the plane of the membrane can be sufficient, and it is an advantage if the crystal is only one unit cell thick, since the electrons do not penetrate more than a few hundred ångströms of organic matter.

### 11.2.2 METHODS FOR CRYSTALLIZATION OF MEMBRANE PROTEINS

There are three main categories of two-dimensional crystals of membrane proteins in lipid bilayers. Firstly, some proteins are crystalline *in vivo*, as is frequently found for bacterial cell walls. In this case the problems are to isolate the cell walls from the bacteria, and prepare the specimen in a form suitable for electron microscopy, without destroying the order of the crystals. This is achieved by mild treatment of the specimen, which includes such measures as avoiding conditions of low or high pH, and avoiding drying out and heating of the specimen. Most of the crystalline cell walls studied so far have shown hexagonal symmetry with unit cell dimensions of about 100 Å. The crystals are often highly ordered only in small patches, separated by dislocations, giving for example, a local fivefold symmetry. These dislocations are necessary – if the crystal was perfect it would not cover a small closed body, but become an infinitely large cylinder or flat sheet.

Secondly, in some cases it has been possible to crystallize a membrane protein by selectively extracting all other proteins from a membrane by detergents. This is the case with ($Na^+ + K^+$)-ATPase (Hebert *et al.*, 1982).

Thirdly, the most general approach for crystallizing membrane proteins is to purify the protein, and then crystallize it. In principle this method should be applicable to all proteins. The first example of such a procedure was the purification and crystallization of cytochrome reductase from *Neurospora crassa* mitochondria (Wingfield *et al.*, 1979). Whole mitochondria were solubilized in a high concentration of the non-ionic detergent Triton X-100 and applied to an affinity chromatography column with cytochrome *c* bound to it. Only cytochrome reductase and cytochrome oxidase bound to the column. These two membrane proteins were eluted together and separated on the basis of their different molecular weights using a Sephadex column. During all these steps the concentration of the detergent was kept above the critical micelle concentration (CMC) to ensure that the Triton was mainly present in the form of micelles. Only in this form will Triton solubilize membrane proteins. It is thought that one such micelle with a molecular weight of 90 000 always solubilizes one membrane protein complex independent of the size of the complex.

In this way the protein was purified such that its spectrum showed only the expected cytochrome reductase peaks and such that SDS-gels showed only those subunits known to be present in cytochrome reductase. Phospholipids, solubilized with Triton X-100, were then added to the purified protein and excess detergent was removed, either by adding Bio Beads, which selectively adsorb the detergent molecules, or by dialysis. The latter procedure is much slower but may be more easily controlled. A similar procedure was used to crystallize a subcomplex of cytochrome reductase from a purified preparation (Hovmöller *et al.*, 1981).

Other examples of crystallization from purified, monodisperse membrane proteins are the crystallization of cytochrome oxidase from cholate (Goldfarb *et al.*, 1979) and from deoxycholate (Fuller *et al.*, 1979). Using this method there is no upper limit to the size of the crystals. Crystals of cytochrome reductase as large as 20 × 20 μm have been prepared. A part of such a crystal is seen in Fig. 11.1.

A summary of the effect of different variables on the crystallization of cytochrome reductase and its subcomplex has been published (Hovmöller *et al.*, 1983). The most important variables were found to be pH (5.5–6.0 optimal), temperature (25–37° C), protein to lipid ratio (0.5–2.0 mg of protein per mg of phospholipid) and duration of dialysis (must exceed 48 h).

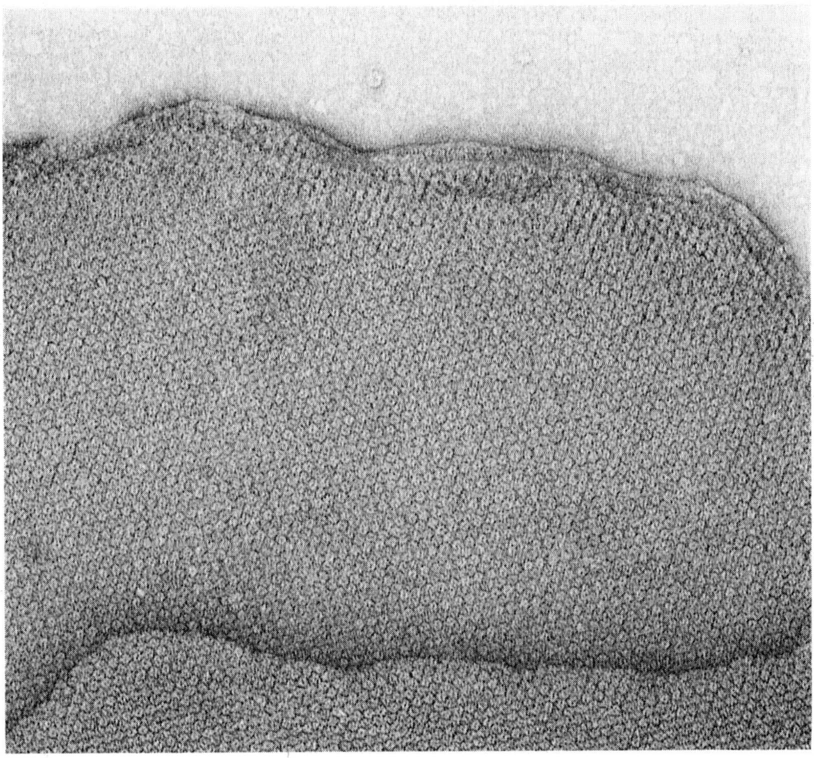

**Fig. 11.1** Part of a very large membrane crystal of cytochrome reductase.

### 11.2.3 DIFFERENT CRYSTAL FORMS

Membrane proteins have been found to crystallize in three different forms – as membrane sheets, as cylindrical tubes and as fibres. All three forms have been found for cytochrome reductase (Figs 11.1, 11.2 and 11.3). Although the fibres may stack up in bundles of some order, they are not suitable for structure determination. Cylindrical tubes collapse when they dry down on to the microscope grid, and electron micrographs taken of such collapsed tubes show the two crystalline lattices from the top and bottom of the tube superimposed. This gives rise to a moiré pattern, but the two lattices can be separated by optical or computer filtering of the diffraction. The best ordered crystals are usually the single or multiple layered sheets. Hovmöller *et al.* (1983) found that crystals of cytochrome reductase grown at pH 5.5 mainly formed sheets; at pH 6.0 over 90% of the crystals were in the form of tubes; and at pH 7.0 small amounts of both crystal forms were found, together with very many small membrane fragments, unsuitable for structure analysis because of their poor order.

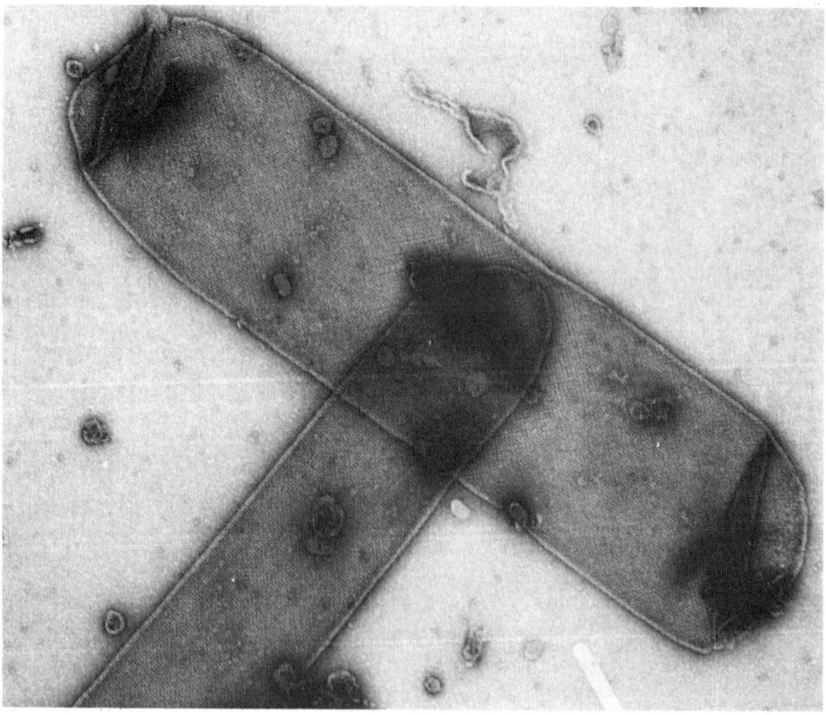

**Fig. 11.2** Tubular double-layer vesicles of cytochrome reductase. These tubes are probably cylindrical in solution, but collapse when dried down on the grid.

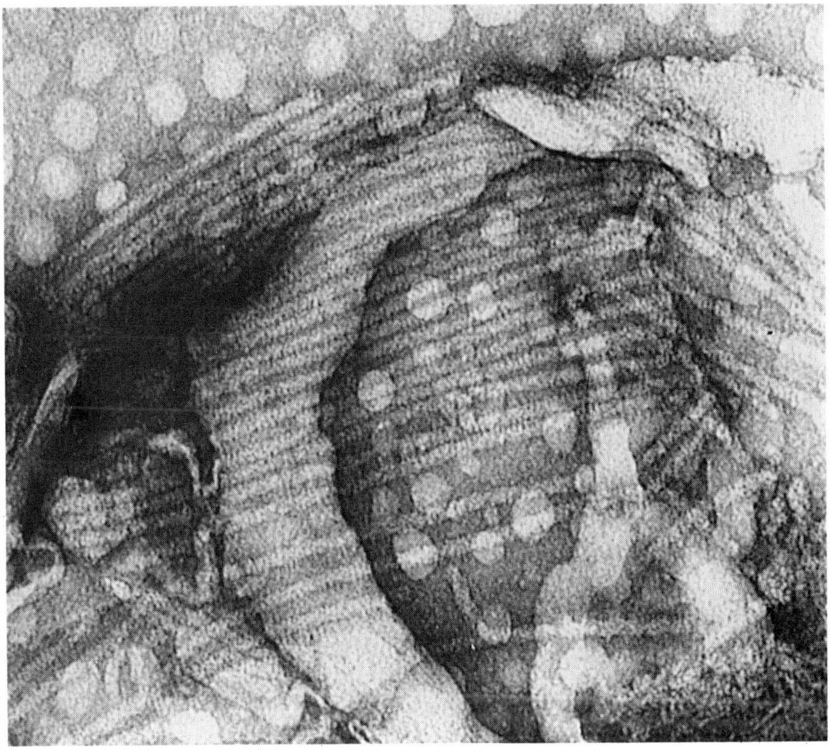

Fig. 11.3 Fibrous aggregates of cytochrome reductase.

It is interesting to note that the membrane sheets frequently grow as stacks, a few layers high. These layers are perfectly in register, and the crystal is then a very thin but truly three-dimensional crystal (Fig. 11.4). Multilayered stacks of membrane crystals have also been found for cytochrome oxidase (Vanderkooi *et al.*, 1972; Goldfarb *et al.*, 1979) and bacteriorhodopsin (Henderson and Shotton, 1980). The pore protein from the outer mitochondrial membrane (Mannella, 1982) forms cylindrical tubes.

### 11.2.4 SPECIMEN PREPARATION

The three main methods for studying membrane proteins by transmission electron microscopy are: (1) the use of negative stain, (2) plastic embedding followed by ultrathin sectioning and (3) embedding in sugars. Negative staining is the most widely used method since it is fast and easy and gives an easily interpretable result. However, the attainable resolution is limited to about 25 Å. A drop of the sample, containing about 1 mg/ml of protein is

**Fig. 11.4** A large multilayer membrane crystal of cytochrome reductase. The preparation was first negatively stained, dried and then shadowed with platinum. Note that all layers of the crystal are in register. (Reproduced from Hovmöller *et al.*, 1983, with permission.) Scale bar is 1.0 μm.

placed on a piece of plastic on the laboratory bench. The grid is allowed to float on the drop for about 20 s, after which the drop is shaken off from the grid. The grid may be washed for a few seconds by floating on a drop of distilled water to remove particles which are not tightly sticking to the grid, followed by shaking off excess water. The grid is stained by floating on a drop of a 1–2% solution of the staining material in a suitable buffer. Common stains used are uranyl acetate, ammonium molybdate or phosphotungstic acid. Finally, the grid is dried by blotting with a piece of filter paper, followed by some minutes of air-drying. The value of the pH and concentration of the staining solution may be important for good preservation of the crystalline order, and for a good contrast, and it has been found that each particular membrane protein crystal is best stained at some particular combination of pH and concentration.

Plastic embedding followed by sectioning with an ultramicrotome is more laborious but a standard technique in most EM laboratories. This may be the only way to get a view of the protein from a direction parallel to the plane of the membrane. See Fig. 11.5.

The best results in the study of the three-dimensional structure of membrane

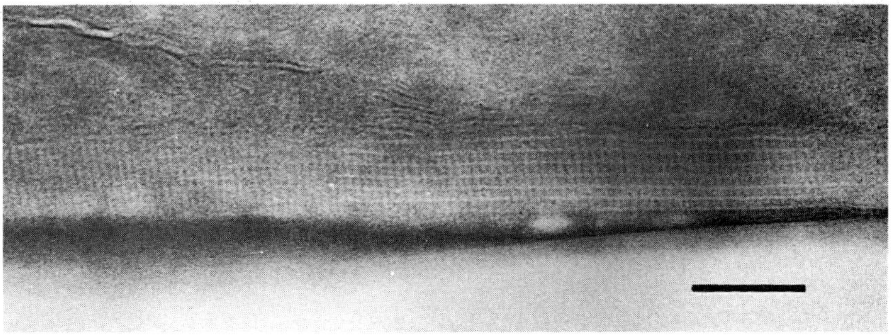

**Fig. 11.5** Thin transverse section of epon-embedded multilayer cytochrome reductase membrane crystals, showing that several layers are present. Scale bar 1000 Å. (Reproduced from Hovmöller *et al.*, 1983, with permission.)

proteins have been obtained by embedding crystals of bacteriorhodopsin in glucose instead of a heavy metal stain, and using very low electron dose ($<1\ e^-/\text{Å}^2$) microscopy to minimize radiation damage (Unwin and Henderson, 1975). The pictures are so underexposed in these conditions, and the contrast is so low, that nothing can be seen by eye on the micrograph, but by computerized image processing of a large area the average of many hundreds of molecules can be calculated, and the signal will rise clearly above the background noise. The technique can only be used when a large proportion of the membranes are of good crystalline order.

### 11.2.5 TWO-DIMENSIONAL AVERAGING OF A MEMBRANE CRYSTAL

By averaging the information from many molecules the signal-to-noise ratio can be improved and thus the possibility of seeing finer details of the protein molecule is increased. The averaging is easily done if all protein molecules are oriented in the same way (or in a limited number of symmetry-related ways), as is the case in a crystal. The first averaging studies were performed by optical filtering but now computers are used. The optical diffractometer is still used for the initial screening of the micrographs, to determine whether a membrane is crystalline, and if so, how well ordered it is. Crystalline material can be recognized in the microscope as a pattern much like the weave of a piece of fabric, but the quality of a crystal is not easily determined by eye. In an optical diffractometer the quality is immediately clear from the number of diffraction points in the back focal plane of a positive lens placed in front of or behind the micrograph. An example of a simple but sufficient optical diffractometer for these investigations is shown in Fig. 11.6.

It is also possible to determine the degree of defocus and astigmatism of a picture in the optical diffractometer. A series of dark rings shows the contrast

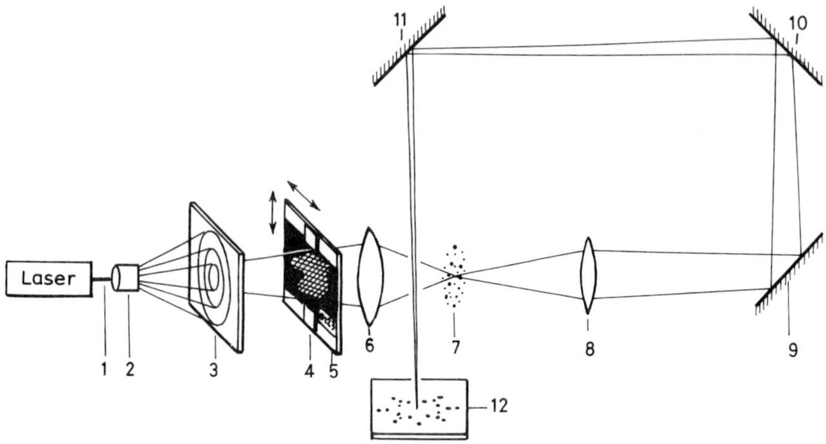


**Fig. 11.6** Schematic drawing of an optical diffractometer, used to analyse diffraction patterns of electron micrographs. The laser beam (1) is directed into a microscope lens (2), which focuses the parallel laser light on to a pinhole, after which the light diverges. The light from the pinhole gives a circular diffraction pattern. Only the central circular bright portion is used, the outer rings of light being stopped by an aperture (3). By adjusting the size of this aperture, smaller or larger areas of the crystal can be analysed. Small magnets (4) hold the electron micrograph (5) on to a plate which can be translated up–down and left–right when screening for the best area of the micrograph. A positive lens with a focal length of about 200 mm focuses the laser light at (7), where a small diffraction pattern is seen. This image is magnified about ten times by another positive lens (8), with a focal length of about 500 mm. It is very convenient to use three mirrors (9, 10 and 11) whereby the enlarged diffraction pattern at (12) can be displayed near the micrograph. With this arrangement the operator can translate the micrograph and at the same time see the diffraction pattern. The mirrors must be high-quality planar metal plates, with evaporated silver on. The diffraction pattern (12) can be displayed on opaque glass or preferably on simple half-transparent drawing film.

transfer function. The first dark ring should occur at a radius outside the spots furthest away from the centre which represent the highest resolution in the micrograph. A detailed discussion on the effect of the contrast transfer function may be found in Unwin and Henderson (1975). If the picture is astigmatic the rings become ellipses. An example of a good focus can be seen in Fig. 11.7.

When a good area has been found in the optical diffractometer it is digitized with a microdensitometer. For a typical magnification of 40 000 times and a resolution of 25 Å the image needs to be scanned with a raster size of between 25 and 50 $\mu$m, corresponding to 6.25 to 12.5 Å of the crystal. Depending on the size of the crystal and the power of the computer, an area from $256 \times 256$ to $1024 \times 1024$ points may be recorded. The Fourier transform of the digitized area is calculated, and the result is similar to the optical diffraction pattern, except that the calculated Fourier transform contains both the phase and the amplitude information, whereas the optical diffraction pattern contains only amplitude information. The diffraction pattern may be displayed on a

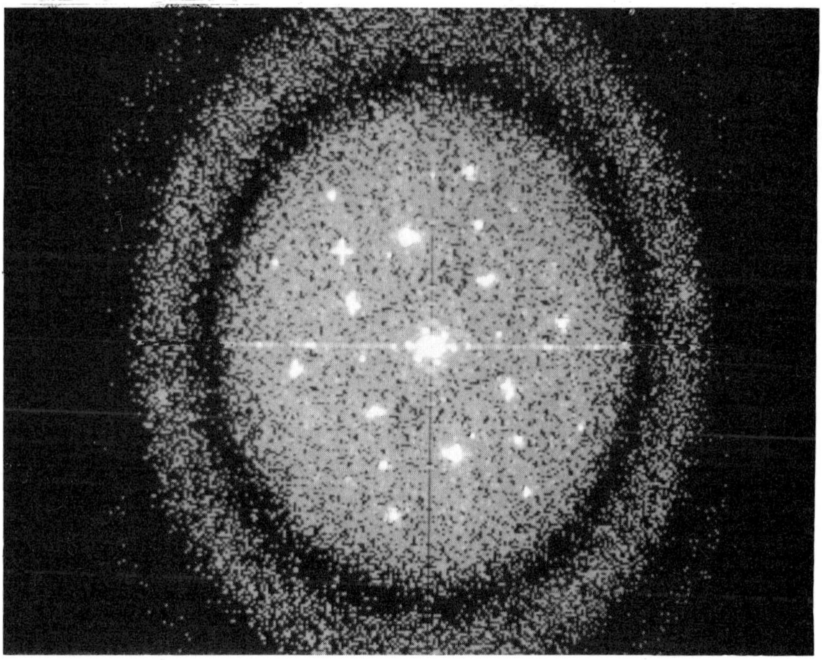

**Fig. 11.7** Diffraction pattern of a crystal of cytochrome oxidase. The crystal quality is acceptable, as judged by the presence of many and sharp diffraction spots. The pattern shows mirror symmetry across two perpendicular axes, so-called mm-symmetry. The unit cell dimensions a and b are not identical, the angle $\gamma$ between the axes is 90°. This and the absence of axial reflections with $h$ odd for ($h$00) or $k$ odd for (0$k$0) indicates that the two-sided plane group is $p22_12_1$ (see Fig. 11.9). The contrast transfer function is visualized in the background noise, with a central bright area and outside this alternating dark and bright rings. All reflections within the central bright area have correct phase values, those in the first bright ring are shifted by 180 degrees, the next ring being correct etc. As this picture is slightly astigmatic, the rings are not circular but elliptical. By changing the focus the radii of the bright rings are changed.

computer graphics terminal or on paper. The exact positions and indices of a few diffraction spots are found and used to determine the positions of all diffraction peaks in the lattice. The amplitudes and phases are then read off at the predicted lattice points in the Fourier transform. The list of these reflections, perhaps 25 to 50, is used to calculate the inverse Fourier transform. By taking Fourier amplitudes and phases only from the predicted lattice points, most of the irregular noise in the original micrograph has been removed, since this falls mainly in the regions of the Fourier transform between the diffraction points. The inverse Fourier transform will correspond to the average over all the unit cells that were scanned. An example of the result is seen in Fig. 11.8.

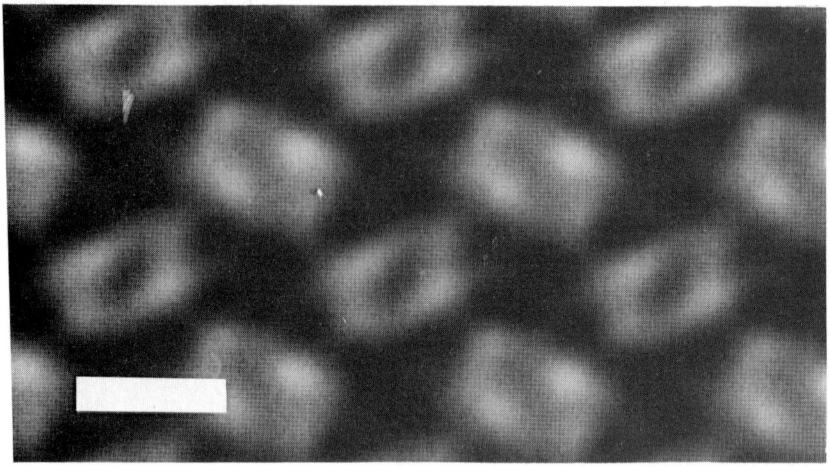

**Fig. 11.8** Computer-averaged two-dimensional projection of cytochrome reductase. The packing of the molecules as dimers is easily observed. Alternate rows of dimers are directed towards or away from the reader. This packing corresponds to the two-sided plane group $p22_12_1$ (see Fig. 11.9). Dimers rotated to the left are facing one way through the membrane, and dimers rotated to the right are facing the opposite way through the membrane. Two such dimers are related by $2_1$ screw axes along $x$ and $y$. The scale bar is 100 Å.

### 11.2.6 SYMMETRY IN THE CRYSTAL AND WHAT IT TELLS US ABOUT THE PROTEIN

A crystal of a membrane protein, just as other crystals, displays symmetry. For a general molecule there are 230 different symmetries possible for three-dimensional crystals; these are called the 230 space groups. In contrast, the possible symmetries for a membrane crystal only one unit cell thick are limited by two factors. Screw axes perpendicular to the membrane are not possible. This criterion limits the possible symmetries to the 80 so-called two-sided plane groups. Most of these contain mirror planes, and this is not allowed in protein crystals, since all proteins are chiral and contain only L-amino acids, and not the mirror-related D-amino acids. This leaves only 17 possible two-sided plane groups for membrane proteins. These are listed in Table 11.1 together with their characteristics. Diagrams of the 17 possible ways of packing proteins in a membrane crystal are shown in Fig. 11.9.

Once a membrane protein is obtained in crystalline form, its unit cell dimensions and, if the crystal is of a reasonable quality, its symmetry, can be determined. From the lengths of the two cell axes and the angle between them in the plane of the membrane ($a$, $b$ and $\gamma$) the area of the unit cell, given by $a \times b \times \sin\gamma$, can be calculated. If the number of protein molecules per unit cell and thickness of the crystal are known, we can determine the volume in $Å^3$ per

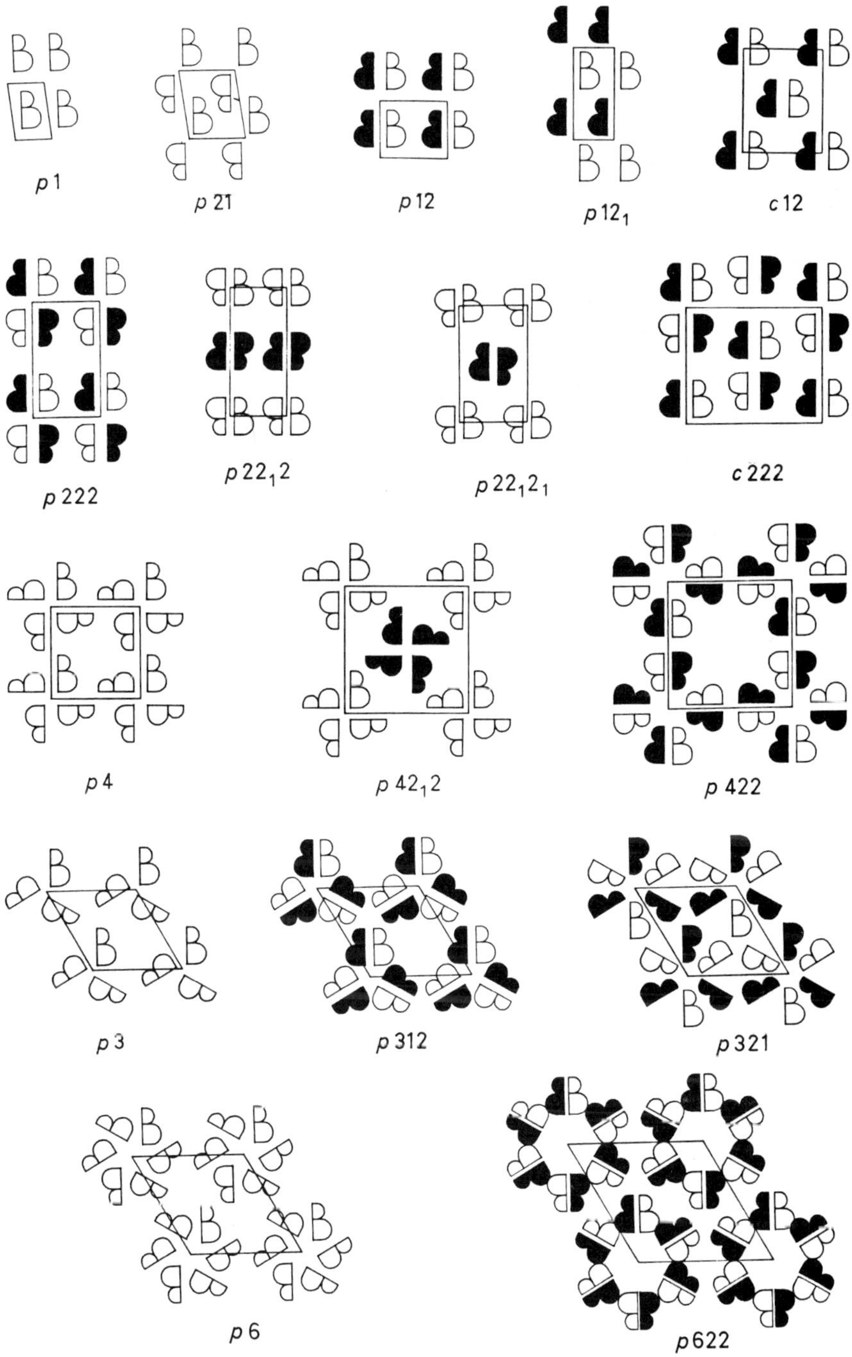


**Fig. 11.9** Diagrams of the 17 two-sided plane groups. These are the only possible symmetries for membrane crystals, one unit cell thick, of chiral molecules such as proteins. The B symbolizes a protein molecule. The filled B's face the other way from the unfilled B's. Notice that the area of the unit cell increases roughly proportionally with the number of molecules per unit cell. The plane groups are named according to Holser (1958). The cell type is indicated by a small letter, *p* or *c* (primitive or centred). The axis perpendicular to the membrane plane is always chosen as the *z* axis. The symmetry along the *z* axis is always described by the first number following the cell type. All figures are drawn as seen along the *z* axis, with *y* vertical and *x* horizontal. Membrane crystals found *in vivo* are expected to belong to one of the plane groups *p*1, *p*21, *p*3, *p*4 or *p*6, since only in these plane groups are all molecules facing the same way out of the membrane. Membrane crystals crystallized *in vitro* may form any of the 17 symmetries.

**Table 11.1** Characteristics of the 17 two-sided plane groups

| *No.* | *Symbol* | *Relations between* a *and* b *axes* | *Equivalent positions* | *Systematically absent reflections* | *Symmetry of diffraction pattern* | *Reflections with phase 0° or 180°* |
|---|---|---|---|---|---|---|
| 1 | $p1$ | — | $(x, y, z)$ | — | $\|F(hkl)\| = \|F(\bar{h}\bar{k}\bar{l})\|)$ | — |
| 2 | $p21$ | — | $(x, y, z)$, $(\bar{x}, \bar{y}, z)$ | — | as 1 and $\|F(hkl)\| = \|F(\bar{h}\bar{k}l)\|$ | $(hk0)$ |
| 3 | $p12$ | $\gamma = 90°$ | $(x, y, z)$, $(\bar{x}, y, \bar{z})$ | — | as 1 and $\|F(hkl)\| = \|F(\bar{h}k\bar{l})\|$ | $(h0l)$ |
| 4 | $p12_1$ | $\gamma = 90°$ | $(x, y, z)$, $(\bar{x}, \frac{1}{2}+y, \bar{z})$ | $(0k0)$, $k$ odd | as 3 | $(h0l)$ |
| 5 | $c12$ | $\gamma = 90°$ | $(x, y, z)$, $(\frac{1}{2}+x, \frac{1}{2}+y, z)$<br>$(\bar{x}, y, \bar{z})$, $(\frac{1}{2}-x, \frac{1}{2}+y, \bar{z})$ | $(hkl)$, $h+k$ odd | as 3 | $(h0l)$ |
| 6 | $p222$ | $\gamma = 90°$ | $(x, y, z)$, $(\bar{x}, \bar{y}, z)$<br>$(\bar{x}, y, \bar{z})$, $(x, \bar{y}, \bar{z})$ | — | mm-symmetry, i.e. $\|F(hkl)\| = \|F(\bar{h}kl)\| = \|F(h\bar{k}l)\| = \|F(hk\bar{l})\| = \|F(\bar{h}\bar{k}l)\| = \|F(\bar{h}k\bar{l})\| = \|F(h\bar{k}\bar{l})\| = \|F(\bar{h}\bar{k}\bar{l})\|$ | $(hk0)$<br>$(h0l)$<br>$(0kl)$ |
| 7 | $p22_12$ | $\gamma = 90°$ | $(x, y, z)$, $(\bar{x}, \frac{1}{2}+y, \bar{z})$<br>$(\bar{x}, \bar{y}, z)$, $(x, \frac{1}{2}-y, \bar{z})$ | $(0k0)$, $k$ odd | as 6 | $(hk0)$<br>$(h0l)$ |
| 8 | $p22_12_1$ | $\gamma = 90°$ | $(x, y, z)$, $(\bar{x}, \bar{y}, z)$<br>$(\frac{1}{2}-x, \frac{1}{2}+y, \bar{z})$, $(\frac{1}{2}+x, \frac{1}{2}-y, \bar{z})$ | $(h00)h$ odd<br>$(0k0)k$ odd | as 6 | $(hk0)$ |
| 9 | $c222$ | $\gamma = 90°$ | $(x, y, z)$, $(\bar{x}, \bar{y}, z)$<br>$(\bar{x}, y, \bar{z})$, $(x, \bar{y}, \bar{z})$<br>$(\frac{1}{2}+x, \frac{1}{2}+y, z)$, $(\frac{1}{2}-x, \frac{1}{2}-y, z)$<br>$(\frac{1}{2}-x, \frac{1}{2}+y, \bar{z})$,<br>[illegible] | $(hkl)$, $h+k$ odd | as 6 | as 6 |

| | | | | | | |
|---|---|---|---|---|---|---|
| | | $\gamma=90°$ | $(\bar{x}, \bar{y}, z)$, $(y, \bar{x}, z)$ | | $\lvert F(hk\bar{l})\rvert=\lvert F(\bar{h}\bar{k}\bar{l})\rvert=$<br>$\lvert F(\bar{k}hl)\rvert=\lvert F(k\bar{h}l)\rvert=$<br>$\lvert F(k\bar{h}\bar{l})\rvert=\lvert F(\bar{k}h\bar{l})\rvert$ | |
| 11 | $p422$ | $a=b$<br>$\gamma=90°$ | $(x, y, z)$, $(\bar{y}, x, z)$<br>$(\bar{x}, \bar{y}, z)$, $(y, \bar{x}, z)$<br>$(y, x, \bar{z})$, $(\bar{x}, y, \bar{z})$<br>$(\bar{y}, \bar{x}, \bar{z})$, $(x, \bar{y}, \bar{z})$ | — | as 10 and<br>mm-symmetry;<br>$\lvert F(hkl)\rvert=\lvert F(\pm h \pm k \pm l)\rvert=$<br>$\lvert F(khl)\rvert=\lvert F(\pm k \pm h \pm l)\rvert$ | $(hk0)$<br>$(h0l)$<br>$(0kl)$<br>$(h\bar{h}l)$<br>$(hhl)$ |
| 12 | $p42_12$ | $a=b$<br>$\gamma=90°$ | $(x, y, z)$, $(\bar{y}, x, z)$<br>$(\bar{x}, \bar{y}, z)$, $(y, \bar{x}, z)$<br>$(\frac{1}{2}+y, \frac{1}{2}+x, \bar{z})$, $(\frac{1}{2}-x, \frac{1}{2}+y, \bar{z})$, $(\frac{1}{2}-y, \frac{1}{2}+x, \bar{z})$, $(\frac{1}{2}+x, \frac{1}{2}-y, \bar{z})$ | $(h00)$ $h$ odd<br>$(0k0)$ $k$ odd | as 11 | $(hk0)$<br>$(h\bar{h}l)$<br>$(hhl)$ |
| 13 | $p3$ | $a=b$<br>$\gamma=120°$ | $(x, y, z)$, $(\bar{y}, x-y, z)$<br>$(y-x, \bar{x}, z)$ | — | 6-fold symmetry<br>$\lvert F(hkl)\rvert=\lvert F(k, \bar{h}+\bar{k}, l)\rvert=$<br>$\lvert F(\bar{h}+\bar{k}, h, l)\rvert=\lvert F(\bar{h}\bar{k}\bar{l})\rvert=$<br>$\lvert F(\bar{k}, h+k, \bar{l})\rvert=$<br>$\lvert F(h+k, \bar{h}, \bar{l})\rvert$ | — |
| 14 | $p312$ | $a=b$<br>$\gamma=120°$ | $(x, y, z)$, $(\bar{y}, x-y, z)$<br>$(y-x, \bar{x}, z)$, $(\bar{y}, \bar{x}, \bar{z})$<br>$(x, x-y, \bar{z})$, $(y-x, y, \bar{z})$ | — | as 13 and mirror symmetry<br>across (110)-diagonal;<br>$\lvert F(hkl)\rvert=\lvert F(khl)\rvert$etc. | $(hhl)$<br>$(h, 2\bar{h}, l)$<br>$(2\bar{k}, k, l)$ |
| 15 | $p321$ | $a=b$<br>$\gamma=120°$ | $(x, y, z)$, $(\bar{y}, x-y, z)$<br>$(y-x, \bar{x}, z)$, $(y, x, \bar{z})$<br>$(\bar{x}, y-x, \bar{z})$, $(x-y, \bar{y}, \bar{z})$ | — | as 13 and mirror symmetries<br>across h and k axes;<br>$\lvert F(hkl)\rvert=\lvert F(kh\bar{l})\rvert$ etc. | $(h\bar{h}l)$<br>$(h, 2h, l)$<br>$(2k, k, l)$ |
| 16 | $p6$ | $a=b$<br>$\gamma=120°$ | $(x, y, z)$, $(\bar{y}, x-y, z)$<br>$(y-x, \bar{x}, z)$, $(\bar{x}, \bar{y}, z)$<br>$(y, y-x, z)$, $(x-y, x, z)$ | — | as 13 and<br>$\lvert F(hkl)\rvert=\lvert F(hk\bar{l})\rvert$ etc. | $(hk0)$ |
| 17 | $p622$ | $a=b$<br>$\gamma=120°$ | $(x, y, z)$, $(\bar{y}, x-y, z)$<br>$(y-x, \bar{x}, z)$, $(\bar{x}, \bar{y}, z)$<br>$(y, y-x, z)$, $(x-y, x, z)$<br>$(y, x, \bar{z})$, $(\bar{x}, y-x, \bar{z})$<br>$(x-y, \bar{y}, \bar{z})$, $(\bar{y}, \bar{x}, \bar{z})$<br>$(x, x-y, \bar{z})$, $(y-x, y, \bar{z})$ | — | as 16 and mirror symmetries<br>across $h$ and $k$ axes and across<br>(110)-diagonal<br>$\lvert F(hkl)\rvert=\lvert F(khl)\rvert=$<br>$\lvert F(hk\bar{l})\rvert=\lvert F(k, \bar{h}+\bar{k}, l)\rvert$ etc. | $(hk0)$<br>$(hhl)$<br>$(h\bar{h}l)$<br>$(h, 2h, l)$<br>$(h, 2\bar{h}, l)$ |

protein molecule. If, for example, $a = 120$ Å, $b = 80$ Å and the thickness $= 100$ Å, we can calculate that the unit cell volume is 960 000 Å$^3$. If the symmetry is $p22_12_1$, for which there are four molecules per unit cell, each protein molecule, together with the lipids and water that surround it, then occupies 960 000/4 Å$^3$, or 240 000 Å$^3$. Three-dimensional crystals of water-soluble proteins contain an average percentage of protein of about 50% of the volume, the rest being solvent. Not enough data on membrane crystals are available yet to determine whether membrane crystals deviate appreciably from this value, but it seems likely that most membrane protein crystals will contain between 40 and 60% protein. Assuming 50% in the example above, we may estimate the volume of one protein molecule to be 240 000/2 Å$^3$, or 120 000 Å$^3$. The density of proteins is 1.35 g/cm$^3$, which is equivalent to 1.2 Å$^3$/dalton. The molecular weight of the protein is then given by 120 000/1.2, or 100 000. This value is uncertain owing to the approximation of the fraction of the volume occupied by protein and may be 20% higher or lower. This usually is the molecular weight of the monomer – in some plane groups we may find dimers, trimers, tetramers or hexamers, the molecular weights of which will then be 2, 3, 4 or 6 times greater, respectively.

The determination of the plane group symmetry follows the same path as the determination of the space group of a three-dimensional crystal studied by X-ray diffraction, except that electron microscopy data provide fewer reflections. This disadvantage is compensated for by the availability of phase information, which is not experimentally available in X-ray diffraction experiments.

The diffraction pattern of the crystal is examined in the optical diffractometer. If the pattern shows 6-fold symmetry, the symmetry of the crystal belongs to a trigonal or a hexagonal plane group ($p3$, $p312$, $p321$, $p6$ or $p622$). To which of these plane groups a crystal belongs can be determined by analysis of the relationships of amplitudes and phases between certain combinations of reflections. These relations are listed in Table 11.1. If the pattern shows 4-fold symmetry ($a = b$, the $\gamma$ angle $= 90°$ Å and the intensities of the diffraction points show 4-fold symmetry), the possible symmetries are $p4$, $p422$ or $p42_12$. The latter may be distinguished by the fact that all reflections along the $a$ and $b$ axes with odd indices are absent. The next lower symmetry is the orthorhombic class. Here the angle $\gamma$ must also be 90°, but the $a$ and $b$ axes need not be equal. The symmetry of the diffraction pattern must show mirror symmetries across both axes, so-called mm-symmetry. If all reflections for which the sum of the indices $(h + k)$ is odd are absent, then the plane group is centred, with $c21$ or $c222$ as the two possibilities. If only the axial reflections are absent the symmetry may be $p22_12_1$ (absences along both $a$ and $b$ axes) or $p12_1$ or $p222_1$ (absences only along one axis). Some plane groups may be distinguished only if the phases of the reflections are known, for which it is necessary to process the image by computer.

One serious difficulty with the plane group determination of membrane protein crystals is the small number of unit cells, and thus the low quality of the diffraction pattern, in contrast to X-ray diffraction in which many more reflections are observed. The plane groups $p12$ and $p12_1$ are distinguished by the systematic absences of $(0k0)$; $k$ odd in $p12_1$ whereas $p12$ does not have any systematically absent reflections. If the unit cell is 100 Å × 100 Å, there will be four reflections on each of the two crystallographic axes $h$ and $k$, within the typical resolution in negative stain, 25 Å. If only four reflections are theoretically present along an axis, it may be hard to tell if the two odd reflections (1 and 3) are systematically absent or just weak. The opposite situation may also be found – the symmetry of the diffraction pattern may be lower than the real symmetry of the crystal due to uneven stain distribution between the top and bottom sides of the membrane. This is a considerable risk for all plane groups with 2-fold or $2_1$ screw axes in the plane of the membrane.

If one of the symmetry elements, a 2-, 3-, 4-, or 6-fold axis, passes through a protein molecule, we can immediately say that the molecule is a dimer, trimer, tetramer or hexamer, respectively.

### 11.2.7 THREE-DIMENSIONAL DATA COLLECTION – TILT SERIES

A protein molecule is a three-dimensional object, and thus a projection, although informative, is not sufficient. A three-dimensional model can be obtained by tilting the membrane in the microscope, and thus obtaining views of the protein molecules over a range of angles. In order to achieve a resolution of 25 Å, a tilt series taken every 10° in the range −60° to +60° along each of two roughly perpendicular tilt axes is sufficient. In practice several tilt series are collected, and those images with the best preserved crystalline order and taken at the most suitable defocus conditions are selected for further processing. The amplitudes and phases of all reflections taken from the different images in the tilt series are then combined in the computer using exact information about the tilt angle and the direction of the axis about which the crystal was tilted. In this way a full three-dimensional data-set is built up. From this data set the inverse Fourier transform is calculated, the result of which is a three-dimensional map of the protein (Fig. 11.10). For full information on the technical details of the procedure, see Amos *et al.* (1982).

### 11.2.8 THE HEIGHT OF THE PROTEIN MOLECULE

It is difficult to get a good estimate of the height of the molecule by the techniques of electron microscopy and image processing. The difficulty arises because of the shrinking that occurs both in the protein and in the stain, as the sample is boiled in the electron beam. This is a severe problem in the structure determination of membrane proteins by this method. One method which is

**Fig. 11.10** Three-dimensional structure of cytochrome reductase, obtained by combining series of tilted views of membrane crystals of the enzyme (Leonard *et al.*, 1981). Each section is 10 Å thick. The hydrophobic part embedded in the lipid bilayer is painted white.

often used to supplement information obtained by observing the specimen in negative stain is metal shadowing. A very thin film of metal atoms is deposited on to the grid on which the crystals are lying. If the metal is deposited at a fairly low angle, 5° to 20°, the membrane crystal will throw a long shadow, and from the length of this shadow the height of the protein can be determined, if the angle is accurately known. A standard such as tobacco mosaic virus, with a known diameter of 180 Å, may be applied to the grid (see Fig. 11.11).

Another method used is to embed the crystals in plastic followed by ultrathin sectioning. This gives a well-defined repeat distance, especially for stacks of

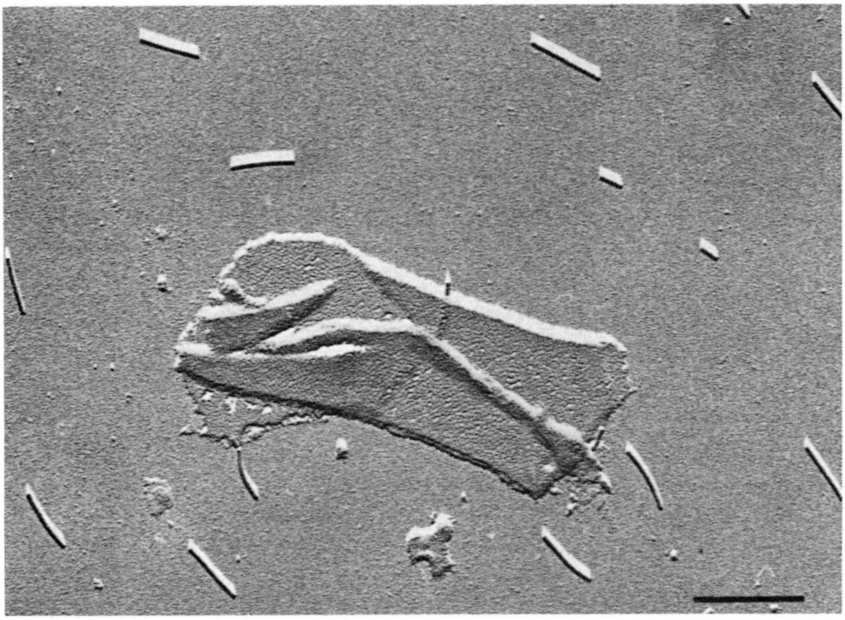

**Fig. 11.11** Example of height determination of a membrane crystal by metal shadowing. One large membrane crystal of cytochrome reductase and several tobacco mosaic virus (TMV) particles are shown. The TMV has a well-known diameter of 180 Å and is used as an internal thickness standard. (Reproduced from Karlsson *et al.*, 1983, with permission.) Scale bar is 0.5 $\mu$m.

membranes, from which the length of the unit cell perpendicular to the plane of the membrane can be measured. In order to ensure that the membranes are sectioned perpendicular to the membrane, and not at an oblique angle, one method is to section the specimen after it has been applied to a grid which then acts as a support and holds the crystal in a fixed orientation to the knife edge. A very elegant version of this technique has been developed by Jésior, in which he first observes a crystal in the conventional electron microscope, and then under a light microscope sections through the very same crystal (Jésior, 1982).

The height of a membrane protein is unlikely to be less than the thickness of the membrane it traverses. The thickness of a lipid bilayer varies with the type of lipid, but for $C_{18}$ lipids which are commonly found in natural membranes the hydrophobic region is $2 \times 22$ Å. The hydrophilic head groups add a further 6–10 Å on each side of the membrane (Pearson and Pascher, 1979), making a total of 55–65 Å for the thickness of a lipid bilayer including the head groups (Fig. 11.12). The heights of membrane proteins studied so far range from about 45 Å for bacteriorhodopsin to about 150 Å for cytochrome reductase.

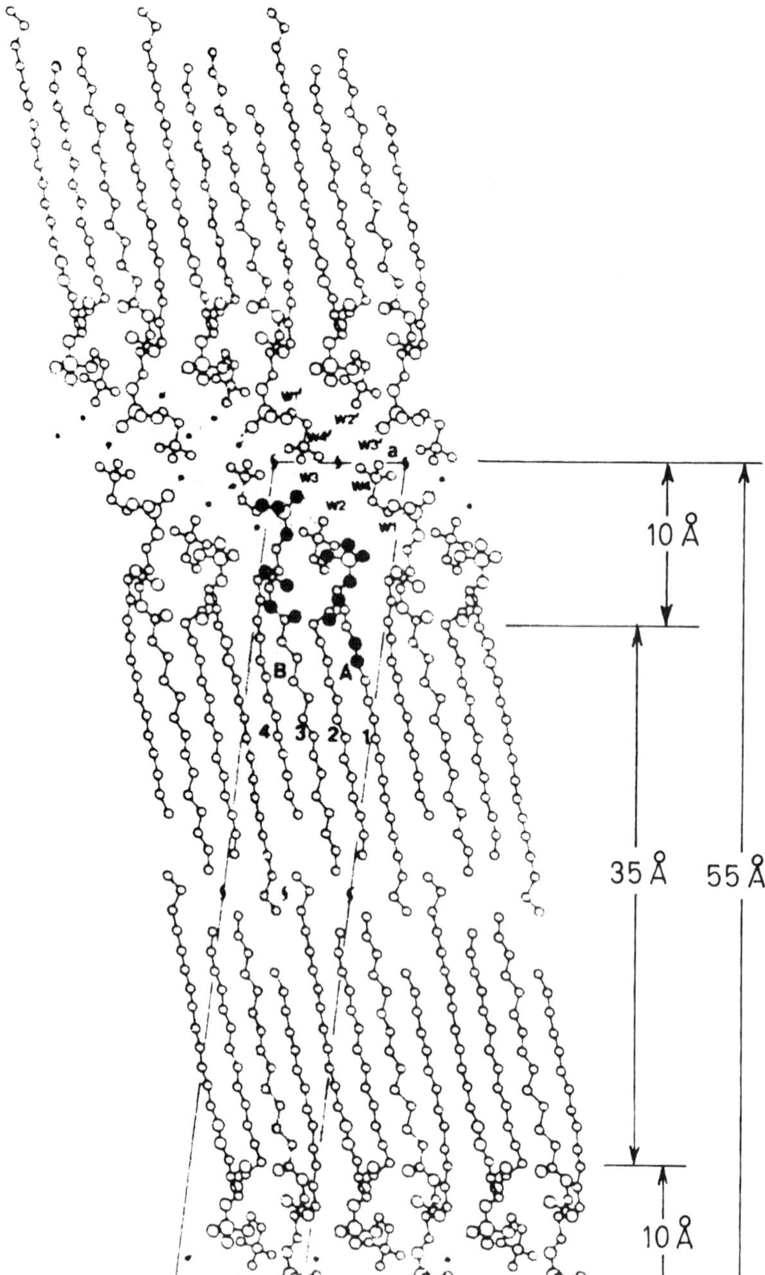


**Fig. 11.12** The structure of a lipid bilayer as determined by X-ray crystallography. One unit cell is indicated. The thickness of the hydrophobic portion of the membrane is 35 Å in this case, where the lipids are C-14. The thickness of the membrane increases with 2 × 1.25 Å per $-CH_2-$ of the hydrophobic chain. The hydrophilic head groups are each 10 Å thick for this lecithin structure. After Pearson and Pascher (1979).

### 11.2.9 LOCALIZATION OF INDIVIDUAL SUBUNITS IN MULTISUBUNIT COMPLEXES

Many membrane proteins are very large multisubunit complexes. This is, for example, the case for all the enzymes of oxidative phosphorylation, where the complexes contain from seven different subunits up to perhaps 20. These complexes are larger than any of the water-soluble proteins studied by X-ray crystallography, except for viruses, which show high non-crystallographic symmetry. An effect of non-crystallographic symmetry is to considerably reduce the size of the unique protein structure which is to be determined. A feasible way forward for high-resolution structure determination of multi-subunit complexes may be to study the intact complex to low resolution by electron microscopy in negative stain, and then split it up into more manageable subcomplexes or single polypeptides, which can be studied again by electron microscopy, or by X-ray crystallography. By this method it may be possible to localize precisely the different subunits within the whole complex, using a three-dimensional model of the whole complex and a series of three-dimensional models obtained by specific removal of subunits. The high-resolution structures obtained from X-ray crystallography of the individual subunits may then be placed at their appropriate positions, and a complete high-resolution electron-density map of the whole complex will emerge.

The first steps along this path have been taken for cytochrome reductase, of which the whole complex has been studied by electron microscopy (Leonard *et al.*, 1981) a hydrophobic subcomplex has been prepared and studied by electron microscopy (Karlsson *et al.*, 1983), and other hydrophilic subunits have been crystallized as three-dimensional crystals.

### 11.2.10 EXAMPLES OF THREE-DIMENSIONAL STRUCTURE DETERMINATIONS OF MEMBRANE PROTEINS

The first well studied of all membrane proteins, at least from a structural point of view, is bacteriorhodopsin from the purple membrane of *Halobacterium halobium.* Henderson and Unwin showed by low-dose electron microscopy techniques that the protein is made up of seven rods, possibly $\alpha$-helices, all of which are close to perpendicular to the plane of the membrane. In a later study by Leifer and Henderson on a different crystal form, this arrangement of seven rods was confirmed, showing that the internal structure of the protein molecule was unaffected by packing forces in the crystals. The structures determined from the two different crystal forms are shown in Fig. 11.13.

Cytochrome reductase has already been mentioned, and the structure shown in Fig. 11.10. Another protein complex of the oxidative phosphorylation pathway, cytochrome oxidase, has been investigated in different crystal forms, and three-dimensional structure determinations from crystals obtained

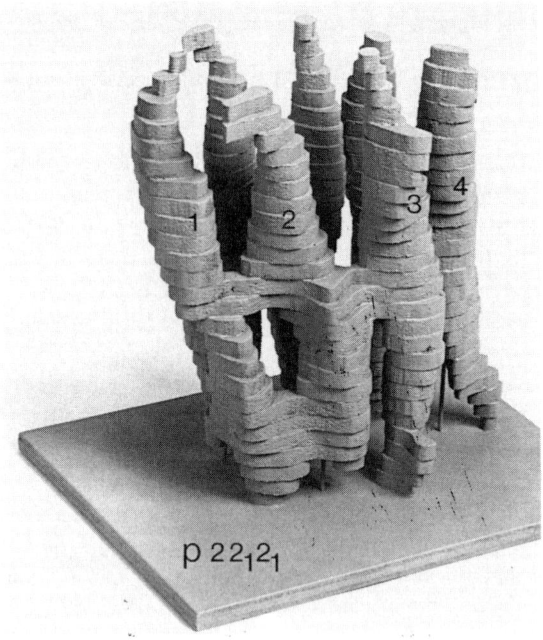


(a)

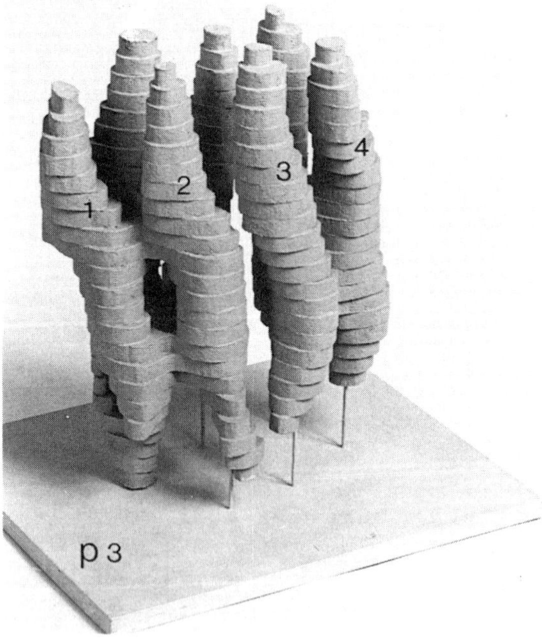


(b)

**Fig. 11.13** Three-dimensional models of bacteriorhodopsin at 7 Å resolution. (a) Structure in $p22_12_1$ crystal form (crystallized *in vitro*). (b) Structure in $p3$ as the protein crystals appear *in vivo*. At this resolution the structure is not affected by packing forces in the different crystal forms. The rod-like features may be α-helices and are roughly perpendicular to the plane of the membrane. (Reproduced from Leifer and Henderson, 1983, with permission.)

with the different detergents Triton X-100 and deoxycholate have been published (Deatherage *et al.*, 1982; Fuller *et al.*, 1979).

A very large number of bacteria contain regular arrays of subunits as the outermost component of their cell envelopes. These surface layers (S-layers) have been found to form crystals *in vivo* for more than 140 different organisms. A review of this field, including a list of all these bacteria can be found in Sleytr and Messner (1983).

Four three-dimensional structures of bacterial S-layers have been published. These are the S-layers of *Sulpholobus acidocaldarius* (Deatherage *et al.*, 1983), *Chlamydia thrachomatis* (Chang *et al.*, 1982), *Synechocystis* CLII (Karlsson *et al.*, 1984) and *Bacteroides buccae* (Sjögren *et al.*, 1985) (see Fig. 11.14). All three membranes have 6-fold symmetry, although the organization of the protein subunits in the membrane is very different.

The connexons or gap junctions between communicating cells comprise cylinders made up of six subunits tilted about their axes. The connexons are organized in a hexagonal lattice, with unit cell dimensions 85 Å in the plane and a thickness of about 75 Å (Unwin and Zampighi, 1980).

The luminal plasma membrane protein from urinary bladder has a rather similar architecture to that of the gap junctions, except that the cell dimensions are very different; $a=b=160$ Å in the hexagonal unit cell, and a thickness of at least 50 Å (Brisson and Wade, 1983).

Several other membrane proteins have been obtained in crystalline form but their three-dimensional structures remain to be studied. These include the ($Na^+$, $K^+$)ATPase from kidney (Hebert *et al.*, 1982), the pore-forming protein of mitochondrial outer membrane (Mannella, 1982), the channel-forming protein porin from *E. coli* outer membranes (Dorset *et al.*, 1983) rhodopsin from frog retina (Corless *et al.*, 1982) and the acetylcholine receptor from *Torpedo californica* (Kistler *et al.*, 1982).

## 11.3 X-ray diffraction

### 11.3.1 MEMBRANE PROTEINS STUDIED BY X-RAY DIFFRACTION

X-ray crystallography has always been the best method for three-dimensional structure determinations. Today the structures of about two hundred water-soluble proteins have been determined by X-ray crystallography. However, this method has not until recently been used in the study of membrane proteins, owing to the difficulty in making three-dimensional crystals of suitable size and quality for X-ray diffraction studies. Recently, Michel (1982) has obtained crystals of bacteriorhodopsin and photosynthetic reaction centres and Garavito *et al.* (1983) have obtained crystals of the pore protein of *Escherichia coli*. These crystals are suitable for X-ray diffraction studies. The

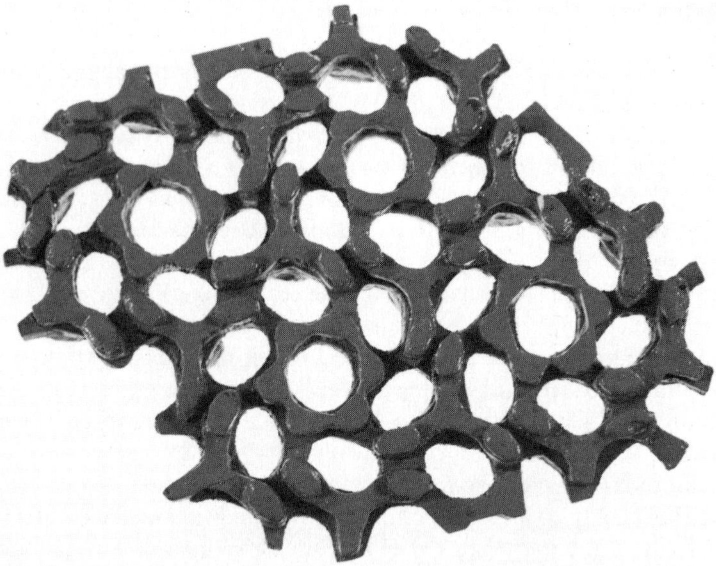

(a)

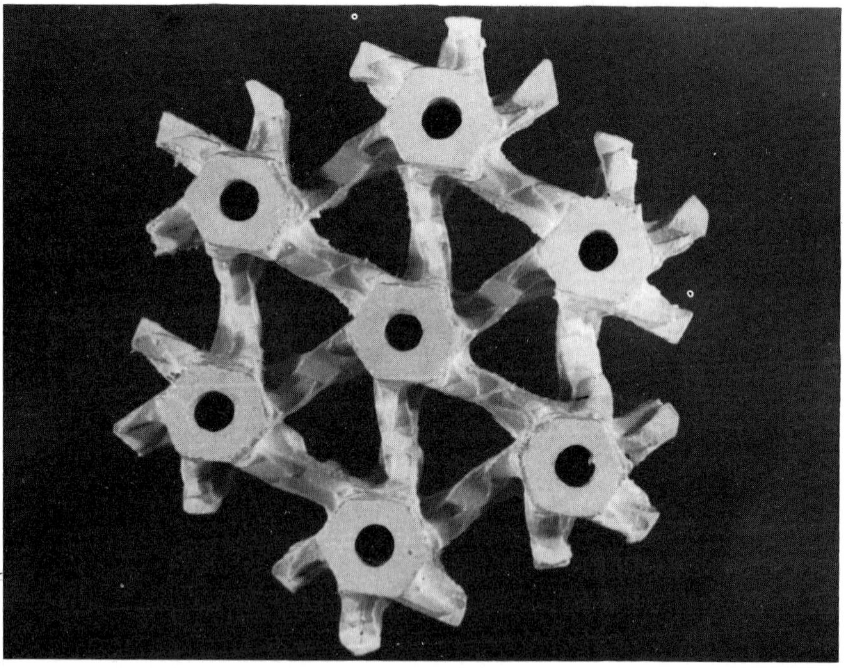

(b)

**Fig. 11.14** Three-dimensional structures of bacterial cell walls/surface layers. (a) *Bacteroides buccae* (Sjögren *et al.*, 1985); (b) *Synechocystis* CLII (Karlsson *et al.*, 1984). These crystals are found *in vivo* and both have *p*6 symmetry, the most common symmetry for bacterial cell walls and surface layers. Although the crystallographic symmetry is the same the structures are distinctly different. (Reproduced with permission.)

structure determinations, including the search for heavy atom derivatives, are now underway.

The principles used for obtaining three-dimensional crystals of membrane proteins have been to render hydrophobic membrane proteins hydrophilic by solubilizing them in detergent, and then to crystallize the whole protein–detergent complex by salting out with ammonium sulphate, as is done for water-soluble proteins. It has been found an advantage to use small amphiphilic molecules as detergents, and several suitable new such detergents have been synthesized during the last few years.

It is outside the scope of this book to give a detailed description of the methods for data collection and structure determination of proteins using X-ray diffraction, and there are already a number of excellent reviews and textbooks in this field.

Yet there should be little doubt that during the next few years we shall see spectacular results coming from X-ray diffraction studies of crystals of membrane proteins. We may expect to see for the first time the structures of membrane proteins at atomic resolution, and with this eventually understand the molecular mechanisms of such fundamental phenomena as oxidative phosphorylation, photosynthesis and membrane pumps. The first membrane protein structure determined to atomic resolution was the photosynthetic reaction centre (Deisenhofer, Michel and Huber, 1985).

## 11.4 References

Amos, L. A., Henderson, R. and Unwin, P. N. T. (1982) *Proc. Biophys. Mol. Biol.*, **39**, 183–231.

Brisson, A. and Wade, R. H. (1983) *J. Mol. Biol.* **166**, 21–36.

Chang, J. J., Leonard, K., Arad, T., Pitt, T., Zhang, Y. X. and Zhang, L. H. (1982) *J. Mol. Biol.*, **161**, 579–90.

Corless, J. M., McCaslin, D. R. and Scott, B. L. (1982) *Proc. Natl. Acad. Sci. USA*, **79**, 1116–20.

Deatherage, J. F., Henderson, R. and Capaldi, R. A. (1982) *J. Mol. Biol.*, **158**, 487–99.

Deatherage, J. F., Taylor, K. A. and Amos, L. A. (1983) *J. Mol. Biol.* **167**, 823–52.

Deisenhofer, J., Michel, H. and Huber, R. (1985) *Trends Biochem. Sci.*, **10**, 243–8.

Dorset, D. L., Engel, A., Häner, M., Massalski, A. and Rosenbusch, J. P. (1983) *J. Mol. Biol.*, **165**, 701–10.

Fuller, S. D., Capaldi, R. A. and Henderson, R. (1979) *J. Mol. Biol.*, **134**, 305–27.

Garavito, M., Jenkins, J., Jansonius, J. N., Karlsson, R. and Rosenbusch, J. P. (1983) *J. Mol. Biol.*, **164**, 313–27.

Goldfarb, W., Frank, J., Kessel, M., Hsung, J. C., King, C. H. and King, T. E. (1979) in *Cytochrome Oxidase* (eds T. E. King *et al.*) Elsevier, Amsterdam, pp. 161–66.

Hebert, H., Jörgensen, P. L., Skriver, E. and Maunsbach, A. B. (1982) *Biochim. Biophys. Acta*, **689**, 571–4.

Henderson, R. and Shotton, D. (1980) *J. Mol. Biol.*, **139**, 99–109.

Henderson, R. and Unwin, P. N. T. (1975) *Nature (London)*, **257**, 28–32.
Holser, W. T. (1958) *Z. Kristallographie*, **110**, 266–81.
Hovmöller, S., Leonard, K. and Weiss, H. (1981) *FEBS Lett.*, **123**, 118–22.
Hovmöller, S., Slaughter, M., Berriman, J., Karlsson, B., Weiss, H. and Leonard, K. (1983) *J. Mol. Biol.*, **165**, 401–6.
Hovmöller, S., Sjögren, A., Farrants, G., Sundberg, M. and Marinder, B.-O. (1984) *Nature (London)*, **311**, 238–41.
Jésior, J.-C. (1982) *EMBO J.*, **1**, 1423–8.
Karlsson, B., Hovmöller, S., Weiss, H. and Leonard, K. (1983) *J. Mol. Biol.*, **165**, 287–302.
Karlsson, B., Vaara, T., Lounatmaa, K. and Gyllenberg, H. (1983) *J. Bacteriol.*, **156**, 1338–43.
Kistler, J., Stroud, R. M., Klymkowsky, M. W., Lalancette, R. A. and Fairclouch, R. H. (1982) *Biophys. J.*, **37**, 371–83.
Leifer, D. and Henderson, R. (1983) *J. Mol. Biol.* **163**, 451–66.
Leonard, K., Wingfield, P., Arad, T. and Weiss, H. (1981) *J. Mol. Biol.*, **149**, 259–74.
Mannella, C. (1982) *J. Cell Biol.*, **94**, 680–7.
Michel, H. (1982) *J. Mol. Biol.*, **158**, 567–72.
Pearson, R. H. and Pascher, I. (1979) *Nature (London)*, **281**, 499–501.
Perkins, S. J. and Weiss, H. (1983) *J. Mol. Biol.*, **168**, 847–66.
Sjögren, A., Hovmöller, S., Farrants, G., Ranta, H., Haapasalo, K., Ranta, K. and Lounatmaa, K. (1985) *J. Bacteriol.*, **164**, 1278–82.
Sleytr, U. B. and Messner, P. (1983) *Annu. Rev. Microbiol.*, **37**, 311–39.
Unwin, P. N. T. and Henderson, R. J. (1975) *J. Mol. Biol.*, **94**, 425–40.
Unwin, P. N. T. and Zampighi, G. (1980) *Nature (London)*, **283**, 545–9.
Vanderkooi, G., Senior, A. E., Capaldi, R. A. and Hayashi, H. (1972) *Biochim. Biophys. Acta*, **274**, 38–48.
Wingfield, P., Arad, T., Leonard, K. and Weiss, H. (1979) *Nature (London)*, **280**, 696–7.

# 12

# Lateral motion of membrane proteins

HANS-GEURG KAPITZA and KENNETH A. JACOBSON



## 12.1 Lateral motion in membranes

### 12.1.1 INTRODUCTION

Living cells are highly complicated and active systems. They divide, change their morphology, move and, in general, perform a seemingly infinite number of functions. Therefore one expects that the system components of these systems are organized into dynamic structures rather than in rigid ones. This concept includes the possibility that many components of the cells are quite mobile, at least temporarily or under certain conditions.

Biomembranes are a case in point of this controlled mobility concept. The most common feature of biological membranes is the bilayer structure of lipid molecules which appears, in many cases, to be a two-dimensional fluid. Bilayer assembly is an entropy-driven effect based on removing the water molecules structured around the aliphatic chains of lipids in aqueous solution (Tanford, 1973). The lipid molecules so assembled into the bilayer are rather loosely associated and can freely move laterally with respect to each other. Lipid bilayers provide both high flexibility and the diffusion barriers between the outer and membrane-enclosed aqueous medium. By definition, integral proteins are inserted into the core of membrane while peripheral proteins are bound to the surface of the membrane. Over a decade ago, the 'Fluid Mosaic' model (Singer and Nicholson, 1972) was postulated to describe the dynamic

Techniques for the Analysis of Membrane Proteins. Edited by C. I. Ragan & R. J. Cherry. Published in 1986 by Chapman and Hall, 11 New Fetter Lane, London EC4P 4EE

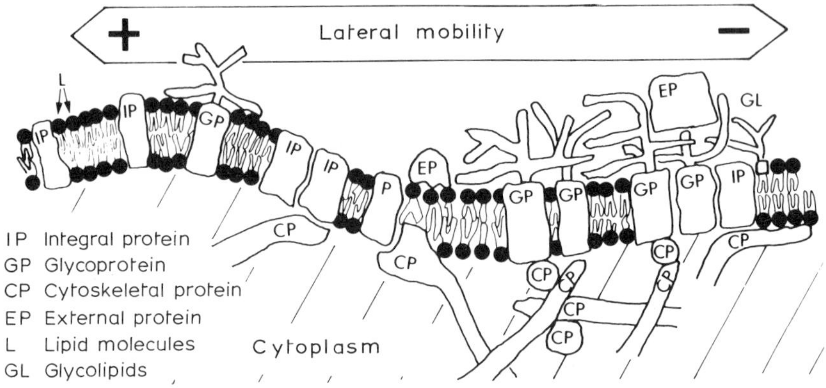


**Fig. 12.1** The degree of lateral motion of membrane proteins expected in the plasma membrane of a eukaryotic cell.

*Left side: rapid motion*
- Low protein density (dilute system)
- Weak or no interaction with peripheral proteins and other membrane proteins
- Large mobile fraction of membrane; proteins displaying rapid lateral diffusion

*Right side: Restricted motion*
- High protein density ('crowded' system)
- Strong intramembrane interaction and binding to peripheral components
- Large 'immobile' fraction

structure of biological membranes. The range of lateral mobility of membrane proteins expected within the general constraints of this model is depicted in Fig. 12.1.

This chapter focuses on the measurement of lateral motion of membrane proteins. It is intended to give practical instructions on how to perform measurements on membrane proteins rather than presenting a review of the state of today's knowledge. Measurements of lateral motion have been made on the proteins of the plasma membranes of cells, of organelle membranes, and on proteins reconstituted into model bilayer membranes made either from synthetic or extracted lipids. Today, it is widely accepted that the lateral motion of membrane proteins is obligatory in many biological reactions although the specific relationships have been shown in only a few cases. Axelrod (1983) has discussed the problem of defining the relationship between lateral motion of membrane proteins and biological function considering the following examples: acetylcholine receptors in the formation of synapses, rhodopsin in visual transduction, cell surface immunoglobulins in immunological responses, $\beta$-catecholamine receptors in response to hormones, and the cytochromes in oxidative phosphorylation.

In addition, correlations have been established between lateral motion of membrane proteins and biological function in the following more general processes (for review, see Axelrod, 1983): ligand-binding receptor down-regulation, cell–cell contact specificity, and viral infection of cells. In addition to functional correlations, the measurement of the diffusion rates in membranes can also provide a dynamic view of the molecular environment immediately surrounding the diffusant, revealing structural interactions which may limit translational diffusion.

There has been a substantial biophysical effort to establish methods to measure the lateral motion of membrane proteins (LMMP). The initial goals have been: (1) to find 'baseline values' for LMMP in artificial membranes and compare the values to those measured in cell membranes and (2) to understand the physical laws valid for LMMP, primarily employing membrane proteins reconstituted into artificial membranes. In this regard, a description of membrane protein mobility has been derived (Saffman and Delbrück, 1975). The membrane protein is assumed to be a right circular cylinder with a smooth surface embedded in the lipid bilayer and its long axis oriented normal to the plane of the membrane. The cylinder undergoes thermally driven lateral motion; its velocity is determined by the viscous drag arising from the surrounding lipid continuum. The Saffman–Delbrück model treats both the lateral diffusion and the rotational diffusion in terms of the Einstein–Stokes law for viscous friction adapted to two dimensions. The lateral diffusion coefficient $D_L$ of a cylinder with radius $a$ and height $h$ spanning the membrane is:

$$D_L = (k_B T/4\pi\eta h)\,[\ln(\eta h/\eta_w a) - \gamma] \tag{12.1}$$

where $k_B$ = Boltzmann's constant ($1.3806 \times 10^{-23}$ joule/kelvin), $T$ = temperature, $\eta$ = viscosity of the 'fluid' membrane, $\eta_w$ = viscosity of the surrounding aqueous phase and $\gamma$ = Euler's constant ($\approx 0.5772$).

The conditions found in the reconstituted membranes (single bilayer, low protein concentration; left side of Fig. 12.1) approximate the situation described by the Saffman–Delbrück equation (Equation 12.1). In such membranes above the gel to liquid crystalline phase transition temperature, protein diffusion is rapid ($D \sim 10^{-8}$ cm²/s) approaching the rates of diffusion of lipid probe molecules [$10^{-8} \leqslant D \leqslant 10^{-7}$ cm²/s (Cherry and Peters, 1982; for reviews, see Vaz *et al.*, 1982a; Jacobson, 1983)], and the remarkably weak dependence of $D$ on protein size has been experimentally verified (Vaz *et al.*, 1982b; W. L. C. Vaz, personal communication).

Thus, with at least a semiquantitative understanding of lateral diffusion accomplished in these simpler systems, many membrane biologists have turned their attention to investigating the types and role of LMMP in specific biological functions. In fact, new, more sophisticated techniques are being introduced to assist in this task (for example, see end of this chapter and Koppel and Sheetz, 1983).

### 12.1.2 GENERAL TYPES OF LATERAL MOTION OCCURRING IN MEMBRANES

Five states of lateral motion may occur in membranes: (a) lateral diffusion; (b) lateral flow; (c) forced lateral motion; (d) no observable motion; (e) removal/reinsertion of membrane proteins. The first three states of motion are two-dimensional in nature, the last one is three-dimensional but may appear as lateral motion in experiments.

(a) *Lateral diffusion*
This type of motion has been described as Brownian motion resulting from thermally driven molecular collisions (Einstein, 1905; Smoluchowski, 1906). This type of motion shows no predictability for the direction of each single 'step' but step length can be described by known probability distributions.

(b) *Lateral flow*
In contrast to diffusion, the lateral flow of membrane components can be described by vector fields varying in time. For pure flow, the integral flow into an area should compensate the total outflow out of this area during a given observation time. If this balance does not occur sources and sinks must exist which require, in the case of membranes, net transport from and to the third dimension (bulk medium).

(c) *Forced lateral motion*
In many cases one could imagine that membrane components are tightly bound to peripheral membrane structures (glycocalyx, cytoskeleton). If these structures have a component of motion parallel to the membrane, the linked membrane components will also move. If the structures are static, the bound membrane components will not move (case d).

(d) *No observable motion*
If the labelled membrane proteins are bound to others forming a two-dimensional network or bound to peripheral structures, there may be no observable motion at all.

(e) *Removal/reinsertion*
In living cells, membrane components may be removed from the membrane and reinserted at another point. The net effect may result in apparent lateral motion.

## 12.2 Measurements of lateral motion in membranes

Small systems require the application of microscopic methods since the dimensions of typical cells range between 5 μm and 200 μm while cellular organelles with the exception of the nucleus have dimensions less than a few microns. The determination of the lateral mobility of proteins in these systems requires the direct observation of their motion or the measurement of time-dependent changes in their local concentrations. The methods to be reviewed here are the photobleaching techniques, mixing of membrane components following the fusion of two cells, and the lateral electrophoresis of membrane constituents. We have confined ourselves, in this chapter, to measurements on single membrane entities using microscopic methods; thus, other fluorescence and magnetic resonance techniques which average over many membranes are not discussed. The principles, applications and key citations are listed in Table 12.1.

### 12.2.1 THE FLUORESCENCE RECOVERY AFTER PHOTOBLEACHING (FRAP) TECHNIQUE

(a) *Experimental*

This method, introduced some ten years ago (Peters *et al.*, 1974), has proven to be a powerful technique for measurements of lateral motion in membranes. Most of this chapter will therefore be spent on describing this technique to give guidance to the interested investigator on how to perform the measurements. Extensions of this technique are still being developed (Koppel and Sheetz, 1983) and one of the more recent improvements, laser photobleaching combined with time-resolved spatial photometry (TRSP), will be introduced at the end of the chapter.

Classical experiments for the measurement of diffusion usually require the generation of a well-defined concentration gradient for the species observed. Then the time-dependent decay of the concentration distribution is measured and, with knowledge of the boundary conditions, the diffusion coefficient can be calculated.

The basic principle of the FRAP technique is the combination of the methods of microfluorescence for the observation of a given species in the membrane and the generation of a well-defined spatial pattern by bleaching fluorophores bound to membrane components. Lateral motion is detected by the decay and the movement of this pattern with time. The concept of the simplest instrumental format, spot photobleaching, is illustrated in Fig. 12.2. First, the emission, $F_k(t<0)$ produced by a fluorescence label on or in a region of a membrane, surface, or thin volume is measured. Second, an intense short photobleaching pulse of light is directed to the same region of the specimen, rapidly destroying a substantial amount of the fluorescence in that region. After this photobleaching pulse, the end of which is defined as time zero, the recovery of fluorescence within the photobleached region is measured as a function of time with a beam coincident with the photobleaching beam and of the same intensity as that used to measure $F_k(t<0)$. This fluorescence intensity increases from $F_k(0)$ at time zero to $F_k(\infty)$ at the time when the recovery is judged complete. The kinetics of recovery can be empirically characterized by the time ($\tau_{\frac{1}{2}}$) required to reach 50% of complete recovery. The

**Table 12.1** Methods for the measurement of lateral motion of membrane proteins

| *Technique* | *Fluorescence recovery after photobleaching (FRAP)* | *Fluorescence redistribution after fusion (FRAF)* | In situ *electrophoresis* |
|---|---|---|---|
| Principle | Time-dependent decay/movement of a photobleached spatial pattern | Intermixing of fluorescence-labelled membrane components of two individual cells after fusion | Re-randomization of a non-homogeneous distribution of membrane proteins after a previously applied electrical field has been removed |
| Application | Measurement of lateral motion in any larger membrane, artificial or natural. Also applicable to fluids, surface films, liquid crystals and cytoplasmic transport in cells | Measurement of lateral motion in cell plasma membranes | Measurement of lateral motion in plasma membranes (a) and inner mitochondrial membranes (b) |
| Author(s) | Peters *et al.* (1974)<br>Edidin *et al.* (1976)<br>Axelrod *et al.* (1976)<br>Jacobson *et al.* (1976)<br>Koppel *et al.* (1976) | Frye & Edidin (1970)<br>Fowler & Branton (1977)<br>Schindler *et al.* (1980) | (a) Poo (1981)<br>(b) Sowers & Hackenbrock (1980) |

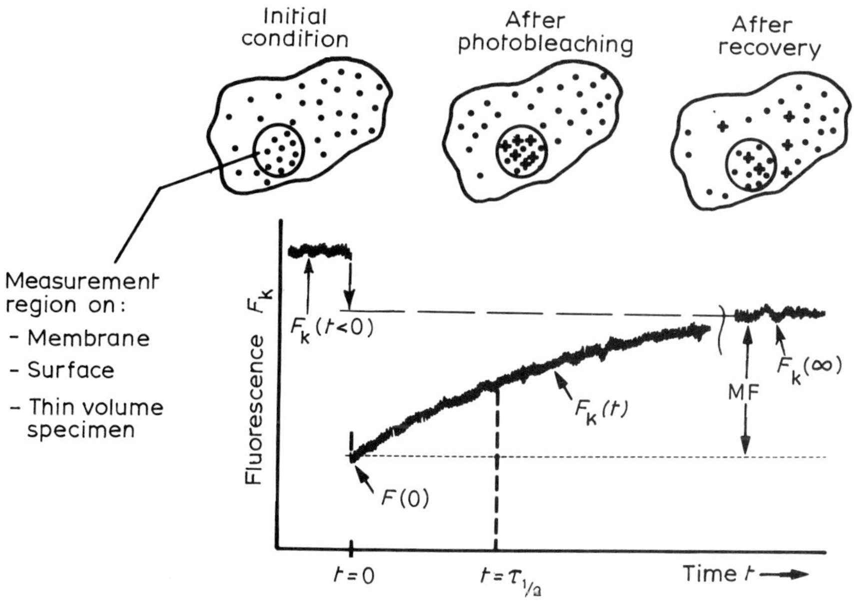


**Fig. 12.2** The concept of a photobleaching measurement. The measurement gives the diffusion coefficient (or flow velocity) of fluorescence labelled components in a surface, membrane or solution sample. Unbleached fluorophores on the cell surface are shown as dots (●) and the bleached fluorophores as crosses (×). The area irradiated by laser is indicated by the circular outline on the specimen. $F_k(t<0)$ = fluorescence intensity before photobleaching. $F_k(0)$ = fluorescence intensity immediately after photobleaching. $F_k(t)$ = increasing fluorescence intensity during recovery. $F_k(\infty)$ = final intensity after recovery has been completed. $\tau_{1/2}$ is the characteristic time of recovery.

mobile fraction $\mathrm{MF} \equiv [F_k(\infty) - F_k(0)]/[F_k(t<0) - F_k(0)]$ characterizes the extent to which the initial fluorescence $F_k(t<0)$ is regained. Fluorophores immobile on the time scale of the measurement are indicated by an MF of less than 100%. If spontaneous recovery of fluorescence does not occur and the probe beam does not itself induce photobleaching, the recovery kinetics can be related to the rate of lateral transport of neighbouring unbleached fluorophores into the previously bleached region in a manner described below (Section 12.2.1.d).

(b) *Classes of fluorescence labels for the investigation of lateral motion of membrane components*

(1) Small hydrophobic or amphiphilic fluorescent molecules that are unspecifically dissolved in or bound to artificial and natural membranes. These probes, such as anilinonaphthalene sulphonate (ANS) and diphenylhexatriene (DPH) are useful to investigate the changes in 'fluidity' or 'microviscosity' of lipid bilayers but have not been used for FRAP studies, thus far.

(2) The second class of probe molecules is made of amphiphilic molecules resembling the membrane's lipid molecules as closely as possible (Derzko and Jacobson, 1980). Synthetic or natural lipid molecules are labelled with a fluorophore or certain probes are conjugated to long acyl chains with the expectation that the probe molecules will approximate the behaviour of native lipid molecules in the membrane.

(3) Membrane peptides and proteins can be labelled using small fluorophores covalently linked to selected amino acids, such as those containing a free amino function (e.g. lysine) or a sulfhydryl group (e.g. methionine). For photobleaching studies, reactive fluorescein, eosin and rhodamine derivatives have been employed. The NBD (*N*-4-nitrobenzo-2-oxa-1,3-diazole) and dansyl moieties (Tank *et al.*, 1982a) have been used to label cyclic and linear gramicidin, respectively. Sometimes, surprising specificity has been achieved in labelling natural membranes with the small molecular probes (Peters *et al.*, 1974; Cherry *et al.*, 1976; Schindler *et al.*, 1980a).

Alternatively, a generally higher degree of specificity may be achieved by using a fluorescent ligand which recognizes a given membrane protein. Small ligands such as the $\beta$-adrenergic analogue (Henis *et al.*, 1982) have been employed to label specific receptors. Fluorescent derivatives of polypeptide hormones such as epidermal growth factor (EGF) and insulin (Schechter *et al.*, 1979) have also been used. In the case of EGF, the fluorescence signal was amplified by conjugating a heavily labelled lactalbumin to EGF without appreciable loss of activity (Schechter *et al.*, 1978).

The most general method available for providing specific labels is the raising of antibodies towards specific membrane antigens. Many measurements of plasma membrane protein diffusion have been described in the literature (for review, see Jacobson and Wojceiszyn, 1981) and recently the technique has also been used on the proteins of the inner mitochondrial membrane (Gupte *et al.*, 1985; Hochman *et al.*, 1982). In general one wants to avoid cross-linking by the antibody meaning that either univalent $F_{ab}$ fragments of polyclonal antibodies or monoclonal antibodies (which usually act as a univalent reagent) should be employed (for discussion, see Jacobson *et al.*, 1984). Indirect immunofluorescence technology, involving amplification of the stain with a second labelled antibody directed toward the primary antibody, is not used because antigen redistribution is usually induced by the second antibody in unfixed cells.

The fluorescence-labelled plant lectins (e.g. concanavalin A, wheat-germ agglutinin etc.) have been used for the earlier measurements of the lateral motion of their corresponding receptors. These generally multivalent ligands recognize certain sugar groups present on components of the cell membrane and the glycocalyx. The interpretation of data using such probes must be approached cautiously because the lectins will induce cross-linking and internalization and, depending on type and concentration, can be toxic to the cells. Thus, the results will, in general, be dependent on the time after labelling the cells. Nevertheless, probes like the divalent succinyl derivative of Con A which shows a reduced tendency to cross-link can be a useful preliminary probe for measuring the lateral diffusion of membrane glycoproteins.

(4) There are a number of covalent protein labels available today, but not all are suitable for FRAP experiments. First there are restrictions due to the optical properties of the light source (usually, a laser), the microscope and the detector devices. The excitation wavelengths of the fluorophores are usually restricted to the range from 350 nm (UV) to below 600 nm (yellow). Emission wavelengths range from roughly 380 nm (UV) to 700 nm (red). For practical purposes, one tries to work with wavelengths from 400 nm to 580 nm for excitation and from 480 nm to 650 nm for emission. Shorter excitation wavelengths require both UV optics and UV lasers while measurement of longer wavelength emission requires special, cooled photo detectors. Currently, there are few probe molecules for the 'red end' of the visible spectrum although such labels could be advantageous (see next paragraph).

The specimen also imposes restrictions in terms of reducing the cellular absorption of excitation energy. Absorption of the excitation light by the cell has at least two undesired impacts: possible harmful photochemical reactions and background contributions to the measured emission intensity (autofluorescence). The autofluorescence of most biological materials appears especially strong upon UV, violet and blue excitation, but is remarkably reduced with excitation in the green and yellow part of the spectrum. Thus, one tries to select probes using the longest wavelength possible for the excitation. Finally, the ideal fluorescent label for photobleaching studies will not alter normal membrane protein interactions between themselves or other membrane structures either before or after photobleaching.

(c) *The equipment required to perform FRAP measurements*

Photobleaching measurements generally employ three major components: the laser as the light source for both photobleaching and excitation of fluorescence, the microscope equipped for epifluorescence, and the detector for the fluorescence emission combined with electronics for recording the time-dependent change in emission intensity following photobleaching. Figs 12.3 and 12.4 show schematic diagrams for the optical and electronic elements, respectively. Spot photobleaching systems have been described in detail by Koppel *et al.* (1976) and Jacobson *et al.* (1977).

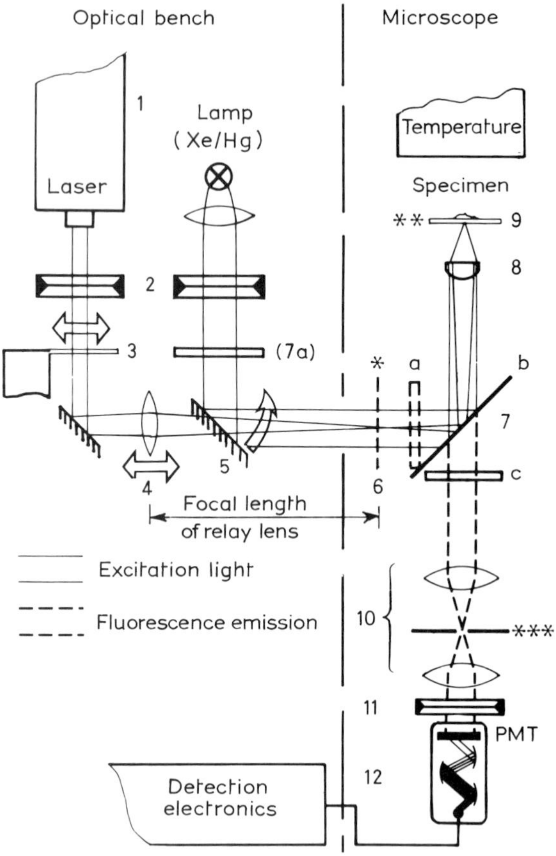


**Fig. 12.3** Schematic diagram of the experimental set-up for FRAP measurements. 1, laser; 2, electromagnetic shutter for laser and lamp; 3, device for laser beam attenuation: solenoid-actuated neutral density filter or switchable beam splitter; 4, relay lens to focus laser beam into the intermediate image plane (*); 5, swing-back mirror or beam splitter for lamp light excitation; 6, intermediate image plane. Objects inserted here are imaged into the specimen plane; 7, filters for wavelength separations: (a) excitation filter – for use with lamp only, (b) dichroic (or neutral) mirror, (c) emission filter; 8, microscope objective lens; 9, specimen (e.g. cell on coverslip); 10, optics forming intermediate image plane in front of the detector. Adjustable pinholes are inserted here to select area (volume) seen by the detector; 11, detector protection shutter; 12, detector (e.g. photomultiplier tube, PMT) and associated electronics.

*Table or bench.* Unless the investigator plans a multiple use facility (e.g. FRAP, fluorescence correlation spectroscopy, image analysis, etc.) normally there is no need for expensive vibration isolation optical tables like those used for interferometric optics experiments provided the building housing the laboratory is free from excessive vibration. High-density surface-coated particle board (3 cm thickness) mounted to a steel frame or a sturdy large table or laboratory bench will give satisfactory support. Money saved by not

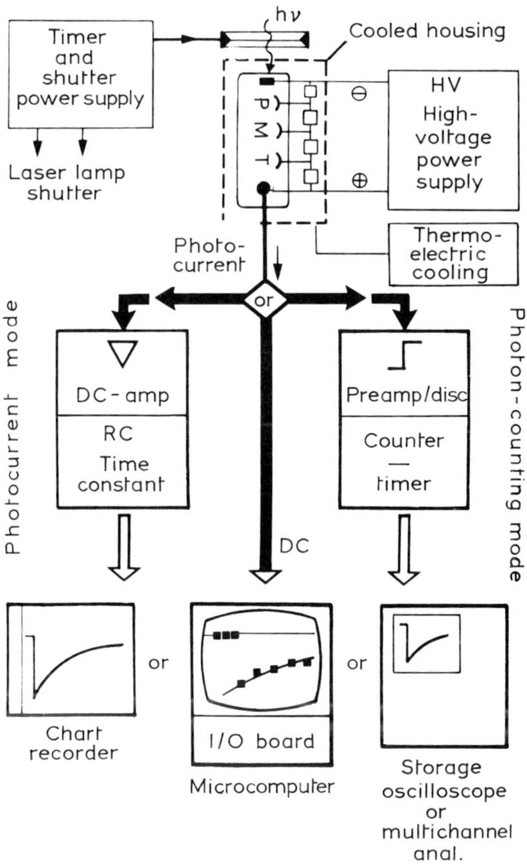


**Fig. 12.4** Configuration for recording fluorescence intensities in the FRAP experiment. From top to bottom: the photoelectric signal coming from the phototube is processed either in the photocurrent mode using amplifiers with high input resistance and following time-constant circuitries for signal-smoothing (left-side) or the photon-counting mode (right side). The display and the documentation of the time-dependent fluorescence intensity require only one of the options shown at the bottom.

purchasing an expensive optical table is better spent on a microscope of excellent mechanical and optical quality.

*Laser.* The expected diffusion coefficients define the needs for laser power in that faster diffusion requires the bleaching dose to be delivered in a shorter time. Single-line-power laser power can be estimated from the following considerations. The bleach energy required depends on the spot diameter used (usually 1.0 to 5 $\mu$m for focused laser beams) and the bleaching characteristics of particular fluorophore selected, subject to the constraint that the bleach energy is delivered in a time $\leqslant 0.1\ \tau_{\frac{1}{2}}$. Empirically, variations in the bleaching quantum yield for the currently available probes appear to vary at least by an order of magnitude. Experience shows that roughly $10^{-4}$ to $10^{-2}$ Ws optical input energy in the specimen plane is required to get reasonable bleaching ($\geqslant$50%). Since energy is the product of power and time, it is possible to calculate the laser power needed from $\tau_{\frac{1}{2}}$, which is approximately equal to $w_s^2/4\,D$ where $w_s$ is the radius of the focused laser beam in the specimen plane. An upper limit of power required is therefore:

$$P \leqslant 4 \times 10^6 \times D(\text{Watt});\ D \text{ in cm}^2\,\text{s}^{-1}. \qquad (12.2)$$

For example, to measure diffusion coefficients of $10^{-8}$ cm$^2$ s$^{-1}$, the laser power required at one line (wavelength) should be 40 mW. If there is any doubt, the higher powered laser is the safer choice. Note that if only slow protein diffusion is to be measured ($D \leqslant 10^{-9}$ cm$^2$ s$^{-1}$) and excitation wavelengths in the blue or green part of the spectrum are required, a small air-cooled Argon ion laser ($\leqslant$100 mW@488 nm/514 nm) or HeCd gas laser ($\leqslant$40 mW@442 nm) will be sufficient.

Higher-powered ion lasers provide a range of emission lines not available on smaller models (e.g. argon 350 nm to 514 nm; krypton 350 nm to 752 nm). These lasers are less convenient in that they are water-cooled and require high-current outlets for the power supplies. It should be noted that laser power generally decreases during the lifetime of the laser discharge tube. Usually one can expect a several thousand hour operating life for the laser tube and a slow decrease of maximum output power. The above lasers provide continuous wave (cw) operation, i.e. a constant intensity output in time. Properly attenuated flash-lasers, like pulsed dye lasers, could be used but their beam profile is usually poor.

*Beam attenuation.* Attenuation of the photobleaching beam for interrogation can be accomplished by rapidly shifting a neutral density filter into the beam path. In this case, extreme care must be taken to obtain, from manufacturers of specialized optics, neutral density filters polished to have nearly parallel sides; otherwise, the filter is a wedge and the interrogation beam deviates from the optical axis, causing systematically longer recovery times than should be obtained (Barisas, 1980). Various beam-splitting arrangements obviate this problem (Barisas, 1980; Koppel, 1979).

*Microscope.* FRAP measurements require the use of microscopes with superior mechanical and optical quality. The mechanical stability means that the microscope stand or housing is sturdy and made with high precision. This is usually the case with the larger research microscopes of the leading manufacturers. The use of an inverted microscope is desirable because they can be outfitted with excitation light sources and the detectors mounted close to the benchtop. The upright microscope has been widely used, but usually needs support structures to hold the detector(s) (usually at the top of the system). The inverted microscope has advantages for the observation of biological specimens in addition, because both normally mounted specimens and specialized vessels containing the specimen can be viewed.

The microscope should be equipped for epifluorescence excitation through the objective lens and provide phase contrast optics to examine the specimen. Other optics can be useful depending on the project. The best objective lenses are those for normal epifluorescence having high aperture, low stray light and low autofluorescence. There is no problem in the use of lenses with built-in phase rings for laser light excitation unless the laser spot desired is much larger than the minimal beam waist achievable. In this case the 'shadow' of the phase ring would appear in the image plane. It is highly desirable to have the 'intermediate' image planes accessible; holders for diaphragms and grids placed on this plane in the excitation light path should allow not only reproducible adjustment within the plane ($x$, $y$ axis) but also along the optical axis for focusing of the desired pattern on the specimen.

Pinholes and other apertures placed at image planes in the emission light path allow out-of-focus fluorescence to be largely eliminated (Koppel *et al.*, 1976). This is extremely useful in reducing, for example, tissue culture substrate fluorescence or in confining the FRAP measurement to one of two membranes when they are separated by more than 5 $\mu$m (Petersen and McConnaughey, 1981; Dragsten *et al.*, 1979).

*Detection electronics.* Photobleaching instruments detect fluorescence using photomultipliers (PMT) either to measure photocurrent or to count photons. The photomultiplier should be selected to have high sensitivity and low dark current. A multistage PMT with a reduced bialkali cathode (typically 10 mm diam.) is a good choice for the detection of emission from the near-UV ($\sim$ 380 nm) to the yellow or orange part of the spectrum ($\sim$ 600 nm). For the frequent measurement of yellow or red emission wavelengths, a multialkali or, even better, a GaAs photocathode is recommended, but these PMT require cooled housings to reduce their high number of thermal dark counts. Cooling from room temperature to 0° C or to about $-25°$ C reduces noise by 10- and 100-fold respectively. Beyond that limit, cooling is not as effective and the quantum efficiency drops somewhat. The cooling required is easily achieved with thermoelectric elements. The water-cooled thermoelectric elements are preferable to fan-driven air-cooled elements because the fan vibrations are transmitted through the microscope.

To record the photocurrent (photon counts), any amplifier (photon counter) is suitable, provided that the time constant (counting interval; including start, stop, reset) is much less than $\tau_{\frac{1}{2}}$ in order not to greatly distort the recovery curve. Finally, the signal needs to be displayed on a chartwriter, storage oscilloscope or multichannel analyser operated in the multiscaling mode.

The most advanced instruments will utilize an on-line microcomputer, which will not only perform the data acquisition (DC or photon counting) but will also handle the calculation of results and the timing of the experiment.

*Control electronics.* Some electronic hardware is necessary to open and close the two electromagnetic shutters and to actuate the beam attenuator. It is necessary to close the PMT protection shutter for the duration of the bleaching pulse and at least for the first 20 ms after that because of strong long-lived emission from the specimen and optical elements.

The other main function is to switch on and off the attenuated laser beam in certain intervals (for example, 1 s 'on', 50 s 'off'). This prevents undesired bleaching of the fluorophores during the prebleach/recovery measurement in longer-lasting experiments.

To perform these functions one normally will specify the control electronics required and have these built in an electronics shop. The other approach is to find a suitable combination of pulse generators in a master/slave organization driving the electronic shutter power supplies. Illumination times ('on') are selected on these power supplies and the pulse generators define the time intervals ('off').

*Other equipment.* Often the microscope stage needs to be thermostatted to perform diffusion measurements as a function of temperature. Heating/cooling stages or 'air curtain incubators' are commercially available.

*Operating precautions.* Novice operators should familiarize themselves with laser safety rules. In addition, the laser should not enter the microscope without the proper blocking filter protecting the operator's eyes from the strong reflections coming back from the specimen or optical elements in the fluorescence illuminator. During adjustment, care should be taken to avoid having the laser spot focused on surface coatings or mirror surfaces inside the microscope. Damage to these elements could occur.

*Beam size calibration.* A spot photobleaching set-up is especially dependent on a good estimate of the spot size of the laser beam focused on the specimen plane. A preliminary estimate can be obtained by comparing the size of the focused laser beam exciting a fluorescent specimen with the ocular micrometer. A suitable test specimen can consist of fluorophores dissolved in an aqueous solution which is drawn into a microcapillary or squeezed between a microscope slide and a coverslip. The illuminating laser beam is focused to the first glass/water interface or to the median plane between the two glass/water interfaces. The rapid diffusion of fluorophores in water and a low intensity for the probing beam will maintain the fluorescence profile excited by the beam. For quantitative purposes, a series of microphotographs is taken of the fluorescent spot and then compared with another one – using the same optics – displaying the bars of a calibrated stage micrometer. The negatives may be scanned with a microdensitometer or prints enlarged at identical magnification may be evaluated with a ruler. The diameter for a beam exhibiting a rectangular intensity profile is easily measured by direct comparison of the apparent diameter with the distances between the micrometer bars, because of the clearly visible boundary of the spot. The case of the Gaussian beam profile requires the densitometer to find the $e^{-2}$ value defining the radius. This value is easily found by inserting filters to attenuate the beam to its $e^{-2}$ value or about 13.5% and taking a second exposure at this intensity. It is important to take the photographs of these spots at a high overall magnification (high zoom factor, high magnification eyepiece) to make the spot size in the plane of the film material large compared with the resolution-limiting grain size.

More sophisticated techniques to measure the radii of small microbeams are based on convolution, integrating and scanning methods (Schneider and Webb, 1981). These methods will require mounting high-precision translation stages on the conventional microscope stage. In the convolution method a spot is bleached into a thin film of immobilized fluorophores and subsequently laterally scanned at a constant velocity by the same but highly attenuated laser beam. This results in a flow type recovery and, since the flow velocity is well defined, the radius of the beam may be calculated from the recovery half time using the flow recovery formula given by Axelrod *et al.* (1976). The method can be performed without a motorized scanning stage by simply using an accurate translating stage manually driven by a differential micrometer to a point where one-half of the bleached fluorescence is recovered (Derzko and Jacobson, 1980). The integration method uses a 'knife-edge' to scan the beam-waist laterally to integrate in a stepwise manner the intensity seen behind the edge. From the resulting curve the intensity distribution of the beam may be derived (Firester *et al.*, 1977). The well-defined translational movement of fluorescent 'point sources' (small fluoresceinated polystyrene beads or other polymer beads coated with fluoresceinated macromolecules) along the diameter of the beam-waist has also been used to give directly the intensity distribution across the beam-waist.

Diffusion coefficients known by other methods should be measured on a new machine to ensure its accuracy and the lack of systematic sources of error. The diffusion coefficients for molecules and macromolecules in fluid mixtures have been calculated and measured using the Einstein–Stokes formula and the FRAP experiment, respectively (Jacobson *et al.*, 1976; Axelrod *et al.*, 1976; Barisas and Leuther, 1979). If agreement is not achieved, the beam diameter estimate should be checked as well as the superimposition of the measuring beam on the bleached hole. One should also check that the dissolved test macromolecules have not aggregated.

(d) *The calculation of diffusion coefficients and the mobile fraction*

Lateral transport coefficients can be evaluated from the kinetics of fluorescence recovery after photobleaching under certain assumptions. (1) Bleaching is assumed to be first order in the concentration of unbleached fluorophores. This allows the concentration of unbleached fluorophore to be calculated as a function of position following photobleaching, forming the initial condition for the solution of the transport equation. The bleached profile is characterized by the parameter $K$ and the effect of increasing $K$ is shown in Table 12.2 for both the Gaussian and rectangular beam intensity profiles. (2) Initial treatments considered there to be an infinite reservoir of fluorophore; that is, as a boundary condition, the membrane was assumed to be an infinite plane. Recent analytical work by Koppel and Sheetz (1983) has lifted this restraint making it possible to analyse the data from small membrane fragments. (3) In the case of diffusion, Fick's second law is assumed to hold:

$$\frac{\delta C}{\delta t} = D\nabla^2 C \tag{12.3}$$

where $C$ is the time-dependent local density of observed fluorophores and $\nabla^2$ is the Laplace-operator in a plane applied to $C(x, y, t)$.

Equation 12.3 is solved subject to the above boundary conditions; a spatial integration yields the fluorescence recovery at any time $t$ (Table 12.2) which then can be fitted to the experimental recovery data using $D$ as the adjustable parameter.

The 'three point fit' is the simplest way to extract diffusion coefficients and mobile fractions from the recovery kinetics. It is valid when the investigator is confident that the mechanism of recovery is single-component diffusion. Three points on the recovery curve are considered (Fig. 12.2): $F_k(t=0)$, the prebleach intensity, $F_k(t=0)$, the intensity recorded immediately after photobleaching is completed, and $F_k(\infty)$, the intensity after recovery process is completed. The characteristic time, $\tau_{\frac{1}{2}}$, for 50% recovery of the mobile fluorophores is the time when the fluorescence recovers to half of its full extent, i.e.

$$F_k(\tau_{\frac{1}{2}}) = F_k(0) + \frac{F_k(\infty) - F_k(0)}{2} \tag{12.4}$$

$D$ can then be calculated from the simple formulae given in Table 12.2. The mobile fraction, MF, is calculated as:

$$\text{MF}(\%) = \frac{F_k(\infty) - F_k(0)}{F_k(t<0) - F_k(0)} \times 100 \tag{12.5}$$

The complement of this value (100% − MF) describes the fraction of fluorophores which appear immobile on the time scale of the experiment.

**Table 12.2** Mathematical formulation for the FRAP experiment (Axelrod *et al.*, 1976)

| | | |
|---|---|---|
| Circular intensity profile $I(\mathrm{r})$ of excitation light beam; $I_0$: intensity on optical axis; $w$ = beam 'radius' | 'Gaussian': $I(\mathrm{r}) = I_0 \exp(-2r^2/w^2)$ | 'Rectangular': $I(\mathrm{r}) = \begin{cases} 0 & r > w \\ I_0 & r \leqslant w \end{cases}$ |
| Local concentration $C_k(r, t)$ of active fluorophore immediately after photobleaching (general): | $C_k(r, t) = C_0 \exp[-\alpha T I(\mathrm{r})]$<br>Bleaching parameter: $K = \alpha T I(0)$<br>$C_0$ prebleach concentration; $\alpha$ overall absorption;<br>$T$ bleach time | |
| Bleaching results in different concentration profiles $C_k(r, t=0)$ dependent on $I(\mathrm{r})$ and $K$: | 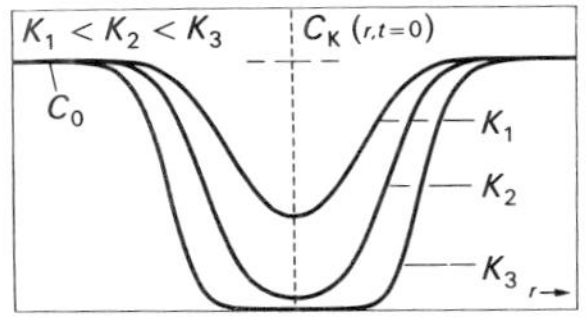  | 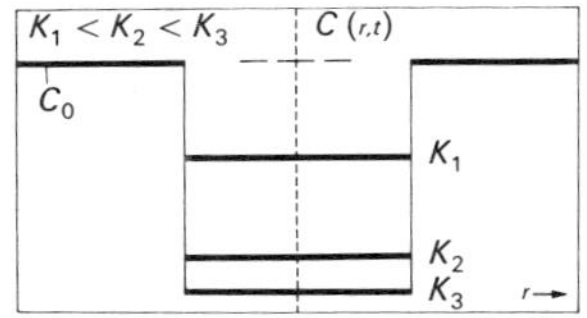  |
| Time-dependent fluorescence intensity (recovery curve) observed with the same, but attenuated (factor: $a$) light beam: | $F_k(t) = (q/a)\int I(\mathrm{r}) C_k(r, t)\,d^2r$<br>$q$ = overall quantum efficiency | |
| Differential equation for lateral motion including isotropic diffusion in a plane and constant flow $v_0$ along $x$: | $\dfrac{\partial C_k(r, t)}{\partial t} = D\nabla^2 C_k(r, t) - v_0 \dfrac{\partial C_k(r, t)}{\partial x}$<br>$D$ = diffusion coefficient | |

This differential equation needs to be solved for the appropriate initial conditions (see above). Assuming that $v_0 = 0$, this leads to:

Series solutions for recovery curves:
$I_{01} I_2$: Bessel functions of first kind and order zero and two

Gaussian:

$$F_k(t) = F(t<0) \sum_{n=0}^{\infty} [(-K)n/n!]\left[1 + n\left(1 + \frac{2t}{\tau D}\right)\right]^{-1}$$

Rectangular: Fractional recovery $f(t)$

$$f(t) = 1 - \left\{\left(\frac{\tau_D}{t}\right)\exp\left(\frac{-2\tau_D}{t}\right)\left[I_0\left(\frac{2\tau_D}{t}\right) + I_2\left(\frac{2\tau_D}{t}\right)\right] + 2\sum_{n=0}^{\infty} \frac{(-1)n(2n+2)!(n+1)!}{(n!)^2[n+2)!]^2}\left(\frac{\tau_D}{t}\right)^{n+2}\right.$$

Connect the characteristic time $\tau_{\frac{1}{2}}$ of the recovery and the characteristic time $\tau_D$ of diffusion using this formula:

$$\tau_{\frac{1}{2}} = \gamma_D \tau_D;\ \gamma_D = \begin{cases} \gamma(k);\ \text{example } \gamma_D = 1.2 \text{ for } k = 2.3 \\ \text{for the 'Gaussian' beam profile} \\ 0.88 \text{ for the 'rectangular' beam profile} \end{cases}$$

These calculations are only valid in the case of one diffusing species in the absence of significant flow.

More sophisticated methods for the evaluation of recovery curves involve the use of calculators or computers. The series solution for the fluorescence recovery (Axelrod *et al.*, 1976) can be programmed into hand calculators and the calculated recovery compared to the actual recovery for various values of $D$ or $V$, the flow velocity. Computers allow the choice of the adjustable parameter, $D$ or $V$, to be optimized in accord with the non-linear least-square fit criteria. The lateral diffusion coefficients for two or more diffusing species co-existing in the same sample may be evaluated by computer modelling the data according to the theory of Axelrod *et al.* (1976) as exemplified by the analysis of Derzko and Jacobson (1980).

Several 'shortcut' procedures have been useful. One elegant method (Axelrod *et al.*, 1976) relates to the fact that in the equations describing the recovery curves the diffusion coefficient $D$ is always a multiplicative factor of the time $t$ and the mobile fraction MF is always a scaling factor for the value on the ordinate $[F_k(t) - F_k(0)]$. The expression

$$f_k(t) = \frac{F_k(t) - F_k(0)}{F_k(t<0) - F_k(0)}$$

'normalizes' the values of the recovery curve $[0 \leqslant f_k(t) \leqslant 1]$. If the 'normalized' recovery curves are plotted in a log–log representation, then these multiplicative scaling factors will show up as orthogonal displacements along the time axis as a measure for $D$ and along the intensity axis as a measure for MF.

Recently a simpler method for the evaluation of recovery data from spot-photobleaching experiments has been developed. This method first linearizes the recovery curve and then performs a linear regression fit to derive the diffusion coefficient and the mobile fraction. This method depends on the fact that $\mathrm{r}(t)$, a reciprocal function of the recovery given by:

$$\mathrm{r}(t) = \frac{F_k(\infty)}{F_k(\infty) - F_k(t)} \tag{12.7}$$

is linear over a large range of times, $t$, when there is only one diffusion coefficient and no flow involved and a Gaussian beam is employed (Yguerabide *et al.*, 1982; van Zoelen *et al.*, 1983). Multiple diffusion coefficients and/or flow will both cause deviations from quasilinear behaviour of the reciprocal function. The accuracy of this method depends critically on the inherent signal/noise ratio in the data gathered from the experiment. This method is very convenient for the visual inspection of recovery curves since deviations from simple diffusion show up as non-linearities.

### (e) *Uncertainties in the calculations of transport parameters from spot photobleaching*

*Instrumental sources.* The major uncertainty in the calculation of lateral diffusion coefficients arises from uncertainty in beam radius, $w_s$, since this quantity is squared in the formula for $D$ (Table 12.2). Further systematic error may be incurred if the measuring beam is not exactly centred on the axis of the photobleached hole (Barisas, 1980). Another systematic error is normally found in the determination of $F_k(0)$, the fluorescence intensity measured immediately after the photobleaching process has been finished. As mentioned, recording the fluorescence emission following bleaching normally starts after some delay time ($\leqslant 20$ ms). Therefore as diffusion becomes more rapid the measured $F_k(0)$ will become systematically higher than the true $F_k(0)$ due to some 'filling in' of the bleached hole before the first interrogation.

*Photobleaching.* The validity of photobleaching results has been questioned because of the possible harmful effects of the photobleaching light pulse (Bretscher, 1980). One potential avenue for membrane photodamage, protein cross-linking, has been demonstrated in conventional photochemical experiments carried out under conditions different from the photobleaching measurements. This putative photodamage is not easy to detect on the single cell level. For example, no detectable alteration in cell surface morphology at the scanning electron microscope level or the Trypan Blue permeability barrier was induced by spot photobleaching. Also, within experimental limitations neither repeated bleaching on the same spot nor increased bleaching dose affected the measurements of fluorescein wheatgerm agglutinin receptor diffusion on the fibroblast cell surface (Jacobson *et al.*, 1978). Most importantly, recent experiments indicate that there is no appreciable photodamage artifact in the measurement of lateral mobility. When two spectrally distinguishable fluorescent labels were directed towards the same cell surface determinant, extensive photobleaching of one of the surface labels over the entire cell did not alter lateral diffusion coefficients subsequently determined by spot photobleaching of the second label (Wolf *et al.*, 1980). Also, measurements of the membrane diffusion coefficient of rhodopsin in the visual receptor membrane both by fluorescence photobleaching and by a photobleaching method which uses the natural photochemistry of rhodopsin yield the same values within experimental error (Wey and Cone, 1981). Finally, careful comparison shows that lateral diffusion coefficients inferred from rates of antigen intermixing on heterokaryons and obtained independently by photobleaching are consistent (Edidin and Wey, 1982). In an elegant extension of this point, Koppel and Sheetz (1981) have shown that diffusion coefficients derived from membrane protein intermixing following red cell–red cell fusion agree with those obtained by subsequently photobleaching the identical fused cell pair.

Photobleaching (fading) during the recovery phase of the experiment will distort the measurement (Jacobson *et al.*, 1984) by reducing the mobile fraction and shortening the apparent $\tau_{\frac{1}{2}}$ because recovery no longer reaches its full extent. Thus, calculated lateral diffusion constants are systematically too high. The fading effect depends strongly on the fluorophore used and its molecular environment. It is difficult to correct the recovery data for fading and thus best to reduce the interrogation beam intensity to the minimum still allowing an adequate signal to noise ratio.

*Properties of the specimen.* Surface roughness and spurious flow-like behaviour within the specimen are two potential sources of difficulty in a typical FRAP experiment. Surface irregularity has long been suspected of causing difficulty in the calculation of diffusion coefficients from FRAP data. However, theoretical calculations (Aizenbud and Gershon, 1982) suggest that surface 'roughness' will not affect the calculated *D* too much and this is borne out by measurements on villous and non-villous regions of the egg membrane (Wolf *et al.*, 1982). The calculations dealt with the relatively early phases of recovery; it is certainly possible that a villous surface could give rise to an apparent immobile fraction.

In experiments designed to calculate diffusion constants, the investigator is interested in detecting possible non-diffusive behaviour (e.g. flow) to eliminate its impact on the calculation of diffusion constants. Flow-like behaviour will contaminate pure diffusion results when the observed membrane moves appreciably during the time of observation due to either its own motion or microscope stage movement. Such translation can be verified by observing the location and shape of the object before and after the photobleaching measurements have been carried out. Without the need to fit the data to calculated recovery curves (Axelrod *et al.*, 1976), flow can sometimes be recognized in the FRAP kinetics by observing either discontinuities in the slope of the recovery curve or fast completion of the recovery process not showing the typical smooth asymptotic approach to the final value.

Flow-like behaviour can be easily detected when the effective flow velocities are large compared to the rates of diffusive recovery. It requires a much higher signal to noise ratio to detect flow when its contribution to recovery is comparable to diffusion.

### 12.2.2 VARIATIONS OF THE PHOTOBLEACHING TECHNIQUE

During the past few years several investigators have developed variations of the basic spot photobleaching technique.

(a) *Multipoint and 'flying spot' analysis of fluorescence redistribution after photobleaching*

Koppel (1979) has introduced a multipoint analysis of the basic FRAP experiment by monitoring recovery after a localized bleach at several points within the bleached and surrounding area. This 'flying spot' approach allows diffusion and flow to be readily recognized (Koppel *et al.*, 1981).

(b) *Pattern photobleaching*

This technique employs the bleaching of periodic patterns on the membrane (Smith and McConnell, 1978). The lateral transport of the observed fluorophores causes the gradual decay of these bleached patterns in time. The evaluation of this decay may be done in a very direct fashion by photographing the decaying pattern (via an image intensifier) and calculating the time constant for the decrease in the amplitude. If the bleached pattern consists of parallel stripes of constant width, scanning across this pattern will deliver a 'square wave' if done immediately after the bleaching. The 'square wave' may be decomposed into a sum of sinusoidal functions (Fourier decomposition), each characterized by a spatial frequency. For practical purposes, it is sufficient to follow only the decay of the first harmonic in the Fourier series. This amplitude decays as a simple exponential function, characterized by a time constant which contains the diffusion coefficient. Bleaching a two-dimensional pattern (a 'checkerboard' pattern) followed by a two-dimensional Fourier transform of the decaying bleached patterns allows detection of anisotropic lateral motion, i.e. different diffusion coefficients along different directions in the plane (Smith *et al.*, 1979).

(c) *Variations of basic pattern photobleaching*

Patterns have usually been produced by Ronchi rulings but recently they have also been generated by the interference of two coherent laser beams for fringe pattern photobleaching (Davoust *et al.*, 1982). Lanni and Ware (1982) demonstrated that scanning the pattern during the recovery phase could be used to improve signal to noise. Koppel and Sheetz (1983) used this principle in combination with a localized pattern bleach to study recovery over both the pattern repeat distance (small) and the dimension of the entire pattern (large), allowing simultaneous analysis of a larger range of diffusion rates.

(d) *Photobleaching using the effect of total internal reflection (TIR) on membranes or interfaces*

The Fresnel formula describing the reflection of light beam incident on interfaces between two media having different refraction indices predicts that a light beam propagating in a medium with a higher refraction index is totally reflected at the boundary between a medium with lower refraction index if the angle of incidence equals or exceeds a critical angle (for quartz/water: 65.4°).

However, a so-called 'evanescent wave' enters the second medium to a depth of roughly 0.1 $\mu$m and is capable of exciting only those fluorophores which are located at the interface or within this small depth. This special type of illumination has been used for the measurement of adsorption/desorption and surface diffusion rates of proteins at the quartz/aqueous buffer interface (Burghardt and Axelrod, 1981). It has also been used to study only molecules in the ventral (substrate attached) membrane of tissue culture cells (Axelrod, 1981).

(e) *Continuous fluorescence microphotolysis (CFM)*
The kinetics of photobleaching the fluorescence from a selected area of the specimen are governed by the intrinsic photochemical kinetics of the fluorophore and the rate of lateral transport into the bleached area. The latter effect acts to retard the depletion of active fluorophores in the area. Peters *et al.* (1981) have developed the mathematical description of the kinetics of 'fading' of the fluorescence from selected areas in a method termed continuous fluorescence microphotolysis (CFM). This method allows the use of a reduced concentration of fluorophores, makes use of the problem of undesired bleaching found in conventional FRAP, and works well for measurements of relatively high values of diffusion coefficients. The CFM curve represents the sum of both effects, the loss of fluorescence due to the photobleaching event and any recovery of fluorescence due to lateral motion. Typically, the curves are dominated by a rapid initial loss of fluorescence followed by an approach to a steady state at long times where the loss of fluorescence due to photobleaching is balanced by the diffusive influx of unbleached fluorophores. With computer-based data fitting and assumptions about the bleaching kinetics (usually assumed to be of first order), the lateral diffusion coefficient or the flow velocity can be extracted.

(f) *Photoinactivation localized by a glass micropipette*
This experiment differs from the other photobleaching techniques in that it does not rely on fluorophore bleaching. The UV irradiation of a small area on an electrically active frog muscle cell surface is defined by an attached micropipette normally used for measurements of transmembrane currents (Stühmer and Almers, 1982); this irradiation results in the irreversible inactivation of sodium channels. Thus, after radiation inactivation no sodium current is observed. The re-entry of active channel molecules into the observed area via lateral motion restores the current. The recovery kinetics of this increasing current directly indicate the time constant for lateral motion in the membrane. However, in the first measurement of this type no lateral motion of the sodium channels was observed. Recent control experiments measuring lateral diffusion of fluoresceinated lectin–receptor complexes in these membranes demonstrated that there is no difference observed between results taken

before and after UV irradiation. This argues against the conclusion that the observed lateral diffusion is generally affected by the intense UV irradiation (Weiss *et al.*, 1984).

### 12.2.3 FLUORESCENCE REDISTRIBUTION AFTER FUSION (FRAF)

One of the key observations for the formulation of the Fluid-Mosaic model was that of antigen intermixing following fusion of heterokaryons (Frye and Edidin, 1970). Their experiments showed that membrane antigens could diffuse laterally throughout the cell membrane. The method was extended to the diffusion of labelled proteins in the erythrocyte membrane (Fowler and Branton, 1977). Koppel *et al.* (1981) showed that the time course of the redistribution could be followed by a flying spot method.

Two samples containing equal amounts of cells are prepared for the determination (Fig. 12.5). One sample is treated with fluorescently labelled molecules to visualize the desired component of the membrane; the other is not. After this step the samples are mixed and a cell-fusing agent is added (for example, polyethyleneglycol (PEG) or Sendai virus). Within a short time, pairs of fused cells become apparent and roughly 50% of these cells are expected to

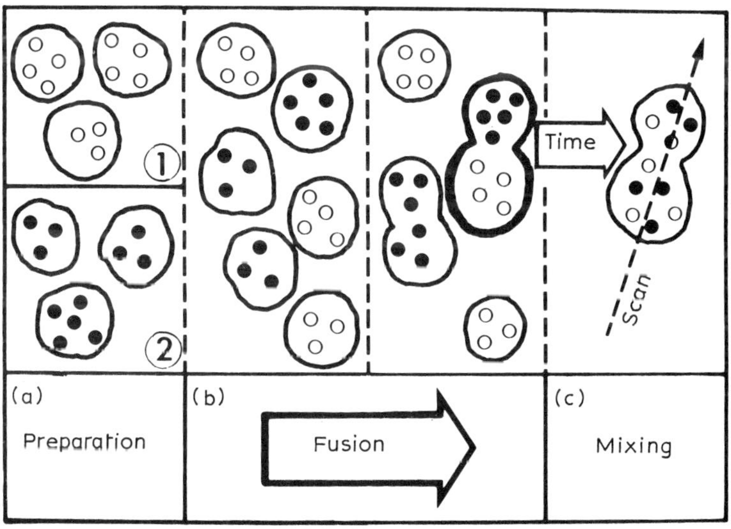


**Fig. 12.5** Principle of the fluorescence recovery after fusion (FRAF) experiment. a. Two separate samples containing cells; one is fluorescence-labelled (1), the other not (2). b. The two samples are combined to one volume and the fusing agent is added; pairs of fused cells develop. c. Fused cell pairs consisting of one labelled cell and an unlabelled one exhibit redistribution of fluorophores over the initally unlabelled membrane. It is possible to take intensity profiles by scanning along the long axis in defined time intervals and to calculate the diffusion coefficient for the labelled species (Schindler *et al.*, 1980a).

show the combination of an unstained cell fused with a stained one. Visual observation shows that fluorescence gradually redistributes into the initially unstained cell of the fused cell pair. The scanning beam device can take the emission intensity profiles along a given axis at certain time intervals, and the time-dependent changes of this profile will reflect the lateral motion in these membranes. A method of moments analysis can be used to extract lateral diffusion coefficients from these profiles (Koppel *et al.*, 1980).

### 12.2.4 *IN SITU* ELECTROPHORESIS OF MEMBRANE COMPONENTS

Electric fields have been applied to cell and organelle membranes to create asymmetric concentration distributions of membrane components by electrophoresing charged components to one side of the cell or the other. After the field is removed, the membrane components will relax toward an equilibrium distribution by diffusion. Fluorescence microscopy (Poo, 1981) and freeze-fracture electron microscopy (Sowers and Hackenbrock, 1980) have been used to observe this relaxation process. In the fluorescence approach cells are scored according to whether their distributions have relaxed while in the EM approach, the intramembranous particle redistribution after membrane electrophoresis can be observed directly (Fig. 12.6). Such quantification

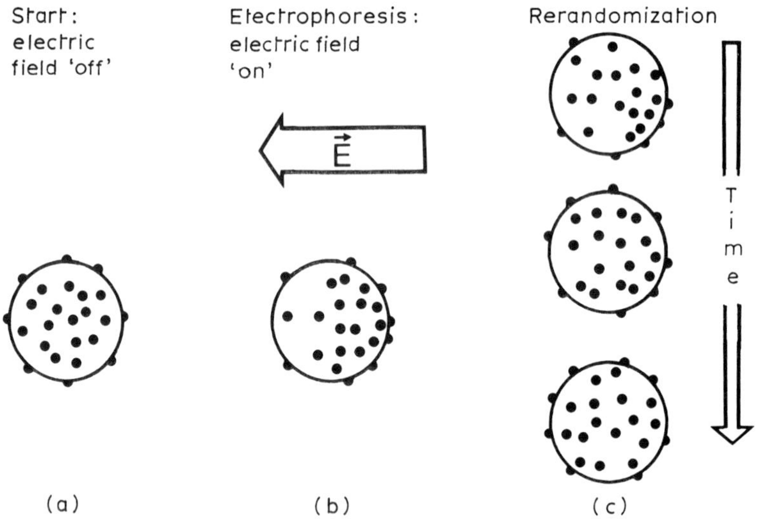


**Fig. 12.6** *In situ* electrophoresis: measurements of lateral diffusion by electrophoretic displacement and rerandomization. a, vesicles containing integral proteins ('particles') in a random array. b, asymmetric distribution of particles after an electric field has been applied to the sample but before relaxation. c, The asymmetric distribution relaxes after the field is turned off due to lateral diffusion. Freezing at various times during the relaxation process allows a diffusion coefficient to be estimated.

results in estimates of the lateral diffusion coefficients of the electrophoresed membrane components.

### 12.2.5 SOME TECHNIQUES USED FOR MEASUREMENTS OF FAST LATERAL DIFFUSION ARE NOT SUITABLE FOR MEMBRANE PROTEIN DIFFUSION

Some techniques have been developed for the measurement of the fast lateral diffusion of lipid probe molecules in membranes. However, techniques such as electron spin resonance (ESR), nuclear magnetic resonance (NMR), excimer-fluorescence, triplet-triplet-annihilation and the fluorescence correlation spectroscopy (FCS) require either a high number of intermolecular collisions of probe molecules in a given time interval (high concentration, fast diffusion) or many large concentration fluctuations in a given area and time (very low concentration, very fast diffusion) in order to yield valid measurements. Therefore these techniques are not easily applied to measurements of the relatively slow lateral motion of proteins in the membranes of cells.

## 12.3 Benchmarks for the interpretation of results

It is useful to interpret the lateral diffusion coefficients of membrane proteins in a comparative manner. All values for diffusion coefficients measured so far are located between the extremes of macromolecules diffusing in viscous solutions on the one hand, and in the 'solid state' on the other.

Table 12.3 shows a selection of absolute values for diffusion coefficients found in non-living matter and in four classes of membranes investigated so far. For comparison the corresponding range of diffusion coefficients for lipid probes is also given.

In most experiments the investigator is interested in finding out what relative changes in diffusion coefficients occur as the sample is treated in different ways (e.g. temperature variation, drugs altering cytoskeleton, etc.) or when diffusion is monitored in different regions of the same specimen (e.g. different regions of a single cell surface; Jacobson *et al.*, 1984; Edidin, 1982). Such comparative studies require valid 'base-line' values to be established before the significance of departures from these values can be interpreted. Of course, a complete interpretation of the results from biomembranes is always enhanced by information from other spectroscopic, biochemical and ultra-structural techniques.

The fact that lateral motion of proteins in plasma membranes is nearly always slower than what one would expect qualitatively from the fluid mosaic model requires finding mechanisms to explain possible 'restraints' to lateral diffusion (Jacobson, 1983; Jacobson and Wojcieszyn, 1981; see Fig. 12.1, right side). Such restraints fall in the following classes. (1) Self-association of proteins to oligomeric forms in the membrane may reduce mobility by

**Table 12.3** Comparison of absolute values of coefficients of lateral diffusion found on membranes and those found in condensed matter

| *Type of membrane* | *Artificial membranes containing reconstituted proteins* | *Vesicles derived directly from cell membranes* | *Cell plasma membranes* |
|---|---|---|---|
| Lateral diffusion of lipid probe molecules in membranes ($cm^2\ s^{-1}$) | $4 \times 10^{-8} \leqslant D \leqslant 2 \times 10^{-7}$ for a review, see: Vaz *et al.* (1982a) | $D \sim 3 \times 10^{-8}$<br>Wu *et al.* (1982)<br>Tank *et al.* (1982) | $5 \times 10^{-9} \leqslant D \leqslant 10^{-8}$<br>Kapitza & Sackmann (1980)<br>Peters (1981) |
| Lateral diffusion of proteins in membranes ($cm^2\ s^{-1}$) | $10^{-9} < D < 4 \times 10^{-8}$ dependent on lipid/protein ratio: Peters & Cherry (1982); for a review, see: Vaz *et al.* (1982a) | $D \sim 4 \times 10^{-9}$<br>Wu *et al.* (1982)<br>Tank *et al.* (1982) | $> 10^{-12} < D < 5 \times 10^{-10}$<br>for an overview, see: Jacobson & Wojcieszyn (1981)<br>Peters (1981)<br>Jacobson (1983) |

For comparison: some typical diffusion coefficients found in condensed matter:

| *Fluids* | *Polymers* | *Solids* |
|---|---|---|
| $H_2O$@20° C: $D \sim 10^{-5}\ cm^2\ s^{-1}$<br>Rare gas@ −189° C: $D \sim 10^{-5}\ cm^2\ s^{-1}$ (van der Waals fluid) | Above glass transition temperature $< 10^{-7}\ cm^2\ s^{-1}$ (Organic vapours) | $\sim 10^{-20} < D < 10^{-7}\ cm^2\ s^{-1}$ depends on material, defect structures and amount of chemical impurities |

increasing the size of the diffusion entity. However, it has been shown that the size-dependence of lateral diffusion is weak (Vaz *et al.*, 1982a) as predicted by the Saffman–Delbrück model. (2) Extensive lateral association of membrane components could be expected to reduce lateral motion of proteins in membranes gradually up to the point where a closed meshwork (percolation transition) arises. In this case only a small mobile fraction would be observed with the long range diffusion coefficient reduced. For example one could envisage a 'network-type' cluster, separating the membrane into many small, more 'fluid' areas. Biomembranes have also been described in terms of polymer networks to understand some results from diffusion measurements (Schindler *et al.*, 1980b). (3) A high lateral density of non- or weakly interacting proteins, producing 'crowding' in the membrane, may decrease the lateral motion by increasing the apparent viscosity of the membrane. This has been shown for model membranes containing increasing amounts of protein (Cherry and Peters, 1982) and also when cholesterol is present in membranes. (4) Structures that are peripheral to the lipid bilayer of the membrane, the cortical cytoskeleton on the inner and the glycocalyx on the outer surface of a typical plasma membrane, may weakly or strongly bind integral membrane proteins reducing the rate of lateral diffusion or immobilizing the proteins, respectively (Koppel, 1981, Tank *et al.*, 1982b). Multiple binding between different components would strongly enhance this immobilization effect.

Indeed, protein immobilization is the extreme case of mobility restraint and is indicated by the immobile fraction often observed in FRAP measurements on the proteins of biomembranes. This fraction (100% − mobile fraction in %) describes the amount of protein that is bound to peripheral structures or 'trapped' into clusters, networks or other aggregates for times long compared to the $\tau_{\frac{1}{2}}$ for recovery, meaning that observable diffusion of this fraction of the labelled species does not occur on the time scale of the measurement. (Operationally, for spot photobleaching, the diffusion coefficient of the immobile fraction can be assumed to be $\sim 10^{-12}$ cm$^2$ s$^{-1}$ or less.) However, it is well to remember that since the living cell has only a limited lifetime as an entity and is subject to substantial morphological and structural changes, these membrane components cannot be considered to be in a true solid state.

As a general interpretative principle, it is useful to realize that the cell and its organelles are today considered to be highly organized structures in three dimensions (Marx, 1983). As a consequence, the boundaries of any compartment of a cell including the outer boundary of the cytoskeleton where the plasma membrane is found are also highly organized. It is probable, then, that the lateral motion of proteins in membranes reflects, at least in part, structural changes in the three-dimensional environment. Thus, association/dissociation processes from structures within or immediately adjacent to the membrane are likely to be the rate-limiting steps for the lateral motion of membrane proteins. The hydrodynamic model is only applicable in the limiting case, where proteins are completely released from peripheral binding sites and therefore able to diffuse freely in the lipid bilayer.

## 12.4 A new development for the measurement of lateral motion in biomembranes: time-resolved spatial photometry (TRSP)

The development of the computer-based techniques for image acquisition and analysis and the ongoing increase in the performance of micro- and mini-computers allow today the analysis of digitized two-dimensional intensity distributions with time. At the same time, ultra-high-sensitivity video cameras are now commercially available, allowing observation of even weak fluorescence patterns seen in the fluorescence microscopic images of membranes and cells. These techniques have been adapted to measure lateral motion of membrane components (Kapitza *et al.*, 1985; Salmon *et al.*, 1984). There are at least two different ways to observe and analyse the motion. The first one is to combine the conventional FRAP experiment with the quantitative image analysis, while the second one records simply the time evolution of a fluorescence pattern as it changes due to inhomogeneous local accumulation or depletion of labelled material as in, for example, 'patching' and 'capping'. In the usual photobleaching experiment, the changes of intensity within the observed area (spot) are recorded by integrating the emission intensity arising from this area. The video-based measurement 'reads' the whole intensity pattern, picture element by picture element, from a membrane area, recording each intensity as a digitized value. Subsequent frames show the time-dependence of these intensities and certain changes in the intensities will reflect lateral motion of the labelled components. Frame averaging and filtering techniques have been used to improve the signal to noise ratio. Fig. 12.7 shows the results of an experiment in which the intensity map after photobleaching is shown in terms of isointensity contours; local diffusion coefficients and flow velocities are extractable from such data. For immediate observation of the data on a monitor, conversion of intensity level ranges to arbitrary colours (pseudocolour) allow the human eye to see patterns which may remain undetected in a representation showing only grey-level distributions. In addition, the whole experiment including the timing for data acquisition and sample illumination, as well as parts of the image analysis, is under the control of the microcomputer. These techniques should allow the examination of subtle redistributions of membrane components as the living cell performs its many functions. At this time also, rapid Fourier transform image analysis methods are being adapted to correct the raw photometric data for the fluorescence emission from out-of-focus planes in the cytoplasm, opening the way for *in situ* measurements of protein traffic inside the cell (Fay and Fogarty, 1983) in three dimensions.

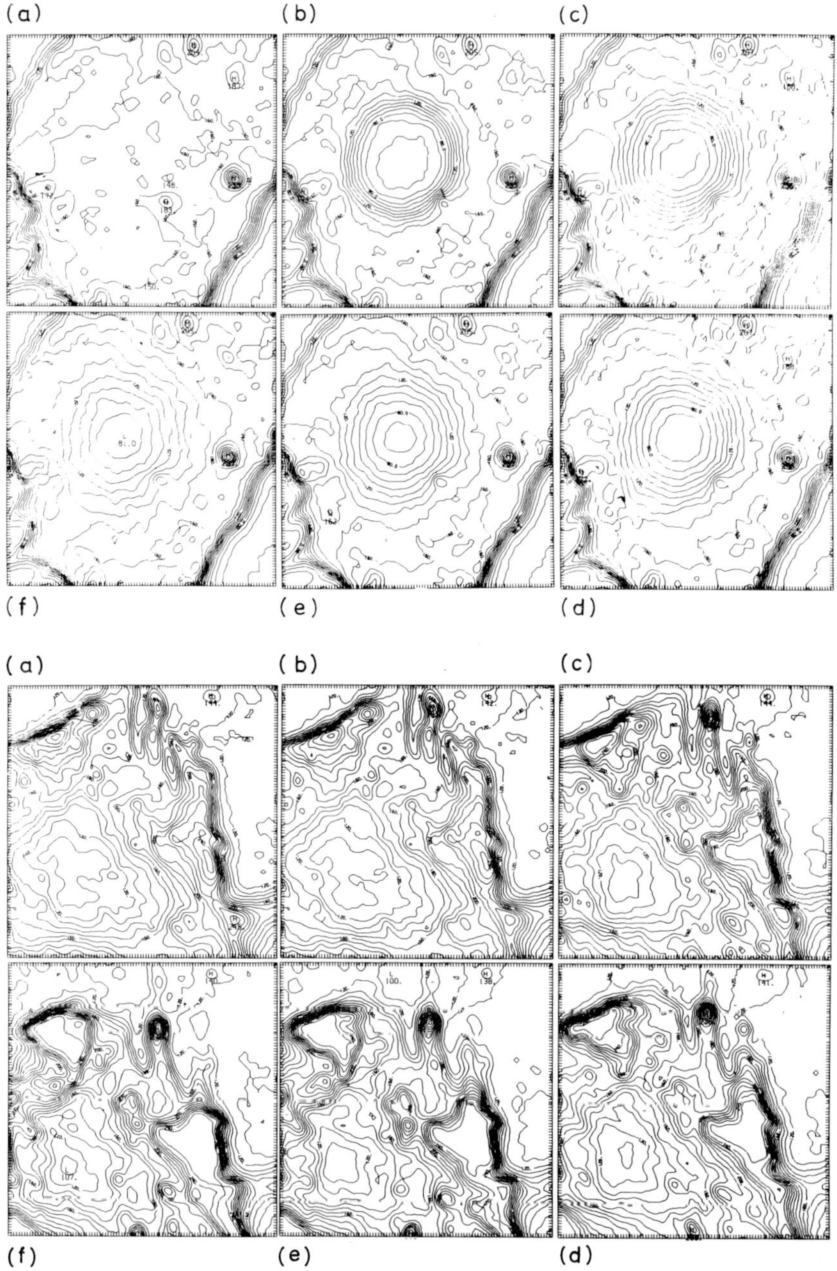


**Fig. 12.7** The application of the time-resolved spatial photometry (TRSP) to the measurement of lateral motion on cell membranes. Example 1 (above): the lateral diffusion of lipid probe molecules (diI) in a fibroblast plasma membrane is demonstrated. The maps a to f show the development of the intensity pattern during the photobleaching experiment. Isointensity plot a shows the intensity distribution in the (flat) membrane before the bleaching; b gives the situation 2 s after the bleaching; c, d and e exhibit the isotropic diffusional recovery. The delay time between maps is 10 s. Between e and f: 30 s. Example 2 (below): patching of the rhodamine-labelled lectin wheatgerm agglutinin (WGA) bound to the surface of a fibroblast. Note the growth of areas exhibiting high fluorescence intensity in time during 1 h of observation (a–f). (Acknowledgement: We gratefully acknowledge the help of Mr George McGregor with the development of software for TRSP.)

## 12.5 References


Aizenbud, B. M. and Gershon, N. D. (1982) *Biophys. J.*, **38**, 287–93.

Axelrod, D. (1981) *J. Cell Biol.*, **89**, 141–5.

Axelrod, D. (1983) *J. Membr. Biol.*, **75**, 1–10.

Axelrod, D., Koppel, D. E., Schlessinger, J., Alson, E. and Webb, W. W. (1976) *Biophys. J.*, **16**, 1055–69.

Barisas, B. G. (1980) *Biophys. J.*, **29**, 545–8.

Barisas, B. G. and Leuther, M. D. (1979) *Biophys. Chem.*, **10**, 221–9.

Bretscher, M. S. (1980) *Trends Biochem. Sci.*, **5**, 6.

Burghardt, T. P. and Axelrod, D. (1981) *Biophys. J.*, **33**, 455–68.

Cherry, R. J., Burkli, A., Busslinger, M. and Schneider, G. (1976) *Nature (London)*, **263**, 389–93.

Cherry, R. J. and Peters, R. (1982) *Proc. Natl. Acad. Sci. USA*, **79**, 4317–21.

Davoust, J., Devaux, P. F. and Leger, L. (1982) *EMBO J.*, **1**, 1233–8.

Derzko, Z. and Jacobson, K. A. (1980) *Biochemistry*, **19**, 6050–7.

Dragsten, P., Henkart, P., Blumenthal, R., Weinstein, J. and Schlessinger, J. (1979) *Proc. Natl. Acad. Sci. USA*, **76**, 5163–7.

Edidin, M. and Wey, T. (1982) *J. Cell Biol.*, **95**, 458–62.

Edidin, M., Zagyansky, Y. and Lardner, T. J. (1976) *Science*, **191**, 466–8.

Einstein, A. (1905) *Ann. D. Physik*, **IV/17**, 549–60.

Fay, F. S. and Fogarty, K. (1983) *J. Cell Biol.*, **97**, 294a.

Firester, A. H., Heller, M. E. and Sheng, P. (1977) *Appl. Optics*, **16**, 1971.

Fowler, V. and Branton, D. (1977) *Nature (London)*, **268**, 23–6.

Frye, L. D. and Edidin, M. (1970) *J. Cell Sci.*, **7**, 319–35.

Gupte, S., Wu, E. S., Hoechli, L., Hoechli, M., Jacobson, K. A., Sowers, A. E. and Hackenbrock, C. R. (1985) *Proc Natl. Acad. Sci. USA*, **81**, 2606–610.

Henis, Y. I., Hekman, M., Elson, E. L., Helmreich, E. J. M. (1982) *Proc. Natl. Acad. Sci. USA*, **79**, 2907–11.

Hochman, J., Schindler, M., Lee, J. G. and Ferguson-Miller, S. (1982) *Proc. Natl. Acad. Sci. USA*, **79**, 6866–70.

Jacobson, K. A. (1983) *Cell Motil.*, **3**, 367–73.

Jacobson, K. A., Derzko, Z., Wu, E. S., Hou, Y. and Poste, G. (1977) *J. Supramol. Struct.*, **5**, 565–76.

Jacobson, K. A., Hou, Y., Derzko, Z., Wojcieszyn, J. and Organisciak, D. (1981) *Biochemistry*, **20**, 5268–75.

Jacobson, K. A., Hou, Y. and Wojcieszyn, J. (1978) *Exp. Cell Res.*, **116**, 179–89.

Jacobson, K. A., O'Dell, D. S. and August, J. T. (1984) *J. Cell Biol.*, **99**, 1613–23.

Jacobson, K. A. and Wojcieszyn, J. (1981) *Comments Mol. Cell Biophys.*, **1**, 189–99.

Jacobson, K. A., Wu, E. S., Poste, G. (1976) *Biochim. Biophys. Acta*, **433**, 215–22.

Kapitza, H. G., McGregor, G. and Jacobson, K. A. (1985) *Proc. Natl. Acad. Sci., USA*, **82**, 4122–126.

Kapitza, H. G. and Sackmann, E. (1980) *Biochim. Biophys. Acta*, **595**, 56–64.

Koppel, D. E. (1979) *Biophys. J.*, **28**, 281–91.

Koppel, D. (1981) *J. Supramol. Struct. Cell Biochem.*, **17**, 61–7.

Koppel, D. E., Axelrod, D., Schlessinger, J., Elson, E. L. and Webb, W. W. (1976) *Biophys. J.*, **16**, 1315–29.

Koppel, D. E. and Sheetz, M. P. (1983) *Biophys. J.*, **43**, 175–81.

Koppel, D. E. and Sheetz, M. P. (1981) *Nature (London)*, **293**, 159–61.

Koppel, D. E., Sheetz, M. P. and Schindler, M. (1980) *Biophys. J.*, **30**, 187–92.

Koppel, D. E., Sheetz, M. P. and Schindler, M. (1981) *Proc. Natl. Acad. Sci. USA*, **78**, 3576–80.

Lanni, F. and Ware, B. R. (1982) *Rev. Sci. Instrum.*, **53**, 905–8.

Marx, J. L. (1983) *Science*, **222**, 1109–11.

Peters, R. (1981) *Cell Biol. Int. Rep.*, **5**, 733–60.

Peters, R., Bruenger, A. and Schulten, K. (1981) *Proc. Natl. Acad. Sci. USA*, **78**, 962–6.

Peters, R., Peters, J., Tews, K. H. and Bähr, W. (1974) *Biochim. Biophys. Acta*, **367**, 282–94.

Petersen, N. O. and McConnaughey, W. B. (1981) *J. Supramol. Struct. Cell Biochem.*, **17**, 213–21.

Poo, M. (1981) *Annu. Rev. Biophys. Bioeng.*, **10**, 245–76.

Saffman, P. G. and Delbrück, M. (1975) *Proc. Natl. Acad. Sci. USA*, **72**, 3111–13.

Salmon, E. D., Saxton, W. M., Leslie, R. J., Karow, M. L. and McIntosh, J. R. (1984) *J. Cell Biol.*, **99**, 2165–74.

Schechter, Y., Hernaez, L., Schlessinger, J. and Cuatrecasas, P. (1979) *Nature (London)*, **278**, 835–8.

Schechter, Y., Schlessinger, J., Jacobs, S., Chang, K. J. and Cuatrecasas, P. (1978) *Proc. Natl. Acad. Sci. USA*, **75**, 2135–9.

Schindler, M., Koppel, D. E. and Sheetz, M. P. (1980a) *Proc. Natl. Acad. Sci. USA*, **77**, 1457–61.

Schindler, M., Osborn, M. J. and Koppel, D. E. (1980b) *Nature (London)*, **283**, 346–50.

Schneider, M. B. and Webb, W. W. (1981) *Appl. Optics*, **20**, 1382–8.

Singer, S. J. and Nicholson, G. L. (1972) *Science*, **173**, 720–31.

Smith, B. A., Clark, W. R. and McConnell, H. M. (1979) *Proc. Natl. Acad. Sci. USA*, **76**, 5641–4.

Smith, B. A. and McConnell, H. M. (1978) *Proc. Natl. Acad. Sci. USA*, **75**, 2759–63.

Smoluchowski, M. V. (1906) *Annal. Phys.*, **IV/21**, 756–80.

Sowers, A. and Hackenbrock, C. R. (1980) *Proc. Natl. Acad. Sci. USA*, **78**, 6246–50.

Stühmer, W. and Almers, W. (1982) *Proc. Natl. Acad. Sci. USA*, **79**, 946–50.

Tanford, C. (1973) *The Hydrophobic Effect: Formation of Micelles and Biological Membranes*, John Wiley, New York and London.

Tank, D. W., Wu, E. S., Meers, P. R. and Webb, W. W. (1982a) *Biophys. J.*, **40**, 129–35.

Tank, D. W., Wu, E. S. and Watt, W. W. (1982b) *J. Cell Biol.*, **92**, 207–12.

Van Zoelen, E. J. J., Tertoolen, G. J. and de Laat, S. W. (1983) *Biophys. J.*, **42**, 103–8.

Vaz, W. L. C., Criado, M., Madeira, V. M. C., Schoellmann, G. and Jovin, T. M. (1982b) *Biochemistry*, **21**, 5608–12.

Vaz, W. L. C., Derzko, Z. and Jacobson, K. A. (1982a) *Cell Surf. Rev.*, **8**, 84–128.

Weiss, R. E., Stühmer, W. and Almers, W. (1984) *Biophys. J.*, **45**, 269a.

Wey, C. L. and Cone, R. A. (1981) *Biophys. J.*, **33**, 225–32.

Wolf, D. E., Edidin, M. and Dragsten, P. R. (1980) *Proc. Natl. Acad. Sci. USA*, **77**, 2043–5.

Wolf, D. E., Handyside, A. H. and Edidin, M. (1982) *Biophys. J.*, **38**, 295–7.

Wu, E. S., Tank, D. and Webb, W. W. (1982) *Proc. Natl. Acad. Sci. USA*, **79**, 4962–6.

Yguerabide, J., Schmidt, J. A. and Yguerabide, E. E. (1982) *Biophys. J.*, **39**, 69–75.

# 13
# Rotational diffusion of membrane proteins

DAVID D. THOMAS



## 13.1 Introduction

### 13.1.1 BACKGROUND AND PREVIEW

The quantitative description of molecular mechanisms involved in membrane processes requires direct information about the motions of membrane proteins. The correlation of protein motions with the functional state of the membrane can provide evidence for motions directly involved in such processes as active transport, transmembrane signalling and electron transport. More generally, protein mobility is a sensitive indicator of protein–protein and protein–lipid associations that may be important for function. Measurements of *rotational* motions are particularly important, because of their high sensitivity to the shape and size of the protein (or assembly of proteins) and to the effective viscosity of the surrounding lipid environment.

Measurements of membrane protein rotational motion are made primarily with the use of molecular probe methods, based on both optical and magnetic resonance spectroscopy, that have been developed over the past decade. Until the mid-1970s, nearly all studies of rotational motion in membranes focused on the lipid components, with very few studies of protein motion, despite the central role of proteins in membrane function. This can be blamed in part on a shortage of information about the structures of membrane proteins, but the main problem was that the spectroscopic techniques used to study molecular dynamics – conventional electron paramagnetic resonance (EPR), transient

Techniques for the Analysis of Membrane Proteins. Edited by C. I. Ragan & R. J. Cherry.
Published in 1986 by Chapman and Hall, 11 New Fetter Lane, London EC4P 4EE

fluorescence anisotropy (TFA) and nuclear magnetic resonance (NMR) – were primarily sensitive only to motions in the nanosecond time range or faster. This is the appropriate time range for the rapid rotational motions of lipid hydrocarbon chains and protein side chains, or for the tumbling of small proteins in aqueous solution. However, protein–lipid and protein–protein interactions in membranes produce an environment much more viscous than water, with the result that the large-scale rotational motions of proteins tend to occur in the microsecond time range or slower. Although nanosecond motions within proteins, presumably corresponding to small-scale motions (e.g. motions of amino acid side chains), can be detected in membranes, the lack of high-resolution structural information usually prevents a clear interpretation. In contrast, the increasing amount of low-resolution structural information available for a number of membrane proteins often permits microsecond and millisecond rotational motions of membrane proteins to be plausibly interpreted in terms of specific large-scale motions (e.g. overall protein rotation, motions of protein assemblies, or segmental motions of large domains within proteins), as illustrated in Fig. 13.1 and discussed in detail in section 13.1.2. This review will be restricted primarily to a discussion of methods for measuring these large-scale (slow) protein motions.

The goal of detecting microsecond rotational motions within membranes has helped motivate the parallel development of two areas of spectroscopic technology, one in EPR (saturation transfer EPR, ST-EPR, using nitroxide spin labels) and the other in optical spectroscopy (transient optical anisotropy, TOA, using long-lifetime chromophores). Both of these methods use molecular probes having (a) orientation-dependent spectroscopic signals and (b) long (microsecond-to-millisecond) excited-state lifetimes (usually referred to as relaxation times in EPR). The measurements of rotational motion are based on (a) orientation-dependent photoselection, i.e. the selective excitation of molecules on the basis of orientation, and (b) the observation of the effects of molecular reorientation during the excited-state lifetime. In the case of ST-EPR, the same nitroxide spin labels are used as in conventional EPR; the methodological innovations are based mainly on instrumental conditions (requiring new theoretical approaches) that optimize the sensitivity to excited-state processes. In the optical techniques, the instrumental and theoretical methods are very similar to those used previously to study excited-state processes using fluorescence; the main innovation is the development and use of long-lifetime probes (having microsecond-to-millisecond excited-state lifetimes, in contrast to the nanosecond lifetimes of most conventional fluorescent probes).

The primary purpose of the present review is to discuss the methods currently used in EPR and optical spectroscopy to measure the microsecond and millisecond motions of integral membrane proteins. The remainder of the introduction will discuss concepts common to both classes of techniques.

Subsequent sections will discuss these two classes separately but will emphasize the comparison of the principles, instrumental methods and sensitivities of different methods. I will discuss a few selected applications to illustrate the methodology, without attempting to provide a comprehensive list of references to applications. A number of related reviews have appeared discussing these methods for measuring slow rotational motions: (a) a more general discussion of ST-EPR and optical methods, discussing both principles and applications (Thomas, 1978); (b) a brief discussion of the principles of both methods and their application to muscle proteins (Thomas *et al.*, 1985); (c) applications of both methods to membranes (Cherry, 1979); (d) a general review of ST-EPR (Hyde and Thomas, 1980); (e) principles and applications of ST-EPR in membranes (Thomas, 1982, 1985; Hemminga, 1983); (f) optical methods in membranes (Cherry, 1978; Jovin *et al.*, 1981); and (g) minireviews of optical applications in membranes (Cherry, 1982; Garland, 1982).

### 13.1.2 MATHEMATICAL DESCRIPTION OF ROTATIONAL DIFFUSION IN A MEMBRANE

The interpretation of measurements requires (a) the prediction of probable modes of membrane protein motions and (b) a mathematical framework in which to characterize these motions. Fig. 13.1 illustrates the general types of large-scale rotational diffusion expected for membrane proteins. The simplest

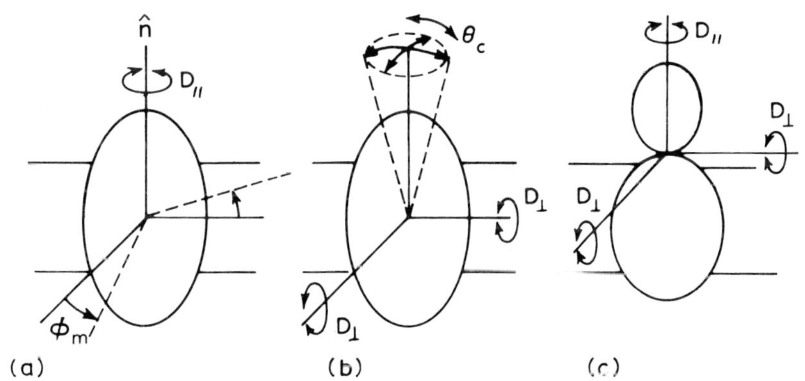


**Fig. 13.1** Schematic illustration of large-scale rotational motions of membrane proteins that are expected to occur in the microsecond-to-millisecond time range. (a), Rigid-body uniaxial rotation about the membrane normal ($\hat{n}$), corresponding to random reorientations $\phi_m$ of molecule-fixed axes in the membrane plane. The rate of this motion is characterized by the axial diffusion coefficient $D_\parallel$. (b), Rigid-body off-axis rotation (wobble) about axes in the membrane plane, corresponding to reorientation of the protein-fixed axis which has its average orientation normal to the membrane. The rate of this motion is characterized by the diffusion coefficient $D_\perp$, also referred to as $D_w$, and the amplitude is characterized by the cone half-angle $\theta_c$. (c), Segmental motions (both axial and wobble) of a large domain of the protein relative to the remainder of the protein, which is assumed to be fixed in the membrane. See text for further discussion.

is rigid-body rotation, the rotation of the entire protein with respect to the membrane in which it is embedded. Considering the structure of biological membranes, in which amphipathic proteins are embedded in a planar lipid bilayer, the most likely kind of rigid-body motion would seem to be *uniaxial rotation*, the rotation of a protein about an axis parallel to the membrane normal ($\hat{\mathbf{n}}$ in Fig. 13.1a), corresponding to reorientation of molecule-fixed axes in the plane of the membrane by the angle $\phi_m$ (Fig. 13.1a). (In general, I will use $\phi$ to represent a rotation *about* a symmetry axis and $\theta$ to represent the *reorientation* of a symmetry axis. Bold-face symbols are vectors, and ^indicates a unit vector.) As in the case of most macromolecular rotations in which the solvent molecules (in this case lipid) are much smaller than the solute (in this case protein), the motion is assumed to correspond to Brownian rotational diffusion, i.e. a series of small-amplitude, random rotations that are properly described by a (diffusion) differential equation:

$$\mathrm{d}p(\phi_m,t)/\mathrm{d}t = -D_{\|}(\mathbf{d}^2/\mathrm{d}\phi_m{}^2)p(\phi_m, t) \tag{13.1}$$

$p(\phi_m, t)$ is an ensemble probability distribution and $D_{\|}$ is the rotational diffusion coefficient, which characterizes the rate of motion in rad$^2$ s$^{-1}$. The quantitative significance of $D_{\|}$ is best understood by imagining that all molecules have the same initial orientation $\phi_m(t=0)$, then solving the equation for $p$ and calculating the mean (ensemble average) square rotation occurring within a time $t$:

$$\langle[\Delta\phi_m(t)]^2\rangle = 2D_{\|}t \tag{13.2}$$

The brackets indicate an average over all molecules, i.e. weighting by the solution for $p$. This ensemble average is equivalent to the most probable value for an individual molecule. Thus, the most probable angle through which the protein rotates within a time $t$ is $(2D_{\|}t)^{\frac{1}{2}}$ rad. $D_{\|}$ is inversely proportional to the effective viscosity of the membrane, and if this viscosity remains constant, $D_{\|}$ decreases as the size of the protein (or protein assembly) increases. For a cylindrical protein of radius $a$, diffusing in a membrane of thickness $h$ and viscosity $\eta$, it has been predicted theoretically that (Saffman and Delbrück, 1975):

$$D_{\|} = kT/(4\pi\eta a^2 h) \tag{13.3}$$

In the above example, a protein-fixed principal axis of rotation is assumed to remain strictly perpendicular to the membrane plane. The other type of rigid-body motion expected for an integral membrane protein is the reorientation of this axis, corresponding to rotation about axes in the membrane plane (Fig. 13.1b). Electrostatic and thermodynamic arguments predict that this type of 'off-axis' motion should be *restricted in amplitude* (boundary conditions must be imposed on Equation 13.1) and that diffusion should be similar about all axes in the plane, resulting in the prediction that the protein's principal axis will *wobble in a cone.* Thus this motion is characterized by two parameters: a wobbling diffusion coefficient $D_{\perp}$ (often referred to as $D_w$, as discussed below) and a cone half-angle $\theta_c$ (Fig. 13.1b), which describes the limits for rotations $\theta_m$ of the molecule-fixed symmetry axis relative to its average orientation $\hat{n}$. Because the rotation occurs formally about two axes, the expression for the mean square angular displacement (Equation 13.2) becomes $\{[\Delta\theta_m(t)]^2\} = 4D_{\perp}t$, and is valid only for small rotations $\Delta\theta_m$. Similarly, for unrestricted isotropic ($D_{\parallel} = D_{\perp} = D_{iso}$, $\theta_c = 180°$) rotational motion (e.g. that of a spherical protein in solution),

$$\langle[\Delta\theta_m(t)]^2\rangle = 4D_{iso}t \tag{13.4}$$

Although knowledge of overall protein motions (rigid-body rotations) may provide information about viscosity and aggregation (i.e. about protein–lipid and protein–protein interactions), segmental motions *within* proteins may be more direct indicators of protein conformational changes involved in function. Indeed, the current level of knowledge about membrane structure suggests that most membrane proteins contain sizeable aqueous domains that are not in contact with lipid, suggesting the possibility of segmental motions of these aqueous portions of proteins. As illustrated in Fig. 13.1(c), both axial (torsional) and wobbling motions are possible. Whereas axial motions are expected to dominate the rigid-body (overall) motions (Fig. 13.1a), wobbling might be more important for segmental motions occurring in the aqueous regions of membrane proteins, where the constraints to off-axis wobbling are expected to be less severe.

### 13.1.3 GENERAL SPECTROSCOPIC PRINCIPLES

The challenge to the spectroscopist is to (a) introduce probes with fixed orientations relative to the protein axes in Fig. 13.1, (b) perform a spectroscopy experiment that is sensitive to the probes' rotational diffusion, and (c) interpret the probes' motions in terms of the proteins' motions. This section discusses in general terms the principles behind step (b). The methods most commonly used to measure membrane protein rotational motion, in optical and EPR spectroscopy, depend on the principles of (1) orientation-sensitive excitation (photoselection), (2) an orientation-dependent absorption measurement that monitors reorientation of the photoselected molecules.

Consider a spectroscopy experiment (either optical or EPR) in which the sample consists of a random orientation distribution of membrane fragments (e.g. vesicles), with probes (dyes or spin labels) attached to proteins. The absorption of this sample is continuously monitored with a weak detection beam (early times in Fig. 13.2). Now let us introduce a brief excitation pulse of radiation at $t = 0$, intense enough to perturb the Boltzmann distribution, i.e. to deplete the ground state population of probes, so that the absorption decreases.

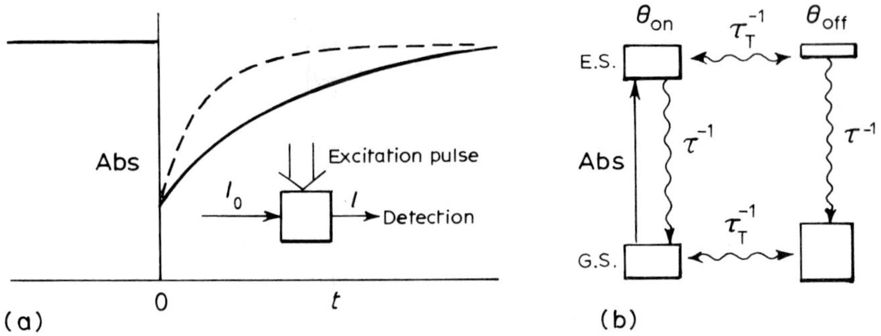


**Fig. 13.2** Schematic illustration of an orientation-dependent transient absorption experiment, illustrating the basis for sensitivity to rotational motion for both optical and EPR spectroscopy. (a), Hypothetical experimental data (discussed in text). (b), Time constants that determine the rate of recovery of the system to its initial absorption level: $\tau$ is the excited-state lifetime (designated $T_1$ in EPR), and $\tau_T$ is the *transfer* time required for rotation $\theta_T$ from an orientation where the probability of excitation is high ($\theta_{ON}$) to one where it is low ($\theta_{OFF}$). In optical spectroscopy, $\tau_T$ is identical to the rotational correlation time $\tau_r$, whereas $\tau_T \ll \tau_r$ for EPR. See text for further discussion. If $\tau_T < \cong \tau$, the motion is fast enough to make the recovery more rapid, and to decrease the level of saturation (ground-state depletion) in a steady-state experiment.

If either the excitation or detection do not depend on molecular orientation (e.g. if unpolarized light is used in the optical experiment), or if the excited molecules do not change orientation, the absorbance will decay back to its original value with an exponential time constant equal to the excited-state lifetime (usually designated $\tau$ in optical spectroscopy, $T_1$ in magnetic resonance), as shown in the solid curve in Fig. 13.2. However, if (a) the absorption is dependent on molecular orientation (e.g. if vertically polarized light is used for both excitation and detection), so that the initial absorption is anisotropic (i.e. the distribution $p(\theta)$ in Equation 13.1 is not random), and (b) the molecules rotate enough during the excited-state lifetime to change the probability of probe absorption (thus decreasing the anisotropy), the absorbance will recover more rapidly, as shown in the dashed curve in Fig. 13.2. The time-dependence of this recovery will have two exponential decay components, with the long-time decay dominated by the same rate $\tau^{-1}$, and the early decay dominated by the rate $(\tau^{-1} + \tau_T^{-1})$. The *rotational transfer time* $\tau_T$ is the time required for rotational motion through an angle $\theta_T$ sufficient to reduce the excitation probability by a factor of 2. This angle is about 45° for polarized optical spectroscopy and about 4° for EPR (discussed in Section 13.3.1), so $\tau_T$ is much less in EPR for a given rate of motion, as discussed below. The discussion of $\tau_T$ in the present review closely parallels that of Dalton (1985), where $\tau_{SD}$ was used.

(a) *Correlation times*

Although diffusion coefficients are generally used to describe the rotational motions of proteins, correlation times $\tau_r$ are also used, particularly in connection with the analysis of spectroscopic data. A correlation time is properly defined as an exponential decay constant for a *time autocorrelation function* that characterizes fluctuations in a spectroscopic observable. For example, in time-resolved optical experiments, the function detected (polarization anisotropy, $r(t)$) is given by

$$r(t) = \langle P_2(\hat{\boldsymbol{\mu}}_{\mathbf{ob}} \cdot \hat{E}) \rangle \equiv \langle P_2(\cos\theta_{\mathrm{ob}}) \rangle \equiv \langle (3\cos^2\theta_{\mathrm{ob}} - 1)/2 \rangle, \qquad (13.5)$$

where $\theta_{\mathrm{ob}}$ defines the orientation of the probe's observed transition moment $\hat{\boldsymbol{\mu}}_{\mathbf{ob}}$ relative to the polarization direction $\hat{E}$ in space (discussed below). In this case, the exponential decay constants detected ($\tau_T$ in Fig. 13.2) are defined as the correlation times $\tau_r$. Using this definition of correlation time, an important result for isotropic rotational diffusion, for both optical and magnetic resonance spectroscopy, is

$$\tau_r = 1/(6D_{\mathrm{iso}}), \qquad (13.6)$$

so Equation 13.4 becomes

$$\langle \Delta\theta^2(t) \rangle = 2t/3\tau_r, \text{ and } \tau_T/\tau_r = (3/2)(\theta_T)^2 \qquad (13.7)$$

Thus the root-mean-square rotation in one $\tau_r$ is $(2/3)^{\frac{1}{2}}$ radian $\cong 47°$, which is $\equiv \theta_T$ in optical spectroscopy but is $\cong 12\theta_T$ in EPR. Thus, while the detected $\tau_T$ is identical to $\tau_r$ in optical spectroscopy, $\tau_T \cong 0.007\tau_r$ in EPR; i.e. the detected signal in EPR decays much more rapidly than the correlation function in Equation 13.5. Nevertheless, Equation 13.5 is used to define rotational correlation times $\tau_r$ in magnetic resonance, as well as in optical spectroscopy.

Other definitions of rotational correlation times are sometimes used, resulting in some confusion. Although a correlation time properly refers to the *reorientation of a probe's axis*, as defined by the exponential decay of $r(t)$ (Equation 13.5), some workers (particularly in EPR) associate a 'correlation time' with the rate (diffusion coefficient) of rotation *about a macromolecular diffusion axis*. For example, although axial diffusion (Equations 13.1–13.3) in general results in a multiexponential decay (corresponding to multiple correlation times) of $r(t)$ (Equation 13.5), a single 'correlation time' is sometimes defined to be $\tau_\parallel \equiv 1/(6D_\parallel)$ (Polnaszek *et al.*, 1981; Robinson and Dalton, 1980). This agrees with the standard definition of $\tau_r$ only for isotropic motion (Equation 13.6). Alternatively, a 'relaxation time' $\equiv 1/D_\parallel$ is sometimes defined (Cherry, 1979).

(b) *Data analysis*

If the time-resolved absorption is detected directly, as shown in Fig. 13.2 (the usual case in optical spectroscopy), the amplitudes and time constants of the rotation-dependent absorption recovery are detected directly (determined by fitting the data to a sum of exponentials). In a steady-state experiment (the usual case in EPR), theoretical simulations or empirical model systems are required to estimate correlation times and amplitudes. In either case, model-dependent theoretical simulations are then used to relate these observed amplitudes and correlation times to the amplitudes and rates (diffusion coefficients) of molecular motion. This relationship can be complex, since the probe axis $\hat{\boldsymbol{\mu}}_{ob}$ that determines absorption (and the angle $\theta_{ob}$ in Equation 13.5) may not be perpendicular to the protein's diffusion axis, in which case $\theta_{ob}$ in Equation 13.5 is not the same as $\phi_m$ in Equations 13.1 and 13.2, or $\theta_m$ in Equation 13.4.

The optical and magnetic resonance techniques differ in the mechanism and resolution of orientation-selection (photoselection), and there is considerable variation in the experimental methods used to detect the absorption recovery. The rest of this review will describe these methodological variations, illustrated by selected applications.

## 13.2 Transient optical anisotropy

### 13.2.1 PRINCIPLES

As discussed above, the two methods under discussion (transient optical anisotropy and saturation transfer EPR) share many common principles. Since the relationship between these principles and the interpretation of data is more straightforward for the optical methods, I will discuss these methods first; analogies with EPR principles will be pointed out in the second half of this review. This section introduces the theoretical principles shared by the various optical methods. A more detailed discussion has been published by Jovin *et al.* (1981).

(a) *Photoselection*

The various optical methods discussed below differ in the mode of detection, but photoselection is the same in all cases: excitation with a pulse of polarized light. Although most studies are done on suspensions of randomly oriented membranes, a pulse of vertically polarized light selectively excites those chromophores having their absorption transition moments $\hat{\boldsymbol{\mu}}_a$ parallel to the polarization direction $\hat{E}$, producing an orientation distribution of excited molecules $p(\theta_{ex}, t=0) = \cos^2 \theta_{ex} \sin \theta_{ex}$, where $\cos \theta_{ex} = \hat{\boldsymbol{\mu}}_a \cdot \hat{E}$. It should be noted that this is a fairly broad distribution, so that a transition moment must rotate through an angle $\theta_T$ of about 45° within the excited-state lifetime to *transfer* it from a high to a low detection probability (or vice versa), i.e. in order for its rotational motion to be easily detected. Note that this $\theta_{ex}$ refers to the orientation of the chromophore's transition moment, whereas the $\theta_m$ discussed above refers to the angle of macromolecular (protein) rotation about a molecule-fixed diffusion axis. These angles are the same only if the transition moment is perpendicular to the diffusion axis.

(b) *Detection*

Following excitation at $t=0$, the photoselected population of probes is monitored, by detecting either their absorption ($\hat{\boldsymbol{\mu}}_{ob}=\boldsymbol{\mu}_a$) or emission ($\hat{\boldsymbol{\mu}}_{ob}=\hat{\boldsymbol{\mu}}_e$) of polarized light, to ask how far they rotate per unit time. For example, if vertically polarized excitation is followed by the detection of vertically polarized absorption, a signal like that shown in Fig. 13.2 is detected. As described in detail in the section below on Methods, signals are measured that permit the calculation of an anisotropy parameter $r(t)$, whose theoretical relationship with the time-dependent orientation distribution of those chromophores that were excited at $t=0$ is given above in Equation 13.5. As discussed in detail in the section below on Data Analysis (13.2.4), the rate and amplitude of the decay of $r(t)$ are closely related to the rate ($D$) and amplitude ($\Delta\theta$) of molecular rotational motion.

The variety of signals for which $r$ (and hence $\theta_{ob}$) can be detected is illustrated in Fig. 13.3 and Table 13.1. All of these methods require that the excited molecule does not return to the ground state ($S_0$) before its orientation can be detected by probing $\hat{\boldsymbol{\mu}}_{ob}$. Since measurements in the microsecond and millisecond time range are to be made, the probe must have an excited-state lifetime much longer than the nanosecond times typical of excited singlet states ($S_1$). The most common method of achieving a long lifetime is to choose a probe having a stable *triplet state* ($T_1$) that is populated efficiently through *intersystem crossing* from $S_1$. Alternatively, another kind of long-lived state may be produced, e.g. a photochemical product, as discussed below.

*Absorption methods.* The most commonly used method of detection is absorption. Molecules that absorb light during the excitation pulse are no longer available to absorb light. This results in *ground-state depletion* and is detectable as a *decrease* in absorption in the same absorption band used for excitation. Excitation with polarized light results in *transient absorption anisotropy* (transient dichroism), which is reported as a time-dependent decay of *absorption (polarization) anisotropy*:

$$r(t) \equiv \frac{\Delta A_{\parallel}(t) - \Delta A_{\perp}(t)}{\Delta A_{\parallel}(t) + 2\Delta A_{\perp}(t)} \tag{13.8}$$

where $\Delta A_{\parallel}$ and $\Delta A_{\perp}$ are the changes in absorbance of light polarized parallel and perpendicular to the excitation polarization $\hat{E}$. The form of $r$ (the difference absorption divided by the total absorption) removes the dependence of the decay on the excited-state lifetime $\tau$ (Fig. 13.2), so that the time constants for the decay of $r$ are simply the rotational correlation times $\tau_r$, as discussed below. In this case, $\hat{\boldsymbol{\mu}}_{ob}=\hat{\boldsymbol{\mu}}_{ex}=\hat{\boldsymbol{\mu}}_a$, so $r(t)=\langle P_2(\cos\theta_{ob})\rangle=\langle P_2(\hat{\boldsymbol{\mu}}_a\cdot\hat{E})\rangle$. Although $\Delta A$ is usually observed as a decrease in absorption in the excited absorption band, transient absorption anisotropy can also be observed by detecting an *increase* in absorption in another wavelength band, in which the excited state (or

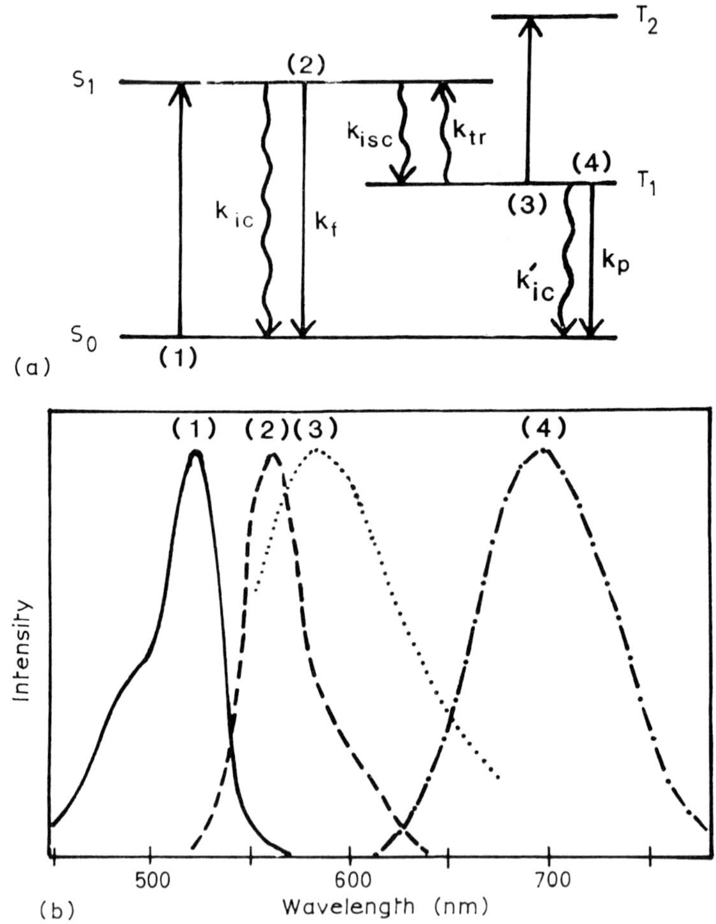


**Fig. 13.3** (a) Schematic illustration of energy levels and transitions for a typical triplet-state optical probe (e.g. eosin). Radiative transitions are indicated by straight arrows, and non-radiative transitions by wavy arrows. $S_0$ and $S_1$ are singlet states, and $T_1$ and $T_2$ are triplet states. The radiative transitions are (1) singlet absorption, (2) fluorescence, (3) triplet–triplet absorption and (4) phosphorescence. (b), Spectra of eosin-5-iodoacetamide (Eo5IA, see Fig. 13.4) bound to myosin (Eads *et al.*, 1984), numbered as in (a). Spectra are normalized to the same maximum height. The absorption coefficient for singlet absorption is about $8 \times 10^4$ $\text{M}^{-1}$ $\text{cm}^{-1}$, about twice the value for triplet absorption. The intensities of phosphorescence and delayed emission are roughly equal, and are about 100 times less than that of prompt fluorescence.

photochemical product) has a higher extinction coefficient (e.g. *triplet–triplet absorption:* $T_1 \rightarrow T_2$, illustrated in Fig. 13.3).

*Emission methods.* Several methods involving emission are also used. Direct emission from the long-lived excited state is most often detected as

**Table 13.1** Spectroscopic properties of some triplet-state probes

| *Probe* | *Abs(nm)* | $\tau$*(ms)* | $Q_t$ | $Q_f$ | $Q_p$ | *Abs** | *Phos* | *DF* | *FD* |
|---|---|---|---|---|---|---|---|---|---|
| Eosin | 530 | 1–2 | 0.71 | 0.19 | 0.0005 | ++ | ++ | + | + |
| Erythrosin | 530 | 0.3–0.4 | 0.98 | 0.02 | 0.002 | ++ | +++ | – | – |
| Di-iodofluorescein | 502 | 0.5 | | | | + | + | ++ | ++ |
| Dibromofluorescein | 514 | 0.9 | 0.49 | | | + | + | | |
| Fluorescein | 490 | 1 | 0.05 | 0.95 | | – | – | + | ++ |
| Rhodamine | 559 | 3–5 | $\leq$0.01 | 0.99 | | – | – | – | +++ |
| Pyrene | 350 | 1 | | | | ++ | | | |
| Tryptophan | 280 | 1000 | | | | | + | | |

These are typical values; there is considerable variation, especially due to the wide range of possible binding environments. References include Garland and Moore, 1979; Johnson and Garland (1982) and Austin *et al.* (1979).

*'+' and '–' signs indicate the relative usefulness of each probe in each technique (Abs = absorbance, Phos = phosphorescence, DF = delayed fluorescence, FD = fluorescence depletion, $Q_t$ = quantum yield for triplet state formation, $Q_f$ = quantum yield for fluorescence emission, $Q_p$ = quantum yield for phosphorescence emission).

*phosphorescence* ($\tilde{\boldsymbol{\mu}}_{ob}=\hat{\boldsymbol{\mu}}_p$: $T_1 \rightarrow S_0$) (Austin *et al.*, 1979; Moore *et al.*, 1979), which typically occurs at wavelengths 100 nm (or more) higher than absorption (see Fig. 13.3). The results are reported as the time-dependent decay of *emission (polarization) anisotropy*

$$r(t) \equiv \frac{I_{\parallel}(t) - I_{\perp}(t)}{I_{\parallel}(t) + 2I_{\perp}(t)} \tag{13.9}$$

where $I_{\parallel}$ and $I_{\perp}$ are the phosphorescence emission intensities polarized parallel and perpendicular to the excitation polarization $\hat{E}$. Thus Equation 13.5 becomes $r(t) = \langle P_2(\cos\theta_{ob})\rangle = \langle P_2(\hat{\boldsymbol{\mu}}_p \cdot \hat{E})\rangle$. Fluorescence ($\hat{\boldsymbol{\mu}}_{ob}=\hat{\boldsymbol{\mu}}_f$: $S_1 \rightarrow S_0$) is usually not useful for detecting slow molecular processes, due to the nanosecond lifetimes of most excited singlet states. However, a long-lived *delayed fluorescence* may be detected after the return of the photoselected molecule from its long-lived excited state, most typically due to the reverse of the intersystem crossing process that populated the excited (triplet) state in the first place (Greinert *et al.*, 1979). The polarization anisotropy of this delayed emission is defined as in Equation 13.9, where $r(t) = \langle P_2(\hat{\boldsymbol{\mu}}_f \cdot \hat{E})\rangle$. Alternatively, fluorescence may be observed by exciting a population of probes a second time. If this is done after photoselection but before the ground state has been fully repopulated, *fluorescence depletion* is detected, corresponding to a *decrease* in fluorescence due to ground-state depletion. If the total fluorescence is then detected, using a wide-aperture lens, the *fluorescence depletion anisotropy* is equivalent to the absorption anisotropy, substituting $\Delta F$ for $\Delta A$ in Equation 13.8 (Johnson and Garland, 1981).

### 13.2.2 SAMPLE PREPARATION

(a) *Probes*

Some of the probes commonly used to study membrane protein rotational motion using transient optical anisotropy are shown in Fig. 13.4, and properties are listed in Table 13.1.

*Extrinsic probes.* The most generally applicable probes, both in terms of the proteins that can be studied and the detection methods that can be used, are organic dyes having easily populated and stable triplet states. These molecules, often referred to as 'triplet probes', are usually attached covalently to proteins by means of such reactive groups as isothiocyanate (reacting primarily with amino groups), iodoacetamide (reacting primarily with SH groups), maleimide (reacting primarily with SH groups, but also with amino groups).

The most commonly used of these are derivatives of *eosin* (2,4,5,7-tetrabromofluorescein), in which the heavy Br atoms apparently facilitate intersystem crossing, increasing the quantum yield for excitation to the long-

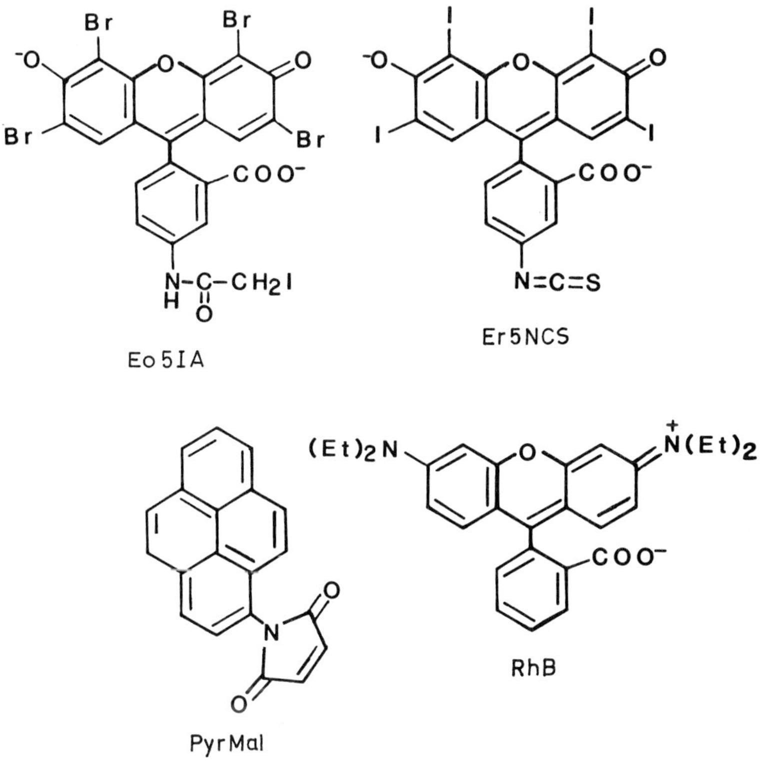


**Fig. 13.4** Examples of triplet-state probes used to study membrane proteins. The chromophore components include eosin (Eo), erythrosin (Er), pyrene (Pyr), and rhodamine B (RhB). Reactive groups include iodoacetamide (IA, specific for sulphhydryl groups), isothiocyanate (NCS, specific for amino groups) and maleimide (Mal, specific for sulphhydryl groups).

lived triplet state (Cherry and Schneider, 1976). Eosin's spectral properties make it a versatile probe that can be studied using most of the detection methods discussed in this review. Upon excitation, about 70% of the excited molecules are converted to the triplet state, and the triplet-state lifetime, as measured by the decay of ground-state depletion (absorption around 530 nm), triplet–triplet absorption (absorption around 630 nm), phosphorescence emission (around 700 nm), fluorescence depletion (emission around 560 nm), or delayed fluorescence (emission around 560 nm), is often as long as several milliseconds. The wavelength-dependence of these detection methods is illustrated by the spectra in Fig. 13.3.

A closely related triplet probe is *erythrosin* (2,4,5,7-tetraiodofluorescein). Erythrosin's absorption and emission occurs at similar wavelengths to eosin's, but some of the quantum yields are significantly different. Most importantly, the quantum yield for triplet formation is at least 98%, prompt fluorescence is negligible (<2%) and the quantum yield for phosphorescence ($2 \times 10^{-3}$, Garland and Moore, 1979; Moore *et al.*, 1979) is higher than that of eosin. As a result, the absolute sensitivity of phosphorescence (of the order of $10^{-11}$ moles; Garland and Moore, 1979) and the ratio of phosphorescence to fluorescence are much greater than for eosin. The latter is important because of the difficulty in preventing interference from intense fluorescence signals at early times after the exciting flash (discussed below in the instrumental section). Thus erythrosin is a nearly ideal probe for phosphorescence studies. Its high triplet quantum yield also makes it an excellent absorption probe. On the negative side, there is a reciprocal relationship between the quantum yield for excitation and the emission lifetime, with the result that the maximum excited-state lifetime of an erythrosin derivative is several hundred microseconds, about 10 times less than that of eosin, thus restricting erythrosin to studies in the submillisecond time range. In addition, the low fluorescence quantum yield prevents the use of erythrosin in fluorescence depletion studies (Johnson and Garland, 1981).

Another useful triplet probe is pyrene, used previously with excitation at 353 nm and detection of triplet absorption at 421 nm (Lavalette *et al.*, 1977). There are no published reports of triplet studies of pyrene probes on membrane proteins.

In analysing the criteria for sensitivity in fluorescence depletion experiments, Johnson and Garland (1982) concluded that it is desirable to maximize the ratio $Q_f/Q_t$, where $Q_f$ is the quantum yield for fluorescence emission and $Q_t$ is the quantum yield for triplet state formation (intersystem crossing). Note that these are essentially the opposite of the criteria for sensitivity in phosphorescence. Thus probes such as rhodamine, which have high ratios and are thus useless for phosphorescence or absorbance experiments, are potentially quite useful for optimizing the sensitivity of this method.

A number of triplet probes, including most of the eosin and erythrosin derivatives mentioned here, are available commercially, chiefly from Molecular Probes (Junction City, Oregon, USA).

*Intrinsic probes.* The most ubiquitous intrinsic triplet probe in membrane proteins is tryptophan. It absorbs around 280 nm, fluoresces around 350 nm and phosphoresces from 410 to 520 nm. Its phosphorescence lifetime in proteins can be as long as hundreds of milliseconds, permitting phosphorescence anisotropy to detect even slower rotational motions than are measurable with eosin and most other triplet probes (Strambini and Galley, 1976, 1980). Triplet absorption of tryptophan in proteins has also been reported (Hicks *et al.*, 1978). The principal disadvantage of tryptophan as a probe is that most proteins have several tryptophan residues, making it difficult to study one site selectively. No applications have been published on slow tryptophan rotation in membrane proteins.

Intrinsic chromophores that undergo a photochemical reaction have proven quite useful in studies of membrane protein rotational motion, using transient absorption (ground-state depletion) dichroism. An important example is the retinal chromophore of rhodopsin (Cone, 1972) and bacteriorhodopsin (RaziNaqvi *et al.*, 1973; Heyn *et al.*, 1977; Kouyama, *et al.*, 1981). The absorption change observed when haem–CO is photolysed has also been exploited to study cytochrome *c* oxidase (Kawato *et al.*, 1981, 1982a) and cytochrome *P*-450 (Richter, *et al.*, 1979; Kawato *et al.*, 1982b; Gut *et al.*, 1982).

Analogues of intrinsic chromophores, designed to have appropriate triplet-state properties, have been used to study rotational motions of membrane proteins. For example, Vaz *et al.* (1979) replaced the haem of cytochrome $b_5$ with rhodium(III)–protoporphyrin IX, and observed transient dichroism. Dixit *et al.* (1982) substituted Zn for the Haem Fe in cytochrome *c*, producing a fluorescent and phosphorescent probe that permitted the study of cytochrome *c* motions on mitochondrial membranes.

(b) *Labelling*

As in any extrinsic probe experiment, achieving and demonstrating appropriate labelling is often the most difficult and crucial phase of the experiment. Most of the desired objectives (specificity, complete labelling, and preservation of protein function) are not unique to triplet or membrane studies and will not be discussed in detail here. Specific labelling of a single protein component in a membrane with complex protein composition (e.g. mitochondrial membranes, plasma membranes) often requires labelling a purified protein followed by reconstitution. Proteins in membranes with simple protein composition (e.g. sarcoplasmic reticulum, purple membranes, rod cell disc membranes) are conveniently studied either in native membranes or in reconstituted systems. It is also desirable to know the site of labelling within the protein structure, but it is only very recently that significant structural information has been available for a few membrane proteins.

There is an additional labelling requirement that is more strict than in most other applications: if the probe is to be used in measuring microsecond motions, its microsecond and submicrosecond motions relative to the protein must be considerably restricted. As discussed below, the time resolution available in optical spectroscopy can, in principle, be used to separate rapid probe motions from slower (presumably protein) motions, but this becomes more difficult as the amplitude of the rapid motions increases and the remaining anisotropy decreases.

(c) *Concentration and size*

The sensitivities of various optical and EPR methods, in terms of the concentration and amount of probes needed, are summarized in Table 13.2.

**Table 13.2** Sensitivities of various methods

| | *Concentration (mol/l)* | *Volume (l)* | *Moles* | *Dead time (ns)* | *Time resolution*[a] *(ns)* |
|---|---|---|---|---|---|
| OPTICAL | | | | | |
| Absorption, beams $\perp$ | $10^{-6}$ | $10^{-3}$ | $10^{-9}$ | 5[b] | 50 |
| Absorption, beams $\parallel$ | $10^{-6}$ | $10^{-4}$ | $10^{-10}$ | 5[b] | 50 |
| Phosphorescence | $10^{-8}$ | $10^{-4}$ | $10^{-12}$ | 100[c] | 50 |
| Delayed fluorescence | $10^{-8}$ | $10^{-4}$ | $10^{-12}$ | 1000[d] | 50 |
| Fluorescence depletion | | | $10^{-19}$ | >10 000[c,d] | 50 |
| EPR | | | | | |
| Saturation recovery | $5 \times 10^{-4}$ | $2 \times 10^{-5}$ | $10^{-8}$ | 100[e] | 20 |
| Pulsed ELDOR* | $10^{-3}$ | $2 \times 10^{-5}$ | $2 \times 10^{-8}$ | 100[e] | 20 |
| Continuous wave ELDOR* | $10^{-3}$ | $2 \times 10^{-5}$ | $2 \times 10^{-8}$ | | |
| $V_2'$ | $10^{-5}$ | $2 \times 10^{-5}$ | $2 \times 10^{-10}$ | | |
| $U_1'$ (cavity) | $10^{-3}$ | $2 \times 10^{-5}$ | $2 \times 10^{-8}$ | | |
| $U_1'$ (loop-gap resonator)[f] | $10^{-6}$ | $10^{-6}$ | $10^{-12}$ | | |

Values are approximate minimum achieved so far, not necessarily absolute limits.
*Electron-electron double resonance.
[a]Limited by the bandwidth of the amplifier and the channel width of the digitizer (Eads *et al.*, 1984; Forrer *et al.*, 1980).
[b]Limited by the laser pulse width (Eads *et al.*, 1984).
[c]Limited by prompt fluorescence and light-scattering, usually requiring PM gating (Austin *et al.*, 1979).
[d]Limited by laser pulse width needed for significant ground-state depletion (Johnson and Garland, 1981).
[e]Limited by microwave pulse width and cavity (resonator) ringing time (Hyde, 1979; Kusumi *et al.*, 1982).
[f]Thomas *et al.* (1983).

These are approximate values that depend on the details of the particular instrumentation and probe used, but there are certain definite principles to be mentioned. For example, optical absorption techniques require sufficient concentration and optical path length to result in an absorbance of the order of 0.1 or greater, usually requiring that the product of the concentration and path length is at least $10^{-6}$ M.cm. The required volume can be minimized by using co-linear excitation and detection beams (discussed in Section 13.2.3 below). Phosphorescence and delayed fluorescence, like other emission methods, have much more intrinsic sensitivity than absorption methods, making them applicable to smaller and more dilute samples, although sensitivity is limited by typically low quantum yields. The fluorescence depletion method takes advantage of the high quantum yield of fluorescence to obtain even higher sensitivity, permitting even experiments on a localized region of the surface of a single cell (Garland and Johnson, 1985).

(d) *Turbidity*
Light-scattering causes serious problems for both adsorption and emission detection, decreasing the applicability of optical techniques to turbid samples.

(e) *Oxygen removal*
All of the triplet probes used in transient optical anisotropy experiments are quenched by oxygen. The concentration of $O_2$ in aqueous solution at room temperature (about 0.4 mM) is enough to reduce eosin's triplet lifetime from about 1 ms to about 10 $\mu$s, severely limiting the sensitivity of any detection method to slow motion. The 1 ms lifetime is easily regained by bubbling the solution for a few minutes with high-purity argon or nitrogen. However, this kind of treatment is often too harsh for proteins or membrane suspensions, so gentler methods have been developed, including (a) stirring the sample while directing a stream of argon at its surface (Cherry, 1979), (b) dialysing the sample against argon-bubbled buffer (Eads *et al.*, 1984), (c) separating the sample from a dithionite solution with a gas-permeable membrane (Garland and Johnson, 1985), and (d) adding an oxygen-consuming enzyme system (Horie and Vanderkooi, 1981; Johnson and Garland, 1982; Eads *et al.*, 1984).

### 13.2.3 INSTRUMENTATION FOR TRANSIENT OPTICAL ANISOTROPY

An instrument used for measuring the transient anisotropy of absorption or emission of eosin-labelled proteins, with time resolution in the range from 50 ns to several ms, is shown in Fig. 13.5 (Eads *et al.*, 1984). Figs. 13.5(a) and 13.5(b) show the configurations used for the detection of absorption and emission respectively.

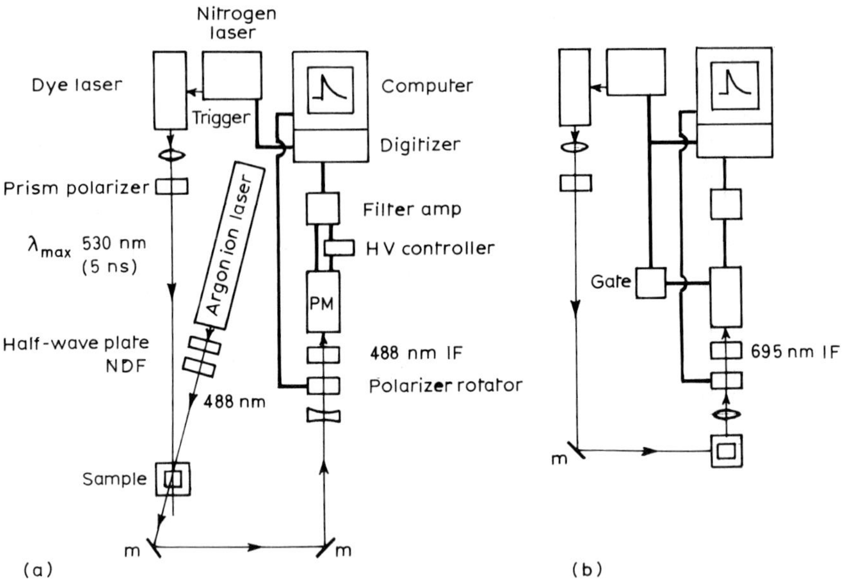


**Fig. 13.5** Schematic drawing of instruments used to measure slow (microsecond and millisecond) rotational motions of optical probes (e.g. eosin derivatives). Details are discussed in the text. (a), Absorption. (b), Emission. (From Eads *et al.*, 1984.)

(a) *Excitation*

For both types of experiments, excitation is accomplished with a brief vertically polarized pulse from a dye laser. Pulsed excitation ('pumping') of the dye laser is provided by a nitrogen laser (National Research Group 0.5-5-150/1B) with a nominal pulse width of 5 ns, and a repetition rate of 1–60 Hz. The nitrogen laser's wavelength (337 nm) would be ideal for exciting pyrene, but eosin and most other triplet probes are optimally excited at much longer wavelengths (see Fig. 13.3). The dye laser (NRG DL-0.03) is operated as a tuned (less than 1 nm bandwidth, using a grating) or broadband (about 10 nm bandwidth, using a mirror) source, using Coumarin 485 or 500 (Exciton) at 10 mM in ethanol, to excite eosin in its visible absorption band near 520 nm. Other excitation light sources that could be used include a flash lamp (Strambini and Galley, 1980), a flash-lamp-pumped dye laser (Cherry, 1978), a dye laser pumped by a XeCl excimer laser (Corin *et al.*, 1985) or a frequency-doubled Nd-YAG laser (530 nm, Mühlebach and Cherry, 1982).

Pulsed laser excitation offers the advantage of high light energy density (needed to produce a detectable signal) in a narrow wavelength band (needed for selective excitation) in a time short compared to the microsecond and millisecond motions being measured. The nitrogen, excimer and Nd-YAG lasers all have pulse widths of 20 ns or less with virtually no residual 'tail'.

Most flash lamps (and flash-lamp-pumped dye lasers) that can deliver enough energy for triplet studies have pulse widths of the order of 1 $\mu$s or more, with small but significant tails lasting for many $\mu$s, preventing the accurate measurement of the more rapid $\mu$s motions. High excitation energy is more important for absorption detection than for emission detection, since a detectable change in absorption requires the excitation and conversion of a significant fraction (preferably about 10%) of the probed molecules into the triplet (or other long-lived state). Even if the absorption coefficient (nearly $10^5$ for eosin) and the efficiency (quantum yield) of triplet conversion (as high as 70% for eosin) are high, a ground-state depletion of 10% requires an energy density per pulse of about 0.1 mJ/cm$^2$. (This value is inversely proportional to the product of the absorption coefficient and the triplet yield.) The nitrogen-pumped dye laser shown in Fig. 13.5 produces a pulse energy of the order of 0.01 to 0.05 mJ. Thus, in order to obtain a sufficient energy density for absorption experiments, it is necessary to focus the excitation beam to a diameter considerably less than 1 cm. Excimer-pumped dye lasers and Nd-YAG lasers produce more than 10 times as much light as the nitrogen-pumped dye laser, making it much easier to obtain a substantial absorbance change without the need to focus the beam to such a small diameter. The use of high excitation energy density in absorption can produce saturation artifacts that are usually not present in emission experiments (Kawato and Kinosita, 1981). In addition, photochemical reactions can result in bleaching of the probe (reducing the signal/noise ratio) and even functional damage to the labelled protein, although the functional damage can usually be minimized by deoxygenation (Bürkli and Cherry, 1981).

(b) *Absorption detection*

Figure 13.5(a) shows the configuration used to detect singlet absorption (ground-state depletion). The excitation (pump) laser is operated in the broadband mode and focused to a diameter of 1–2 mm to maximize energy density. The pulse-induced depletion and regeneration of the ground singlet state is monitored by changes in absorbance of a low-power (50 $\mu$W) continuous-wave probe beam at 488 nm from an argon-ion laser (Spectra-Physics 165), which is arranged to be very nearly co-linear with the pulse beam within the cuvette. Most other investigators have used a probe beam perpendicular to the pump beam, requiring that the pump beam excite the entire cross-section of the sample. The co-linear arrangement permits the use of small-diameter beams, thus permitting the use of a smaller sample (using a 2 mm wide cuvette in this case) and a lower pulse energy to achieve the same absorbance and ground-state depletion. The sample is contained in a box equipped for temperature and atmosphere control, and the sample may be stirred magnetically during experiments. The polarization of the probe beam is adjusted to give equal intensities (in the absence of excitation) for analyser orientations parallel and perpendicular to the plane of excitation, by rotation of a half-wave plate. Fine control of the average (DC) signal level during experiments is obtained by electronically controlling the photomultiplier (PM) high voltage to produce a constant output. The controller is slow (1 s response time) and does not affect the very rapid (ms) polarized transmittance transients.

The time-dependent transmittance is detected after passing through a $488 \pm 1$ nm interference filter to reject prompt fluorescence and light-scattering due to the excitation pulse. These latter artifacts are much less intense (relative to the desired transmittance signals) than in emission experiments, due to the long (2 m) path length from the sample to the PM, and to the much higher intensity of the observed signal (permitting the PM to be operated at lower voltage, i.e. lower sensitivity). As a result, PM gating is usually not necessary in absorption experiments, and the dead time at the beginning of the experiment is reduced to the width of a single data channel (as low as 50 ns for the digitizer used in this instrument) or the laser pulse width (5 ns in this case), whichever is greater.

The transmittance signal from the PM can be filtered to limit the bandwidth and reduce high-frequency noise over the approximate range 65 kHz to 3 MHz. The signal is visualized on an oscilloscope, and digitized by a LeCroy 2256A 20 MHz waveform Digitizer (minimum period is 50 ns per point). Polarizer rotation, data acquisition, and signal averaging are under control of a LeCroy 3500M Multichannel Analyzer. The polarizer is rotated every 50–200 pulses, and a total of 1000 to 10 000 transients are added, with $I_{\parallel}$ and $I_{\perp}$ being stored separately. An alternative to this rotating polarizer scheme is to detect $I_{\parallel}$ and $I_{\perp}$ simultaneously, using a polarizing beam splitter and two PMs (Cherry, 1978). Although two parallel detection systems are more expensive and usually decrease time resolution by a factor of two, they decrease the time required for an experiment by at least a factor of two and are effective in reducing errors due to the variation of excitation intensity and systematic electronic noise. The absorbance changes are calculated from intensity transmittance records obtained with ($I(t)$) and without ($I_0(t)$) pulsed excitation, obtained in separate runs on the same sample, as follows:

$$\Delta A_{\parallel}(t) = -\log \frac{I_{\parallel}(t)}{I_{0\parallel}(t)}, \quad \Delta A_{\perp}(t) = -\log \frac{I_{\perp}(t)}{I_{0\perp}(t)}$$

Then the anistropy $r(t)$ is calculated as in Equation 13.8. The sum, corresponding to total absorption recovery, is calculated as $\Delta A(t) + 2\Delta A_{\perp}(t)$.

Triplet absorption anisotropy signals from eosin samples are detected and analysed in a similar way (Eads *et al.*, 1984), except that the probe beam is provided by a HeNe laser, since its wavelength (630 nm) is more appropriate for the $T_1 \rightarrow T_2$ transition (see Fig. 13.3).

Absorption methods generally have the advantage that signals are intense and do not suffer from artifacts due to intense spikes of prompt fluorescence and light scattering during the excitation pulse, with the result that the signals can be more easily detected at short times. Detection dead times as short as 10 ns have been reported (Eads *et al.*, 1984). Another advantage is that the detected signal is a relatively strong probe beam, eliminating or reducing the requirements for high-sensitivity photomultipliers (photodiodes can be used), amplifiers (that can reduce time resolution), and extensive signal averaging to achieve acceptable signal/noise. The major drawbacks of absorption methods, in comparison to other methods discussed here, are that they require (a) a sufficient number (concentration times path length times beam area) of probes to absorb a detectable fraction ($>=5\%$) of the probe beam intensity, resulting in a requirement for a minimum of about $10^{-9}$ probe molecules, and (b) a sufficiently energetic excitation light pulse (at least 100 $\mu J/cm^3$) to excite a detectable fraction ($>=5\%$) of the probes. As discussed above, the intense excitation ('pump') beam can cause optical and photochemical artifacts.

(c) *Emission detection*

Figure 13.5(b) shows the configuration used to detect eosin *phosphorescence*. Emission detected at 90° to the excitation beam passes through a polarizer oriented to analyse the components parallel ($I_{\parallel}$) and perpendicular ($I_{\perp}$) to the vertical polarization axis of the excitation beam. Triplet emission in the red (phosphorescence) is selected, and the contribution to the signal from scattering of pulsed excitation and from the eosin singlet emission band centred near 550 nm (prompt fluorescence) is reduced, by (a) using a $695 \pm 30$ nm interference filter (Spectrofilm, Inc.), and by (b) gating the photomultiplier (PM, Hamamatsu R928, having a red-sensitive S20 photocathode) on (turning on the high voltage) at about 800 ns after the pulse. With eosin, both of these measures must be used to prevent the spike of prompt fluorescence and light scattering from causing electronic artifacts that can last as long as hundreds of microseconds after the excitation pulse. Because of erythrosin's higher phosphorescence yield and much lower fluorescence yield, high-quality phosphorescence signals can be detected at short times after the pulse, without the need for PM gating, assuming that light scattering is not too severe (Moore *et al.*, 1979). A monochromator may be used to select the detected wavelength range, in place of the interference filter, reducing the signal intensity but permitting the recording of spectra (see Fig. 13.3). Detection of the two polarizations is accomplished in emission as for absorption, using a computer-controlled rotating polarizer. An alternative method is to use a T format and two PMs to detect the two signals simultaneously (Moore *et al.*, 1979). The phosphorescence emission anisotropy is calculated from Equation 13.9.

The detection of *delayed fluorescence* is accomplished simply by using the appropriate interference filter (centred at 560 nm rather than 695 nm, for eosin and related probes). Artifacts due to prompt fluorescence are much more severe, suggesting the need not only for PM gating but also for the use of a probe with a very low prompt fluorescence yield, such as di-iodofluorescein (Greinert *et al.*, 1979, 1982).

*Fluorescence depletion* signals are detected with a configuration similar to that used for absorption (an intense pulsed excitation beam and a weak continuous detection beam), except for the addition of fluorescence detection at 90° (Johnson and Garland, 1982). If the total fluorescence is detected, the decrease in fluorescence intensity is proportional to the decrease in absorbance and can be analysed in essentially the same way (Equation 13.8). This method can be used to achieve extremely high sensitivity, in terms of the number of molecules needed to obtain a detectable signal. Measurements on single cells have been reported (Johnson and Garland, 1981, 1985). Detection dead times are typically greater than with other techniques, due to the long pulse width that is often required to achieve significant depletion.

### 13.2.4 DATA ANALYSIS

(a) *Empirical analysis: curve-fitting*

In most cases, the first step in data analysis is the relatively model-independent fitting of the observed anisotropy decay function to a sum of exponential terms:

$$r(t) = \sum_{i=1}^{n} r_i(t)\exp(-t/\tau_{ri}) + r_\infty \qquad (13.10)$$

where $\tau_{ri}$ are the *rotational correlation times* and $r_\infty$ ($\equiv r(\infty)$) is the *residual anisotropy*. For purposes of theoretical analysis (discussed below), this equation is usually normalized by dividing by the *initial (fundamental) anisotropy* $r_0(\equiv r(0) = \sum_{i=1}^{n} r_i + r_\infty)$. As discussed below, the amplitudes ($A_i = r_i/r_0$) and correlation times ($\tau_{ri}$) are related to the angular amplitudes and time constants of rotational motions, but there is not always a one-to-one correspondence between the *i*th decay term and a particular rotational degree of freedom (e.g. rotation about an axis). The empirical fitting of the data to Equation 13.10 is usually carried out by a non-linear least-squares procedure. The number of parameters is usually limited by the quality of the data and/or the assumed motional model. The quality of the fit is assessed by overlaying the data and fit (see Fig. 13.6), and sometimes also by calculating and plotting the residual (data minus fit) and/or the autocorrelation function of this residual. Perhaps the most rational procedure is to start with a single-exponential function ($n = 1$ in Equation 13.10), and increase $n$ until no further improvement is observed in the fit (as judged by the chi-squared value). Most analyses stop at $n = 1$ or 2.

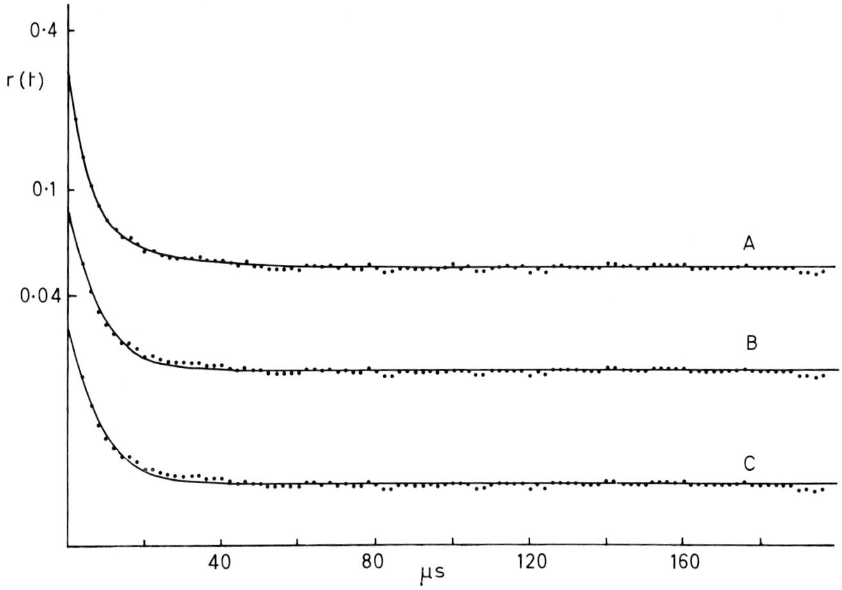


**Fig. 13.6** Data from time-resolved absorption anisotropy experiments on reconstituted bacteriorhodopsin (212 mol DMPC/mol BR, 28°C), along with the best-fit curves from least-squares procedures (from Cherry and Godfrey, 1981). The anisotropy decay is plotted correctly in (A); the same curve (fitted to different functions) is displaced vertically in (B) and (C) for clarity. (A), best fit to Equation 13.12 (uniaxial rotation), resulting in $\theta_{\mu n} = 77°$ and $D_{\parallel} = 7.8 \times 10^4$ rad$^2$ s$^{-1}$. An equally good fit, with essentially equivalent results, was obtained by fitting to Equation 13.10, with $n = 2$ (not shown). (B), An alternative fit to Equation 13.12, with $\theta_{\mu n} = 36°$. (C), The best fit to a single exponential decay plus constant (Equation 13.10, with $n = 1$). Note that, in both (B) and (C), the fit to the residual anisotropy is as good as in (A), but the fit to the early decay is not as good. See text for further discussion.

(b) *Theoretical modelling*

Once the anisotropy decay function is analysed and the parameters $r_0$, $r_\infty/r_0$, $A_i = r_i/r_0$, and $\tau_{ri}$, are determined, further analysis usually requires the assumption of a model for the type of rotational motion. The choice of model depends on the types of motion predicted for a membrane protein (discussed above in Section 13.1.2), and on the number of parameters needed to fit the data.The theoretical analysis required to make this connection has been discussed in detail (Cherry, 1979; Lipari and Szabo, 1980; Kawato and Kinosita, 1981; Kinosita *et al.*, 1984). This section will present some of the important results of the theoretical analysis and illustrate how it is applied to the analysis of experiments on membrane protein motions.

*Isotropic rotational diffusion.* In the simple case of isotropic rotational diffusion, as expected for the rotation of a probe rigidly fixed to a spherical protein tumbling freely in an isotropic medium, the anisotropy decay is given by a single-exponential decay with a time constant $\tau_r = 1/6D$ (Equation 13.6) and no residual anisotropy ($r_\infty = 0$). In an anisotropic medium such as a membrane, rotational diffusion is expected to be anisotropic, resulting in a more complex decay.

*Initial and residual anisotropies; model-independent conclusions.* Although the theoretical analysis of anisotropy decay is usually based on a motional model, there are certain relatively model-independent conclusions that can be derived, particularly from the values of $r_0$ and $r_\infty$. The theoretically maximum possible value of the initial anisotropy $r_0$ is 0.4 for absorption and 0.4 $P_2(\hat{\boldsymbol{\mu}}_\mathbf{a} \cdot \hat{\boldsymbol{\mu}}_\mathbf{e})$ for emission. The absolute value of $r_0$ can be decreased by (a) excitation so intense that a significant fraction of the probes are excited, (b) light scattering or (c) rotational motion on a time scale too fast to be detected. The first two effects are not very important, since they multiply $r(t)$ by the same factor for all $t$, and most analysis depends on $r(t)/r(0)$, as discussed below. A non-zero value of $r_\infty$ implies that, on the time scale of the detection, the probes' transition moments do not all rotate isotropically without restriction through 180°. One possible explanation is that some molecules rotate isotropically and freely, while others rotate very slowly (are effectively rigid) on the observed time scale. However, if there is reason to believe that all probes are in equivalent environments (e.g. all attached to the same site on a protein), so that the entire anisotropy decay is representative of the motion of a single probe, the value of $r_\infty/r_0$ can be analysed directly in terms of the extent to which the motion is restricted in angular amplitude on the observed time scale (Lipari and Szabo, 1980):

$$r_\infty/r_0 = \langle P_2(\hat{\boldsymbol{\mu}}_\mathbf{ob} \cdot \hat{\mathbf{n}})\rangle^2 = \langle (3\cos^2\theta_{\mu n} - 1)/2\rangle^2 \qquad (13.11)$$

where $\theta_{\mu n}$ is the angle between the observed transition moment $\hat{\boldsymbol{\mu}}_\mathbf{ob}$ and the system's symmetry axis $\hat{\mathbf{n}}$ (e.g. the membrane normal) in Fig. 13.1.

*Uniaxial rotation.* If a chromophore is rigidly fixed to a membrane protein that rotates as a rigid body about the membrane normal, characterized by a rotational diffusion coefficient $D_\parallel$ (Fig. 13.1(a)), its anisotropy decay takes the form:

$$\begin{aligned} &r(t)/r_0 = A_1 \exp(-t/\tau_{r1}) + A_2 \exp(-t/\tau_{r2}) + A_3, \\ &\tau_{r1} = 1/4D_\parallel,\ \tau_{r2} = 1/D_\parallel, \\ &A_1 = (3/10)\sin^4\theta_{\mu n},\ A_2 = (6/5)\sin^2\theta_{\mu n}\cos^2\theta_{\mu n}, \\ &A_3 = r_\infty/r_0 = [P_2(\cos\theta_{\mu n})]^2 = (1/4)(3\cos^2\theta_{\mu n} - 1)^2, \end{aligned} \qquad (13.12)$$

where $\theta_{\mu n}$ is the fixed angle between the probe's absorption transition moment and the membrane normal (Cherry, 1978; Kinosita *et al.*, 1984). (This is a different $\theta$ from the one depicted in Fig. 13.1.) Note that there is only one diffusion coefficient but there are two correlation times. Some workers assume that there is uniaxial diffusion and fit the data to Equation 13.12, thus determining values for $D_\parallel$ and $\theta_{\mu n}$ (Cherry, 1979; Cherry and Godfrey, 1981; see Fig. 13.6). Equation 13.12 assumes that the sample is a randomly oriented suspension of membrane fragments. If the experiment is done on oriented membranes, the analysis is much simpler and the conclusions are much less model-dependent (Cone, 1972). In particular, steady-state dichroism experiments on oriented membranes can be used to determine whether there is indeed a single chromophore orientation $\theta_{\mu n}$ with respect to the membrane normal.

The most thorough test (and application) of Equation 13.12 has been provided by transient dichroism (absorption anisotropy) measurements on the intrinsic retinal chromophore of the purple membrane protein bacteriorhodopsin (BR) reconstituted in dimyristoylphosphatidylcholine (DMPC) vesicles (Cherry and Godfrey, 1981; Fig. 13.6). As long as the temperature was above the lipid phase transition and the lipid/protein ratio was high, corresponding to conditions favouring monomeric proteins, reasonable agreement was observed with Equation 13.12. The least model-dependent observation was a non-zero constant residual anisotropy, $r_\infty/r_0 = 0.191 \pm 0.026$, implying that $\langle P_2(\hat{\boldsymbol{\mu}}_{ob} \cdot \hat{\mathbf{n}})\rangle = (0.191)^{\frac{1}{2}} = 0.44 \pm 0.06$. The data were analysed further by fitting to Equation 13.10. Better fits were obtained by assuming two exponentials ($n = 2$ in Equation 13.10) than one. The ratio of the two correlation times $\tau_{r2}/\tau_{r1}$ was found to be approximately 6, in reasonably close agreement with the predicted value of 4 for uniaxial rotation (Equation 13.12). Assuming that Equation 13.12 applies, the observed residual anisotropy corresponds to a chromophore orientation $\theta_{\mu n}$ of either $78° \pm 3°$ or $38° \pm 1.5°$. In order to distinguish between these two values, the data were fitted directly to Equation 13.12, varying only the parameters $D_\parallel$ and $\theta_{\mu n}$. The best fit was obtained for $\theta_{\mu n} = 78°$ and $D_\parallel = 7.5 \times 10^{-4}\ \text{rad}^2\ \text{s}^{-1}$ (i.e. $\tau_{r2} = 4\tau_{r1} = 1/D_\parallel = 1.33 \times 10^{-5}$ s) at 28° (Fig. 13.6). This value of $D_\parallel$ is quite consistent with that predicted from Equation 13.3, based on the known structure of the protein (Cherry and Godfrey, 1981). The observed form of the anisotropy decay cannot be easily distinguished from that predicted for other types of anisotropic motion (e.g. wobbling in a cone, as discussed below). However, two other pieces of evidence argue strongly for the interpretation of uniaxial rotation in this case. First, steady-state linear dichroism measurements on oriented membranes provide fairly direct support for the uniform orientation of the chromophores at a similar angle to that obtained in this transient measurement (Heyn *et al.*, 1977). A wobbling chromophore (discussed below), having a wide range of orientations relative to the membrane normal, would have resulted in a much lower steady-state dichroism. Second, the value of $r_\infty/r_0$ was observed to be constant over a wide range of temperatures and lipid/protein ratios, a result more clearly consistent with a fixed chromophore orientation than with a wobbling motion, the amplitude of which might be expected to depend on temperature or protein aggregation. The dependence of the correlation times ($\tau_{r1}$) and amplitudes ($A_i$) on the temperature and lipid/protein ratio in this system provided direct insight into protein–protein (aggregation) and protein–lipid interactions (Cherry and Godfrey, 1981). In general, poor agreement with Equation 13.12 (often observed at low temperatures or lipid/protein ratios) is considered diagnostic for protein aggregation in this system and others, including eosin-labelled band 3 in erythrocyte membranes (Nigg and Cherry, 1979, 1980), cytochrome oxidase (Kawato *et al.*, 1981) and cytochrome *P*-450 (Richter *et al.*, 1979).

*Restricted rotation (wobbling in a cone).* Instead of a complete inhibition of a protein's off-axis motion, this wobbling motion may occur with a restricted but non-zero amplitude. If we assume that the observed transition moment $\hat{\boldsymbol{\mu}}_{\mathbf{ob}}$ is parallel to the molecular symmetry axis $\hat{\mathbf{m}}$, then the expression for $r_\infty/r_0$ (Equation 13.11) is equivalent to the familiar *order parameter* S used to characterize the amplitude of rotations in partially ordered systems. A value of $S=1$ implies complete 'order' (no rotational motion) on the observed time scale and a value of $S=0$ implies complete 'disorder' (unrestricted rotational motion) on the observed time scale. It is important to emphasize that the above definition of S depends on the time window being observed and is unrelated to disorder due to motions in a time range too fast to affect $r_0$ or too slow to affect $r_\infty$. For example, if $\hat{\mathbf{m}}$ (and hence $\hat{\boldsymbol{\mu}}_{\mathbf{ob}}$) wobbles with diffusion coefficient $D_{\mathrm{w}}$ ($D_\perp$), within the walls of a cone having half-angle $\theta_{\mathrm{c}}$ (Fig. 13.1b), its anisotropy decay takes approximately the form (Kinosita *et al.*, 1977; Kawato and Kinosita, 1981):

$$\begin{aligned} r(t)/r_0 &= A\exp(-t/\tau_r) + r_\infty/r_0, \\ \tau_r &= \langle\sigma\rangle/D_{\mathrm{w}} \\ r_\infty/r_0 &= S^2 = [(1/2)\cos\theta_{\mathrm{c}}(1+\cos\theta_{\mathrm{c}})]^2 \end{aligned} \qquad (13.13)$$

Plots of $\langle\sigma\rangle$ *vs* $\theta_{\mathrm{c}}$ have been published by Kinosita *et al.* (1977), and an analytical expression for $\langle\sigma\rangle$ has been published by Lipari and Szabo (1980); $\langle\sigma\rangle=6$ for isotropic motion ($\theta_{\mathrm{c}}=180°$), $\langle\sigma\rangle=4$ for $\theta_{\mathrm{c}}=90°$, and $\langle\sigma\rangle\cong 0.292(\theta_{\mathrm{c}})^2$ for $\theta_{\mathrm{c}}\leqq 30°$ (Lipari and Szabo, 1980). This result is only slightly sensitive to the precise type of orientational distribution assumed (e.g. a strict cone or a Gaussian distribution, Kinosita *et al.*, 1982, 1984). This model would apply to the case of a rigid protein wobbling about the bilayer normal (Fig. 13.1b) or to segmental motion within the protein that results in a similar motion (Fig. 13.1c). The shapes of the anisotropy decay curves expected for uniaxial (Equation 13.12) and wobbling (Equation 13.13) motions can be distinguished only if very high signal/noise is obtained. As discussed above, measurements on oriented membranes are the best means of distinguishing between these two models. An additional criterion that is often applied is based on the plausible (but usually not very rigorous) expectation that submicrosecond-to-microsecond motions having temperature-dependent correlation times and/or amplitudes ($\tau_{ri}$ and $r_i$ in Equation 13.10) are likely to be due to segmental (wobbling?) motions *within* the protein (Fig. 13.1c), rather than overall protein motions. This kind of evidence for internal wobbling motions has been obtained and analysed for cytochrome $b_5$ (Vaz *et al.*, 1979, using absorption) and for the Ca-ATPase in sarcoplasmic reticulum membranes (Bürkli and Cherry, 1981, using absorption of Eo5IA; Spiers *et al.*, 1983, using phosphorescence of Er5IA).

## 13.3 Saturation transfer electron paramagnetic resonance (ST-EPR)

### 13.3.1 PRINCIPLES OF ST-EPR

As in optical absorption spectroscopy, ST-EPR is sensitive to rotational motion because reorientation of the probe changes the probability of excitation and thus decreases saturation (transfers it away). This saturation transfer becomes detectable if it happens at a rate that is competitive with the intrinsic (rotation-independent) relaxation processes.

(a) *Spectrally resolved photoselection*

The EPR experiment (either conventional or saturation transfer) begins with a photoselection process that produces a non-random orientation distribution of excited probes. However, this orientation-dependent excitation does not depend on the orientation of the absorption transition moment, since these moments (spins) are all polarized either parallel or anti-parallel to the applied magnetic field and thus perpendicular to the polarization of the excitation field. Instead, the orientation dependence arises from the *orientational resolution within the spectrum*. That is, the anisotropic magnetic interactions make the Hamiltonian, and hence the separation between the energy levels of the two spin states, and hence the position of the EPR absorption line, dependent on the orientation of the spin label relative to the applied (DC) magnetic field $\boldsymbol{H}_o$. To a good approximation, the position at which a nitroxide spin label contributes to the absorption spectrum ($H_{res}$, the value of $H_0$ at which resonance occurs) depends on only two variables: the nitrogen nuclear quantum number $m_I$ ($-1$, 0, or $+1$) and $\langle \hat{\mathbf{z}} \cdot \boldsymbol{H}_o \rangle^2 = \cos^2(\theta_{zH})$ where $\theta_{zH}$ is the angle between $\boldsymbol{H}_o$ (which replaces $E$ in Equation 13.5) and the spin label's *principal axis* $\hat{\mathbf{z}}$ *(which replaces* $\hat{\boldsymbol{\mu}}_{ob}$ in Equation 13.5):

$$H_{res}(\theta_{zH}, m_I) = h\nu/g(\theta_{zH})\beta - m_I T(\theta_{zH}),\ (m_I = +1, 0, -1)$$
$$g(\theta_{zH}) = g_{\parallel} \cos^2\theta_{zH} + g_{\perp} \sin^2\theta_{zH}, \qquad (13.14)$$
$$T(\theta_{zH}) = [T_{\parallel} \cos^2\theta_{zH} + T_{\perp} \sin^2\theta_{zH}]^{\frac{1}{2}}$$

This orientation-dependence is illustrated by simulated conventional EPR ($V_1$) spectra in Fig. 13.7, assuming typical values for $g$ and $T$ tensors ($T_{\|}=35$ gauss, $T_{\perp}=7$ gauss, $g_{\|}=2.00241$, $g_{\perp}2.00741$), with $hv/2\beta=3400$ gauss. Thus for a uniformly oriented population of spin labels having a single value of $\cos^2\theta_{zH}$ (e.g. in a single crystal, in a stack of perfectly crystalline membranes oriented normal to $H_0$, or in a bundle of perfectly helical fibres oriented parallel to $H_0$), one observes a three-line spectrum in which the position of the centre line is determined by $g(\theta_{zH})$ and the splitting between the three lines is determined by $T(\theta_{zH})$ (Fig. 13.7) (Thomas and Cooke, 1980). Fig. 13.7 shows that each of these lines is so narrow that spectra corresponding to only slightly different orientations are easily resolved. As explained in the legend to Fig. 13.7, this orientational resolution varies throughout the spectrum and has a minimum value (smallest resolved rotation) of $\theta_T \cong 4°$ (Equation 13.7).

If there is only partial orientational order of the spin labels within a sample, the EPR spectrum provides a direct and unambiguous read-out of the orientational distribution $p(\theta_{zH})$, providing independent information on the average orientation, the orientational disorder, and the number of discrete preferred orientations in the population (Thomas and Cooke, 1980; Barnett and Thomas, 1984; Barnett *et al.*, 1986). Polarized optical absorption, lacking this orientational resolution in the frequency domain of the spectrum, can only characterize an orientational distribution with a single parameter, the order parameter (see Equations 13.11 and 13.13). An additional advantage of EPR is that it is not only sensitive to the orientation of the principal axis, but also is slightly sensitive to rotation *about* this axis (not shown in Equation 13.14 or Fig. 13.7, but discussed in connection with Fig. 13.15). Although EPR experiments usually involve steady-state measurements, resolution in the frequency domain compensates for the lack of time resolution and permits analysis in terms of multiple components and anisotropic motions.

In the absence of orientational order of the supramolecular assemblies being studied (e.g. in a solution or in a randomly oriented dispersion of membranes), orientational information cannot be derived directly from the EPR spectrum, but the orientational resolution remains in the spectrum (see Fig. 13.7, bottom), so that a very narrow distribution of orientations is excited at one spectral position. Thus a change in the orientation of a probe by as little as $\theta_T=4°$ can change its position in the spectrum significantly (decrease its excitation probability by a factor of 2), and *rotational motion transfers spins between excited and unexcited parts of the spectrum.* The orientational resolution of this photoselection is *more than 10 times better than in optical spectroscopy,* where $\theta_T \cong 45°$. This highly resolved *rotation-dependent transfer of spins* is the basis of motional, as well as orientational, sensitivity for both conventional and saturation transfer EPR. Substituting $\theta_T=4°=0.07$ rad into Equation 13.7, the transfer time $\tau_T$ (Fig. 13.2), in which a molecule undergoes a detectable rotation ($\theta_T$), is 0.007 $\tau_r$, more than 100 times less than in optical spectroscopy.

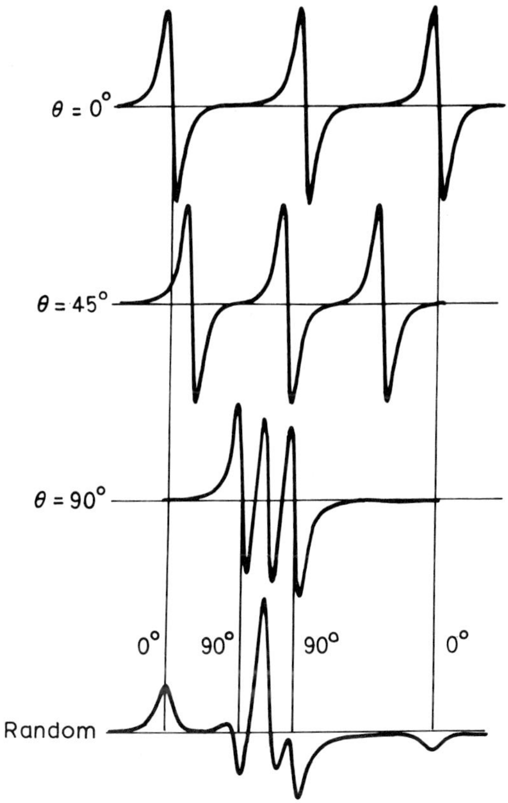


**Fig. 13.7** Computer simulations illustrating the orientation-dependence of nitroxide EPR spectra. $V_1$, the 'first harmonic' absorption spectrum, is plotted against the applied DC magnetic field strength $H_0$. $V_1 = dV_0/dH_0$, where $V_0$ is the amplitude of microwave absorption. The baseline in each spectrum is 100 gauss wide. The top three spectra correspond to uniformly oriented populations of spin labels, each having a single value of $\cos^2\theta_{zH}$, where $\theta_{zH}$ is the angle between the nitroxide spin label's principal axis ($\hat{\mathbf{z}}$) and the applied field $\hat{\mathbf{H}}$. For these spectra $V_0(H_0, \theta_{zH}, m_I) = L[H_0 - H_{res}(\theta_{zH}, m_I)]$, where $H_{res}$ is given by Equation 13.14 and $L$ is a Lorentzian lineshape function having a linewidth (half-width at half-height) of $1/(\gamma T_2) = 2.35$ gauss, where $\gamma$ is the gyromagnetic ratio ($1.76 \times 10^7$ rad s$^{-1}$ gauss$^{-1}$) and $T_2$ is the transverse relaxation time (about $2.4 \times 10^{-8}$ s for slowly tumbling nitroxides). The slight dependence of the spectrum on $\psi_{zH}$, i.e. rotation about the principal axis (deviation from axial symmetry) is neglected (discussed in Fig. 13.15). The bottom spectrum corresponds to a randomly oriented population of spin labels, immobile in the nanosecond time range (a 'powder' spectrum). It is obtained by summing $V_1$ spectra corresponding to all values of $\theta_{zH}$ from 0 to 90°, with weighting factors proportional to $\sin\theta_{zH}$. Lines at 0° and 90° illustrate that, although there is no preferred orientation in this population, the orientational resolution (anisotropy) remains in the spectrum: excitation at a particular value of $H_0$ corresponds to selective excitation of a narrow range of values of $\cos^2\theta_{zH}$. The sensitivity of the spectral position to the orientation ($dH_{res}/d\theta_{zH}$, evaluated from Equation 13.14) varies throughout the spectrum and has a maximum value of about 35 gauss rad = 0.61 gauss/degree (Thomas *et al.*, 1976). Thus the angular half-width of excitation, and thus the rotation required for transfer, is given by $\theta_T = (2.35 \text{ gauss}) \times (dH_{res}/d\theta_{zH})^{-1}$, and has a minimum value of 4°.

(b) *Conventional EPR: motional averaging*
If $\tau_T$ is comparable to or less than the inverse of the frequency resolution in the spectrum (roughly equal to $T_2$, the transverse relaxation time, which is typically 24 ns for nitroxides), the anisotropy will be averaged and the spectrum will be narrowed by the rotational motion. Therefore, this *motional averaging* can be detected only when $\tau_r$ is less than a microsecond. Conventional EPR is performed under conditions where the shape of the nitroxide spectrum, in the absence of preferred orientation, is determined almost exclusively by motional averaging, and is thus sensitive only to *submicrosecond* rotational motions. The methods of ST-EPR are designed to detect slower motions.

(c) *Saturation, ground-state depletion and relaxation*
The term 'conventional EPR' indicates an experiment that probes the linear response of a spin system to the exciting microwave radiation, implying that the spin system is never significantly perturbed from equilibrium, with a constant excess population in the ground state, as determined by the Boltzmann distribution. If the microwave field strength $H_1$ is intense enough, the spin system becomes saturated, i.e. the excess population in the ground state becomes significantly depleted, and the net steady-state absorption is no longer proportional to $H_1$. The amount of saturation depends on the competition between the rates of excitation and relaxation back to the ground spin state (see Fig. 13.2). In the absence of rotational motion, the rate constant for recovery from saturation is $1/T_1$, where $T_1$ is the longitudinal relaxation time and is quite analogous to the excited-state lifetime in optical spectroscopy.

(d) *Rotational diffusion and saturation transfer*
Rotational motion can increase this rate of recovery by causing saturation transfer, according to the following argument (see Fig. 13.2 and Thomas *et al.*, 1976): In a steady-state EPR experiment, only a small fraction of the probes, corresponding to a narrow angular range (as little as $\theta_T = 4°$), are at resonance and are therefore subjected to saturating radiation at any one time during the spectral scan. Therefore, rotational diffusion that transfers these probes out of this saturated angular range (and transfers other probes into it) during the excited-state lifetime $T_1$ will *transfer saturation* to other spectral regions, thus decreasing the saturation (ground-state depletion) at resonance. Thus the condition for effective saturation transfer is

$$\tau_T < \cong T_1$$

Since $\tau_T$ can be as little as 0.007 $\tau_r$, and $T_1$ is about 7 $\mu$s for slowly rotating nitroxides (Huisjen and Hyde, 1974), saturation transfer can be detected whenever the rotational correlation time

$$\tau_r < \cong T_1/0.007 \cong 1 \text{ ms}.$$

Optimum sensitivity to $\tau_r$ occurs in the range of 10–100 $\mu$s, and saturation transfer becomes maximal when $\tau_r$ is less than 1 $\mu$s. In optical spectroscopy $\tau_T = \tau_r$, so correlation times $\tau_r$ can only be measured if they are $< \cong$ the excited-state lifetime, which is about 1 ms for eosin. Since $T_1$ is much greater than $T_2$, virtually any nitroxide EPR experiment becomes more sensitive to microsecond rotational motion when saturation is imposed. The development of ST-EPR instrumentation and methodology has involved a search for an EPR experiment that is optimally sensitive to saturation, hence optimally sensitive to microsecond rotational motion that causes saturation transfer. These methods will be discussed in Section 13.3.3.

### 13.3.2 SAMPLE PREPARATION

(a) *Probes*

ST-EPR experiments are performed on the same types of nitroxide spin labels used in conventional EPR. For detailed discussions, including methods of synthesis, see Gaffney (1976), Morrisett (1976) and Keana (1979). Examples of spin labels used to study membrane proteins are shown in Fig. 13.8. Most have a nitroxide group in a six-membered (piperidinyl) or five-membered (pyrrolidinyl) ring, connected to a group for covalent attachment. The piperidinyl and pyrrolidinyl rings are much smaller than the chromophoric parts of most optical probes, and thus are more readily incorporated into spin-labelled analogues of substrates and other non-covalent protein ligands. The

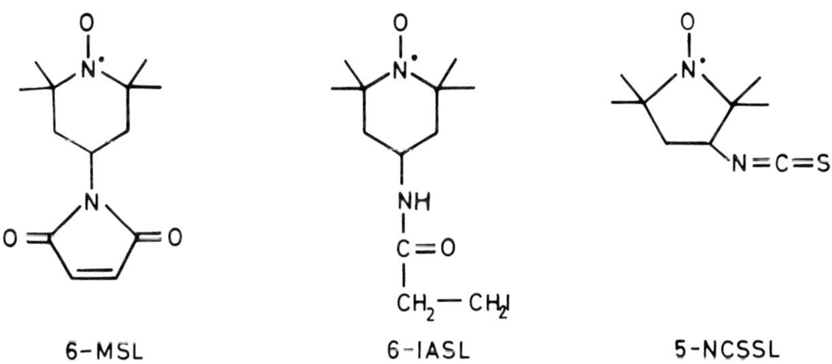


**Fig. 13.8** Examples of nitroxide spin labels used in ST-EPR studies of membrane proteins. The probe most often used is the maleimide derivative MSL (*N*-[1-oxyl-2,2,6,6-tetramethyl-4-piperidinyl]maleimide). In the figure, it is designated 6-MSL, indicating that it is based on the 6-membered piperidinyl ring; the five-membered ring pyrrolidinyl derivative (5-MSL) is also used. Also shown is another commonly used SH-directed label, the iodoacetamide derivative IASL (or 6-IASL), (*N*-[1-oxyl-2,2,6,6-tetramethyl-4-piperidinyl]iodoacetamide). At or below pH 7, both MSL and IASL react specifically with SH groups, and the varying reactivities of SH groups (due to varying surface accessibility, p*K*, etc.) often make it possible to label selectively a single SH group or class of SH groups. Amino groups can be labelled with isothiocyanate spin labels (NCSSL), including the pyrrolidinyl derivative shown.

spectroscopic properties that determine the spectral sensitivity to microsecond rotational motion ($T_1$, $T_2$, $T_\parallel$, $T_\perp$, $g_\parallel$, $g_\perp$) are similar for all nitroxides and are themselves insensitive to microsecond rotational motions. A number of nitroxide spin labels are available commercially, chiefly from Aldrich (Milwaukee, Wisconsin, USA). In principle, ST-EPR could be used to measure rotational motion of paramagnetic probes other than nitroxides, including some intrinsic ones (Hyde *et al.*, 1970), but these applications have been rare.

(b) *Labelling*

The procedures for designing probes and attaching them to membrane proteins, and the control experiments necessary to characterize the labelling, are similar to those for optical probes. However, significant practical differences arise from the fact that the spectroscopic portion of the spin probe is much smaller than that of most optical probes, especially triplet-state probes such as eosin (compare Figs. 13.4 and 13.8). Besides offering a smaller perturbation to the protein's structure, the smaller size of spin labels, coupled with the tendency of the nitroxide group to form hydrogen bonds with proteins, tends to result in a more complete immobilization of the probe within the protein's structure. This is particularly true of the SH-directed probe 6-MSL (Fig. 13.8). Thus spin labels tend to reflect more faithfully the large-scale motions of proteins, as discussed below in the case of the Ca-ATPase, although the time resolution of optical anisotropy data can sometimes be used to resolve probe motions from protein motions. In any case, it is important (in both EPR and optical experiments) to carry out control experiments on immobilized (e.g. cross-linked) proteins, to determine whether the probe motions reflect overall or internal protein motions (discussed below in Fig. 13.14). Site-specific labelling with spin labels is sometimes facilitated, relative to that of optical probes, by (a) spin labels' lower tendency to bind non-covalently to membranes, and (b) the sensitivity of the nitroxide group to chemical reduction and oxidation, which can sometimes be used to eliminate signals from probes non-rigidly bound to surface residues (Graceffa and Seidel, 1980; Swanson *et al.*, 1980). The latter sensitivity, however, sometimes makes spin labels unstable in metabolically active preparations, particularly in intact cells.

(c) *Sample concentration and size*

EPR experiments on aqueous samples are generally limited to the study of volumes less than 0.1 ml (see Table 13.2), because of the unacceptably high microwave loss (dissipation) at higher volumes due to the high dielectric constant of water. The result is a generally higher required concentration than in optical experiments. However, the most commonly performed steady-state ST-EPR experiment ($V_2'$) requires about the same amount of material (at least $10^{-10}$–$10^{-9}$ moles of probe) as does an optical absorption experiment. EPR techniques do not approach the sensitivity of optical emission techniques, with the possible exception of the $U_1'$ experiment performed with a loop-gap resonator (discussed below). As in optical spectroscopy, present pulsed EPR methods are significantly less sensitive than steady-state methods, although the improved quality of the data, resulting in less ambiguous interpretation, justifies the extra material and/or time required.

(d) *Turbidity*

EPR experiments are unaffected by optical turbidity, eliminating many of the troubling artifacts due to turbidity in optical spectroscopy and permitting experiments on some samples that would be virtually inaccessible to optical techniques, e.g. membrane pellets.

(e) *Oxygen removal*

Molecular oxygen can decrease the excited-state lifetime ($T_1$) in nitroxide EPR, but the effects are much less than in triplet-state optical spectroscopy, partly because intrinsic optical lifetimes are longer. Thus, ST-EPR signals are usually easy to obtain in oxygenated samples. But deoxygenation can increase the sensitivity and decrease the potential ambiguity caused by the fact that increases in $O_2$ collision rates can decrease saturation and hence mimic increased rotational motion. In addition to the techniques discussed above for $O_2$ removal (see 13.2.2.e), gas-permeable EPR cells are sometimes used (Plachy and Windrem, 1977; Kusumi *et al.*, 1982).

### 13.3.3 INSTRUMENTATION

(a) *Time-resolved ST-EPR*

Since time-resolved experiments provide, potentially, the most direct information about the rates and amplitudes of rotational motion, and since time-resolved EPR is most directly analogous to time-resolved optical anisotropy, this section will begin with a discussion of this developing EPR technology. However, this method is in its infancy, and applications to membrane protein motions have not yet been published, so this discussion will be brief in comparison with that in the section on steady-state methods.

*Saturation recovery.* Figure 13.9(b) shows a schematic diagram of a pulsed EPR spectrometer used to detect saturation transfer in the time domain (see Hyde, 1979; Kusumi *et al.*, 1982; Forrer *et al.*, 1980), and Fig. 13.10 shows saturation-recovery data from a spin-labelled membrane protein (P. Fajer, T. Squier and D. Thomas, unpublished). This spectrometer measures the transient EPR absorption following a saturating pulse of radiation, and the experiment is thus referred to as saturation recovery. The experiment is essentially the same as the transient absorption experiment described in

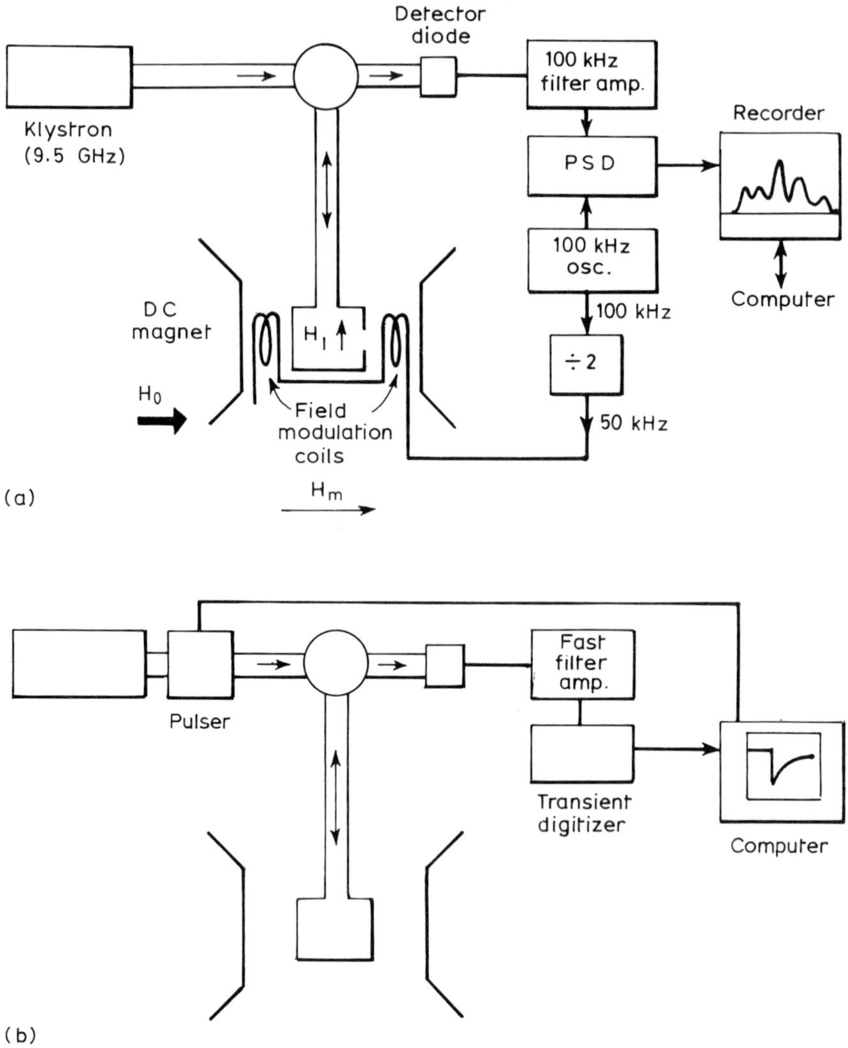


**Fig. 13.9** Schematic drawings of instruments used to perform saturation transfer EPR experiments. (a), Steady-state EPR, using 'modulation' or 'passage' techniques. The configuration shown is the one most often used in ST-EPR, namely, the $V'_2$ display, in which the second harmonic (100 kHz) response to the 50 kHz modulation is detected. Conventional first harmonic ($V_1$) EPR is performed with the same commercially available instrument (IBM Instruments/Bruker) by bypassing the ÷2 circuit. (b), Time-resolved saturation recovery. In this instrument, the same microwave source (klystron) is used for excitation as for detection, although a separate source for pulsed excitation can also be used. This simplified drawing omits many details, including the arrangement of microwave bridge components used to minimize artifacts due to free-induction decay.

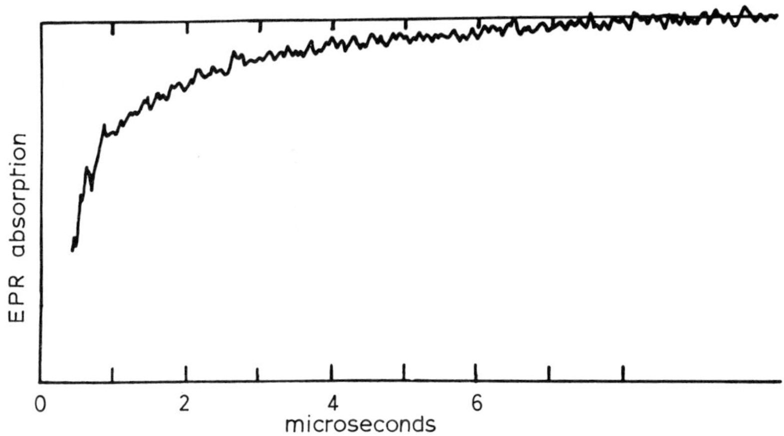


**Fig. 13.10** Saturation-recovery EPR data from a maleimide spin label (MSL) attached to the Ca-ATPase in sarcoplasmic reticulum membranes at 4° C. Data were obtained from the instrument shown in Fig. 13.9(b). The absorption ($V_0$), at a field value $H_{res}$ near the centre of the spectrum, is plotted against the time after a 200 ns saturating pulse at the same spectral position.

Fig. 13.2, and it is quite analogous to the measurement of $A_{\parallel}(t)$ in transient optical absorption. The radiation (microwave) source for excitation is a klystron. As in optical absorption, the intense (saturating) pulsed (usually submicrosecond) radiation used for excitation ('pump') and the low-intensity continuous radiation used for detection ('observe') can either be provided from the same source (as shown in Fig. 13.9b) or from separate sources. Microwaves are transmitted through a waveguide or a co-axial cable and are absorbed by the sample in a resonator (e.g. a resonant cavity or a loop-gap resonator). Absorption is detected by measuring a decrease in the microwave intensity reaching a detector diode.

The signal is amplified, digitized and averaged essentially as described above for optical experiments. Because signal/noise is intrinsically much lower in magnetic resonance than in optical spectroscopy, many more transients must be averaged in EPR (typically $10^6$) than in optical spectroscopy (typically $10^3$). To compensate for this, pulsed microwave sources and signal averagers used in EPR experiments operate at repetition rates as high as 100 kHz (Forrer *et al.*, 1980), at least 1000 times faster than in most pulsed optical spectrometers used for studying microsecond motions. Higher repetition rates are possible in EPR partly because of the availability of high-repetition-rate pulsed microwave sources (most intense pulsed lasers have maximum repetition rates much less than 1 kHz), but also because the excited-state lifetimes ($T_1$ in EPR) are much shorter (1–10 $\mu$s for nitroxide spin labels vs. 0.1–1 ms for eosin). Despite this difference in excited-state lifetime (and, hence, the rate at which data can be collected), the correlation times $\tau_r$ accessible to the two techniques are similar

(0.1 $\mu$s–1 ms), due to the higher orientational resolution of EPR, which results in a shorter $\tau_T$ (time required for a detectable rotation) for the same $\tau_r$ (see Equation 13.7). As indicated in Fig. 13.2, in the absence of rotational motion, the saturation recovers with an exponential time constant $T_1$; while in the presence of motion, the recovery will have an additional component $(T_1^{-1}+\tau_T^{-1})^{-1}$. Thus the motion will be detectable only if $\tau_T(\cong 0.007\tau_r)$ is $< \cong T_1 \cong 7$ $\mu$s, i.e. as long as $\tau_r$ is $< \cong 1$ ms. To analyse the data in terms of the effects of rotational motion, it is necessary to measure $T_1$ independently by either (a) performing the same experiment on an immobilized sample or by (b) applying a very long and intense excitation pulse, so that the entire spin system (all orientations) is saturated and rotational motion can transfer no saturation (Smigel *et al.*, 1974a; Fajer *et al.*, 1986). A detailed theory remains to be developed, but it seems clear that both rate and amplitude information can be extracted from the decay kinetics, much as in optical spectroscopy.

*Other developments in pulsed EPR.* In addition to saturation-recovery, spin-echo techniques can also be used to detect the time-resolved transfer of spins due to rotational motion (Brown, 1979; Millhauser and Freed, 1984). Since the radiation fields are turned off during the detection of the echo, and one observes a decay of the echo rather than a recovery of absorption, this technique is analogous to emission techniques (e.g. phosphorescence) in optical spectroscopy. (However, strictly speaking, spontaneous emission is not detectable in magnetic resonance.) Due to the superior orientational resolution of EPR, it is clear that more detailed (less ambiguous) information should be obtainable than in optical spectroscopy. To maximize the information obtained, it is essential to vary the spectral position (orientation), and even to vary the relative position of excitation and detection (usually referred to as electron–electron double resonance, ELDOR). The ELDOR experiment is somewhat analogous to varying the relative polarizer orientations in optical spectroscopy, but the resolution of photoselection is much greater in EPR. The pulsed variation in spectral position can be done either by (a) using different microwave frequencies for the excitation and detection sources (Hyde *et al.*, 1984) or by (b) quickly changing the DC magnetic field after the excitation pulse (Millhauser and Freed, 1984). Maximal information can be obtained if the experiment is performed on an ordered sample, such as a stack of membranes oriented normal to the magnetic field (Kar *et al.*, 1984).

A number of technical problems remain to be solved in order for pulsed EPR experiments to find wide application in biological studies using nitroxide spin labels. These include the difficult problems in rejecting artifacts due to intense excitation pulses. One of the most commonly used approaches to this problem, the use of 'bimodal' resonant cavities to isolate excitation from detection, is very difficult to apply with aqueous samples (Huisjen and Hyde, 1974; Mailer *et al.*, 1980). As a result, most pulsed EPR experiments are performed on nonaqueous or frozen samples. Despite these and other problems, a commercially manufactured instrument usable for pulsed EPR experiments on spin-labelled systems should be available within a few years. Meanwhile, the majority of ST-EPR experiments on membrane proteins will continue to be performed using steady-state methods, as described in the following section.

(b) *Steady-state ST-EPR*

In principle, in both optical and EPR spectroscopy, the steady-state (CW) absorption is sensitive to the competition between relaxation and rotational diffusion (see Fig. 13.2). For example, if the microwave radiation is intense enough to cause saturation, the conventional EPR spectrum is less saturated (more intense) when $\tau_T < \cong T_1$. In contrast to optical spectroscopy, this saturation transfer affects not only the intensity of the spectrum but also its *lineshape* (relative intensities within the spectrum, which are much more precisely measurable than absolute overall intensity in EPR), as explained by the following argument (Thomas *et al.*, 1976): The sensitivity to rotational motion varies with the spectral position. That is, different positions ($H_{res}$ values) in the spectrum have different values of $dH_{res}/d\theta_{zH}$ (see Equation 13.14), so different regions have different values of $\tau_T$ (inversely proportional to $dH_{res}/d\theta_{zH}$), the time in which saturation is transferred, for the same value of $\tau_r$. The sensitivity of the spectrum to rotational motion is proportional to $dH_{res}/d\theta_{zH}$, which has its minimum values at $\theta_{zH}=0°$ or $90°$ (*turning points*, designated L, C, H below) and maximum values at intermediate spectral positions (designated L″, C′, H″ below).

The conventional ($V_1$) EPR spectrum, which shows no lineshape sensitivity to μs rotation in the absence of saturation, shows significant sensitivity in both intensity and lineshape when the microwave power is sufficient to cause saturation. However, these changes are too small to provide much precision, because (a) only partial saturation is easily achieved, so the intensity change due to motion is usually only a small fractional increase, and (b) the $V_1$ spectrum has most of its intensity near the turning points (Fig. 13.7, bottom), minimizing its sensitivity to both intensity and lineshape changes.

*Modulation spectroscopy.* In order to obtain more sensitivity to saturation transfer, detection methods have been developed that exploit the principles of *modulation spectroscopy* (Thomas *et al.*, 1976). The principles and technology involved are analogous to those involved in modulation techniques in optical spectroscopy (Weber, 1977), but the added orientational resolution of EPR results in increased motional effects not only on signal *intensity*, but also on spectral *lineshape*. Even in the conventional EPR experiment, the detected signals are not strictly continuous wave (CW). In addition to the large DC field $H_0$ that determines the orientation of the excited spins, and the absorbed microwave radiation field $H_1$ (perpendicular to $H_0$, with magnitude $2H_1 \cos wt$, with $w/2\pi$ usually $= 9.5$ GHz), there is a *modulation field* ($H_m = \hat{z}0.5\, H_m \cos w_m t$, with $w_m/2\pi$ usually $= 100$ kHz), parallel to $H_0$, that modulates the spin energy levels and therefore causes spins to go in and out of resonance at $w_m$. The orientations of these fields are illustrated in Fig. 13.9(a). This modulation is a standard feature in virtually all steady-state EPR spectrometers, and it is used in conjunction with a phase-sensitive detector primarily for purposes of high-frequency ($> w_m$) noise rejection.

If the modulation frequency $w_m$ is much less than $T_1^{-1}$, the modulation amplitude $H_m \ll 1/\gamma T_2$ (about 5 gauss), and the microwave field intensity is low enough to avoid saturation ($\gamma^2 H_1{}^2 T_1 T_2 \ll 1$), the EPR absorption signal follows the modulation precisely (in-phase) and the result is simply the derivative $V_1 = dV_0/dH_0$ (see legend to Fig. 13.7). However, if (a) $H_1$ increases enough to cause saturation ($\gamma^2 H_1{}^2 T_1 T_2 > \cong 1$), and (b) $w_m$ increases to become $> \cong T_1^{-1}$ (typical values: $w_m = 2\pi \times 10^5\ s^{-1}$, $T_1 = 1.5 \times 10^5\ s^{-1}$), *the signal lags behind the modulation* and a signal can be observed even when the reference phase on the phase-sensitive detector is set 'in quadrature', i.e. *90°-out-of-phase* with respect to the modulation. This method of using saturation and rapid field modulation to produce out-of-phase EPR signals has sometimes been called 'rapid passage'. The advantage of this out-of-phase signal is that it is present *only* in the presence of saturation and is thus selectively sensitive to saturation transfer which decreases its intensity. That is, saturation transfer causes a larger fractional change in an out-of-phase signal than in an in-phase signal. The conventional detection mode is designated $V_1$ (V for absorption, 1 for first harmonic), and the out-of-phase signal is designated $V_1'$. The intensity of $V_1'$ is more sensitive to saturation transfer than that of $V_1$, but its lineshape is not (Thomas *et al.*, 1976).

*The $V_2'$ spectrum.* A signal that has been found to have good sensitivity to saturation transfer, in terms of *both intensity and lineshape*, is $V_2'$ (V for absorption, 2 for second harmonic, and ′ for out-of-phase) (Hyde and Thomas, 1973; Thomas *et al.*, 1976). This detection method, performed essentially as described by Thomas *et al.* (1976b), remains the technique of choice for biological ST-EPR, and most of the rest of the present discussion of ST-EPR techniques will focus on it. For more detailed discussions of the methodology, including refinements developed since 1976, see Hyde and Thomas (1980) and Squier and Thomas (1986). The instrumental configuration used to detect $V_2'$ is shown in Fig. 13.9(a). The modulation frequency is typically set at 50 kHz and the phase-sensitive detector at 100 kHz. A significant second harmonic response to the modulation is obtained whenever the modulation amplitude $H_m$ is $> \cong$ the intrinsic linewidth $1/\gamma T_2$. The advantage of the $V_2'$ signal (as compared with $V_1'$) is that, in the presence of saturation and in the absence of microsecond rotational motion, its lineshape has significant intensity *throughout* the spectrum (Fig. 13.11, top right), including spectral

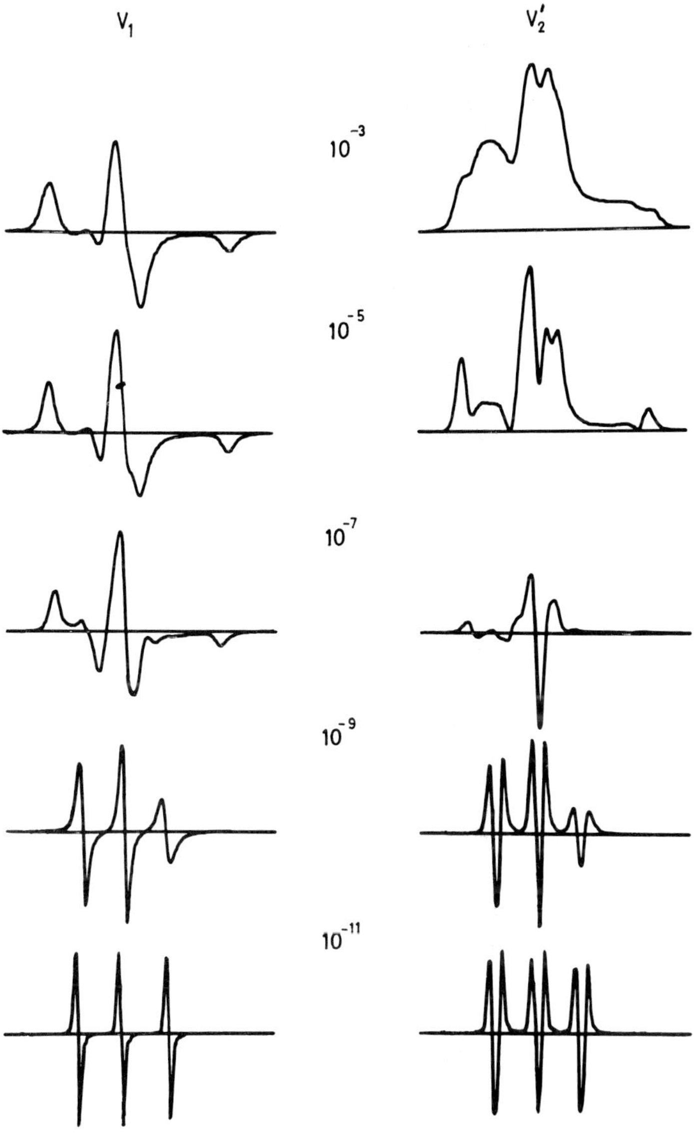


**Fig. 13.11** Reference EPR spectra obtained from experimental model systems, corresponding to isotropic Brownian rotational diffusion at known rotational correlation times (Squier and Thomas, 1986). Each row shows a conventional ($V_1$) and saturation transfer ($V'_2$) EPR spectrum from the same sample. The bottom two rows of spectra were obtained from solutions of small spin labels, and the rest were obtained from spin-labelled haemoglobin. The viscosity was varied by varying the glycerol concentration and temperature, and the rotational correlation time $\tau_r$ for isotropic Brownian rotational diffusion was calculated from Equation 13.16. The baseline is 100 gauss wide. Phase-sensitive detection was at 100 kHz. Field amplitudes for $V_1$: $H_1=0.032$ gauss, $H_m=2$ gauss; for $V'_2$: $H_1=0.25$ gauss, $H_m=5$ gauss.

positions between the turning points that have maximum sensitivity to rotation (maximum values of $dH_{res}/d\theta_{zH}$, minimum values of $\tau_T$). These most rotation-sensitive regions of the spectrum are clearly revealed in the model system spectra in Figs. 13.11 and 13.12: These are the regions whose intensities decrease most dramatically when $\tau_r$ decreases (decreasing saturation). Spectra are typically characterized by measuring the line-height ratios of two positions in the spectrum, one having maximal sensitivity (e.g., L″, C′, or

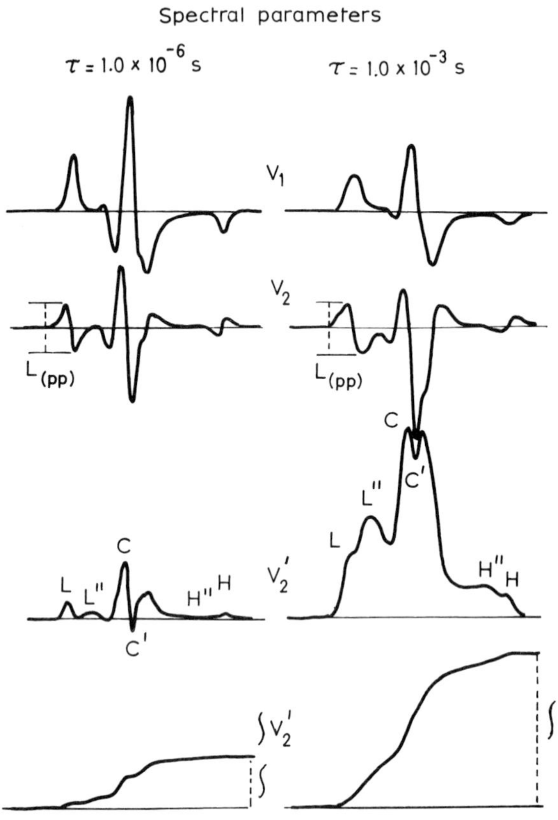


**Fig. 13.12** Parameters used to quantify ST-EPR ($V_2'$) spectra, illustrated with conventional ($V_1$) and ST-EPR ($V_2'$) reference spectra obtained from MSL-labelled haemoglobin in aqueous glycerol solutions. The spectra on the left are from a sample in 50% glycerol at −12° C, corresponding to a $\tau_r$ value of $10^{-6}$ s. The spectra on the right are from a sample in 90% glycerol at −32°, corresponding to a $\tau_r$ value of $10^{-3}$ s. Power was adjusted so that $H_1$ was 0.25 G. $V_2'$ (out-of-phase) spectra are normalized to correspond to the same number of spins (by dividing by the double integral of $V_1$) so that the effects of motion on both shape and absolute intensity are illustrated. $V_1$ and $V_2$ (in-phase) spectra are shown at a gain ten times less than $V_2'$. The most commonly used line-shape parameter is L″($V_2'$)/L($V_2'$) (usually written as L″/L); C′/C and H″/H are also used (Thomas *et al.*, 1976). Parameters that are sensitive both to lineshape and absolute intensity changes are L″($V_2'$)/$L_{pp}$($V_2$) and the normalized integral of $V_2'$ (bottom) (Squier and Thomas, 1986).

H″ in Fig. 13.12) and the other having minimal sensitivity (L, C, or H), as discussed quantitatively in Section 13.3.4.

$V_2'$ experiments are performed using the same commercial EPR spectrometers used in conventional ($V_1$) experiments (Fig. 13.9a), and the only difference in instrumental configuration is the '÷2' circuit that permits the detection of the second harmonic. However, the same experimental conditions (saturating microwave power, high modulation amplitude, out-of-phase detection) that make $V_2'$ more sensitive to slow rotational motion also make it more sensitive to changes in other physical and instrumental variables, such as intrinsic spin relaxation times, $H_m$, $H_1$, and the phase setting on the phase-sensitive detector. The latter two instrumental variables, in particular, must be set with considerably more care than in conventional EPR.

*$H_1$ determination.* The extent of saturation, which determines the intensity and shape of the $V_2'$ spectrum, depends primarily on the intensity of microwave radiation that depletes the ground state, and only secondarily on saturation transfer that reduces the effects of this radiation by transferring saturation to other (not irradiated) spectral positions. In the absence of saturation transfer, steady-state saturation (defined as the fractional depletion of the ground state) is $1-(1+SAT)^{-1}$, where $SAT=\gamma^2 H_1{}^2 T_1 T_2$. $H_1$ is the microwave field amplitude at the sample and is proportional to $P^{1/2}$, where $P$ is the power incident upon the sample. Thus, in order to ensure that the extent of saturation depends only on saturation transfer, it is desirable to know $SAT$ and keep it constant for spectra that are to be compared. It is often safe to assume that $T_1$ and $T_2$ are nearly constant (although these should be directly measured with time-resolved EPR if possible), leaving $H_1$ as the most important variable to control in the experiment. The most reliable means of measuring $H_1$ is to measure $SAT$ for a standard sample, for which $T_1$ and $T_2$ are precisely known. This procedure is described elsewhere in detail (Thomas *et al.*, 1976; Fajer and Marsh, 1982; Squier and Thomas, 1986). The sample of choice is a deoxygenated solution of peroxylamine disulphonate ('PADS' or 'Fremy's salt'), 0.9 mM in 50 mM $K_2CO_3$ at 20°C. In this sample, $T_1$ and $T_2$ are equal and are easily determined from the linewidth of the $V_1$ spectrum in the absence of saturation, resulting in a straightforward calculation of $H_1(1/2)$, the value of $H_1$ at which $SAT=1$ (fractional saturation is 1/2). $H_1(1/2)$ is about 0.1 gauss for the standard PADS sample. $V_1$ spectral intensities are proportional to $H_1 SAT$, and $H_1$ is proportional to $P^{1/2}$, so $V_1/P^{1/2}$ is plotted against $P^{1/2}$. When the plotted value is 1/2, $SAT=1$ and $H_1=H_1(1/2)$. Since $H_1$ is proportional to $P^{1/2}$, $H_1$ is thus known for all values of $P$. A further correction must be applied if the $Q$ of the microwave cavity changes due to changes in the sample geometry or dielectric constant of the sample (Fajer and Marsh, 1982; Squier and Thomas, 1986). The standard value of $H_1$ used by most workers is 0.25 gauss, this can correspond to a power setting anywhere from 1 to 100 mW,

depending on the resonator and sample characteristics. Unless the sample is very small ($\leq$5 mm in length), the value of $H_1$ may be heterogeneous over the sample, so $H_1$ is an average value (Fajer and Marsh, 1982).

*Modulation phase setting.* In $V_2'$ experiments, the reference phase on the 100 kHz phase-sensitive detector is set 90° away from the conventional in-phase setting. The signal detected under these conditions (lagging in phase because of saturation) is typically much less intense than the in-phase signal that is to be rejected, so a much more precise setting is necessary than in conventional EPR. For example, if the in-phase signal ($V_2$) is 10 times as intense as the out-of-phase signal (a typical value), and no more than 5% error in the $V_2'$ signal intensity can be tolerated (a typical requirement), the phase must be set with an accuracy of 0.3 degrees. This is possible if the following procedure is followed (Thomas *et al.*, 1976; Hyde and Thomas, 1980; Squier and Thomas, 1986). The most intense (central) feature of the second harmonic signal is recorded at low power, typically corresponding to an $H_1$ value of 0.03 gauss or less, to avoid saturation (and the accompanying phase lag). The phase is varied until an approximate null is achieved. The power is then turned up to correspond to $H_1 = 0.25$ gauss, and the gain is adjusted to give the desired $V_2'$ signal height. For the final (fine) phase setting, the power is reduced again to a low (non-saturating) level and the signal intensity (peak-to-trough) is recorded at two phase settings, approximately 2° on either side of the approximate null phase value determined above. The precise value is determined and set by linear interpolation between these two settings to determine the phase corresponding to a null signal. The power is then turned up to the desired level and the $V_2'$ spectrum is recorded. It is best to perform the final phase adjustment just before each $V_2'$ measurement, with all spectrometer settings (especially gain) except power set at the same values to be used in the $V_2'$ experiment, since changes in gain and other settings can cause phase shifts (Thomas *et al.*, 1976). In order to avoid phase drift during the recording of the spectrum (which can require acquisition times ranging from a few minutes to a few hours), it is advisable to warm up the instrument for several hours before starting experiments and to keep the ambient temperature constant during the experiment to within 1°.

*Proposed modifications in $V_2'$ instrumentation.* Although the above instrumentation and methodology for $V_2'$ experiments continues to be used in most membrane protein studies, a number of possible improvements have been proposed. Several of these proposals centre around the use of digital detection and signal processing to reduce (or eliminate) the time and effort required for the phase-determination step (Hemminga and de Jager, 1981; Watanabe *et al.*, 1980, 1982; Sasaki *et al.*, 1980; Evans, 1981; Vistnes, 1983). The use of microwave frequencies other than 9.5 GHz (Johnson and Hyde, 1981;

Johnson *et al.*, 1982a,b) and modulation frequencies other than 50 kHz (Hyde and Thomas, 1973) may offer additional information about the details of rotational motion. $^2$H and/or $^{15}$N substitution in nitroxide spin labels offers increased spectral resolution (Beth *et al.*, 1981a,b).

*Alternatives to $V_2'$.* Just as in pulsed EPR, more detailed information can be obtained in steady-state EPR if the relative positions (molecular orientations) of excitation and detection are different, corresponding to ELDOR (Smigel *et al.*, 1974b). However, the instrumental complexity and low absolute sensitivity of this technique have severely limited its application to biological samples. The first ST-EPR experiments on nitroxides were performed using the *dispersion* mode (designated U), which is analogous to refraction in optical spectroscopy, and which is available as an option on many commercial EPR spectrometers. The dispersion display found most useful was $U_1'$ (dispersion, first harmonic, out-of-phase), and it was found to have good sensitivity to microsecond rotational motion (Hyde and Dalton, 1972). This experiment offers many potential advantages over $V_2'$ (e.g. higher signal intensities, simpler line-shapes, simpler theoretical analysis, useful at a wide range of modulation frequencies), but noise levels were found to be unacceptable for biological experiments in standard instruments, leading to the development of $V_2'$ (Hyde and Thomas, 1973). More recent studies using alternative cavity structures have reduced these levels significantly (Huisjen and Hyde, 1974; Mailer *et al.*, 1980; Sehr *et al.*, 1983; Froncisz and Hyde, 1982). The most promising of these devices is the 'loop-gap resonator' (Froncisz and Hyde, 1982), which yields $U_1'$ spectra having better signal-to-noise ratios than can be achieved with $V_2'$ spectra (Thomas *et al.*, 1983). These devices could make $U_1'$ the ST-EPR experiment of choice in the near future (Thomas *et al.*, 1983), and they may also prove crucial in improving the sensitivity of pulsed EPR and ELDOR experiments.

### 13.3.4 DATA ANALYSIS

This section will focus exclusively on the analysis of $V_2'$ spectra of $^{14}$N spin labels at 9.5 GHz (X-band), since these conditions are used in most membrane applications. The relationship between the data and rotational dynamics is less direct than in a time-resolved experiment, so that spectra are analysed by comparing them with reference spectra obtained either from experiments on model systems or from theoretical simulations using a computer.

(a) *Reference spectra from experimental model systems*

The most reliable empirical model systems are those corresponding to isotropic rotational diffusion. The most commonly used model system for analysing ST-EPR spectra is maleimide-spin-labelled haemoglobin (MSL-Hb) in aqueous glycerol solutions (Hyde and Thomas, 1973; Thomas *et al.*, 1976), although maleimide-spin-labelled bovine serum albumin has also been used (Kusumi *et al.*, 1980). The piperidinyl maleimide spin label (Fig. 13.8) binds quite rigidly to haemoglobin (and many other proteins), and haemoglobin has been shown to behave hydrodynamically like a rigid sphere, so the correlation time for the isotropic rotational diffusion of the attached probe can be calculated directly from the Stokes–Einstein–Debye equation:

$$\tau_r = 1/6D = V\hat{\eta}/kT, \tag{13.15}$$

where $V$ is a haemoglobin's molecular (hydrated) volume ($cm^3$) and $\eta$ is the viscosity (poise). For haemoglobin, this results in the relationship (McCalley *et al.*, 1972; Thomas *et al.*, 1976)

$$\tau_r = (7.6 \times 10^{-4}) \times \eta/T. \tag{13.16}$$

The correlation time is varied by varying the temperature and glycerol concentration. The viscosity can be measured or looked up in a table (Squier and Thomas, 1986). Examples of spectra (both $V_1$ and $V_2'$) obtained from MSL-Hb are shown in Figs. 13.11 and 13.12.

*Spectral parameters.* The comparison of $V_2'$ spectra is made quantitative by defining parameters, measured from the spectrum, that are (a) particularly sensitive to rotational motion and (b) conveniently, accurately and reproducibly measurable. These criteria are best met by parameters that characterize changes in the *lineshape*. The most commonly used lineshape parameters in $V_2'$ are the three line-height ratios L″/L, C′/C and H″/H from the low-, centre- and high-field regions of the spectrum, introduced by Thomas *et al.* (1976) and illustrated in Fig. 13.12. The amplitudes L, C and H are at spectral positions where the rotational sensitivity ($dH_{res}/d\theta_{zH}$) is low ('turning points', where $\theta_{zH} \cong 0°$ or $90°$); whereas L″, C′ and H″ are at intermediate positions ($\theta_{zH} \cong 45°$) where rotational sensitivity is high. To minimize the ambiguity of experimental measurement, each of these values is measured at an extremum, not at a particular value of $H_{res}$, except for the H″ value (usually not a well-defined peak), which is usually defined to be approximately 10 gauss to the left of H (Squier and Thomas, 1986). Fig. 13.13(a) shows a plot of the most commonly used parameter, L″/L, vs $\tau_r$ (Squier and Thomas, 1986).

The use of reference spectra in analysing membrane protein data is illustrated in Fig. 13.14 (Thomas and Hidalgo, 1978). These spectra were obtained from a maleimide spin label (6-MSL, see Fig. 13.8) covalently and specifically attached to SH groups on the Ca-ATPase of sarcoplasmic reticulum membranes, without inhibiting enzymic activity (Thomas *et al.*, 1982). The conventional ($V_1$) spectrum of a suspension of vesicles (Fig. 13.14A, left), is characteristic of 'strongly immobilized' spin labels, implying the absence of large-amplitude ns rotational motions. Thus the $V_2'$ spectra can be analysed in

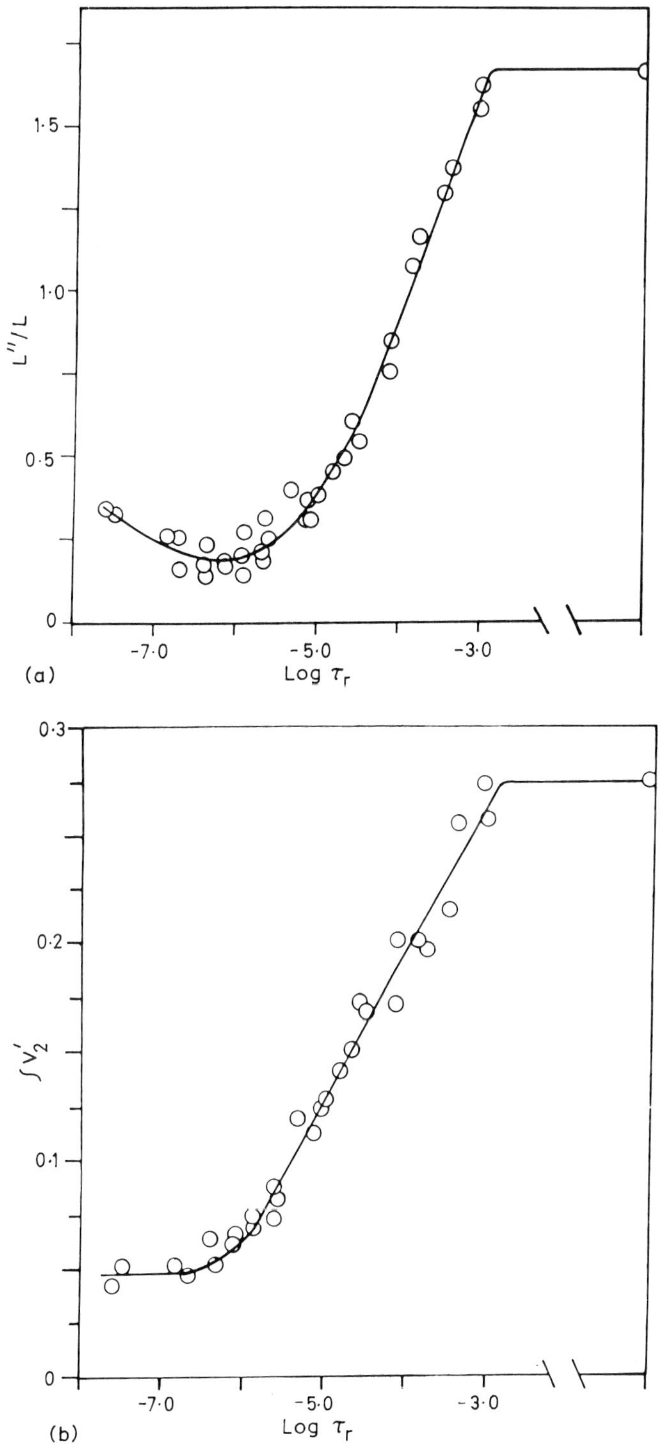


**Fig. 13.13** Experimental parameter plots used to measure rotational correlation times from ST-EPR ($V_2'$) spectra (Squier and Thomas, 1986). Parameters, defined in Fig. 13.11, were measured from $V_2'$ spectra of MSL-haemoglobin in aqueous glycerol solutions (see Fig. 13.11). Correlation times $\tau_r$ for isotropic Brownian rotational diffusion were calculated according to Equation 13.16.

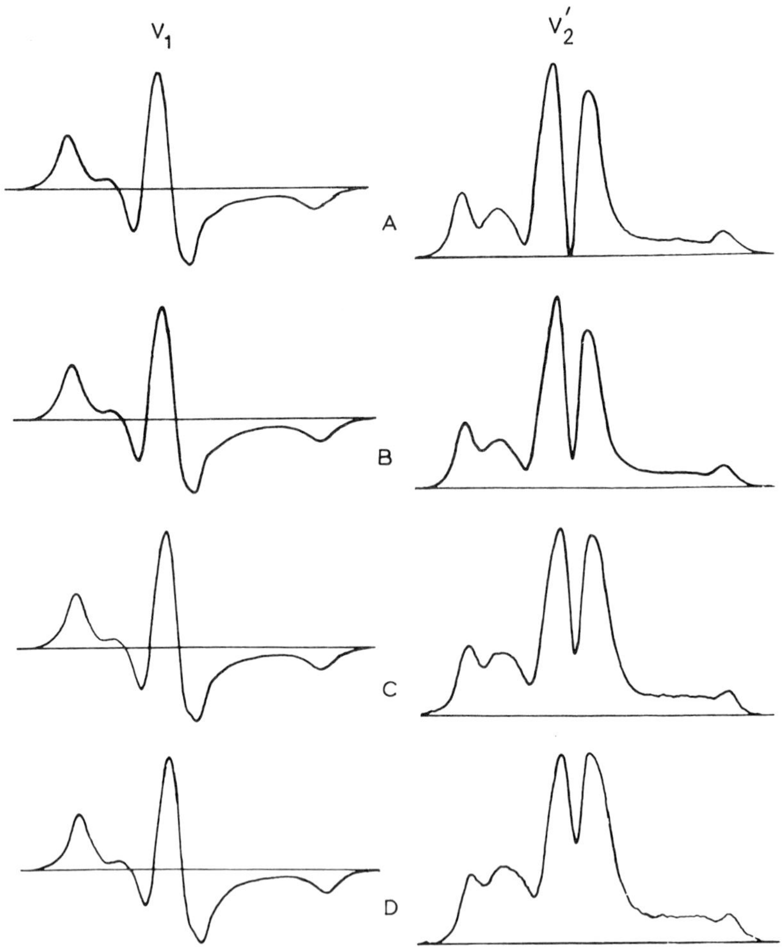


**Fig. 13.14** Conventional ($V_1$) and ST-EPR ($V_2'$) spectra of MSL attached to the Ca-ATPase of SR at 4°C. A, A suspension of membrane vesicles (L″/L = 0.75, $\tau_r = 60\ \mu s$). B, A membrane pellet (L″/L = 0.75). C, Vesicles treated with 80 mM glutaraldehyde for 10 min (L″/L = 0.91). Gels showed that the proteins were partially cross-linked, with about half of the Ca-ATPase remaining monomeric. D, SR vesicles treated with 80 mM glutaraldehyde for 12 h (L″/L = 1.1). Gels showed that no monomers or small oligomers remained (from Thomas and Hidalgo, 1978).

terms of microsecond motions. The $V_2'$ spectrum of this vesicle suspension (Fig. 13.14A, right) has an L″/L value of 0.75, yielding an effective correlation time of 60 $\mu s$ (from Fig. 13.13a). The remaining spectra in Fig. 13.14 are controls designed to further characterize the nature of the motion. The L″/L value is not changed by pellet formation (Fig. 13.14B, right), indicating that the observed motion is not that of the membrane vesicles themselves.

Protein–protein cross-linking with glutaraldehyde does increase L″/L (Fig. 13.14c,d), indicating substantially decreased probe mobility, presumably due to decreased protein mobility. A similar increase in L″/L was observed upon immobilizing the protein in gel-phase lipids (Hidalgo *et al.*, 1978) or reducing the lipid/protein ratio (Thomas *et al.*, 1982). The most likely interpretation is that the spectra are reflecting large-scale protein rotations with respect to the membrane. The fact that spectra from the cross-linked samples do not approach the L″/L value corresponding to a completely rigid system (see Fig. 13.13a) suggests that there may be some motion within the protein molecules. The correlation time obtained in this kind of analysis is termed 'effective' because it is unlikely that these membrane proteins are undergoing the same kind of isotropic motions as are the haemoglobin molecules used to obtain the reference spectra. The fact that the effective correlation time in this system depends significantly on the spectral parameter used (Squier and Thomas, 1986), i.e. that there is no isotropic motion reference spectrum with the same lineshape as the membrane spectrum, indicates directly that the motion is anisotropic and should be analysed by comparison with anisotropic motion reference spectra (discussed in Section 13.3.4.b). However, similar correlation times were later determined in time-resolved triplet anisotropy experiments on this system (Bürkli and Cherry, 1981; Spiers *et al.*, 1983), and similar agreements have been obtained in most other systems studied by both optical and ST-EPR techniques (Thomas, 1985), suggesting that this method of $V_2'$ analysis is useful for determining the time range of the motion.

Although line-height ratios such as L″/L remain the most convenient and reliable means of characterizing $V_2'$ spectra, a number of alternative parameters have been explored, some of which provide complementary information (Squier and Thomas, 1986). The most important of these parameters take into account not only changes in the lineshape but also changes in the *absolute intensity*. All points in the $V_2'$ spectrum decrease in absolute intensity as rotational motion increases, while $V_2$ (in-phase) changes very little with motion (Fig. 13.11). Thus the ratio of line-heights in $V_2'$ is less sensitive to motion than the ratio of a line-height in $V_2'$ to a line-height in $V_2$ (illustrated in Fig. 13.11). This line-height ratio offers the additional advantages of higher signal/noise and lower sensitivity to errors in phase setting (Squier and Thomas, 1986). Another parameter that is sensitive to changes in both lineshape and intensity is the integral of the $V_2'$ spectrum, appropriately normalized (Evans, 1981). A correlation time of 10 $\mu$s or greater results in a spectrum that is entirely above the baseline (has a large integral), while a correlation of 100 ns results in a spectrum that is much reduced in absolute intensity and has nearly equal areas above and below the baseline, resulting in a much smaller integral (illustrated in Figs. 13.12 (bottom) and 13.13). Reduction of the spectrum to a single number (the integral) obscures spectrally resolved lineshape information that is sensitive to the anisotropy or heterogeneity of the motion (discussed below). In this sense, the integral is like the steady-state polarization anisotropy in optical spectroscopy. However, the integral offers the advantage of suppressing contributions from weakly immobilized spin labels ($\tau_r < 10^{-7}$ s), thus permitting the selective study of strongly immobilized labels ($\tau_r > 10^{-7}$ s) (Evans *et al.*, 1981). This suppression principle has been exploited in the study of lipid spin labels (Horvath and Marsh, 1983), but its validity is limited to the suppression of signals in a narrow range of correlation times (Squier and Thomas, 1986). Although $V_2'$ parameters that reflect changes in both lineshape and absolute intensity are more sensitive to motion (show a larger fractional change for the same change in $\tau_r$) than are lineshape parameters alone, intensity parameters are also more sensitive to changes in other physical parameters (e.g. $T_1$, $H_1$). Thus intensity-sensitive parameters are reliable mainly when comparing spectra from similar samples under identical instrumental conditions.

(b) *Theoretical simulation of reference spectra*

Considerable insight into the principles of ST-EPR can be gained by making semiquantitative physical arguments and performing experiments on well-defined model systems, as described above. However, the quantitative analysis of spectral dependence on specific physical properties (e.g. rotational correlation times, relaxation times, motional models) is greatly facilitated by performing theoretical simulations using a digital computer. These simulations involve the numerical solution of the differential equations that describe the coupling of rotational diffusion to the other processes (excitation, relaxation and modulation) that determine the EPR spectrum (Thomas and McConnell, 1974; Thomas *et al.*, 1976). The ultimate purpose of reference spectra, whether obtained by experiments on model systems or by theoretical simulation, is to compare them quantitatively with experimental unknowns and to extract motional information. The experimental and theoretical reference spectra are complementary, because they are subject to different kinds of uncertainty. For example, experimental reference spectra can be performed on the same instrument (and usually the same nitroxide spin label) as are the unknowns, whereas computer simulations are subject to errors in (a) determination of instrumental and physical parameters that are required as input values, and (b) inaccurate approximations in the theory. On the other hand, simulations can be used to study the dependence of spectra on a specific physical parameter (e.g., $T_1$, $\tau_r$, or the anisotropy of motion) that is difficult to vary independently and unambiguously in an experimental model system.

*Isotropic rotational motion.* In the case of isotropic Brownian diffusion, the model systems (especially MSL-Hb) are sufficiently well behaved that the experimental $V_2'$ reference spectra are usually used directly to analyse experimental unknowns and to test the validity of the theoretical formulations (approximations) used in the computer simulations (Thomas *et al.*, 1976; Hyde and Thomas, 1980). For example, reasonably good agreement has been obtained between experimental and simulated plots of the ratio parameters L″/L and H″/H vs $\tau_r$ (Thomas *et al.*, 1976). The accurate simulations of other features in the spectra (e.g. line-height ratios in the centre of the spectrum, absolute spectral intensities) require more detailed and time-consuming simulations, and are more sensitive to errors in instrumental settings or $T_1$ values, supporting the use of the ratio parameters for unambiguous measurements of rotational motion (Thomas *et al.*, 1976; Squier and Thomas, 1986). The availability of more powerful computers, along with more accurate values for relaxation times and other physical parameters, promises to make possible in the near future the accurate simulation of all spectral features (including spectral intensity) of the isotropically tumbling reference samples, without the need for the variation of undetermined physical parameters to obtain good fits (C. Polnaszek and D. Thomas, unpublished). Besides the increased reliability that this will bring to the study of isotropic motion, this will establish that the most difficult and questionable aspects of the simulations – the intrinsic EPR relaxation processes – are being treated correctly, and that this theory can then be extended in a straightforward way to the study of anisotropic motion, where simulations are more crucial.

*Anisotropic rotational motion.* Recent work has focused on extending the theory beyond the model of isotropic Brownian motion (Thomas and McConnell, 1974; Thomas *et al.*, 1976) to consider the types of anisotropic motions more likely to occur in biological systems, particularly in membranes (Fig. 13.1). Simulations are even more important for anisotropic motion than for isotropic, since it is much more difficult to find experimental model systems for generating reliable reference spectra (Gaffney, 1979). In general, as in the case of optical spectroscopy and conventional EPR, data depend not only on the rates (diffusion coefficients), but also on the amplitudes of motions, and on the orientation of the nitroxide relative to the axes of diffusion. In the absence of information about the orientation of the probe and the type of anisotropic motion occurring, or in the absence of reference spectra for the expected type of motion, reference spectra can only be used to determine an *effective correlation time.* This is usually done using parameter plots from isotropic-motion reference spectra. If, for example, the principal nitroxide axis is approximately parallel to the axis of diffusion, the effect on the spectrum will be much less (the effective $\tau_r$ value, using isotropic motion reference spectra, will be greater) than if it is nearly perpendicular. In addition, the effective correlation times measured from different regions of the spectrum will usually be different (i.e. the lineshape does not match that of any spectrum corresponding to isotropic motion), providing direct evidence for anisotropic motion.

*Uniaxial rotation.* Robinson and Dalton (1980, 1981) have verified this principle by simulating ST-EPR spectra (mainly $U_1'$ spectra of $^{15}N$ spin labels) corresponding to the unrestricted motion of rigid ellipsoids of revolution, rotating in an isotropic medium. The simulation of the more experimentally relevant $V_2'$ spectra (of $^{14}N$ spin labels) corresponding to *uniaxial rotation* (Fig. 13.1a) has recently been achieved (Fig. 13.15). These spectra show clear differences in lineshape from those corresponding to isotropic motion. In addition, the spectra are quite sensitive to the orientation of the spin label relative to the axis of diffusion (membrane normal). Thus, despite the lack of time resolution, the spectrally resolved orientational resolution in EPR provides informational detail that is qualitatively similar to that obtained from time-resolved optical anisotropy. For example, uniaxial rotation tends to affect the centre of the spectrum more than it affects the wings of the spectrum, particularly if rotation is about the nitroxide's $\hat{z}$-axis (Marsh, 1980). Thus, when comparing an axial motion spectrum with a series of isotropic motion reference spectra, the effective correlation time is shortest if determined from the centre of the spectrum. A preliminary analysis of Ca-ATPase spectra (e.g. Fig. 13.14A) shows that the lineshape agrees better with reference spectra corresponding to uniaxial motion than with those corresponding to isotropic motion, resulting in a slightly smaller estimate for the rotational correlation time (C. Polnaszek and D. Thomas, to be published).

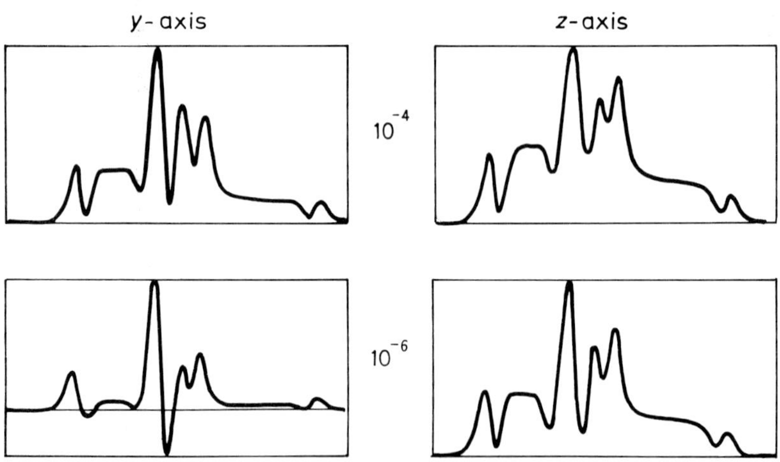


**Fig. 13.15** Theoretical simulations of ST-EPR spectra corresponding to anisotropic (uniaxial) rotational diffusion, like that depicted in Fig. 13.1(a) (C. Polnaszek and D. Thomas, unpublished), with either the nitroxide $z$-axis or $y$-axis fixed parallel to the membrane normal $\hat{\mathbf{n}}$. These simulations take into account not only the major dependence of the line position ($H_{res}$) on the angle $\theta_{zH}$ between the applied field $H$ and the nitroxide's principal ($z$) axis (indicated in Equation 13.14), but also the slight dependence on the angle $\phi_{zH}$ that describes rotation about the $z$-axis (Marsh, 1981; not included in Equation 13.14). Spectra on the left correspond to rotation about the $y$-axis (changing both $\theta_{zH}$ and $\phi_{zH}$, so these are very similar to isotropic motion spectra), and spectra on the right correspond to rotation about the $z$-axis (changing only $\phi_{zH}$).

*Restricted rotation.* Lindahl and Thomas (1982) Lindahl *et al.* (1986) (see also Thomas *et al.* 1985) have considered the effects of *restricted rotation* (wobbling, Fig. 13.1b) on $V_2'$ spectra, reporting that restriction of the angular amplitude of motion, without a change in the rate, can cause a substantial increase in the effective $\tau_r$ (obtained from isotropic reference spectra). As in the case of uniaxial motions (discussed above), extremely anisotropic motions can sometimes be recognized directly from the ST-EPR lineshape; that is, there is no isotropic motion that could give rise to the same shape (the effective correlation time varies with spectral position). However, as in the case of optical anisotropy decays, spectra corresponding to restricted motion are very difficult to distinguish from those corresponding to uniaxial motion.

Whether the anisotropy of motion is due to rotation about a preferred axis or to restricted amplitude, the theoretical simulations indicate that effective correlation times measured from ST-EPR spectra using isotropic motion reference spectra (Figs 13.11–13.13) are most likely to be *greater* than or equal to the actual times, which can only be unambiguously measured in time-resolved experiments. This principle is confirmed by a comparison between the results of steady-state ST-EPR experiments and time-resolved optical experiments on the same systems, e.g. rhodopsin [Cone, 1972 (optical); Baroin *et al.*, 1977 (EPR); Devaux *et al.*, 1982 (EPR)], Ca-ATPase [Thomas and Hidalgo, 1978 (EPR); Bürkli and Cherry, 1981 (optical); Spiers *et al.*, 1983 (optical)], cytochrome oxidase (Swanson *et al.*, 1980 (EPR); Kawato *et al.*, 1981 (optical)] and myosin [Eads *et al.*, 1984 (optical); Barnett and Thomas, 1984 (EPR); Thomas *et al.*, 1985 (both)].

This sensitivity to the anisotropy of motion in a steady-state experiment is much less ambiguous in EPR than in steady-state optical spectroscopy, owing to the orientational resolution of EPR, although the effects are less striking than in conventional EPR (in the ns time range) or in time-resolved experiments. For any of these techniques, EPR or optical, the ambiguity of interpretation can be greatly decreased (e.g. permitting one to distinguish an amplitude change from a rate change, or uniaxial motion from restricted motion) if spectra are obtained from oriented membranes. Conventional EPR can be used to determine the orientation distribution of the probe axis relative to the membrane normal, providing the information needed for an unambiguous interpretation of ST EPR spectra. This approach has proven powerful in analogous studies of oriented muscle fibres (Barnett and Thomas, 1984; Thomas *et al.*, 1985). In the absence of this kind of information, caution must be used when interpreting changes in effective correlation times, and especially when comparing effective correlation times from different systems.

## 13.4 Conclusions

The optical and ST-EPR spectroscopic methods discussed here have matured over the past decade into quantitative techniques for the measurement of microsecond rotational motions in general, and membrane protein rotational diffusion in particular. The two classes of techniques share many physical and experimental principles, but are different enough to provide complementary information. The optical experiments have the primary advantages of time resolution, high absolute sensitivity and, in some cases, intrinsic chromophores, while the steady-state EPR techniques offer the advantages of superior orientational resolution, commercially available instrumentation and smaller probes. Future developments in both fields should increase their power and utility. For example, the development of time-resolved ST-EPR offers the promise of a technique with both time and orientational resolution, which could result in more direct, less ambiguous probing of anisotropic motions than has been previously possible. The further development of fluorescence depletion methods offers the sensitivity to measure rotational motions in a selected region of the surface of a single cell. Both classes of techniques will continue to play important roles in membrane biophysics, providing fundamental information about the dynamics of protein–protein and protein–lipid interactions and about the possible roles of these molecular motions in biological function.

## Acknowledgments

I thank Thomas Squier, Kerry Lindahl and Carl Polnaszek for providing me with unpublished data included in this review. They, along with Richard Ludescher and Piotr Fajer, provided constructive comments on the manuscript. I am grateful to Richard Cherry for his patience and helpful suggestions. This work was supported by grants from the National Institutes of Health (GM 27906, AM32961 and RR01439), the American Heart Association (80-850 and an Established Investigatorship), the National Science Foundation (PCM 8004612) and the Muscular Dystrophy Association of America.

## 13.5 References

Austin, R. H., Chan, S. S. and Jovin, T. M. (1979) *Proc. Natl. Acad. Sci. USA*, **76**, 5650–4.

Barnett, V. A. and Thomas, D. D. (1984) *J. Mol. Biol.*, **179**, 83–102.

Barnett, V. A., Fajer, P., Polzaszek, C. F. and Thomas, D. D. (1986) *Biophys. J.*, **49**, 144–46.

Baroin, A., Thomas, D. D., Osborne, B. and Devaux, P. F. (1977) *Biochem. Biophys. Res. Commun.*, **78**, 442–7.

Beth, A. H., Balasubramanian, K., Wilder, R. T., Venkataramu, S. D., Robinson, B. H., Dalton, L. R., Pearson, D. E. and Park, J. H. (1981b) *Proc. Natl. Acad. Sci. USA*, **78**, 4955–9.

Beth, A. H., Venkataramu, S. D., Balasubramanian, K., Dalton, L. R., Robinson, B. H., Pearson, D. E., Park, C. R. and Park, J. H. (1981a) *Proc. Natl. Acad. Sci. USA*, **78**, 967–71.

Brown, I. M. (1979) in *Time Domain Electron Spin Resonance* (eds L. Kevan and R. N. Schwartz), Wiley, New York, pp. 196–229.
Bürkli, A. and Cherry, R. J. (1981) *Biochemistry*, **20**, 138–43.
Cherry, R. J. (1978) *Methods Enzymol.*, **54**, 447–61.
Cherry, R. J. (1979) *Biochim. Biophys. Acta*, **559**, 289–327.
Cherry, R. J. (1982) in *Membranes and Transport* (ed A. Martonosi), Plenum Press, New York, Vol. 1, pp. 145–52.
Cherry, R. J. and Godfrey, R. E. (1981) *Biophys. J.*, **36**, 257–76.
Cherry, R. J. and Schneider, G. (1976) *Biochemistry*, **15**, 3657–61.
Cone, R. A. (1972) *Nature (London) New Biol.*, **236**, 39–43.
Corin, A. F., Matayoshi, E. D. and Jovin, T. M. (1985) in *Spectroscopy and the Dynamics of Biological Systems* (eds P. M. Bayley and R. E. Dale), Academic Press, London pp. 53–78.
Dalton, L. R. (ed) (1985) *EPR and Advanced EPR studies of Biological Systems*, CRC Press, Boca Raton, Florida.
Devaux, P. F. (1982) *Methods Enzymol.*, **81**, 703–9.
Dixit, B. P. S. N., Waring, A. J., Wells, K. O., Wong, P. S., Woodrow, G. V. and Vanderkooi, J. M. (1982) *Eur. J. Biochem.*, **126**, 1–9.
Eads, T., Austin, R. H. and Thomas, D. D. (1984) *J. Mol. Biol.*, **179**, 55–81.
Evans, C. A. *et al.* (1981) *J. Magn. Reson.*, **44**, 109–16.
Fajer, P. and Marsh, D. (1982) *J. Magn. Reson.*, **49**, 212–24.
Fajer, P., Thomas, D. D., Feix, J. and Hyde, J. S. (1986) *Biophys. J.* (in press).
Forrer, J. E., Wubben, R. C. and Hyde, J. S. (1980) *Bull. Magn. Reson.*, **2**, 441–6.
Froncisz, W. and Hyde, J. S. (1982) *J. Magn. Reson.*, **47**, 515–21.
Gaffney, B. J. (1976) in *Spin Labeling* (ed L. Berliner), Academic Press, New York, pp. 183–238.
Gaffney, B. J. (1979) *J. Phys. Chem.*, **83**, 3345–9.
Garland, P. B. (1982) in *Membranes and Transport* (ed A. Martonosi), Plenum Press, New York, Vol. 1, pp. 153–8.
Garland, P. B. and Johnson, P. (1985) in *Spectroscopy and the Dynamics of Biological Systems*, (eds P. M. Bayley and R. E. Dale), Academic Press, London, pp. 95–118.
Garland, P. B. and Moore, C. H. (1979) *Biochem. J.*, **183**, 561–72.
Graceffa, P. and Seidel, J. C. (1980) *Biochemistry*, **19**, 33–9.
Greinert, R., Finch, S. A. E. and Stier, A. (1982) *Xenobiotica*, **12**, 717–26.
Greinert, R., Staerk, H., Steir, A. and Weller, A. (1979) *J. Biochem. Biophys. Methods*, **1**, 77–83.
Gut, J., Richter, C., Cherry, R. J., Winterhalter, K. H. and Kawato, S. (1982) *J. Biol. Chem.*, **257**, 7030–6.
Hemminga, M. A. (1983) *Chem. Phys. Lipids*, **32**, 323–83.
Hemminga, M. A. and de Jager, P. A. (1981) *J. Mag. Reson.*, **43**, 324–7.
Heyn, M. P., Cherry, R. J. and Müller, U. (1977) *J. Mol. Biol.*, **117**, 607–20.
Hicks, B., White, M. Ghiron, C. A., Kuntz, R. R. and Volkert, W. A. (1978) *Proc. Natl. Acad. Sci. USA*, **75**, 1172–5.
Hidalgo, C., Thomas, D. D. and Ikemoto, N. (1978) *J. Biol. Chem.*, **253**, 6879–87.
Horie, T. and Vanderkooi, J. M. (1981) *Biochim. Biophys. Acta*, **670**, 294–7.
Horvath, L. I. and Marsh, D. (1983) *J. Magn. Reson.*, **54**, 363–73.
Huisjen, M. and Hyde, J. S. (1974) *Rev. Sci. Instrum.*, **45**, 669–77.

Hyde, J. S. (1979) in *Time Domain Electron Spin Resonance* (eds L. Kevan and R. N. Schwartz), Wiley, New York, pp. 1–30.

Hyde, J. S. and Dalton, L. R. (1972) *Chem. Phys. Lett.*, **16**, 568–72.

Hyde, J. S., Eriksson, L. E. G. and Ehrenberg, A. (1970) *Biochim. Biophys. Acta*, **222**, 688–92.

Hyde, J. S., Froncisz, W. and Mottley, C. (1984) *Chem. Phys. Lett.*, **110**, 621–5.

Hyde, J. S. and Thomas, D. D. (1973) *Ann. N.Y. Acad. Sci.*, **222**, 680–92.

Hyde, J. S. and Thomas, D. D. (1980) *Annu. Rev. Phys. Chem.*, **31**, 293–317.

Johnson, M. E. and Hyde, J. S. (1981) *Biochemistry*, **20**, 2875–80.

Johnson, M. E., Lee, L. and Fung, L. W. M. (1982a) *Biochemistry*, **21**, 4459–67.

Johnson, M. E., Thiyagarajan, P., Bates, B. and Currie, B. L. (1982b) *Biophys. J.*, **37**, 553–7.

Johnson, P. and Garland, P. B. (1981) *FEBS Lett.*, **132**, 252–6.

Johnson, P. and Garland, P. B. (1982) *Biochem. J.*, **203**, 313–21.

Jovin, T. M., Bartholdi, M., Vaz, W. L. C. and Austin, R. H. (1981) *Ann. N.Y. Acad. Sci.*, **366**, 176–96.

Kar, L., Millhauser, G. L. and Freed, J. H. (1984) *J. Phys. Chem.*, **88**, 3951–6.

Kawato, S., Gut, J., Cherry, R. J., Winterhalter, K. H. and Richter, C. (1982b) *J. Biol. Chem.*, **257**, 7023–9.

Kawato, S. and Kinosita, K., Jr. (1981) *Biophys. J.*, **36**, 277–96.

Kawato, S., Lehner, C., Müller, M. and Cherry, R. J. (1982a) *J. Biol. Chem.*, **257**, 6470–6.

Kawato, S., Sigel, E., Carafoli, E. and Cherry, R. J. (1981) *J. Biol. Chem.*, **256**, 7518–27.

Keana, J. F. W. (1979) in *Spin Labeling*, Academic Press, New York, Vol. 2, pp. 115–72.

Kinosita, K., Jr., Ikegami, A. and Kawato, S. (1982) *Biophys. J.*, **37**, 461–4.

Kinosita, K., Jr., Kawato, S. and Ikegami, A. (1977) *Biophys. J.*, **20**, 289–305.

Kinosita, K., Jr., Kawato, S. and Ikegami, A. (1984) *Adv. Biophys.*, **17**, 147–203.

Kouyama, T., Kimura, Y., Kinosita, K. Jr. and Ikegami, A. (1981) *J. Mol. Biol.*, **153**, 337–9.

Kusumi, A., Sakaki, T., Yoshizawa, T. and Ohnishi, S. (1980) *J. Biochem. (Tokyo)*, **88**, 1103–11.

Kusumi, A., Subczynski, W. K. and Hyde, J. S. (1982) *Proc. Natl. Acad. Sci. USA*, **79**, 1854–8.

Lavalette, D., Amand, B. and Pochon, F. (1977) *Proc. Natl. Acad. Sci. USA*, **74**, 1407–11.

Lindahl, K. M. and Thomas, D. D. (1982) *Biophys J.*, **37**, 71a.

Lindahl, K. M., Polnaszek, C. F. and Thomas, D. D. (1986) unpublished.

Lipari, G. and Szabo, A. (1980) *Biophys. J.*, **30**, 489–506.

Mailer, C., Thomann, H., Robinson, B. H. and Dalton, L. R. (1980) *Rev. Sci. Instrum.* **51**, 714–21.

Marsh, D. (1980) *Biochemistry*, **19**, 1632–7.

Marsh, D. (1981) in *Membrane Spectroscopy* (ed. E. Grell), Springer-Verlag, Berlin, pp. 51–142.

McCalley, R. C., Shimshick, E. J. and McConnell, H. M. (1972) *Chem. Phys. Lett.*, **13**, 115–19.

Millhauser, G. L. and Freed, J. H. (1984) *J. Chem. Phys.*, **81**, 37–48.

Moore, C., Boxer, D. and Garland, P. (1979) *FEBS Lett.*, **108**, 161–6.

Morrisett, J. D. (1976) in *Spin Labeling* (ed L. Berliner), Academic Press, New York,

pp. 273–338.
Muhlebach, T. and Cherry, R. J. (1982) *Biochemistry*, **21**, 4225–8.
Nigg, E. A. and Cherry, R. J. (1979) *Biochemistry*, **18**, 3457–65.
Nigg, E. A. and Cherry, R. J. (1980) *Proc. Natl. Acad. Sci. USA*, **77**, 4702–6.
Plachy, W. Z. and Windrem, D. A. (1977) *J. Magn. Reson.*, **27**, 237–9.
Polnaszek, C. F., Marsh, D. and Smith, I. C. P. (1981) *J. Magn. Reson.*, **43**, 54–64.
RaziNaqvi, K., Rodriguez, J. G., Cherry, R. J. and Chapman, D. (1973) *Nature (London) New Biol.*, **245**, 249–51.
Richter, C., Winterhalter, K. H. and Cherry, R. J. (1979) *FEBS Lett.*, **102**, 151–4.
Robinson, B. R. and Dalton, L. R. (1980) *J. Chem. Phys.*, **72**, 1312–24.
Robinson, B. R. and Dalton, L. R. (1981) *Chem. Phys.*, **54**, 253–9.
Saffman, P. G. and Delbrück, M. (1975) *Proc. Natl. Acad. Sci. USA*, **72**, 3111–13.
Sasaki, T., Kanaoka, Y. and Watanabe, T. (1980) *J. Magn. Reson.*, **38**, 385–90.
Sehr, P. A., Mailer, C. and Devaux, P. F. (1983) *J. Magn. Reson.*, **52**, 23–30.
Smigel, M. D., Dalton, L. A., Dalton, L. R. and Kwiram, A. L. (1974a) *Chem. Phys.*, **6**, 183–92.
Smigel, M. D., Dalton, L. R., Hyde, J. S. and Dalton, L. A. (1974b) *Proc. Natl. Acad. Sci. USA*, **71**, 1925–9.
Spiers, A., Moore, C. H., Boxer, D. H. and Garland, P. B. (1983) *Biochem. J.*, **213**, 67–74.
Squier, T. C. and Thomas, D. D. (1986) *Biophys. J.*, **49**, 921–35.
Strambini and Galley (1976) *Nature (London)*, **260**, 554–6.
Strambini and Galley (1980) *Biopolymers*, **19**, 383–90.
Swanson, M. S., Quintanilha, A. T. and Thomas, D. D. (1980) *J. Biol. Chem.*, **255**, 7494–502.
Thomas, D. D. (1978) *Biophys. J.*, **24**, 439–62.
Thomas, D. D. (1982) in *Membranes and Transport* (ed A. Martonosi), Plenum Press, New York, Vol. 1, pp. 135–9.
Thomas, D. D. (1985) in *The Enzymes of Biological Membranes* (ed A. Martonosi), Plenum Press, New York, pp. 287–312.
Thomas, D. D., Bigelow, D. J., Squier, T. C. and Hidalgo, C. H. (1982) *Biophys. J.*, **37**, 217–25.
Thomas, D. D. and Cooke, R. (1980) *Biophys. J.*, **32**, 891–906.
Thomas, D. D., Dalton, L. R. and Hyde, J. S. (1976) *J. Chem. Phys.*, **65**, 3006–24.
Thomas, D. D., Eads, T. M., Barnett, V. A., Lindahl, K. M., Momont, D. A. and Squier, T. C. (1985) in *Spectroscopy and the Dynamics of Biological Systems* (eds P. M. Bayley and R. E. Dale), Academic Press, London, p. 239–57.
Thomas, D. D. and Hidalgo, C. (1978) *Proc. Natl. Acad. Sci. USA*, **75**, 5488–92.
Thomas, D. D. and McConnell, H. M. (1974) *Chem. Phys. Lett.*, **25**, 470–5.
Thomas, D. D., Wendt, C. H., Froncisz, W. and Hyde, J. S. (1983) *Biophys. J.*, **43**, 131–5.
Vaz, W. L. C., Austin, R. H. and Vogel, H. (1979) *Biophys. J.*, **26**, 415–26.
Vistnes, A. I. (1983) *Biophys. J.*, **43**, 31–8.
Watanabe, T., Sasaki, T. and Fujiwara, S. (1982) *Appl. Spectrosc.*, **36**, 174–8.
Watanabe, T., Sasaki, T. and Sawatari, K. (1980) *Appl. Spectrosc.*, **34**, 456–60.
Weber, G. (1977) *J. Chem. Phys.*, **66**, 4081–91.

# Index

Italicized page numbers refer to figures and tables